D0998559

ADVANCES IN CHEMICAL PHYSICS

VOLUME XXXIII

EDITORIAL BOARD

C. J. BALLHAUSEN, Kemisk Laboratorium IV, Kobenhaven Universitets-Fysisk-Kemiske Institut, Kopenhagen, Denmark

J. J. M. BEENAKKER, Kamerlingh Onnes Laboratory, Rijksuniversiteit te Leiden, Leiden, Netherlands

RICHARD B. BERNSTEIN, Department of Chemistry, University of Texas at Austin, Austin, Texas, U.S.A.

H. HAKEN, Institut für Theoretische und Angewandte Physik der Technischen Hoschschule, Stuttgart, Germany

YU L. KLIMANTOVITCH, Moscow State University, Moscow, U.S.S.R.

RYOGO KUBO, Department of Physics, University of Tokyo, Tokyo, Japan

M. MANDEL, Chemie-Complex der Rijksuniversiteit, Leiden, Netherlands

PETER MAZUR, Institute Lorentz voor Theoretische Natuurkunde, Leiden, Netherlands

GREGOIRE NICOLIS, Pool de Physique, Faculté de Sciences, Université Libre de Bruxelles, Brussels, Belgium

S. ONO, Institute of Physics, College of General Education, University of Tokyo, Tokyo, Japan

MICHAEL PHILPOTT, IBM Research Center, San Jose, California, U.S.A.

J. C. POLANYI, Department of Chemistry, University of Toronto, Toronto, Ontario, Canada

YVES POMEAU, Commissariat a l'Energie Atomique, Division de la Physique, Centre d'Études Nucleares de Saclay, Gif sur Yvette, France

B. PULLMAN, Institute de Biologie Physico-Chimique, Université de Paris, Paris, France

C. C. J. ROOTHAAN, Departments of Physics and Chemistry, University of Chicago, Chicago, Illinois, U.S.A.

IAN ROSS, Department of Chemistry, Australian National University, Canberra, Australia

JOHN ROSS, Department of Chemistry, Massachusetts Institute of Technology, Cambridge, Massachusetts, U.S.A.

R. SCHECTER, Department of Chemical Engineering, University of Texas at Austin, Austin, Texas, U.S.A.

I. SHAVITT, Battelle Memorial Institute, Columbus, Ohio, U.S.A.

JAN STECKI, Institute of Physical Chemistry of the Polish Academy of Sciences, Warsaw, Poland

GEORGE SZASZ, General Electric Corporate R&D, Zurich, Switzerland

KAZUHISA TOMITA, Department of Physics, Faculty of Science, Kyoto University, Kyoto, Japan

M. V. VOKENSTEIN, Institute of Molecular Biology, Academy of Science, Moscow, USSR.

E. BRIGHT WILSON, Department of Chemistry, Harvard University, Cambridge, Massachusetts, U.S.A.

Advances in

CHEMICAL

PHYSICS

EDITED BY

I. PRIGOGINE

University of Brussels,
Brussels, Belgium
and
University of Texas,
Austin, Texas

AND

STUART A. RICE

Department of Chemistry
and
The James Franck Institute
The University of Chicago
Chicago, Illinois

VOLUME XXXIII

AN INTERSCIENCE® PUBLICATION

JOHN WILEY AND SONS

NEW YORK · LONDON · SYDNEY · TORONTO

QD453
A27
V.33

AN INTERSCIENCE® PUBLICATION

Copyright © 1975 by John Wiley & Sons, Inc.

All rights reserved. Published simultaneously in Canada.

No part of this book may be reproduced by any means,
nor transmitted, nor translated into a machine language
without the written permission of the publisher.

Library of Congress Catalog Card Number: 58-9935
ISBN 0-471-69935-7

Printed in the United States of America

10 9 8 7 6 5 4 3 2 1

INTRODUCTION

To the Memory of Lothar Meyer (1906–1971)

I always find it difficult to describe a friend. It is not merely a matter of being too close to have perspective. The larger problem is to sum up in only a few words the personal traits, the accomplishments, the hopes of a complex being. Where does one begin—and where does one end?

Scientists are fortunate in that their work is a visible creation that stands or falls on its own merits. The very visibility of a man's work is in many senses the best memorial to him. It mirrors his development, shows vividly his strengths and weaknesses—in short, is a part of his humanity. I shall not write about Lothar Meyer's work except to note that he was a pioneer in low temperature physics and chemistry. I think of him as a scientific romantic. His interests were catholic, and he was never one to confine himself for long to a single field. In his constant exploration of new areas nothing seemed to hold back his enthusiasm for a new experiment. It is fitting that this volume of the Advances in Chemical Physics, dedicated to his memory, deals with many different subjects. Lothar Meyer made important contributions to some of the topics discussed and, I am sure, would have been interested in all of them.

What I will always remember about Lothar Meyer, and what is most important to me, was our personal relationship. We first became acquainted when I joined the faculty of the University of Chicago. Even against a background in which the general level of material and psychological support for new young faculty was very high, he was exceptionally generous of his time and effort in behalf of a newcomer. Over the years of our association he was at times my teacher and my student, my collaborator and my antagonist. In a sense he was frequently younger than I, and yet still more experienced and mature. He was always my friend and counselor.

Chicago, Illinois
September 1975

Stuart A. Rice

JAN 2 7 1976

JAN 27 1976

CONTRIBUTORS TO VOLUME XXXIII

M. V. BASILEVSKY, Karpov Institute of Physical Chemistry, Moscow, USSR

A. BEN-REUVEN, Department of Chemistry, Massachusetts Institute of Technology, Cambridge, Massachusetts

MAYNARD A. BRANDT, Department of Chemistry, University of Minnesota, Minneapolis, Minnesota

ROGER CERF, Laboratoire d'Acoustique Moléculaire, Université Louis Pasteur, Strasbourg, France

DOMINIC G. B. EDELEN, Center for the Application of Mathematics, Lehigh University, Bethlehem, Pennsylvania

TARO KIHARA, Department of Physics, Faculty of Science, University of Tokyo, Tokyo, Japan

AKIO KOIDE, Department of Physics, Faculty of Science, University of Tokyo, Tokyo, Japan

C. ALDEN MEAD, Department of Chemistry, University of Minnesota, Minneapolis, Minnesota

K. W. SCHWARZ, Department of Physics and James Franck Institute, University of Chicago, Chicago, Illinois

DONALD G. TRUHLAR, Department of Chemistry, University of Minnesota, Minneapolis, Minnesota

R. G. WOOLLEY, Trinity Hall, Cambridge, England

CONTENTS

ADVANCES IN CHEMICAL PHYSICS

VOLUME XXXIII

MOBILITIES OF CHARGE CARRIERS IN SUPERFLUID HELIUM

K. W. SCHWARZ

The Department of Physics and The James Franck Institute,
The University of Chicago, Chicago, Illinois

CONTENTS

I. INTRODUCTION

A. Early History

Excess charges were first used to probe the properties of superfluid helium in several experiments[1-3] concerned with the motion of such charges through helium under the influence of an applied electric field. One aim of these early investigations was to verify and extend the Landau-Pomeranchuk[4] description of how a foreign particle propagates through a superfluid, the central idea being that for speeds below some critical

1

velocity, a moving impurity cannot create excitations, and thus can only lose momentum through collisions with elementary excitations that are already present. Since the excitation density drops rapidly as the temperature is lowered, it was expected that a foreign impurity such as an excess charge would become increasingly mobile with lower temperature. Not only was such behavior observed, but in fact the temperature variation of the mobility below the λ-point was found to be in good agreement with the predictions of the Landau model.[5]

These early experiments also indicated that whereas in moderately dense helium gas the negative charge carrier is a free electron[6] and the predominant positive carrier a He_2^+ ion,[7] in liquid helium both carriers are localized and are associated with structures large compared with that of a simple ion. Part of the reason for this was elucidated when Atkins[8] pointed out the importance of electrostrictive effects on the liquid near an ion. A strongly bound ion such as He_2^+ will attract the neighboring neutral atoms sufficiently to generate a high-density, permanent cluster of He atoms around the ion. This model, which leads to the idealized structure in Fig. 1a, has in fact provided a quite satisfactory explanation of even the most recently measured properties of the positive charge carriers.

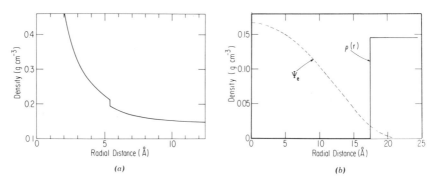

Fig. 1. (a) Idealized structure of the positive charge carrier, calculated for zero pressure at infinity and $\sigma_{ls} = 0.10$ erg/cm². (b) Idealized structure of the negative charge carrier. The electron wavefunction is calculated assuming $V_0 = 1.0$ eV.

Since a stable negative ion of He does not exist, it was not expected that the Atkins explanation would necessarily apply to the negative carrier. Instead, early speculations[9-11] raised the possibility that an excess electron in the liquid would become self-trapped in a region of low density. The properties of such a state were worked out in some detail by Kuper,[12] who concluded that the electron should be localized within a large, well-defined bubble in the liquid (Fig. 1b). Because of the large size of the bubble, electrostrictive effects are insignificant. This "bubble model" of the negative carrier structure received confirmation from the work of Levine and Sanders,[6, 13] who observed the transition from the free to the localized electron state in low-temperature helium gas. Later experiments have furnished additional support for the existence of this novel but simple state, and with the addition of some refinements, the bubble model is now universally accepted.

The systematic measurements of charge carrier mobilities that were carried out by Meyer and Reif[14–16] for various temperatures, pressures, and ^{3}He impurity concentrations demonstrated that such charged probes are indeed very useful for investigating some of the more novel properties of the superfluid state. Subsequent years have witnessed an impressive mushrooming of the experimental application of these probes, and today such applications constitute a rather large and well-defined subfield of superfluid physics. It is not our purpose here to give a complete review of this area: just such a survey has recently been published.[17] However, some of the background is presented in Sections I.B and I.C to put the subsequent discussion into proper context.

B. Charged Probes in Superfluid Helium

The dynamic state of a superfluid may be approximately visualized as being composed of two distinct types of motion.[18] Like any many-body system at very low temperatures, the fluid will contain thermally activated elementary excitations, which may be thought of as a dilute gas of weakly interacting wave packets. The constituents of this gas can exchange momentum with one another and with any boundaries that may be present; thus the elementary excitations have the macroscopic properties of a *normal fluid*. In addition to such incoherent excitations, superfluid helium also exhibits macroscopically phase-coherent motion in the sense that the one-particle reduced density matrix contains a term $\Phi^*(\mathbf{r}, t)$ $\Phi(\mathbf{r}', t)$.[19, 20] Writing $\Phi = f(\mathbf{r}, t)\exp[i\varphi(\mathbf{r}, t)]$ and taking the local averages of the density and the momentum density operators, one finds that this term makes a contribution to the average local density and velocity fields given by

$$\rho_s(\mathbf{r}, t) = f(\mathbf{r}, t)^2$$

$$\mathbf{v}_s(\mathbf{r}, t) = \frac{\hbar}{m} \nabla \varphi(\mathbf{r}, t)$$

(1)

These fields describe a *superfluid* component of the motion of the system, distinct from the normal fluid motion.

The properties of the excitation gas can be described phenomenologically in terms of a kinetic theory of interacting wave packets,[21] but the determination of equations of motion for the fields ρ_s and v_s is a more subtle matter. In principle, it would be desirable to derive the properties of $\Phi(\mathbf{r}, t)$ from microscopic considerations and take the relevant averages. In practice, this is a difficult theoretical project, and here we are content with the usual assumption that mass and momentum are locally conserved. The result of this *ansatz* is, of course, the ideal fluid equations

$$\frac{\partial \mathbf{v}}{\partial t} + (\mathbf{v} \cdot \nabla)\mathbf{v} = -\frac{\nabla p}{\rho}$$

$$\frac{\partial \rho}{\partial t} + \nabla \cdot (\rho \mathbf{v}) = 0$$

(2)

Here ρ and $\rho \mathbf{v}$ refer to the *total* mass and momentum densities. In the low-temperature limit, the elementary excitations will contribute negligibly to such quantities, and (2) then provides a partial, approximate description of the motion of the superfluid component.[22] The range of applicability of these equations is somewhat uncertain, but approximate microscopic calculations[23,24] suggest that they should hold down to a scale of roughly 1 Å. Time variations are assumed to be slow in the sense that (2) represents the quantum-mechanically adiabatic response of the system to time-dependent perturbations.

Let us now consider a structure like one of those in Fig. 1, moving through superfluid helium at a temperature far below the λ-point, where the excitation gas is dilute. In contrast to weakly interacting probes such as neutrons or light waves, which measure properties of the undisturbed, stationary superfluid, the charge carriers may most profitably be viewed as providing a means of investigating the interaction between the superfluid and a *boundary*. That is, the structures sketched in Fig. 1 are large enough to constitute geometrically simple boundaries that are rather well defined on the atomic scale. Moving through the fluid, such a probe will undergo two distinct types of interaction. First, it will interact quantum-hydrodynamically with the fluid to generate nonuniform fields ρ_s, v_s in its immediate neighborhood. To an accuracy of about 1 Å, these fields are the potential flow[25] solutions of (2). Second, the probe will suffer random changes in velocity as elementary excitations scatter off the moving boundary and the associated local velocity field. When the excitation gas is very dilute, it is clear that the two types of interaction can be discussed independently as physically distinct phenomena. A more general criterion for being able to do this is that the normal fluid density be low enough to permit (2) to provide a good description of the local superfluid flow pattern about the probe.

The preceding considerations allow us to roughly classify the main types of experiment performed using charged probes. One line of development has been to extend the early work of Meyer and Reif to gain a detailed understanding of how the elementary excitations scatter off the moving charge carriers. Since one can look at three kinds of excitations (phonons, rotons, and ^{3}He impurities) interacting with the two kinds of charge carrier, a rich variety of scattering processes may be studied by such experiments. It is, of course, of considerable intrinsic interest to see how the various types of excitation interact with the well-defined boundaries represented by the charge carriers. However, it should also be noted that other experiments involving charged probes usually measure a combination of the phenomenon of interest and statistical effects arising from the random fluctuations of the probe velocity due to excitation scattering. Clearly, a quantitative interpretation of such experiments can be obtained only after the scattering processes have been thoroughly understood. Considerable progress has been made in recent years, and the later sections of the chapter review this class of problems.

A quite different line of development has been to use the charge carriers to investigate the quantum-hydrodynamic properties of the superfluid. On the simplest level, one may ask whether the hydrodynamic contribution to the effective mass of the carriers is consistent with our assumption that the local flow field obeys the ideal fluid equations down to a scale of ~ 1 Å. Although measuring these very large effective masses has proved to be difficult,[26] quite accurate values have recently been obtained by Poitrenaud and Williams.[27] In the case of the positive carrier, the hydrodynamic contribution to the effective mass is rather small, and the interpretation is complicated by the effect of electrostriction on the fluid near the charge.[28] Thus while agreement between experiment and calculations based on the electrostrictive model of Atkins is generally very good, this serves more to confirm the model than to test the hydrodynamic properties of the surrounding fluid. The electron bubble has a less complicated structure, and its entire mass is hydrodynamic. Here too, one finds excellent agreement with theoretical predictions and with the results of other experiments (see Section II.C). This represents a simple and accurate confimation both of the bubble model and of the assumptions implicit in the use of the ideal fluid equations to describe the superfluid on the scale of a few ångstrom units.

Turning to more complicated quantum-hydrodynamic phenomena, we find that charged probes have proved to be particularly suitable for the study of vortex structure in helium. Since $\nabla \times \mathbf{v} \equiv 0$ for a superfluid, it cannot (in contrast to an ideal fluid) exhibit local rotation. Early experiments,[29] however, demonstrated conclusively that superfluids can give the macroscopic *appearance* of rotational flow. The resolution of this apparent paradox has been provided by the realization that the helium may contain vortex filaments at the center of which the $\nabla \times \mathbf{v} = 0$ restriction breaks down. The superfluid restrictions apply outside the core region, and it follows at once from (1) that the circulation $\kappa_0 = \oint \mathbf{v} \cdot \mathbf{dl}$ about such a line must be quantized in units of h/m. When averaged over macroscopic distances, a distribution of such quantized vortex lines will look like a rotational flow field.

Since the nonsuperfluid core is estimated[30,31] to have a radius of $\sim 1 \text{Å}$, quantized vortex lines act as line sources of vorticity even on the size scale of the carriers. Thus within the framework of (2), the interaction of a charged probe and a quantized vortex line is to a good approximation identical to the hydrodynamic interaction between a small sphere and a vortex filament of circulation κ_0. One of the features of this interaction is a strong local attraction exerted by the line on the sphere, and in fact such an object passing by a line will be hydrodynamically captured.[32]

The study of vortex structure using charge carriers was initiated by Careri, McCormick, and Scaramuzzi,[33] who looked at the behavior of charges moving across a rotating bucket of helium. Such a rotating system appears macroscopically to be undergoing solid-body rotation ($\nabla \times \mathbf{v} = 2\Omega$), which according. to the quantized vortex-line picture would imply that the liquid is permeated by an array of such lines, directed along the axis of rotation with a density of $2\Omega \, m/h$. The authors did indeed find that a certain fraction of their carriers failed to make it across the bucket, presumably because they became trapped on the lines. The extensive subsequent research[34,35] done on this system has not only verified this interpretation, it has provided detailed information about the trapping processes and about the properties of charges trapped on the lines. Most recently, it has proved possible[36] to observe the appearance and disappearance of single vortex lines in a rotating system, and attempts are currently underway to obtain a "picture" of the vortex-line array by the use of charged probes.[37,38]

Perhaps the deepest questions relate to the discovery by Rayfield and Reif[30] that charge carriers exceeding a certain critical velocity with respect to the superfluid will *create* quantized vorticity in the fluid. Although interesting attempts[39] have been made to explain this phenomenon in terms of the thermal activation of vortex loops on the carrier, recent work[40,41] indicates rather that it is a zero-temperature effect not fundamentally dependent on the presence of the elementary excitations. Under certain special circumstances,[15,42] charged probes will create rotons in the fluid rather than quantized vortex rings; but the two processes appear to be similar in that both are characterized by critical velocities above which there is a rapid rise in the probability per unit time of creating the excitation in question.[43]

In judging the significance of these phenomena, it should be kept in mind that (2) certainly does not provide a complete description of the response of the superfluid to the probe. The situation is analogous to that in the theory of classical liquids, where many of the most interesting questions (including particularly those relating to the stability of simple flow patterns) require a discussion of physical effects that are neglected in the ideal fluid approximation. Similarly here, the appearance of vortex rings or rotons at some critical velocity can only be understood in terms of a more fundamental treatment of the superfluid response than is provided by (2). One notes that (at $T=0$) the moving probe represents a time-dependent boundary condition on the ground state of the system. If such a time-dependent perturbation is slow enough, it will generate parametric changes in the ground-state wavefunction that are approximately described by (2). However, a rapidly moving sphere represents a strongly time-dependent perturbation that will cause the system to become excited. One may speculate that the observed critical velocities represent the transition from one regime to the other. In any case, the creation of vortex rings or rotons by the charged probes shows the breakdown of the usual description of superfluid flow in a uniquely simple and experimentally well-characterized system. It thus bears on important theoretical questions that have been somewhat neglected.

C. Some Recent Developments

The areas of inquiry discussed in the previous section are those which have been pursued the longest and still seem to be the most important. In recent years, however, the use of charged probes has been pursued in a number of ingenious new ways, and this section presents a somewhat disorganized sampling.

An obvious extension of the mobility investigations is to carry them out at normal fluid densities so high that the elementary excitations no longer behave as a weakly interacting gas. The theoretical interpretation of such experiments is rather difficult, and there are now several intriguing observations, whose significance remains to be clarified. One way of increasing the excitation density is, of course, to raise the temperature. The first complication to arise is that because of renormalization effects, both the excitation spectrum and the structure of the charge carriers become functions of temperature above $T \sim 1.2°K$. Although it will affect the magnitude of the scattering cross-sections, this effect alone should not cause any great qualitative changes in the way that the carrier interacts with the excitations. The careful survey of mobilities above $1.2°K$ carried out by Brody[44] seems to bear this out. Above $\sim 1.8°K$, the excitations become sufficiently dense to interact in a very complicated way with one another and with the charged probes. Several authors[45-47] have looked at the carrier mobilities in the temperature region near the λ-point, and Sitton and Moss[47] have made the interesting observation that the drag on the carrier seems to obey a scaling law. No convincing explanation of this phenomenon has been given.

A second way of increasing the excitation density is to add ^3He impurities. Anomalies have been observed in the mobility[48] and in other[49,50] transport properties of ^3He-rich solutions, which were initially interpreted as showing that the presence of ^3He impurities had a strong effect on the roton excitation spectrum. Light scattering experiments[51,52] have failed to show such an effect, in agreement with the predictions of microscopic theory.[53] It has been proposed[54,55] instead that these measurements reflect a property of the ^3He spectrum, and more experimental work along these lines would clearly be of interest.

A further extension of mobility studies has been made possible by the successful injection of various positive impurity ions to serve as excess charge centers.[56,57] The simple Atkins model would predict that the structure of such a carrier should not be affected by the nature of the ion at its center, but systematic variations of the mobility have been observed. All these experiments have been carried out in a temperature region in which the mobility is determined by roton scattering, and since such scattering processes are expected to be particularly sensitive to the detailed features of the carrier structure, it is not clear just how much of a deviation from the Atkins model is implied. Nevertheless, such experiments offer the prospect of gaining greater insight into the microscopic structure of the positive carriers.

In addition to the mobility-related experiments just mentioned, charged probes are also being used to investigate the nature of steady-state turbulence in superfluid helium.[58,59] Such a turbulent state is hypothesized to consist of a random tangle of quantized vortex lines,[60] maintained in dynamic equilibrium by a competition between vortex-creating and vortex-annihilating processes. The actual experiments are analogous to the rotating bucket experiments described in the previous section, in that a current of charge carriers is passed through the turbulent helium and the effects of vortex-line trapping are observed. Results so far seem to confirm the Vinen model[60] and to determine some of the relevant phenomenological parameters; and Moss[61] has recently observed fluctuation effects that appear to be directly related to individual vortex-annihilation events. The topic of quantum turbulence has been somewhat neglected, and further studies will undoubtedly provide valuable new information on the creation and annihilation processes.

As a final example, we note that under the application of suitable external fields both the negative and the positive carriers can form electrostatically bound states just inside the free surface of helium.[62] For typical experimental values of the applied electric field, the carriers will be trapped in a potential well that has its minimum a few hundred ångstrom units from the surface. As mentioned previously, the peculiar features of these states have been exploited by Poitrenaud and Williams[27] to obtain very precise values for the effective masses of the carriers. In addition, recent studies[63] of the field emission of electrons from surface-trapped bubbles into the vacuum raise the possibility of learning more about the microscopic properties of the free helium surface. The reader is referred to Ref. 62 for further details.

II. STRUCTURE OF THE CHARGE CARRIERS

The structures sketched in Fig. 1 constitute local deformations involving about 40 strongly interacting atoms in the case of the positive carrier and several hundred in the case of the negative. If one also remembers that helium at these temperatures must be treated quantum mechanically, it quickly becomes apparent that a genuinely microscopic description of the fluid in the neighborhood of a charge is prohibitively difficult. On the other hand, the relatively large size of the structures, and the considerable quantum-mechanical averaging over interatomic distances that occurs in liquid helium, lead to the natural approximation of treating the fluid as a continuum. As in Section I.B, we therefore describe the state of the fluid in terms of a nonuniform density field $\rho(\mathbf{r})$ and assume that the equations governing this quantity are the macroscopic equations.

Since we wish to discuss densities that are nonuniform on the scale of a few ångstrom units, the local value of $\rho(\mathbf{r})$ must be interpreted as the quantum statistical average of the density operator over a region roughly $1\,\text{Å}$ across. To make the continuum description valid, thermal fluctuations in this local density must be small and rapid enough to be unimportant, and the macroscopic equations must give an adequate description of the long-time behavior of $\rho(\mathbf{r})$. These requirements are difficult to discuss rigorously, since the discussion necessitates a solution for just the problem we are trying to avoid. It is perhaps more to the point to note that the continuum approximation is known to be semiquantitatively successful in describing the properties of systems even smaller than the charged probes,[64] and therefore it should be a phenomenologically sound approach to the present problem. As Section II.C indicates, there is little doubt that (in this case) the end justifies the means.

A. The Positive Carrier

The original paper of Atkins[8] setting out the electrostrictive model of the positive carrier is very complete indeed, and it forms the basis of the following discussion. Later authors[26,65] have preferred to present the same physics in a somewhat different way, focusing more on the local forces acting in the fluid.

The presence of an excess charge at r_0 generates a local displacement field

$$\mathbf{D} = e\frac{\mathbf{r}-\mathbf{r}_0}{|\mathbf{r}-\mathbf{r}_0|^3} \tag{3}$$

in the fluid. This field induces a polarization $\mathbf{P}=\chi\mathbf{D}/(1+4\pi\chi)$ to give a local electric field $\mathbf{E}=\mathbf{D}/(1+4\pi\chi)$, where χ is the dielectric susceptibility per unit volume. Helium is difficult to polarize,[66] and even for solid densities it is sufficient[67] to put $\chi=n\gamma$, where n is the local number density of atoms and γ is the atomic polarizability. The equation of electrostrictive equilibrium $\nabla p=(\mathbf{P}\cdot\nabla)\mathbf{E}$ then takes the simple form

$$\frac{\nabla p}{\rho} = \frac{\beta}{2}\nabla\left[\frac{D^2}{(1+4\pi\beta\rho)^2}\right] \tag{4}$$

where we have written $\chi=\rho\beta$ so that β is the polarizability per unit mass.

To complete the description, we note that below $T\sim1.2°\text{K}$ the equation of state of the liquid is[67]

$$p_l = A_1(\rho-\rho_0) + A_2(\rho-\rho_0)^2 + A_3(\rho-\rho_0)^3 \tag{5}$$

where $\rho_0=0.14513$, $A_1=5.68\times10^8$, $A_2=1.110\times10^{10}$, and $A_3=7.41\times10^{10}$ in cgs units. Inserting (3) and (5) into (4), the resulting expression can be integrated analytically to give

$$B_1\ln\left(\frac{\rho_2}{\rho_1}\right) + B_2(\rho_2-\rho_1) + B_3(\rho_2^2-\rho_1^2)$$

$$= \frac{1}{2}\beta e^2\left[\frac{1}{r_2^4(1+4\pi\beta\rho_2)^2} - \frac{1}{r_1^4(1+4\pi\beta\rho_1)^2}\right] \tag{6}$$

where $B_1 = A_1 - 2\rho_0 A_2 + 3\rho_0^2 A_3$, $B_2 = 2A_2 - 6\rho_0 A_3$, and $B_3 = \frac{3}{2} A_3$. Given the density ρ_1 at r_1, (6) determines the density ρ_2 at any other radius r_2. The local pressure is then derived from (5). Figure 2 shows the computed pressure variations for various values of the pressure at infinity.

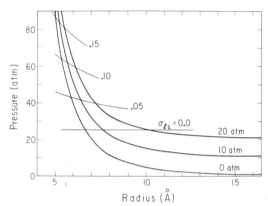

Fig. 2. Electrostrictive variation of the fluid pressure near an excess charge. (After Ref. 65.)

As Fig. 2 indicates, the total pressure near the charge exceeds the melting pressure p_m, which in the low-temperature limit has the value 25.0 atm.[68] This suggests that the helium right around the charge should be solid, with a transition to the liquid state occurring at a radius of roughly 6 Å. More generally, one must include the possibility of a surface energy density σ_{ls} associated with the existence of such a boundary, in which case the condition defining the location of the liquid-solid surface $r = b_+$ is:[69]

$$p_l(b_+) = p_m + \frac{2\sigma_{ls}}{b_+} \frac{v_s}{v_l - v_s} \qquad (7)$$

where p_l is the pressure in the liquid, and v_l and v_s are the molar volumes of the liquid and the solid, respectively, at the melting pressure. Figure 2 also contains plots of $p_m + (2\sigma_{ls}/r)v_s(v_l - v_s)^{-1}$ for various values of σ_{ls}. The intersections of these curves with the $p_l(r)$ curves determine the core radius b_+ for given values of $p_l(\infty)$ and σ_{ls}. Experiments discussed in Section II.C show convincingly that a liquid-solid surface does exist and that the associated surface energy is 0.10 $\pm$ .05 erg/cm^2.

The continuum approximation is not necessarily a very good description of the core region, but one can ignore such quibbles and write an equation of state for solid helium in the form

$$p_s = p_m + A_1'(\rho - \rho_0') + A_2'(\rho - \rho_0')^2 + A_3'(\rho - \rho_0')^3 \qquad (8)$$

where $\rho_0'=0.190$, $p_m=2.53\times 10^7$, $A_1'=1.33\times 10^9$, $A_2'=1.52\times 10^{10}$, and A_3' $=1.99\times 10^{11}$, all in cgs units. This fits the available data[70-72] to within experimental errors up to a pressure of 2000 atm. The equation determining the density variation inside the core is then again (6), with ρ_0 and A_i replaced by their primed values. The value of ρ_1 at $r=b_+$ is determined by the fact that just inside the solid surface, the pressure must be given by $p_s=p_l+2\sigma_{ls}/b_+$, or

$$p_s(b_+)=p_m+\frac{2\sigma_{ls}}{b_+}\frac{v_l}{v_l-v_s} \tag{9}$$

Figure 1a shows the resulting structure of the positive charge carrier for $p_l(\infty)=0$, assuming $\sigma_{ls}=0.10$ erg/cm^2. Table I gives some useful core parameters computed for different values of σ_{ls}. Here m_+ refers to the total mass contained in the solid core (*not* the effective mass).

TABLE I

σ_{ls}	erg/cm^2	0.00	0.05	0.10	0.15
b_+	Å	6.66	5.89	5.37	5.00
$\rho_l(b_+)$	g/cm^3	0.172	0.184	0.194	0.204
m_+	^{4}He masses	47	35	29	24

To conclude this section, we note that the charge carrier structure becomes temperature dependent above $T\sim 1.2°$K. The primary effect is that p_m starts to rise rapidly with increasing temperature; thus from (7) the core radius b_+ will shrink correspondingly. The extra complications introduced by this temperature dependence are of little interest here, but for completeness, Fig. 3 presents the computed values of $b_+(T)$ at the vapor pressure.[73]

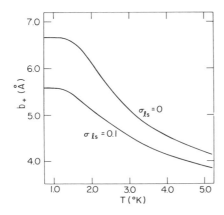

Fig. 3. Variation of the positive carrier core radius with temperature. (After Ref. 73.)

B. The Negative Carrier

A low-energy electron has a net repulsive interaction with a ^{4}He atom, characterized by a positive scattering length[74] $a = 1.13(h^2/m_e e^2)$. The states of an electron in dilute ^{4}He gas will therefore be of a quasifree, extended character, but at higher densities there will be a transition in which the lowest energy states become localized in the statistically occurring low-density regions of the gas.[75] An electron trapped in such a region has time to act back on the gas and will dig itself in deeper by pushing atoms from its immediate neighborhood. This transition has been studied experimentally by several investigators[13,76] and continues to be of interest. Both the experimental observations and calculations[77,78] of the free energy of variously localized electron states make it reasonable to suppose that at liquid densities the electron has become trapped in a spherical cavity from which the ^{4}He atoms have been excluded.

The simplified bubble model is based on the approximation that the density of helium atoms is given by (Fig. 1b)

$$\rho = 0 \qquad r < b$$
$$\rho = \rho_\infty(p, T) \qquad r \geqslant b \tag{10}$$

where $\rho_\infty(p, T)$ is the value of the bulk density far away from the bubble. The equilibrium radius b_- is then determined by minimizing the total energy in the continuum approximation[79]

$$E_{\text{tot}} = E_e(b) + 4\pi\sigma b^2 + \frac{4}{3}\pi p b^3 \tag{11}$$

with respect to b. Here E_e is the energy of the localized electron, σ is the energy density associated with the bubble surface, and p is the pressure in the liquid. From Fig. 2 we see that electrostrictive effects are quite small for $r \geqslant 12\,\text{Å}$, and polarization energy terms have therefore been omitted from (11).

The two problems that arise with (11) are what to use for σ and how to calculate E_e. As to the first of these, the simplest assumption at $p = 0$ is to use the measured surface tension of bulk helium.[80] This ignores corrections arising from the electron wavefunction and from the high curvature of the bubble surface, but the treatment of such corrections also requires a reappraisal of the approximation represented by (10). At nonzero pressure there are no direct measurements of surface tension, and we are forced to use a theoretical estimate of the pressure dependence—for example, the mean-field prediction of Amit and Gross[81] that σ should scale as $c(p)$ $\rho_\infty(p)$, where c is the velocity of sound. The ground-state energy $E_e(b)$ of the localized electron can actually be calculated to a reasonable approximation by assuming that the electron wavefunction ψ_e goes to zero at $r = b$.

However, it has been more fashionable to allow for some leakage of ψ_e into the $r > b$ region by assuming that the liquid presents an effective potential barrier $V_0(\rho_\infty)$ against penetration by the electron (Fig. 1b). The effective barrier height V_0 can be calculated from a Wigner-Seitz argument,[82] and the values so obtained are in good agreement with the electron injection experiments of Sommer[83] and of Woolf and Rayfield.[84]

Given these approximations, it is a simple matter to calculate $b_-(p)$ from (11). The lower curve in Fig. 4 results from scaling σ as $c(p)\rho_\infty(p)$, whereas the upper curve is obtained by using the $p = 0$ bulk value of σ at all pressures. It is clear that the pressure dependence of σ does not introduce a very significant correction. The points in the figure, representing various experimental determinations of $b_-(p)$, are discussed in the next section. For completeness, we show in Fig. 5 the temperature variation of b_- at the vapor pressure.[73] The bubble at first grows with increasing temperature because σ decreases. Eventually it begins to decrease in size as the temperature variation of the vapor pressure becomes the dominant factor.

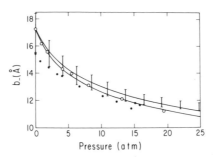

Fig. 4. Variation of the electron bubble radius with applied pressure. The curves show the results of theoretical calculations discussed in the text. The dots are from photoejection experiments,[98] the circles are from vortex-line trapping experiments,[107] and the vertical bars are from measurements of the phonon-limited mobility.[106] (After Ref. 106.)

It should be noted that the model presented here has the virtue of being sufficiently simple to permit the straightforward calculation of other interesting properties of the electron bubble. We refer in particular to treatments of the optical properties[85,86] and of the vibrational modes,[87,88] which have proved to be essential to the interpretation of certain experiments.

Although very successful, the simple bubble model is based on assumptions about the region near the bubble surface which are both phenomenological and to some extent unphysical. Several authors[89–92] have

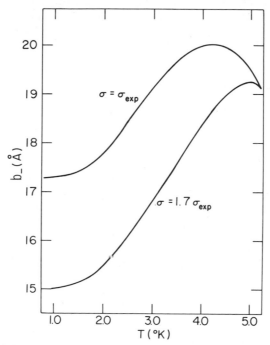

Fig. 5. Variation of the electron bubble radius with temperature. (After Ref. 73.)

therefore tried to motivate this model theoretically and to provide a better description of the surface region. The most recent and probably most realistic of these treatments is that of Padmore and Cole.[78] Their approach is to minimize the free energy F as a functional of the ^{4}He density distribution $\rho(r)$ and the electron wavefunction $\psi_e(r)$. The dependence of F on $\rho(r)$ is chosen to fit known properties of bulk helium, and a variational calculation of the density profile at the free helium surface (using this part of F) successfully reproduces the results of recent microscopic calculations. The effective interaction between the electron and the ^{4}He atoms is again treated in the Wigner-Seitz approximation. Using a trial wavefunction ψ_e, which has the form of a particle in a potential well of finite depth (as in Fig. 1b), and a somewhat more complicated form of $\rho(r)$, the authors obtain the variational solution appearing in Fig. 6.

The calculation of Padmore and Cole provides a great deal of insight into the range of validity of the simple bubble model. The "average" radius they compute is about 17.2 Å, in perfect agreement with the value of b_- predicted by (11). The authors also find that the surface energy of the bubble is nearly identical to the bulk value. On the other hand, the ^{4}He

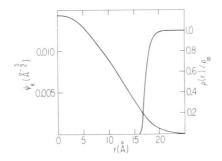

Fig. 6. Result of a variational calculation of the ^{4}He density profile $\rho(r)$ and the electron wavefunction $\Psi_e(r)$ of the electron bubble. (After Ref. 78.)

density profile does not obey the assumption of (10) but rather rises smoothly to the bulk density over a distance of $\sim 4\,\text{Å}$. Thus the effective radius of the bubble may vary by $\pm 2\,\text{Å}$, depending on the phenomenon under consideration.

C. Experimental Evidence

In Table II, we give values of b_+ and b_- at $p = 0$ derived from several independent experiments. The table is arranged roughly according to the order of complexity of the physical process from which the radii are determined. In every case the radius has been extracted more or less indirectly by setting up a theoretical model in which the radius is treated as an adjustable parameter, and then fitting the theory to the data. To the extent that these analyses are oversimplified, the results in Table II should be model dependent. It is therefore gratifying to see the excellent overall agreement between various determinations of the radii, an agreement that is in fact as good as one can expect, given the idealizations inherent in the bubble and electrostrictive models.

Effective masses have been obtained from microwave measurements[26]

TABLE II

Method	b_-	b_+
Effective mass	17.2 ± 0.15	5.8 ± 0.1
Viscosity-limited mobility	$14 \ \pm 2$	5.4 ± 0.4
Photoejection	15.4 ± 0.3	—
^{3}He-limited mobility	$17 \ \pm 1$	—
Phonon-limited mobility	$17 \ \pm 2$	5.4 ± 0.4
Vortex-line trapping	$17 \ \pm 3$	6.4 ± 1

and by measuring the resonant frequency of carriers trapped near the free surface of helium.[27] The latter technique has provided the very precise values $M_+ = (43.6 \pm 2)m^4{}_{He}$ and $M_- = (243 \pm 5)m^4{}_{He}$. The effective mass of a rigid bubble in a fluid of density ρ is just $\frac{2}{3}\pi\rho b^3_-$, permitting the immediate extraction of b_-. The positive carrier mass is the sum of the mass contained in the solid core and a hydrodynamic contribution from the surrounding nonuniform fluid. The relation between core mass and b_+ can be obtained from Table I. Barrera and Baym[28] have shown that the hydrodynamic mass is not significantly affected by the density gradient near the core but is given by $\frac{2}{3}\pi\rho_\infty b^3_+$. Combining these contributions, we are able to extract the value $b_+ = 5.8 \pm 0.1$ Å from the measured effective mass, corresponding to a liquid-solid surface energy of about 0.06 erg/cm^2. Note that our value of b_+ is somewhat lower than that derived in Ref. 27, where the core is treated as a uniform solid.

Another relatively simple approach[73,93,94] is to measure the mobilities of the charge carriers in normal helium and to interpret these in terms of classical, viscosity-limited motion. In practice, the theoretical interpretation[73] involves enough complications to make this a not particularly accurate way of determining the radii. Assuming that the surface of the bubble cannot support tangential stresses (as is the case for a free surface), the modified Stokes law $F = 4\pi v \eta b_-$ determines the mobility in terms of b_- and the independently measured viscosity η. The resulting value $b_- = 14 \pm 2$ Å is somewhat low, perhaps because the surface structure of the bubble in Fig. 6 is not properly taken into account in our simple expression for F. The analysis of the positive carrier mobility is very complicated indeed, since the density variation near the core implies a concomitant viscosity variation. A detailed treatment of the resulting nontrivial hydrodynamics gives $b_+ = 5.4 \pm 0.4$ Å.

A third technique has been to study the photoejection of the electron from the bubble state.[95-98] Experimentally, the mobility of the negative carriers is measured as a function of the wavelength of the applied radiation, and increases in the mobility are observed at certain wavelengths because the electrons are excited up to nonlocalized, continuum states. The resulting resonance structure may be interpreted[99] to yield $b_- = 15.4 \pm 0.3$ Å and $V_0(p=0) \approx 0.95$ eV. This particular measurement probes an "interior" feature of the electron bubble; thus it may very well give a somewhat low value for the effective bubble radius.

Rows 4 and 5 of Table II are obtained from mobility measurements in the regime where the elementary excitations act as a dilute gas. Here the Boltzmann equation is used to derive the mobility from a microscopic model of the carrier-excitation scattering process. In the case of ^{3}He impurities it is found that the negative carrier scatters the impurity quasi-particles like a hard sphere with an effective radius of 21 Å. The value of b_- given in the table is obtained by assigning an effective radius 4 ± 1 Å to the ^{3}He impurity.[100] Phonon scattering can be treated consistently within the continuum approximation, since the thermal wavelengths of the phonons are large on the atomic scale. Results are sensitive to the radius of the carrier, to the pressure response of the carrier surface, and (in the case of the positives) to density variations outside of the core. A complete analysis of ^{3}He and phonon-limited mobilities can be found in Sections IV and V.

The final row of Table II gives values of b_+ and b_- determined from measurements of the lifetime of a charge carrier trapped on a quantized vortex line[35, 101, 102] (see the discussion in Section I.B). The data have been interpreted[34, 103] on the assumption that the region near the vortex line represents an effective potential well whose depth and shape depend on the size of the carrier. The escape of a trapped carrier from the well when an electric field is applied is then treated as the escape of a Brownian particle over a potential barrier. This leads to lifetimes that depend strongly on temperature, with different radii giving very different temperature dependences. Thus measurements of the lifetimes as a function of temperature and applied electric field can be interpreted to yield estimates of the charge carrier radii.

The stochastic model of charge carrier trapping has had considerable success, but it also has some weaknesses. In particular, there is little reason to believe that the close-in interaction of a charge carrier with a vortex line can be described in terms of a simple effective potential, as assumed by Donnelly and Roberts.[34] Later authors[104, 105] have attempted to improve the Donnelly-Roberts theory but have used the same phenomenological *ansatz*. Since the lifetime predictions of the stochastic model are sensitive to the nature of the close-in interaction, it may be well to consider them as mainly of qualitative value. Indeed, the theory does not consistently give all the experimentally observed features: fitting the magnitude of the lifetime[103] of the negative carrier yields $b_- \approx 14.5$ Å, whereas fitting the temperature dependence[35] yields $b_- \approx 20$ Å. This uncertainty is indicated in Table II by taking $b_- = 17 \pm 3$ Å. The radius of the positive carrier has been determined from fitting the lifetime magnitude only, and we have somewhat arbitrarily assigned to it the same fractional uncertainty.

We conclude this section with a brief discussion of the pressure dependence of the carrier radii. The photoejection,[97, 98] the phonon-limited mobility,[106] and the vortex-line trapping measurements[35, 107] have all been extended to measure the electron bubble radius as a function of pressure.

The various experimental results are plotted in Fig. 4, where the results of the vortex-line trapping measurements (open circles) have been normalized to give $b_-(p=0)=17\,\overset{\circ}{A}$. It is clear that the pressure dependence of b_- is in substantial agreement with the prediction of the simple bubble model.

From the discussion of Section II.A and particularly Fig. 2, it is clear that the pressure dependence of b_+ is strongly influenced by the magnitude of σ_{sl}. For $\sigma_{sl}=0$ the core radius grows indefinitely as $p_l(\infty)\rightarrow p_m$, and the absence of such a drastic effect at high pressures[44,94,106] is convincing evidence for the existence of a liquid-solid surface with an associated surface energy density. Although the pressure dependence of b_+ has not been studied as thoroughly as that of b_-, a detailed analysis[73] of mobility measurements in normal helium shows that $b_+(p)$ is satisfactorily given by the electrostriction model, provided we assume $\sigma_{sl}=0.10\ \pm 0.05$ erg/cm^2.

III. THE TRANSPORT PROBLEM

A. Qualitative Aspects

The excitation spectrum[18] of pure superfluid ^{4}He appears in Fig. 7. Since only the regions near $k=0$ and $k=k_0$ are thermally populated, it is convenient to speak of two distinct kinds of elementary excitations. The phonon region is described by $\epsilon=\hbar ck$, where the velocity of sound c at zero pressure is ~ 240 msec^{-1}. Since the wavelengths are considerably greater than interatomic distances, these excitations are simply the quanta of the classical sound field. The roton region may be characterized by

$$\epsilon=\Delta+\frac{\hbar^2(k-k_0)^2}{2m_r} \tag{12}$$

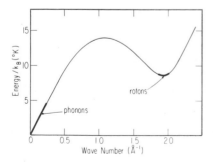

Fig. 7. The low-energy excitation spectrum of superfluid helium. Only the heavily drawn parts of the spectrum are thermally populated.

where at zero pressure $\Delta/k_B = 8.65 \pm 0.04°K$, $k_0 = 1.92 \pm 0.01 Å^{-1}$, and $m_r = 0.16 \pm 0.01 m^4{}_{He}$. The nature of the excitations in the roton region is not perfectly understood, but may be approximately described in terms of a phononlike wavefunction multiplied by a dipolar backflow field.[108] The parameters c, Δ, k_0, and m_r depend on pressure at all temperatures and on temperature when $T \gtrsim 1.2°K$.[109-111] Given the excitation spectrum, it is a simple matter to calculate the number density and the effective mass density of the excitations (Fig. 8). At high temperatures, the normal fluid properties are dominated by rotons because of their large momenta. As the temperature is lowered, the roton density drops exponentially, and the phonons eventually dominate. For the problem of interest to us here, one finds that the motion of the charge carriers is limited by roton scattering for $T \gtrsim 0.6°K$ and by phonon scattering at lower temperatures. Thus both regimes can be studied more or less independently by varying the temperature.

In addition to the thermal excitations, ^{3}He impurities can be added to obtain a third type of excitation. Such impurities are characterized by the

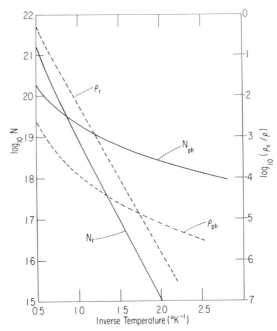

Fig. 8. Number and effective-mass densities of rotons and phonons as a function of temperature.

quasiparticle spectrum

$$\epsilon = -\epsilon_{03} + \frac{\hbar^2 k^2}{2m_3^*} \qquad (13)$$

with $\epsilon_{03}/k_B = 2.8°K$ and $m_3^* = 2.4m^3{}_{He}$.[112] Because of its smaller mass, the ³He atom exerts an excess quantum pressure on its neighboring ⁴He atoms, and therefore inhabits a somewhat larger volume than a ⁴He atom. For our purposes, the ³He quasiparticle can be visualized as a low-density region with a radius of about 4 Å and a fractional volume excess $\alpha = (V_3 - V_4)/V_4$ of about 0.29. The impurity-limited mobility of the charge carriers can be studied experimentally by merely adding a sufficient concentration of impurities to dominate the other scattering mechanisms. From Fig. 8 it is obvious that this requires only a small concentration when $T \lesssim 1.0°K$.

Consider now a gas of charge carriers moving under the combined influence of excitation scattering and the force $e\,\mathscr{E}$ exerted by an external electric field. The standard quantity measured in experiments of the Meyer-Reif type is the mobility $\mu(\mathscr{E}) \equiv \bar{v}/\mathscr{E}$, where $\bar{v}$ is the average drift velocity of the carriers. The bulk of our discussion is concerned with the zero-field limit (denoted simply by μ), but there is a considerable amount of interesting data on the field dependence in various scattering regimes. To determine the steady-state transport properties of a gas, we must calculate the momentum distribution function f_p of the carriers. The density of carriers is assumed so low that they do not affect one another— space charge fields and excitation gas drag effects are neglected. Let us denote by $\Gamma(\mathbf{p} \rightarrow \mathbf{p}')$ the probability per unit time that a carrier with momentum $\mathbf{p}$ be scattered to a different state $\mathbf{p}'$, where Γ will of course depend on the number and character of elementary excitations in the neighborhood as well as on the properties of the carrier. The distribution function is then determined by the Boltzmann equation

$$e\,\mathscr{E} \cdot \frac{\partial f_p}{\partial \mathbf{p}} = \sum_{\mathbf{p}'} \left[f_{\mathbf{p}'} \Gamma(\mathbf{p}' \rightarrow \mathbf{p}) - f_{\mathbf{p}} \Gamma(\mathbf{p} \rightarrow \mathbf{p}') \right] \qquad (14)$$

The use of this equation assumes that the charged probe spends only a small fraction of its time interacting with the excitations. If we estimate the average time of interaction with a single excitation as $b/\langle |\mathbf{v}_g - \mathbf{v}| \rangle$, where b is the size of the probe, $\mathbf{v}$ its velocity, and $\mathbf{v}_g$ the group velocity of the excitation, this requirement becomes

$$\frac{b}{\langle |\mathbf{v}_g - \mathbf{v}| \rangle} \times \text{total scattering rate} \ll 1 \qquad (15)$$

The total scattering rate is roughly the total excitation density N times $\langle |\mathbf{v}_g - \mathbf{v}| \rangle$ times the cross-section. Writing $\sigma \approx b^2$, we obtain the very reasonable condition $Nb^3 \ll 1$ for the applicability of (14). To be conservative, we require $Nb^3 \leqslant 0.1$. Then from Fig. 8 we must restrict our discussion to $T \lesssim 1.7°$K for positive carriers ($b \approx 6$ Å) and to $T \lesssim 1.1°$K for negative carriers ($b \approx 17$ Å). The limits on the relative ^{3}He concentration are N_3/N_4 $\lesssim 2.5\%$ for positives and $N_3/N_4 \lesssim 0.1\%$ for negatives. For temperatures or concentrations much higher than these, the response of the probe to one excitation will depend on the fact that it is simultaneously interacting with many others.

In general the presence of the probe will modify the distribution function of the excitations in its neighborhood, which in turn will affect $\Gamma(\mathbf{p} \rightarrow \mathbf{p}')$. To avoid having to solve such a complicated coupled problem, we further want to confine ourselves to the case in which the momentum relaxation mean free paths of the excitations are much greater than the size of the probe.[113] It is not necessary to engage in a detailed discussion of such mean free paths.[18,21] Rather, we note that the relevant parameter is the distance λ which enters into the kinetic theory formula for the viscosity

$$\eta = \frac{1}{3}\rho_n \bar{v}_g \lambda \tag{16}$$

and which can therefore be estimated from experimental values of η. Consider the data[114,115] plotted in Fig. 9. In the interpretation of such curves it is necessary to remember that phonon scattering is a relatively inefficient momentum relaxation process. Thus the flat part of the pure helium curve at high temperatures arises from the momentum flux carried by rotons whose mean free path is limited entirely by roton-roton scattering. Inserting $\rho_n = \rho_r$ and $\bar{v}_r = \sqrt{k_B T/m_r}$ into (16) yields $\lambda_r \approx 10^{-9} \exp(\Delta/k_B T)$. The requirement $\lambda_r \gtrsim 10b$ then leads to the restrictions $T \lesssim 1.2°$K for negatives and $T \lesssim 1.4°$K for positives. According to Fig. 9, the effect of adding moderate amounts of ^{3}He is to reduce λ_r somewhat, but this effect is small enough not to impose any strong restrictions on N_3/N_4.

The sharp rise in η below $\sim 1.2°$K is due to momentum flux carried by phonons, whose viscosity mean free path is also limited by roton scattering. Writing $\rho_n = \rho_{ph}$ and $\bar{v} = c$ in (16), we can estimate from Fig. 9 that $\lambda_{ph} \approx 2 \times 10^{-8} T^{-9/2} \exp(\Delta/k_B T)$. The condition that $\lambda_{ph} \gtrsim 10b$ then does not represent any additional restriction, since $\lambda_{ph} \approx 200$ Å at $1.4°$K. Figure 9 also shows that the addition of 1% of ^{3}He reduces λ_{ph} by about a factor of 5 at $1°$K, and by a lower factor at higher temperatures. Since $\lambda_{ph}/5$ ≈ 2000 Å at $1°$K, the presence of a few percent impurities clearly causes no problems.

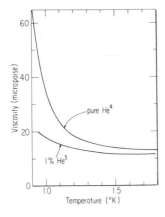

Fig. 9. Viscosity of pure ⁴He (upper curve) and of a 1% solution of ³He in ⁴He.

Finally, we consider the ³He viscosity mean free path, which will be limited by roton or ³He scattering. The scattering cross-section for impurities incident either on rotons or other impurities is roughly 50 Å.² Hence $\lambda_3 \approx 2 \times 10^{14}/N$, where N is the density of rotons and ³He impurities. The restriction $\lambda_3 \gtrsim 10b$ then implies roughly the same temperature limits as in the case of λ_r, and a concentration limit of order 1%.

To summarize, one can use the Boltzmann equation with confidence if on the average there is less than one excitation in the neighborhood of the charge carrier. The effect of the probe on the local distribution function of the excitations can be neglected if the viscosity mean free path of the dominant excitation species is large compared with the size of the carrier. If we require that both these conditions be fulfilled by a factor of 10, our discussion is restricted to $T \lesssim 1.1°$K for negative carriers and to $T \lesssim 1.4°$ for positive carriers. The limits on the impurity concentration are $N_3/N_4 \lesssim 0.1\%$ and $N_3/N_4 \lesssim 1\%$, respectively. It may be argued that a factor of 10 is somewhat excessive and that these conditions could be relaxed. However, since the application of pressure can increase the roton density by a factor of up to 5, our seemingly ultraconservative limits are in fact quite reasonable if we wish to discuss pressure as one of the variables.

B. Formal Aspects

Given a description of the microscopic processes in which individual excitations scatter off the charge carriers, we can readily find $\Gamma(\mathbf{p} \rightarrow \mathbf{p}')$ by summing over the equilibrium distribution n_k of the excitations. The microscopic description can then be tested by calculating $\bar{v}/\mathcal{E}$ from (14) and comparing the result against the experimentally measured mobilities. In the general case of finite $\mathcal{E}$, the solution of (14) becomes complicated, and it is perhaps wisest to eschew the analytical approach in favor of the less elegant Monte Carlo method. In the zero-field limit, however, this equation can be linearized, and some useful analytical results can be obtained.[28, 116–120] Sections of the following discussion are based on the papers of Baym and co-workers.

Detailed balance requires $\Gamma(\mathbf{p}' \to \mathbf{p}) = \Gamma(\mathbf{p} \to \mathbf{p}') f_{\mathbf{p}}^0 f_{\mathbf{p}'}^{0-1}$, where $f_{\mathbf{p}}^0$ denotes the equilibrium Maxwell-Boltzmann distribution function of the carriers normalized to one carrier per unit volume. In (14) we write $\Phi_{\mathbf{p}} = k_B T (f_{\mathbf{p}} - f_{\mathbf{p}}^0) / f_{\mathbf{p}}^0$ and linearize by replacing $f_{\mathbf{p}}$ on the left with $f_{\mathbf{p}}^0$. Then

$$\frac{e \, \mathscr{E} \cdot \mathbf{p}}{M} f_{\mathbf{p}}^0 = \sum_{\mathbf{p}'} (\Phi_{\mathbf{p}} - \Phi_{\mathbf{p}'}) f_{\mathbf{p}}^0 \Gamma(\mathbf{p} \to \mathbf{p}') \tag{17}$$

The function $\Phi_{\mathbf{p}}$ can be determined from the following variational principle:[121] given an equation $X = \theta \Lambda$, where θ is a linear, self-adjoint, and positive definite operator, then of all the functions that satisfy $\langle \Lambda, \theta \Lambda \rangle = \langle \Lambda, X \rangle$, the solution is that which maximizes $\langle \Lambda, \theta \Lambda \rangle$. Thus for any trial function ψ,

$$\frac{\langle \psi, \theta \psi \rangle}{\langle \psi, X \rangle^2} \geqslant \frac{1}{\langle \Lambda, X \rangle} \tag{18}$$

which in the case of (17) gives

$$\frac{\sum_{\mathbf{p}} \psi_{\mathbf{p}} \sum_{\mathbf{p}'} [\psi_{\mathbf{p}} - \psi_{\mathbf{p}'}] f_{\mathbf{p}}^0 \Gamma(\mathbf{p} \to \mathbf{p}')}{\left[\sum_{\mathbf{p}} \frac{e \, \mathscr{E} \cdot \mathbf{p}}{M} f_{\mathbf{p}}^0 \psi_{\mathbf{p}} \right]^2} \geqslant \frac{1}{\sum_{\mathbf{p}} \frac{e \, \mathscr{E} \cdot \mathbf{p}}{M} f_{\mathbf{p}}^0 \Phi_{\mathbf{p}}} \tag{19}$$

Let $\mathscr{E}$ define the z-direction. Since the excitation gas is isotropic, both $\Phi_{\mathbf{p}}$ and $\psi_{\mathbf{p}}$ must be isotropic in the x–y plane. It is clear that the expression on the right-hand side reduces simply to $(e \mathscr{E} k_B T \bar{v})^{-1}$. Inserting the trial function $\psi_{\mathbf{p}} = \bar{v} \cdot \mathbf{p}$, corresponding to an $f_{\mathbf{p}}$ having the form of a shifted Maxwell-Boltzmann distribution, we obtain the simple inequality

$$\frac{e}{\mu} \leqslant \frac{1}{3 k_B T} \sum_{\mathbf{p}} f_{\mathbf{p}}^0 \mathbf{p} \cdot \mathbf{R}_{\mathbf{p}} \tag{20}$$

where

$$\mathbf{R}_{\mathbf{p}} = - \sum_{\Delta \mathbf{p}} \Delta \mathbf{p} \Gamma(\mathbf{p} \to \mathbf{p} + \Delta \mathbf{p}) \tag{21}$$

is the average rate at which a carrier of velocity $\mathbf{p}/M$ loses momentum by excitation scattering.

If in a typical scattering event the carrier momentum changes by only a small fraction, we can take the alternative approach of solving (14) by making a Focker-Planck expansion in powers of $\Delta \mathbf{p} = \mathbf{p}' - \mathbf{p}$. It is easy to show that in this case (20) is satisfied as an *equality* in the zero-field limit. Since both phonons and ^{3}He impurities typically have momenta small compared with the rms carrier momentum, for these (20) provides an exact expression for the zero-field mobility. This simplification does not apply when roton scattering is important.

Since the excitation gas is assumed to be isotropic, $\mathbf{R_p}$ must take the form $\hat{\mathbf{p}}y(p)$. We can expand $y(p)$ in powers of p and insert in (20). Provided the width of $f_\mathbf{p}^0$ is less than the range of p over which $\mathbf{R_p}$ is linear, it is only necessary to keep the linear term in this expansion. One physical source of nonlinearities in $\mathbf{R_p}$ is a possible velocity dependence in the way that a given type of excitation interacts with a moving carrier. Such a dependence can arise, for example, from Doppler-shift effects in phonon scattering or from scattering by the superfluid backflow around the carrier. The other cause of nonlinearity is the requirement of energy conservation. For the phonon and ^{3}He-limited regimes, it is reasonable to assume that the linear approximation is valid—both types of excitation have high group velocities compared with thermal carrier velocities, and both interact with a carrier as though it had essentially infinite mass. For roton-limited motion it is less obvious that the linear term in $\mathbf{R_p}$ is sufficient to provide a good estimate of μ, but there are heuristic arguments to support such an assumption. By taking the moment of (14), it is easy to show that $e\mathcal{E} = \Sigma_\mathbf{p} f_\mathbf{p} \mathbf{R_p}$. From this it can be argued that if $\mathbf{R_p}$ had a strongly nonlinear behavior for thermal values of $\mathbf{p}$, the field dependence of $\bar{v}$ would already show pronounced nonlinearities for $\bar{v} < \langle p/M \rangle_\mathrm{rms}$. In fact, the field dependences of the roton- and phonon-limited drift velocities are found experimentally to be very similar, which at least suggests that in the thermal range $\mathbf{R_p}$ has a similar $\mathbf{p}$-dependence for rotons and phonons. Although further study is required to decide this question definitely, we continue our discussion on the assumption that e/μ_r is also determined primarily by the linear term in $\mathbf{R_p}$.

If the scattering processes can be described in first order, we may write

$$\Gamma(\mathbf{p} \to \mathbf{p}') = \sum_{\mathbf{k}, \mathbf{k}'} n_\mathbf{k} (1 \pm n_{\mathbf{k}'}) \Gamma(\mathbf{p}, \mathbf{k} \to \mathbf{p}', \mathbf{k}') \qquad (22)$$

where $\Gamma(\mathbf{p}, \mathbf{k} \to \mathbf{p}', \mathbf{k}')$ is the rate at which the carrier scatters one excitation per unit volume of momentum $\hbar\mathbf{k}$ to the state $\hbar\mathbf{k}'$, $n_\mathbf{k}$ is the equilibrium distribution function of the excitations, and the ± 1 term refers to boson (phonons, rotons) or fermion (^{3}He impurities) statistics, respectively. Then the linear term in $\mathbf{R_p}$ looks like

$$\mathbf{R_p^{(1)}} = -\mathbf{p} \cdot \sum_{\Delta \mathbf{p}, \mathbf{k}, \mathbf{k'}} \Delta \mathbf{p} n_{\mathbf{k}} (1 \pm n_{\mathbf{k'}}) \frac{\partial \Gamma}{\partial \mathbf{p}} \bigg|_{\mathbf{p}=0} \qquad (23)$$

Baym et al.[116] have given an ingenious argument that allows the transformation of this expression to one that involves only the scattering rate from an initially stationary carrier. Consider $s(\mathbf{p}) = \bar{n}_{\mathbf{k}}(1 \pm \bar{n}_{\mathbf{k'}})$, where $\bar{n}_{\mathbf{k}}$ is that distribution of excitations which characterizes thermal equilibrium in a reference frame moving with velocity $\mathbf{p}/M$. In the presence of such an excitation gas, a carrier with momentum $\mathbf{p}$ will on the average suffer no momentum loss:

$$\sum_{\Delta \mathbf{p}, \mathbf{k}, \mathbf{k'}} \Delta \mathbf{p} s(\mathbf{p}) \Gamma(\mathbf{p}, \mathbf{k} \rightarrow \mathbf{p} + \Delta \mathbf{p}, \mathbf{k'}) = 0 \qquad (24)$$

Expanding this equation in powers of $\mathbf{p}$, we obtain a series of conditions on Γ:

$$\sum_{\Delta \mathbf{p}, \mathbf{k}, \mathbf{k'}} \Delta \mathbf{p} s(0) \Gamma(0, \mathbf{k} \rightarrow \Delta \mathbf{p}, \mathbf{k'}) = 0$$

$$\sum_{\Delta \mathbf{p}, \mathbf{k}, \mathbf{k'}} \Delta \mathbf{p} \left[\frac{\partial s}{\partial \mathbf{p}} \bigg|_{p=0} \Gamma(0, \mathbf{k} \rightarrow \Delta \mathbf{p}, \mathbf{k'}) + s(0) \frac{\partial \Gamma}{\partial \mathbf{p}} \bigg|_{\mathbf{p}=0} \right] = 0 \qquad (25)$$

and so on. The first equation is trivial. The second, however, combines with (23) to give

$$\mathbf{R_p^{(1)}} = \mathbf{p} \cdot \sum_{\Delta \mathbf{p}, \mathbf{k}, \mathbf{k'}} \Delta \mathbf{p} \frac{\partial s}{\partial \mathbf{p}} \bigg|_{\mathbf{p}=0} \Gamma(0, \mathbf{k} \rightarrow \Delta \mathbf{p}, \mathbf{k'}) \qquad (26)$$

Clearly $\mathbf{R_p^{(1)}}$ is of the form $\mathbf{p} \cdot \mathbf{A}$; thus (20) gives immediately $e/\mu \leq M \operatorname{tr} \mathbf{A}/3$. We neglect the n_k' terms in s. These terms cancel in the infinite mass limit and are not important for rotons. Since in a frame moving with velocity $\mathbf{p}/M$ the excitation spectrum is Doppler shifted to $\epsilon'(\mathbf{k}) = \epsilon(k) - \hbar \mathbf{k} \cdot \mathbf{p}/M$, we have $\partial s / \partial \mathbf{p} = -(\hbar \mathbf{k}/M) \partial n_k / \partial \epsilon$. In first-order scattering theory, we can write

$\Gamma(0, \mathbf{k} \rightarrow \Delta \mathbf{p}, \mathbf{k'})$

$$= \frac{2\pi}{\hbar} |t(k, k', \theta)|^2 \delta[\Delta \mathbf{p} - \hbar(\mathbf{k} - \mathbf{k'})] \delta \left[\frac{\hbar^2 (k - k')^2}{2M} + \epsilon(k') - \epsilon(k) \right] \qquad (27)$$

where θ is the angle between $\mathbf{k}$ and $\mathbf{k}'$ and $|t(k,k',\theta)|^2 = |\langle \Delta\mathbf{p}, \mathbf{k}'|T|0,\mathbf{k}\rangle|^2$. The result of combining all these elements and carrying out the sum over $\Delta\mathbf{p}$ is

$$\frac{e}{\mu} \leqslant -\frac{\hbar^2}{6\pi^2} \int_0^\infty k^3 \frac{\partial n_k}{\partial \epsilon} \, dk \int \frac{d^3\mathbf{k}'}{(2\pi)^3} (k - k'\cos\theta)$$

$$\times \frac{2\pi}{\hbar} |t|^2 \delta\left[\frac{\hbar^2(\mathbf{k}-\mathbf{k}')^2}{2M} + \epsilon(k') - \epsilon(k)\right] \tag{28}$$

The delta function defines a path in k', θ space over which the integrand is to be evaluated. We take the integral over k' to obtain

$$\frac{e}{\mu} \leqslant -\frac{\hbar}{6\pi^2} \int_0^\infty k^3 \frac{\partial n_k}{\partial \epsilon}\left|\frac{\partial \epsilon}{\partial k}\right| dk \int d\Omega \sum_i (k - k_i\cos\theta)\sigma_i(k,\theta) \tag{29}$$

where the k_i's are the θ- and k-dependent roots of the delta function, and the differential cross-sections are defined by

$$\sigma_i(k,\theta) = \frac{|t(k,k_i,\theta)|^2 k_i^2}{(2\pi)^2 \left|\frac{\hbar^2}{M}(k_i - k\cos\theta) + \frac{\partial \epsilon}{\partial k_i}\right|\left|\frac{\partial \epsilon}{\partial k}\right|} \tag{30}$$

For the cases of phonon or ^{3}He scattering, there is only one root. In the infinite mass limit, this root is just k (independent of θ)—the energy delta function reduces to the requirement that $k = k'$. Remembering that in these cases the inequality becomes an equality, we obtain the expression of Baym et al:[116]

$$\frac{e}{\mu} = -\frac{\hbar}{6\pi^2} \int_0^\infty k^4 \frac{\partial n_k}{\partial k} \sigma_T(k) \, dk \cdot \tag{31}$$

where the momentum-transfer cross-section σ_T is defined by

$$\sigma_T(k) = \int d\Omega(1 - \cos\theta)\sigma(k,\theta) \tag{32}$$

Equation (31) furnishes a simple connection between the zero-field mobility and the cross-section for the scattering of an excitation by the stationary carrier. Thus mobility measurements provide a very direct test of microscopic models for phonon and ^{3}He impurity scattering. As Sections IV and V reveal, these processes now appear to be satisfactorily understood. The situation is much more complicated in the case of roton scattering; this fascinating unsolved problem is discussed in Section VI.

IV. ^{3}He-LIMITED MOBILITIES

A. Experimental Results

In general all three types of elementary excitations are present simultaneously, and it is often more convenient to discuss the *drag coefficient* e/μ (so named because the average force exerted by the excitations on the drifting carrier is e/μ times the average drift velocity). In the dilute gas regime the excitations interact independently with the carrier; thus

$$\frac{e}{\mu} = \frac{e}{\mu_{ph}} + \frac{e}{\mu_r} + \frac{e}{\mu_3} \tag{33}$$

each type of excitation making its own additive contribution to the average momentum loss rate of the carrier. The drag due to phonons and rotons decreases rapidly as the temperature is lowered (see Fig. 12), so that for every concentration N_3/N_4 there is a characteristic temperature below which e/μ_3 becomes an important, experimentally measurable contribution. In particular, the dilute gas condition $N_3/N_4 \lesssim 3 \times 10^3$ ppm (discussed in Section III.A) restricts the experimental region to $T \lesssim 0.9°$K.

The earliest measurements on ^{3}He-limited mobilities in superfluid helium were performed by Meyer and Reif[14] for $0.5 \lesssim T \lesssim 0.8°$K. These were later extended to much lower temperatures.[122] More recently, careful investigations involving a wide range of concentrations have been carried out by two groups.[123–125] We first discuss the results obtained in the range $0.3 \lesssim T \lesssim 1.0°$K, where the interpretation turns out to be the simplest. In this region, the impurity contribution e/μ_3 is determined as a function of concentration by first measuring e/μ in pure ^{4}He and subtracting it from the values of e/μ measured after the impurities have been added. It is found experimentally that in this temperature region, e/μ_3 is strictly proportional to the concentration of ^{3}He impurities. This is of course consistent with the supposition that the excitations act as a dilute gas, and it demonstrates that no complicating effects such as changes in the structure of the charge carriers with ^{3}He concentration or terms arising from the interactions between elementary excitations are important. The experimentally determined proportionality constants $e/\mu_3 N_3$ appear as functions of temperature in Fig. 10. Since the number of scatterers is constant, the mobility depends only very weakly on temperature: $e/\mu_3 N_3$

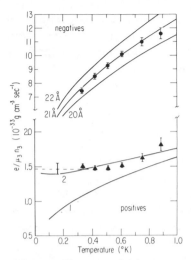

Fig. 10. ^{3}He-limited drag coefficients. The curves in the upper part of the figure give the results of hard-sphere calculations for the negative carrier. Curve 1 shows a similar calculation for the positive carrier; Curve 2 shows the results obtained by using the Bowley-Lekner polarization potential.[132] The dashed line indicates the low-temperature limiting behavior found by Kuchnir et al.[124] (After Ref. 125.)

is seen to vary as $T^{1/2}$ for the negative carriers and to saturate to a constant value at $T \lesssim 0.6°$K for the positives.

The technically more difficult region below $0.3°$K has been investigated by Kuchnir et al.[124] They find that μ_3^+ continues to be temperature independent down to $T = 0.017°$K.[126] A more interesting behavior is exhibited by the negative carriers: in Fig. 11 we have replotted some of the μ_3

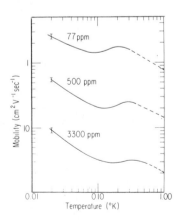

Fig. 11. Low-temperature behavior of the ^{3}He-limited mobility of the negative charge carrier. The solid lines show smoothed data taken from Kuchnir et al.[124]; the dashed lines are $T^{-1/2}$ continuations of these data.

data of Kuchnir et al. up to $0.35°K$ and continued it with the characteristic $T^{-1/2}$ dependence observed at higher temperatures. It is clear that the nature of the ^{3}He-negative interaction undergoes significant changes in the region between 0.1 and $0.3°K$, the effect seeming to shift to slightly higher temperatures as the concentration is increased. The existence of this anomaly is qualitatively understood[127] on the basis of the fact that a ^{3}He atom can reduce its zero point energy relative to a ^{4}He atom by sitting on a free surface. Thus there are bound states of ^{3}He at the surface of the bubble which presumably become occupied at $T \approx 0.25°K$. The surface energy density of the bubble is then reduced substantially,[128, 129] leading by (11) to an increase in the bubble radius. Although the predicted increase in b_- seems to account qualitatively for the observed dip in the mobility curves as T is lowered,[117] the change in μ from the high- to the low-temperature limiting behavior corresponds to about a 50% increase in b_-, which seems to be excessive. This has lead to speculations[124, 130] that the local attraction of the bubble surface for ^{3}He impurities gives rise to a scattering resonance that shows up as the observed anomaly in the mobility.

B. Theory of ^{3}He Scattering

If recoil effects are neglected, the stationary carrier represents a fixed potential $V(r)$ from which the ^{3}He quasiparticle is scattered. Since the de Broglie wavelength of the quasiparticles becomes comparable to the carrier sizes at these low temperatures, it is appropriate to solve Schrödinger's equation for this problem by the method of partial waves. The associated radial equation

$$\frac{d^2 u_l}{dr^2} + \left[k^2 - \frac{2m_3^*}{\hbar^2} V(r) - \frac{l(l+1)}{r^2} \right] u_l = 0 \qquad (34)$$

can be converted to a differential equation for the phase shifts η_l by the method of Levy and Keller[131]

$$\frac{d\eta_l(r)}{dr} = -kr^2 \frac{2m_3^*}{\hbar^2} V(r) [j_l(kr)\cos\eta_l - n_l(kr)\sin\eta_l] \qquad (35)$$

Here j_l and n_l are the spherical Bessel and Neumann functions, respectively, and the quantity $\eta_l(r)$ is the phase shift that would result if the scattering potential were set to zero for radii greater than r. The idea is to start with a suitable boundary condition on $\eta_l(r)$ at some value of r and to integrate (35) out to infinity to find the asymptotic value of η_l. The momentum transfer cross-section is then obtained from

$$\sigma_T(kk) = \frac{4\pi}{k^2} \sum_{l=0}^{\infty} (l+1)\sin^2(\eta_l - \eta_{l+1}) \tag{36}$$

and the drag coefficient e/μ_3 from (31), with

$$n_k = N_3 \left(\frac{2\pi\hbar^2}{m_3^* k_B T}\right)^{3/2} \exp\left(\frac{-\hbar^2 k^2}{2m_3^* k_B T}\right) \tag{37}$$

For want of a better description, we assume that the ^{3}He impurity can only penetrate to within some distance b_{eff} of the center of the charge carrier structure, the quantity b_{eff} playing the role of an effective hard-core collision radius. The boundary condition on $\eta_l(r)$ is then

$$\eta_l(b_{\text{eff}}) = \tan^{-1}\left[\frac{j_l(kb_{\text{eff}})}{n_l(kb_{\text{eff}})}\right] \tag{38}$$

We first discuss scattering by the negative carrier. Since the electron bubble has a well-defined surface and exerts no significant electrostrictive effects on the surrounding liquid, it seems reasonable to set $V(r) = 0$ for $r > b_{\text{eff}}$, absorbing the complicated details of what happens in the $\sim 2\,\text{Å}$ surface region into the parameter b_{eff}. Such a heuristic approach will of course fail if significant resonance effects are introduced by the local attraction that the bubble surface exercises on the ^{3}He impurities. If this were the case, $\sigma_T(k)$ would have a magnitude and k-dependence quite different from the hard-sphere behavior, and it should not be possible to fit $e/\mu_3 N_3$ as a function of T with a reasonable value of b_{eff}. The three curves in the upper half of Fig. 10 show the predictions of the hard-sphere model for various values of b_{eff}. The agreement with experiment is very satisfactory in that the temperature dependence is well represented and $b_{\text{eff}} = 21\,\text{Å}$ is a reasonable collision radius, given $b_- = 17.5\,\text{Å}$. Thus in this temperature range the hard-sphere model seems to be entirely adequate. As we discussed earlier, the anomaly in $e/\mu_3^- N_3$ oserved at lower temperatures is thought to reflect a sudden increase in b_- or possible resonance effects, or both.

A similar calculation for the positive carrier gives curve 1 in Fig. 10. It is evident that a hard-sphere scattering model does not give the correct low-temperature behavior of $e/\mu_3^+ N_3$. This difficulty has been resolved by Bowley and Lekner,[132] who pointed out the importance of electrostrictive effects in the neighborhood of the positive charge carrier. Since the ^{3}He atom occupies a somewhat greater volume than a ^{4}He atom, the electrostrictive force per unit volume is decreased in the neighborhood of a ^{3}He impurity. The result is that a ^{3}He impurity near the positive carrier experiences a repulsive polarization potential given approximately by

$$V(r) = \frac{\alpha\gamma e^2}{2r^4} \qquad (39)$$

where $\alpha = (V_3 - V_4)/V_4$ and γ is the atomic polarizability. It is possible to improve $V(r)$ somewhat[125] over the original Bowley-Lekner form by incorporating the pressure dependence[133] of the volume excess α. Curve 2 in Fig. 15 shows $e/\mu_3^+ N_3$ calculated from this potential with a hard-core cutoff at $b_{eff} = 7.0$ Å. Again the agreement with experiment is good; in particular, it is explained why μ_3 becomes constant at low temperatures.

In summary, the very simple microscopic scattering models outlined above give an adequate description of both the magnitude and the temperature dependence of the ^{3}He-limited mobilities for positive carriers at all temperatures and for negative carriers above 0.3°K. The low-temperature behavior of μ_3^- is not yet perfectly understood, but undoubtedly arises in one way or another from the local attraction exerted by the bubble surface on the ^{3}He impurity. Further work on this problem would clearly be of interest.

V. PHONON–LIMITED MOBILITIES

A. Experimental Results

In pure ^{4}He the mobilities will be limited by collisions with the thermally activated excitations of the phonon-roton spectrum. As discussed in Section III.A, at lower temperatures a changeover is expected from the exponential roton-limited behavior to a phonon-limited regime in which μ is a more gradual function of temperature. This changeover was first observed by Meyer and Reif.[14] Subsequent experimental work[134-137] has provided data sufficiently accurate and extensive to allow a complete theoretical analysis of the phonon-limited behavior.

Figure 12 shows e/μ measured with $p_i(\infty) = 0$. The phonon-dominated regime is clearly displayed by both kinds of carrier. Among the interesting qualitative features of these curves, we note that e/μ_{ph}^- varies as T^3, whereas e/μ_{ph}^+ approaches a T^8 variation. In addition, the magnitude of e/μ_{ph}^- is very large compared with e/μ_{ph}^+. It follows from (31) that if $\sigma_T(k)$ is assumed to vary as k^n, e/μ will go as T^{4+n}. Hence these observations may be interpreted roughly as implying a k^{-1} dependence for phonons

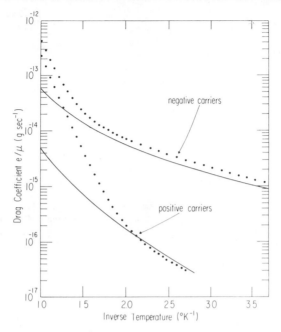

Fig. 12. Drag coefficients in pure ^{4}He at zero pressure. The upper curve shows the prediction of the simplified bubble model when $b_- = 17$ Å, $V_0 = 0.8$ eV. The lower curve gives the prediction of the electrostrictive model when $\sigma_{ls} = 0.10$ erg/cm^2.

scattering off the electron bubble and a k^4-dependence for scattering by the positive carrier. Since typical phonon wavelengths in this temperature regime are greater than 15 Å, the approximate k^4-dependence for the positives can be qualitatively understood as the usual Rayleigh limit for the scattering of a wave from an object that is small compared with the wavelength. The more peculiar scattering behavior of the negative carriers arises because the electron bubble has normal modes of vibration whose characteristic frequencies are comparable to those of the thermally excited phonons: both the large magnitude and the apparent k^{-1}-dependence of σ_T arise from resonance effects.

The pressure dependence of e/μ below 1°K has been studied extensively by Ostermeier.[106] The main effect of applying pressure is to increase the roton population while decreasing the phonon population, thus shifting the roton-phonon changeover region to lower temperatures (Fig. 13). Since the charge carrier structures are sensitive to pressure, one also observes significant changes in $\sigma_T(k)$, especially for the negative carrier, where the resonant frequencies are shifted sharply upward as the electron bubble is compressed. These matters are discussed more fully below.

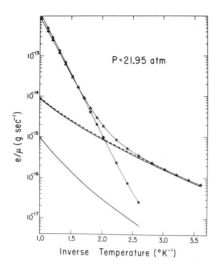

Fig. 13. Drag coefficients in pure ^{4}He at 21.95 atms, again including the theoretically calculated curves. The triangles represent the negative carriers and the circles the positive carriers. (After Ref. 106.)

B. Theory of Phonon Scattering

The importance of resonance effects in determining e/μ_{ph}^- was pointed out by Wang[138] and by Baym et al.[116] Somewhat later it was found[65] that the explanation of the measured positive carrier mobility also involved some interesting complications, in that electrostrictive effects and the response of the liquid-solid phase boundary to the phonon field had to be taken into account.

Figure 14 shows the phonon thermal weight factor $k^4(\partial n/\partial k)$ appearing in (31) for various temperatures. It is clear that we are interested in $\sigma_T(k)$ for wavelengths very long compared with interatomic distances. Given our continuum description of the charge carrier structures, it is entirely consistent to treat the phonon-scattering problem in the same approximation. If we include the electrostrictive force discussed in Section II.B, the ideal fluid equations take the form

$$\frac{\partial \mathbf{v}}{\partial t} + (\mathbf{v} \cdot \nabla)\mathbf{v} = -\frac{\nabla p}{\rho} + \frac{\beta}{2}\nabla\left[\frac{D^2}{(1+4\pi\beta\rho)^2}\right] \tag{40}$$

$$\frac{\partial \rho}{\partial t} + \nabla \cdot (\rho\mathbf{v}) = 0 \tag{41}$$

where $D(r)$ is given by (3) and the relation between p and ρ is given by (5).

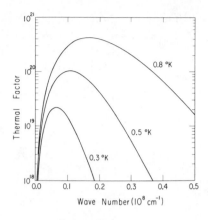

Fig. 14. Thermal weight factor k^4 $(\partial n/\partial k)$ for phonons at temperatures of interest. (After Ref. 65.)

Writing $\mathbf{v}=\mathbf{v}_s$, $p=p_0+p_s$, $\rho=\rho_0+\rho_s$ and linearizing, we obtain

$$\frac{\partial \mathbf{v}_s}{\partial t} = -\nabla\left(\frac{\rho_s}{\rho_0}\right) \tag{42}$$

$$\frac{1}{c_0^2}\frac{\partial p_s}{\partial t} + \nabla\cdot(\rho_0\mathbf{v}_s)=0 \tag{43}$$

where $\rho_s=(\partial\rho/\partial p)_0 p_s=p_s/c_0^2$ has been used to eliminate ρ_s, and some small terms have been neglected. We proceed in the standard way to introduce a velocity potential Φ defined by $\mathbf{v}=\nabla\Phi$. Then the equations governing the sound field are

$$p_s = -\rho_0\frac{\partial\Phi}{\partial t} \tag{44}$$

$$\nabla^2\Phi + \nabla(\ln\rho_0)\cdot\nabla\Phi - \frac{1}{c_0^2}\frac{\partial^2\Phi}{\partial t^2} = 0 \tag{45}$$

Both ρ_0 and c_0 are strong functions of r in the region near the charge, where electrostriction affects the state of the liquid. Far away from the charge, (44) and (45) reduce to the usual sound propagation equations.

As in the theory of ^{3}He scattering, we make use of partial-wave analysis to calculate the cross-section. The associated radial equation

$$\frac{d^2 u_l}{dr^2} + \frac{d(\ln \rho_0)}{dr} \frac{du_l}{dr} + \left[\frac{k^2 c_\infty^2}{c_0^2} - \frac{1}{r} \frac{d(\ln \rho_0)}{dr} - \frac{l(l+1)}{r^2} \right] u_l = 0 \quad (46)$$

can again be transformed into an equation for the phase shifts

$$\frac{d\eta_l}{dr} = k^2 r^2 \frac{d(\ln \rho_0)}{dr} [j_l'(kr) \cos \eta_l - n_l'(kr) \sin \eta_l]$$

$$\times [j_l(kr) \cos \eta_l - n_l(kr) \sin \eta_l]$$

$$- k^3 r^2 \left(1 - \frac{c_\infty^2}{c_0^2} \right) [j_l(kr) \cos \eta_l - n_l(kr) \sin \eta_l]^2 \quad (47)$$

which can easily be integrated on the computer. In (47), c_∞ is the velocity of sound far away from the charge, and the primes denote differentiation with respect to the argument.

Equation (47) explicitly includes electrostrictive effects through the r dependence of ρ_0 and c_0. The resonance effects mentioned earlier, on the other hand, originate with the boundary conditions. The deformation of the bubble or core surface

$$b_s(\theta) = \sum_{l=0}^{\infty} \delta b_l P_l(\cos \theta) \quad (48)$$

in response to the sound pressure field $p_s(\theta) = \sum_{l=0}^{\infty} \delta p_l P_l(\cos \theta)$ can be described by a set of response coefficients: $\delta b_l = \lambda_l \delta p_l$. Using (44) and converting to the phase-shift description, the boundary condition becomes

$$\eta_l(b) = \tan^{-1} \left[\frac{j_l'(kb) + \gamma_l kb j_l(kb)}{n_l'(kb) + \gamma_l kb n_l(kb)} \right] \quad (49)$$

where $\gamma_l = -\lambda_l \rho_0(b) c_\infty^2 / b$. Thus the problem reduces to first characterizing the response of the bubble or core surface to the sound field, then doing the phase shift integration to take care of any electrostrictive effects in the nearby fluid, and finally using (36) and (31) to evaluate e/μ_{ph}.

In the case of the electron bubble, electrostrictive corrections are unimportant even at high pressures where b_- shrinks down to $\approx 12 \, \text{Å}$.[106] The phase shifts are therefore given directly by (49). Celli et al.[88] have given an elegant method of deriving the appropriate γ_l's from the simplified bubble model discussed in Section II.B. They write a more general form of (11) to allow for the deformations of the bubble described by (48). The electron energy E_e is again computed by treating the bubble surface as a sharply defined potential barrier of height $V_0(\rho_\infty)$; but since the shape of the bubble is now distorted, E will be a function of b_0 and the δb_l's, as will the other terms in the total energy. The electron is assumed to adjust its state so fast that at each instant in time the bubble assumes the equilibrium shape consistent with the external pressure field $p_0 + p_s(\theta)$. Then the requirement that the total energy be a minimum with respect to variations in the δb_l's can be used to determine the response coefficients λ_l.[139]

Figure 15 gives $\sigma_T(k)$ as calculated from the theory of Celli et al. for a typical set of bubble parameters. Comparison with Fig. 14 shows that μ_{ph}^- will be completely dominated by the coupling between the phonon field and the resonant modes of vibration of the bubble. The broad s-wave resonance tends to dominate the thermal average, and the qualitative k^{-1} behavior of $\sigma_T(k)$ inferred earlier in our discussion can now be seen to occur because the peak of the thermal weight factor lies to the right of the s-wave resonance. Values of e/μ_{ph}^- calculated with $b_- = 17 \text{Å}$, $V_0 = 0.8 \text{eV}$ are shown in Fig. 12.[140] The agreement with the experimentally observed temperature dependence is excellent, whereas the calculated magnitudes of e/μ_{ph}^- are a consistent 20% below those observed. Given the idealizations of the simple bubble model, this is certainly as good as can be expected.

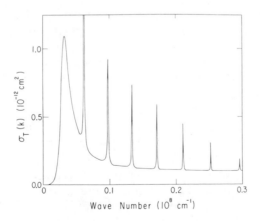

Fig. 15. Momentum transfer cross-section for phonons scattered by an electron bubble. The parameters used to calculate this curve are $b = 16 \text{ Å}$, $V_0 = 0.6$ eV. (After Ref. 65.)

The quality of agreement between theory and experiment in Fig. 12 is somewhat sensitive to the chosen value of b_- and very insensitive to V_0. Fitting the data with b_- as a variable parameter leads to the estimate $b_- = 17 \pm 2\,\text{Å}$ given in Table II. This type of analysis has been applied by Ostermeier[106] to data taken at higher pressures. In Fig. 13, excellent agreement is again obtained with experiment. The derived values of b_- as a function of p are included in Fig. 4.

We now turn to the problems presented by the positive carriers. Here the effects of electrostriction are very large, and the full apparatus of (47) and (49) must be applied. The response of the core to the sound pressure field is easily determined if it is assumed that the liquid-solid phase boundary reacts essentially instantaneously to a local change in the pressure. In the presence of an altered pressure field $p_0 + p_s(\theta)$ the boundary of the core will then be given by [cf. (7)]

$$p_l[b_+ + b_s(\theta)] + p_s(\theta) = p_m + \frac{2\sigma_{ls}}{b_+ + b_s(\theta)} \frac{v_s}{v_l - v_s} \tag{50}$$

Expanding this to first order in the δb_l's and δp_l's yields the simple relation

$$\lambda_l = -\left(\frac{\partial p_l}{\partial r}\bigg|_{r=b_+} + \frac{2\sigma_{ls}}{b_+^2} \frac{v_s}{v_l - v_s} \right)^{-1}, \qquad l = 0, 2, 3, \ldots \tag{51}$$

which may easily be evaluated numerically. It is amusing to note that the core surface has the peculiar property of moving outward in response to an applied pressure.

Using the response coefficients calculated from (51) on the assumption that $\sigma_{ls} = 0.10 \pm 0.05$ erg/cm^2, we obtain the prediction for e/μ_{ph}^+ appearing in Fig. 12. Again the agreement with experiment is remarkably good, such differences as there are falling well within the uncertainty arising from the error limits on σ_{ls}. The data at higher pressures do not provide a significant additional test of the theory, since (as is clear from Fig. 13) it becomes increasingly difficult to separate out the phonon contribution accurately. One can, however, conclude that the measured values of e/μ_{ph}^+ are at all pressures consistent with the electrostriction model and with the assumption that $\sigma_{ls} \approx 0.10$ erg/cm^2.

In summary, the classical sound scattering calculations outlined above provide a surprisingly good description of the magnitude and the temperature dependence of the phonon-limited mobility for both positive and negative carriers. The minor deviations that remain are not significant, given the somewhat idealized nature of the bubble and electrostriction models.

VI. ROTON–LIMITED MOBILITIES

A. Experimental Results

Careful and extensive measurement of μ as a function of temperature and pressure have been made by Brody[44] in the range $1.2 \leqslant T \leqslant 2.2°K$ and by Ostermeier[106] in the range below $1.0°K$. Although the higher temperature data have many interesting features, the discussion of Section III.A indicates that a quantitative treatment of the transport problem in this temperature regime will become rather complicated. It therefore seems more profitable to concentrate on trying to understand the observations at the lower temperatures, where the dilute gas Boltzmann theory outlined in Section III.B is applicable.

The roton-limited drag coefficients e/μ_r are determined from measurements of e/μ by subtracting out the phonon contributions discussed in Section V. For the negative carriers, the best theoretical fit to e/μ_{ph} at a given pressure is multiplied by a fudge factor ranging from 1.2 at zero pressure to 1.04 near the melting pressure, the factor in each case being chosen to produce an exact fit to the low-temperature behavior (Fig. 12). For the positives, the theoretical values of e/μ_{ph}^+ are calculated at each pressure using $\sigma_{ls} = 0.135$ erg/cm^2. This gives an exact fit to the low-temperature behavior when $p = 0$. The resulting phonon contributions are estimated to be reliable to within $\pm 2\%$ up to $1°K$.

A convenient way of separating out the strong effect of the exponential term in the roton number dependence is to present the data in the form

$$\frac{e}{\mu_r} = f(p, T)e^{-\Delta(p,T)/k_B T} \tag{52}$$

where $\Delta(p, T)$ is the pressure- and temperature-dependent roton energy gap.[111] Since any reasonable theory will predict the exponential factor, all the interesting physics is contained in the behavior of the prefactor $f(p, T)$. Figures 16 and 17 show data taken from Refs. 65 and 106. The error bars include the uncertainties arising from estimated errors in the measured mobilities, in the subtracted phonon contribution, and in the roton gap.

One's first impression on looking at these data is that the observed behavior is so reasonable and simple that it should not be difficult to explain. In particular, f increases smoothly with T^{-1} in a manner that seems to be almost independent of pressure and the sign of the charge. Since the roton parameters are independent of temperature in this regime, such behavior must reflect a general property of the roton-carrier interaction or the transport theory. The variation of e/μ_r with pressure also seems to be qualitatively reasonable. For negatives, the size of the carrier shrinks

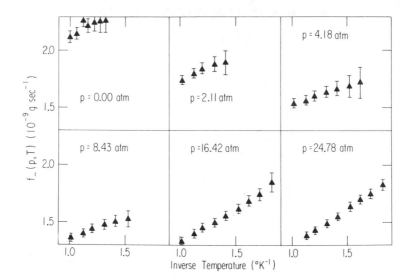

Fig. 16. Roton-scattering prefactor $f_-(p, T)$ for the negative charge carrier. Data taken from Refs. 65 and 106.

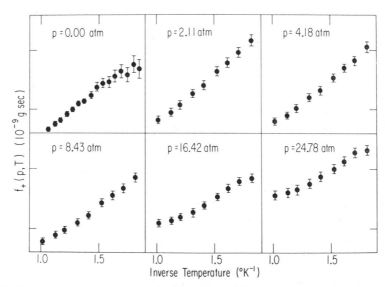

Fig. 17. Roton-scattering prefactor $f_+(p, T)$ for the positive charge carrier. Data taken from Refs. 65 and 106.

with applied pressure, rapidly at first and more slowly at high pressures: a similar variation is evident in e/μ_r^-. For positives, there is a slight increase in carrier size with pressure, and this also seems to be reflected in the variation of the drag coefficient.

Since the rotons have a rather short wavelength compared with the carrier sizes, it is tempting to try to understand the observed behavior in terms of a simple kinetic theory in which the roton-carrier interaction is characterized by an effective cross-section $\bar{\sigma}$ on the order of the geometrical cross-section of the carriers. The resulting prediction

$$\frac{e}{\mu_r} = \frac{\hbar k_0^4}{3\pi^2}\bar{\sigma}e^{-\Delta/k_B T} \tag{53}$$

is unsatisfactory on two counts. First, it gives a prefactor that is independent of temperature, in contrast to the behavior exhibited in Figs. 16 and 17. Second, although it gives roughly the correct variation of e/μ_r with pressure for both carriers, it fails totally to account for the *relative* magnitude of e/μ_r^- and e/μ_r^+: at $p=0$ the ratio of these two quantities is only 1.5, whereas the ratio of the geometrical carrier cross-sections is of order 10. Thus it quickly becomes quite clear that a more sophisticated treatment both of the roton-carrier interaction and of the transport problem is needed to explain the observed behavior. The progress that has been made toward providing such a treatment is discussed next.

B. Theory of Roton Scattering

For a discussion of roton scattering it is necessary to return to (28) to (30), bearing in mind that from a formal point of view the mobilities derived from these equations will be somewhat suspect, because (28) is an inequality and because it contains only the linear term in the momentum loss rate. These technical restrictions should not introduce any qualitative errors into the theoretical predictions, and could easily be overcome by dint of some extra analytical and numerical manipulations if approximate predictions based on (28) are sufficiently encouraging. In this connection it should be pointed out that both Bowley[119] and Barrera and Baym[28] use a somewhat different approach in which the full formal expression for $\mathbf{R_p}$ is inserted into (20) to obtain[141]

$$\frac{e}{\mu} \leqslant \frac{1}{6k_B T}\sum_{\substack{\mathbf{p},\Delta p \\ \mathbf{k},\mathbf{k}'}} \Delta\mathbf{p}^2 f_{\mathbf{p}}^0 n_{\mathbf{k}}(1\pm n_{\mathbf{k}'})\Gamma(\mathbf{p},\mathbf{k}\rightarrow\mathbf{p}+\Delta\mathbf{p},\mathbf{k}') \tag{54}$$

If we now write the transition rate in first order

$$\Gamma = \frac{2\pi}{\hbar}|\langle\mathbf{p},\mathbf{k}|T|\mathbf{p}+\Delta\mathbf{p},\mathbf{k}'\rangle|^2\delta\left[\frac{p^2}{2M}+\epsilon(k)-\frac{(\mathbf{p}+\hbar\mathbf{k}-\hbar\mathbf{k}')^2}{2M}-\epsilon(k')\right] \tag{55}$$

and ignore the p-dependence of the matrix element, we can first take the summation over $\mathbf{p}$ to obtain a Van Hove scattering function $\Sigma_{\mathbf{p}} f_{\mathbf{p}}^0 \delta$ and then carry out the rest of the summations in (54). Although this approach has the advantage of including the nonlinear contributions to $\mathbf{R_p}$ arising from energy conservation, the extra generality so obtained may to some extent be illusory, since it is far from obvious that the p-dependence of the matrix element can be neglected. We prefer the method presented in Section III.B because it leads to analytically simpler results that are based on a consistent order of approximation to $\mathbf{R_p}$.

These points amount to hair-splitting—the real problem is to find the matrix elements $|t|^2$ or, equivalently, the cross-sections σ_i. Before detailing recent attempts to deal with this problem, we briefly discuss the peculiar kinetics implied by the roton dispersion curve. From Fig. 7 it is clear that when $k < k_0$, the group velocity of the incident roton will be antiparallel to its momentum, whereas for $k > k_0$ these will be parallel. The path in k', θ space defined by the energy-conserving delta function has the shape illustrated in Fig. 18a when $(k - k_0)^2 \geqslant 4m_r k k_0 / M$, and that in Fig. 18b when $(k - k_0)^2 \leqslant 4m_r k k_0 / M$. It is seen that if scattering is permitted into a given θ, there are two allowed values of k'. For the branch where k' and k lie on the same side of k_0, the parallel-antiparallel relationship between group velocity and momentum is not altered in the scattering process. The other branch corresponds to scattering across the roton minimum, in which case this relationship is evidently reversed.

Suppose now we have a theoretical model yielding nonpathological matrix elements $|t|^2$. For values of k such that Fig. 18a applies, the integral

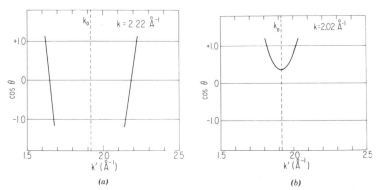

(a) (b)

Fig. 18. Typical locus of final roton states allowed by energy conservation (a) when $(k - k_0)^2 > 4m_r k_0^2 / M$ $(k = 2.22 \text{ Å}^{-1}$ and $m_2 / M = 0.002)$; (b) when $(k - k_0)^2 < 4m_r k_0^2 / M$ $(k = 2.02 \text{ Å}^{-1}$ and $m_r / M = 0.002)$.

over the k-dependent path in k', θ space will not add any strong new k-dependence to the cross-sections beyond what is already contained in $|t|^2$. The situation is different as k approaches k_0, and Fig. 18b applies. Now the *length* of the path in k', θ space (as well as the range of θ sampled) becomes a function of $|k - k_0|$ as an increasingly large number of final states are forbidden by energy conservation. Thus one may expect the d^3k' integration in (28) to produce some peculiar results when k is close to k_0. It is interesting to see whether either of the regimes in Fig. 18 will dominate the k integral in (28) and hence the mobility. The ratio m_r/M is only of order 0.001 to 0.005 because of the large effective masses of the carriers, and this implies that the region of Fig. 18b is defined at the most by $|k - k_0| \lesssim 0.3 \, \text{Å}^{-1}$. On the other hand rotons are concentrated about k_0 with a width of order $|k - k_0| \approx \sqrt{2\pi k_B T m_r/\hbar^2}$, which is also about $0.3 \, \text{Å}^{-1}$. Hence it seems likely that *both* regimes will make significant contributions to the transport properties of the charge carriers.

To calculate e/μ, one must find a model for $|t(k,k',\theta)|^2$, take the integral over paths such as those in Fig. 18 to get the momentum transfer as a function of k, and carry out the k-integral in (28). Two approaches have been taken to implementing this program. One is to simply neglect the k', θ-dependence of $|t|^2$ and to assume some form of the k-dependence. The other is to consider the problem in the limit $m_r/M \to 0$, in which case it is possible to calculate the roton-scattering cross-sections directly. We discuss these in turn.

If $|t|^2$ is a function of k only, the total scattering cross-section is just

$$\sigma(k) = \frac{2\pi}{\hbar} |t|^2 \left| \frac{\partial \epsilon}{\partial k} \right|^{-1} \int \frac{d^3k'}{(2\pi)^3} \delta \left[\frac{\hbar^2(\mathbf{k} - \mathbf{k}')^2}{2M} + \frac{\hbar^2(k' - k_0)^2}{2m_r} - \frac{\hbar^2(k - k_0)^2}{2m_r} \right]$$

(56)

By taking the integral over $z = \cos\theta$ first and dropping negligible terms in m_r/M, we can easily show that

$$\sigma(k) = \frac{m_r M}{\pi \hbar^4} |t(k)|^2 \left[1 - \left(1 - \frac{4m_r k_0^2}{M(k - k_0)^2} \right)^{1/2} \right] \quad \text{if} \quad (k - k_0)^2 \geqslant \frac{4m_r k_0^2}{M}$$

$$= \frac{m_r M}{\pi \hbar^4} |t(k)|^2 \quad \text{if} \quad (k - k_0)^2 \leqslant \frac{4m_r k_0^2}{M}$$

(57)

To derive a k-dependence for $|t|^2$, Barrera and Baym[28] now assume that $\sigma(k)$ should approximately equal the geometrical cross-section πb^2 for all values of k. This leads them to conclude that the matrix element should have a structure

$$|t(k)|^2 = \frac{\pi^2 \hbar^4}{M m_r} \left(b_1^2 + b_2^2 \frac{|k - k_0|^2}{2k_0^2} \frac{M}{m_r} \right) \tag{58}$$

where both b_1 and b_2 are on the order of the carrier radius. Here the first term takes care of the region $|k - k_0| \to 0$, and the second takes care of $|k - k_0| \to \infty$.

While acknowledging the utilitarian spirit in which (58) is derived, one may well question its validity. In particular, it seems unjustified to assume that for k close to k_0 the cross-section should be simply πb_1^2—both $|\partial \epsilon / \partial k|$ and the length of the delta function path in k', θ space go to zero as $k \to k_0$, and there is no obvious reason why in this limit (56) should have anything to do with a geometrical cross-section. Thus both the b_1 and M dependences introduced by the first term in (58) appear to be open to argument.

These reservations are borne out when the behavior calculated from (58) is compared with experiment. If we first assume that e/μ is dominated by rotons for which $(k - k_0)^2 \gg 4m_r k_0^2 / M$, only the second term in (58) is important, and the simple kinetic theory form of (53) is recovered. Since the rotons are clustered too closely about k_0 to make this a realistic picture, the failure of (53) to explain the experimental observations is not surprising. However, the inclusion of the first term in (58) produces other problems. Most notably, e/μ_- is now predicted to increase strongly with applied pressure, whereas it is observed to decrease strongly instead. The difficulty manifestly arises because $b_-^2 / M_- \propto 1/b_-$, so that in (58) $|t(k)|^2$ increases as the electron bubble is compressed. We may conclude that the first term of (58) is not of the correct form to yield good agreement with experiment.

The other line of attack[142, 143] has been to consider the infinite mass limit of (28). In this limit Fig. 18a reduces to two vertical paths at $k' = k$, $2k_0 - k$, and there is no region of the type shown in Fig. 18b. The momentum transfer cross-section in (29) then takes the form

$$\sigma_T(k) = \int d\Omega \left[(1 - \cos\theta) \sigma_s(k, \theta) + \left(1 - \frac{2k_0 - k}{k} \cos\theta \right) \sigma_0(k, \theta) \right] \tag{58}$$

consisting of a part σ_{Ts} which describes scattering into final roton states on the same side of k_0 as k, and a part σ_{T0} describing scattering to the opposite side of the roton minimum. It should be noted that for $k < k_0$, σ_{T0}

is negative—the roton actually gains momentum from the carrier. The differential cross-sections σ_s and σ_0 may be calculated from an effective hamiltonian constructed to fit the roton spectrum [(12)]:

$$\hat{H} = \Delta + \frac{(\hat{p}^2 - p_0^2)^2}{8 m_r p_0^2} + V(\mathbf{r}) \qquad (59)$$

where $\hat{p}$ is the roton momentum operator and $V(\mathbf{r})$ represents the effect of nonuniformities in the fluid near the charge carrier. Ihas[142] has given a simple quasiclassical derivation of the cross-sections, approximating the carrier as a hard sphere in a uniform fluid. A roton incident on the sphere is treated as though it were scattering from a plane tangent to the sphere at the point of incidence. The quantum-mechanical problem of a roton scattering from a hard plane is trivially solved, and it thus becomes very easy to determine the differential cross-sections. The quasiclassical approximation seems to be justified, since the roton wavelength is short compared with the size of the carriers, and indeed a full-dress quantum-mechanical treatment[143] of the same problem gives essentially the same results.

Calculations of this type have also been unsuccessful in describing the experimental observations, yielding values of e/μ that are typically a factor of 5 too low and predicting a wrong temperature dependence for the prefactor $f(p, T)$. The reason for this is easy to see. Insertion of the infinite mass cross-sections into (29) is equivalent to assuming that e/μ is dominated by rotons that have momenta far enough away from k_0 so that the situation of Fig. 18a describes the scattering process, whereas in actuality a large fraction of the rotons are concentrated in the region of k corresponding to Fig. 18b, where the number of allowed final states is strongly restricted by energy conservation. That is, the infinite-mass differential cross-sections certainly do not give an adequate description of the scattering processes when $(k - k_0)^2 \leqslant 4 m_r k_0^2 / M$; and since most of the rotons fall within this limit, it is not surprising that the theory fails.

One obvious next step in trying to construct a realistic theory is to combine the best elements of the two approaches outlined. The idea is to use the infinite-mass differential cross-sections to estimate $|t(k, k', \theta)|^2$ and then to insert these matrix elements into (28) to take account of the energy-conservation restrictions near the roton minimum. The basic assumption in this case is that the matrix elements for scattering from a heavy charge carrier are the same as those for scattering from a fixed carrier. This seems to be a reasonable working approximation, and we are presently investigating whether it leads to a better explanation of the experimental observations.

Acknowledgments

This research has been supported in part by a grant from the National Science Foundation and in part by the Louis Block Fund, The University of Chicago. We have also benefited from support of the Materials Research Laboratory by the National Science Foundation.

Notes and References

1. R. L. Williams, *Can. J. Phys.*, **35**, 134 (1957).
2. L. Meyer and F. Reif, *Phys. Rev.*, **110**, 279 (1958).
3. G. Careri, F. Scaramuzzi, and J. O. Thomson, *Nuovo Cimento*, **13**, 186 (1959).
4. L. D. Landau and I. Pomeranchuk, *Dokl. Akad. Nauk USSR*, **59**, 669 (1948).
5. L. D. Landau, *J. Phys. Moscow*, **11**, 91 (1947).
6. J. Levine and T. M. Sanders, Jr., *Phys. Rev. Lett.*, **8**, 159 (1962).
7. A. V. Phelps and S. C. Brown, *Phys. Rev.*, **86**, 102 (1952).
8. K. R. Atkins, *Phys. Rev.*, **116**, 1339 (1959).
9. R. A. Ferrell, *Phys. Rev.*, **108**, 167 (1957).
10. G. Careri, U. Fasoli, and F. S. Gaeta, *Nuovo Cimento*, **15**, 774 (1960).
11. G. Careri, in *Progress in Low Temperature Physics*, C. J. Gorter, Ed., North-Holland, Amsterdam, 1961, Vol. III, p. 58.
12. C. G. Kuper, *Phys. Rev.*, **122**, 1007 (1961).
13. J. L. Levine and T. M. Sanders, Jr., *Phys. Rev.*, **154**, 138 (1967).
14. L. Meyer and F. Reif, *Phys. Rev. Lett.*, **5**, 1 (1960).
15. F. Reif and L. Meyer, *Phys. Rev.*, **119**, 1164 (1960).
16. L. Meyer and F. Reif, *Phys. Rev.*, **123**, 727 (1961).
17. A. L. Fetter, in *The Physics of Liquid and Solid Helium*, K. H. Bennemann and J. B. Ketterson, Eds., Wiley, New York, 1974.
18. For general discussions of the properties of liquid helium, see J. Wilks, *The Properties of Liquid and Solid Helium*, Clarendon Press, Oxford, 1967, R. J. Donnelly, *Experimental Superfluidity*, University of Chicago Press, Chicago, 1967.
19. O. Penrose and L. Onsager, *Phys. Rev.*, **104**, 576 (1956).
20. P. Nozieres, in *Quantum Fluids*, D. F. Brewer, Ed., North-Holland, Amsterdam, 1966.
21. I. M. Khalantnikov, *Introduction to the Theory of Superfluidity*, Benjamin, New York, 1965.
22. When the average local momentum carried by the elementary excitations become large, (2) and the equations describing the macroscopic behavior of the normal fluid combine to give the familiar two-fluid equations. See Ref. 18.
23. E. P. Gross, *J. Math. Phys.*, **4**, 195 (1963).
24. L. P. Pitaevski, *Sov. Phys. (JETP)*, **13**, 451 (1961).
25. Since the superfluid velocity field v_s is defined as the gradient of a phase, it can have no curl.
26. A. J. Dahm and T. M. Sanders, Jr., *Phys. Rev. Lett.*, **17**, 126 (1966); *J. Low Temp. Phys.*, **2**, 199 (1970).
27. J. Poitrenaud and F. I. B. Williams, *Phys. Rev. Lett.*, **29**, 1230 (1972). Erratum: *Phys. Rev. Lett.*, **32**, 1213 (1974).
28. R. G. Barrera and G. Baym, *Phys. Rev. A*, **6**, 1558 (1972).
29. See Ref. 18 for appropriate references.
30. G. W. Rayfield and F. Reif, *Phys. Rev.*, **136**, A1194 (1964).
31. W. I. Glaberson and M. Steingart, *Phys. Rev. Lett.*, **26**, 1423 (1971).
32. K. W. Schwarz, *Phys. Rev. A*, **10**, 2306 (1974).
33. G. Careri, W. D. McCormick, and R. Scaramuzzi, *Phys. Lett.*, **1**, 61 (1962).

34. For a review, see R. J. Donnelly and P. H. Roberts, *Proc. Roy. Soc. (London) A*, **312**, 519 (1969).
35. W. P. Pratt, Jr., and W. Zimmerman, Jr., *Phys. Rev.*, **177**, 412 (1969).
36. R. E. Packard and T. M. Sanders, Jr., *Phys. Rev. A*, **6**, 799 (1972).
37. R. E. Packard (private communication).
38. J. D. Maynard and T. R. Carver, *Bull. Am. Phys. Soc.*, **19**, 460 (1974).
39. R. J. Donnelly and P. H. Roberts, *Phil. Trans. Roy. Soc. (London) A*, **271**, 41 (1971).
40. K. W. Schwarz and P. S. Jang, *Phys. Rev. A*, **8**, 3199 (1973).
41. R. Zoll and K. W. Schwarz, *Phys. Rev. Lett.*, **31**, 1440 (1973).
42. G. W. Rayfield, *Phys. Rev. Lett.*, **16**, 934 (1966).
43. R. Zoll (to be published).
44. B. A. Brody, thesis, University of Michigan, 1970.
45. G. Ahlers and G. Gamota, *Phys. Lett.*, **38A**, 65 (1972).
46. K. W. Schwarz, *Phys. Rev. A*, **6**, 837 (1972).
47. D. M. Sitton and F. Moss, *Proceedings of the Thirteenth International Conference on Low Temperature Physics* (to be published).
48. B. N. Esel'son, Yu. Z. Kovdrya, and V. B. Shikin, *Sov. Phys. (JETP)*, **32**, 37 (1971).
49. V. I. Saboler and B. N. Esel'son, *Sov. Phys. (JETP)*, **33**, 132 (1971).
50. N. E. Dyumin, B. N. Esel'son, E. Ya Rudavskii, and I. A. Serbin, *Sov. Phys. (JETP)*, **29**, 406 (1969).
51. C. M. Surko and R. E. Slusher, *Phys. Rev. Lett.*, **30**, 1111 (1973).
52. R. L. Woerner, D. A. Rockwell, and T. J. Greytak, *Phys. Rev. Lett.*, **30**, 1114 (1973).
53. A. Bagchi and J. Ruvalds, *Phys. Rev. A*, **8**, 1973 (1973).
54. C. M. Varma, *Phys. Lett.*, **45A**, 301 (1973).
55. M. J. Stephen and L. Mittag, *Phys. Rev. Lett.*, **31**, 923 (1973).
56. G. G. Ihas and T. M. Sanders, Jr., *Phys. Lett.*, **31A**, 502 (1970).
57. W. W. Johnson and W. I. Glaberson, *Phys. Rev. Lett.*, **29**, 214 (1972).
58. D. M. Sitton and F. Moss, *Phys. Rev. Lett.*, **29**, 542 (1972).
59. R. A. Ashton and J. A. Northby, *Phys. Rev. Lett.*, **30**, 1119 (1973).
60. W. F. Vinen, *Proc. Roy. Soc. (London) A*, **243**, 400 (1957).
61. F. E. Moss, *Bull. Am. Phys. Soc.*, **19**, 516 (1974).
62. *Rev. Mod. Phys.*, **46**, 451 (1974).
63. W. Schoepe and G. W. Rayfield, *Phys. Rev. A*, **7**, 2111 (1973).
64. It is well known that the mobility of ions in electrolyte solutions can be obtained to a good approximation from Stokes law. As another example, C. Ebner and D. O. Edwards (Ref. 100) show that the effective mass of a ^{3}He impurity can be approximately obtained from a hydrodynamic argument. We also remind the reader of our discussion of (2), and in particular that mean field calculations on the ground state of helium (Refs. 23 and 24) suggest that the ideal fluid equations hold down to a scale of ~ 1 Å.
65. K. W. Schwarz, *Phys. Rev. A*, **6**, 1958 (1972).
66. The atomic polarizability is ≈ 0.207 Å^3. See M. H. Edwards, *Can. J. Phys.*, **34**, 898 (1956).
67. B. M. Abraham, Y. Eckstein, J. B. Ketterson, M. Kuchnir, and P. R. Roach, *Phys. Rev. A*, **1**, 250 (1970).
68. E. R. Grilly and R. L. Mills, *Ann. Phys.*, **18**, 250 (1962). E. R. Grilly, *Phys. Rev.*, **149**, 97 (1966).
69. L. D. Landau and E. M. Lifshitz, *Statistical Physics,* Addison-Wesley, Reading, Mass., 1958, Chapt XV.

70. J. S. Dugdale and F. E. Simon, *Proc. Roy. Soc. (London) A*, **218**, 291 (1953).
71. J. W. Stewart, *Phys. Rev.*, **129**, 1950 (1963).
72. D. O. Edwards and R. C. Pandorf, *Phys. Rev.*, **140**, A816 (1965).
73. R. M. Ostermeier and K. W. Schwarz, *Phys. Rev.*, **5**, 2510 (1972).
74. T. F. O'Malley, L. Spruch, and L. Rosenberg, *J. Math. Phys.*, **2**, 491 (1961).
75. T. P. Eggarter, *Phys. Rev. A*, **5**, 2496 (1972).
76. H. R. Harrison and B. E. Springett, *Phys. Lett.*, **35A**, 73 (1971); *Chem. Phys. Lett.*, **10**, 418 (1971).
77. H. R. Harrison, thesis, University of Michigan, 1971.
78. T. C. Padmore and M. W. Cole, *Phys. Rev. A*, **9**, 802 (1974).
79. B. E. Springett, M. H. Cohen, and J. Jortner, *Phys. Rev.*, **159**, 183 (1967).
80. K. R. Atkins and Y. Narahara, *Phys. Rev.*, **138**, A437 (1965).
81. D. Amit and E. P. Gross, *Phys. Rev.*, **145**, 130 (1966).
82. J. Jortner, N. R. Kestner, S. A. Rice, and M. H. Cohen, *J. Chem. Phys.*, **43**, 2614 (1965).
83. W. T. Sommer, *Phys. Rev. Lett.*, **11**, 271 (1964).
84. M. A. Woolf and G. W. Rayfield, *Phys. Rev. Lett.*, **15**, 235 (1965).
85. W. B. Fowler and D. L. Dexter, *Phys. Rev.*, **176**, 337 (1968).
86. T. Miyakawa and D. L. Dexter, *Phys. Rev. A*, **1**, 513 (1970).
87. E. P. Gross and H. Tung-li, *Phys. Rev.*, **170**, 190 (1968).
88. V. Celli, M. H. Cohen, and M. J. Zuckerman, *Phys. Rev.*, **173**, 253 (1968).
89. B. Burdick, *Phys. Rev. Lett.*, **14**, 11 (1965).
90. R. C. Clark, *Phys. Lett.*, **16**, 42 (1965).
91. K. Hiroike, N. R. Kestner, S. A. Rice, and J. Jortner, *J. Chem. Phys.*, **43**, 2625 (1965).
92. Y. M. Shih and C.-W. Woo, *Phys. Rev. A*, **8**, 1437 (1973).
93. L. Meyer, H. T. Davis, S. A. Rice, and R. J. Donnelly, *Phys. Rev.*, **126**, 1927 (1962).
94. K. O. Keshishev, Yu. Z. Kovdrya, L. P. Mezhov-Deglin, and A. I. Shal'nikov, *Sov. Phys. (JETP)*, **29**, 53 (1969).
95. J. A. Northby and T. M. Sanders, Jr., *Phys. Rev. Lett.*, **18**, 1184 (1967).
96. J. A. Northby, thesis, University of Michigan, 1966.
97. C. Zipfel and T. M. Sanders, Jr., in *Proceedings of the Eleventh International Conference on Low-Temperature Physics*, J. F. Allen, D. M. Finlayson, and D. M. McCall Eds., University of St. Andrews, Scotland, 1968, Vol. I.
98. C. Zipfel, thesis, University of Michigan, 1969.
99. T. Miyakawa and D. L. Dexter, *Phys. Rev. A*, **1**, 513 (1970).
100. C. Ebner and D. O. Edwards, *Phys. Lett. C*, **2**, 78 (1970).
101. R. L. Douglass, *Phys. Rev. Lett.*, **13**, 791 (1964).
102. A. G. Cade, *Phys. Rev. Lett.*, **15**, 238 (1965).
103. P. E. Parks and R. J. Donnelly, *Phys. Rev. Lett.*, **16**, 45 (1966).
104. T. C. Padmore, *Phys. Rev. Lett.*, **28**, 469 (1972).
105. J. McCauley (to be published).
106. R. M. Ostermeier, *Phys. Rev. A*, **8**, 514 (1973).
107. B. E. Springett, *Phys. Rev.*, **155**, 139 (1967).
108. R. P. Feynman and M. Cohen, *Phys. Rev.*, **102**, 1189 (1956).
109. K. R. Atkins and R. A. Stasior, *Can. J. Phys.*, **31**, 1156 (1953). See also Ref. 67.
110. O. W. Dietrich, E. H. Graf, C. H. Huang, and L. Passell, *Phys. Rev.*, **5**, 1377 (1972).
111. R. J. Donnelly, *Phys. Lett.*, **39A**, 221 (1972).
112. For a review of the properties of ^{3}He impurities, see Ref. 100.
113. S. Chapman and T. G. Cowling, *The Mathematical Theory of Non-Uniform Gases*, Cambridge, 1970.

114. A. D. B. Woods and A. C. Hollis Hallett, *Can. J. Phys.*, **41**, 596 (1963).

115. F. A. Staas, K. W. Takonis, and K. Fokkens, *Physica*, **26**, 669 (1960).

116. G. Baym, R. G. Barrera, and C. J. Pethick, *Phys. Rev. Lett.*, **22**, 20 (1969).

117. L. Kramer, *Phys. Rev. A*, **1**, 1517 (1970).

118. R. M. Bowley, *J. Phys. C*, **4**, 853 (1971).

119. R. M. Bowley, *J. Phys. C*, **4**, 1645 (1971).

120. R. G. Barrera, thesis, University of Illinois, Urbana, 1972.

121. J. M. Ziman, *Electrons and Phonons*, Oxford, 1960, p. 278.

122. D. A. Neeper and L. Meyer, *Phys. Rev.*, **182**, 223 (1969).

123. M. Kuchnir, J. B. Ketterson, and P. R. Roach, *Phys. Lett.*, **36A**, 287 (1971).

124. M. Kuchnir, J. B. Ketterson, and P. R. Roach, *Phys. Rev. A*, **6**, 341 (1972).

125. K. W. Schwarz, *Phys. Rev. A*, **6**, 1947 (1972).

126. The low-temperature value of $e/\mu_3 N_3$ obtained by Kuchnir et al. is about 15% lower than that shown in Fig. 15 (Ref. 125). This is probably due to a small systematic error in the concentration calibration or in the effective drift space distance.

127. A. J. Dahm, *Phys. Rev.*, **180**, 259 (1969).

128. A. F. Andreev, *Sov. Phys. (JETP)*, **23**, 939 (1966).

129. H. M. Guo, D. O. Edwards, R. E. Sarwinski, and J. T. Tough, *Phys. Rev. Lett.*, **27**, 1259 (1971).

130. J. Lekner, *Phil. Mag.*, **22**, 669 (1970).

131. B. R. Levy and J. B. Keller, *J. Math. Phys.*, **4**, 54 (1963).

132. R. M. Bowley and J. Lekner, *J. Phys.*, **3**, L127 (1970).

133. A. E. Watson, J. D. Reppy, and R. C. Richardson, *Phys. Rev.*, **188**, 384 (1970).

134. K. W. Schwarz and R. W. Stark, *Phys. Rev. Lett.*, **21**, 967 (1968).

135. K. W. Schwarz and R. W. Stark, *Phys. Rev. Lett.*, **22**, 1278 (1969).

136. Ref. 46.

137. Ref. 106.

138. S. Wang, thesis, University of Michigan, Ann Arbor, 1967.

139. The p-wave coefficient λ_1 must be considered separately, since this part of the sound wave exerts a net force on the surface. It is easy to show from Newton's law that if m is the mass contained *within* the surface, $\lambda_1 = 4\pi b_0^2/3\omega^2 m$ [i.e., $\gamma_1 = -\alpha/(kb_0)^2$ where α is the ratio of the mass of the displaced fluid to m]. For the negative carrier, $\gamma_1 \to \infty$.

140. Since b_-, V_0, and σ are connected by (11), only two of these may be treated as independent parameters in the phonon-scattering calculation. It is most convenient to choose b_- and V_0.

141. The term $-\Delta p \cdot \mathbf{p}$ obtained when combining (20) and (21) can be replaced by $\Delta p^2/2$.

142. G. G. Ihas, thesis, University of Michigan, Ann Arbor, 1971.

143. I. Iguchi, *J. Low Temp. Phys.*, **4**, 637 (1971).

INTERMOLECULAR FORCES AND CRYSTAL STRUCTURES FOR D₂, N₂, O₂, F₂, AND CO₂

TARO KIHARA AND AKIO KOIDE

Department of Physics, Faculty of Science,
University of Tokyo, Tokyo, Japan

CONTENTS

I. INTRODUCTION

The simplest example of intermolecular forces is provided by the force between two inert-gas atoms, for which the potential energy $U(r)$ is a function only of the distance r between the nuclei. For small distances, U increases steeply for a decrease in distance, corresponding to the repulsive force between the atoms or the mutual "inpenetrability" of the atoms. For larger distances, U increases slowly, asymptotically approaching zero. The increase of U with distance corresponds to the mutual attraction of the atoms, usually called van der Waals attraction. This attractive part of the intermolecular potential has an asymptotic form proportional to r^{-6}. It is usual to approximate the potential energy $U(r)$ between two inert-gas atoms by the Lennard-Jones potential function

$$U(r) = U_0 \left[\left(\frac{r_0}{r} \right)^{12} - 2 \left(\frac{r_0}{r} \right)^6 \right] \tag{1}$$

where U_0 and r_0 are constants to be determined empirically. Extensive studies have been made of this subject regarding both the second virial coefficients[1] and the transport coefficients[2] of gases.

In this chapter we treat, as the next simplest class of intermolecular forces, the force between two such symmetric linear molecules as D_2 (or H_2), N_2, O_2, F_2, and CO_2. The potential energy for the intermolecular force in this case depends on the mutual orientations of the molecules as well as the distance r between the centers of the molecules. Our aim is to reveal the characteristic features of this dependence.

II. PRELIMINARIES

A. Core Potential

For nonpolar polyatomic molecules, Kihara[3] proposed an intermolecular potential function

$$U(\rho) = U_0 \left[\left(\frac{\rho_0}{\rho} \right)^{12} - 2 \left(\frac{\rho_0}{\rho} \right)^6 \right] \tag{2}$$

which is similar to that of Lennard-Jones; however, the variable ρ is set equal to the minimum distance between impenetrable molecular cores. The core may take any shape as long as it is a convex body. If the cores are properly chosen, the sizes and shapes of the molecules can be taken into account in a realistic way.

A convex body is characterized by its three fundamental measures: the volume V, the surface area S, and the measure M, which is the mean curvature integrated over the surface of the convex body.

In terms of V, S, and M of the molecular core, the second virial coefficient $B(T)$ of a one-component gas is given in the form

$$B(T) = \int_0^\infty \left[1 - \exp \frac{-U(\rho)}{kT} \right] \left[\left(S + \frac{1}{4\pi} M^2 \right) + 2M\rho + 2\pi\rho^2 \right] d\rho$$

$$+ V + \frac{1}{4\pi} MS \tag{3}$$

which is calculated to be

$$B(T) = \left(\frac{2\pi}{3} \right) \rho_0^3 F_3(z) + M\rho_0^2 F_2(z)$$

$$+ \left[S + (4\pi)^{-1} M^2 \right] \rho_0 F_1(z) + V + (4\pi)^{-1} MS$$

Here

$$z = \frac{U_0}{kT}$$

and T and k are the temperature and the Boltzmann constant, respectively; U_0 and ρ_0 are the parameters in (2). The functions $F_s(z)$ are the following:

$$F_s(z) = -\frac{s}{12} \sum_{t=0}^{\infty} \frac{1}{t!} \Gamma\left(\frac{6t-s}{12}\right) 2^t z^{(6t+s)/12}, \qquad s = 1, 2, \text{ and } 3$$

If the core of a molecule is properly chosen, it is possible to determine the parameters U_0 and ρ_0 of the molecule by using the observed values of the second virial coefficient.

In previous work a thin rod defined by the atomic nuclei was chosen as the core of a diatomic molecule. Molecular charge distributions calculated by Bader, Henneker, and Cade[4] suggest that somewhat shorter cores are more appropriate for D_2, N_2, and O_2. In this chapter, therefore, we choose the ratio of the core length l to the interatomic distance d as follows:

$$l = 0.65d \qquad \text{for} \quad D_2 \text{ (and } H_2)$$

$$l = 0.85d \qquad \text{for} \quad N_2 \text{ and } O_2$$

$$l = d \qquad \text{for} \quad F_2 \text{ and } CO_2$$

where d for CO_2 indicates the distance between the two oxygen atoms.

The three fundamental measures for a thin rod of length l are

$$V = 0, \qquad S = 0, \qquad M = \pi l \qquad (4)$$

The potential parameters ρ_0 and U_0 determined on the basis of Dymond and Smith's critical compilation of virial coefficients[5] are listed in Table I.

TABLE I

Interatomic Distance d, Core Length l, and Potential Parameters ρ_0 and U_0

	$d(\text{Å})$	$l(\text{Å})$	$\rho_0(\text{Å})$	$U_0/k(°K)$
D_2	0.74	0.48	2.96	41.5
N_2	1.094	0.93	3.60	117
O_2	1.21	1.03	3.20	151
F_2	1.44	1.44	2.90	178
CO_2	2.30	2.30	3.30	316

Here quantum effects have been taken into account for deuterium as a small correction. For hydrogen H_2, the quantum effects are not small, but the ρ_0 and U_0 for H_2 are considered to be the same as those for D_2.

B. Electrostatic Quadrupolar Interactions

Although proper choice of the core can permit us to recognize the size and shape of a molecule in a realistic way, the dependence of the potential depth on the molecular orientation is not accounted for by consideration of the core potential alone. The dependence of the potential depth on the molecular orientation is caused partly by the electrostatic multipole interactions between the molecules.

In the molecule of carbon dioxide O=C=O, the power of the oxygen atom to attract electrons to itself is larger than that of the carbon atom. Hence this spherocylindrical molecule has an electric quadrupole, with negative charge being near the ends and positive charge in the middle.

The quadrupole of such a "uniaxial" molecule is characterized by the moment Q defined by

$$Q = \tfrac{1}{2} \int (2\zeta^2 - \xi^2 - \eta^2)\rho d\tau \tag{5}$$

Here $\rho d\tau$ is the electric charge in the volume element $d\tau$ at the point (ξ, η, ζ), the ζ-axis being on the axis of the molecule. According to this definition, the sign of the quadrupole moment of carbon dioxide is negative. The molecular quadupole moments Q are listed in Table II.

TABLE II
Quadrupole Moments[a]

	$Q(\times 10^{-26} \text{esu})$	$T_c(°K)$	$v_c(\text{Å}^3)$	$Q^2 v_c^{-5/3}(kT_c)^{-1}$
H_2	+0.65	33	108	0.038
D_2	+0.64	38	100	0.036
N_2	−1.4	126	150	0.027
O_2	−0.4	155	123	0.002
F_2	+0.9	144	170	0.008
CO_2	−4.3	304	156	0.097

[a]The values of the electric quadrupole moments are taken from A. D. Buckingham, in *Physical Chemistry*, Vol. IV, Academic Press, New York, 1970, and D. E. Strogryn and A. D. Strogryn, *Mol. Phys.*, **11**, 371 (1966).

The quadrupolar interaction between molecules is characterized by a dimensionless quantity $Q^2 v_c^{-5/3}(kT_c)^{-1}$, in which T_c is the critical temperature and v_c is the volume per molecule at the critical point. These quantities are also included in Table II. The quadrupolar interaction is negligible for oxygen and probably for fluorine, but it is not negligible for nitrogen and hydrogen, and it is quite strong for carbon dioxide.

Owing to quadrupolar interaction, the law of corresponding states should be modified. Figure 1 shows the ratio of the temperature at the triple point T_t to the critical temperature T_c as a function of the dimensionless quantity $Q^2 v_c^{-5/3}(kT_c)^{-1}$; the solid phases are stabilized in a wider temperature range by the quadrupolar interaction.

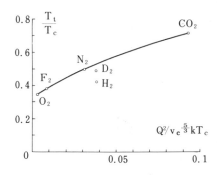

Fig. 1. Ratio of triple-point to critical-point temperatures as a function of the dimensionless quantity for quadrupolar interaction.

A sum of the core potential and the quadrupole interaction energy takes the form

$$U = U_0 \left[\left(\frac{\rho_0}{\rho} \right)^{12} - 2 \left(\frac{\rho_0}{\rho} \right)^6 \right] + \frac{3}{4} \frac{Q^2}{r^5} f(\theta_1, \theta_2, \phi_1 - \phi_2) \qquad (6)$$

in which

$$f(\theta_1, \theta_2, \phi_1 - \phi_2) = 1 - 5\cos^2\theta_1 - 5\cos^2\theta_2 - 15\cos^2\theta_1 \cos^2\theta_2$$

$$+ 2[4\cos\theta_1 \cos\theta_2 - \sin\theta_1 \sin\theta_2 \cos(\phi_1 - \phi_2)]^2 \qquad (7)$$

Here r is the distance between the molecular centers; θ_1 and ϕ_1 are the polar angles of the axis of one molecule with respect to the line connecting the two; θ_2 and ϕ_2 are those of the other (Fig. 2).

The electrostatic quadrupole interaction does not play an important role in gaseous states because an electrostatic multipole interaction almost vanishes when it is averaged with respect to the molecular orientation. In solid states, on the other hand, the crystal structure is governed by the multipolar interaction between molecules;[6] in fact, the crystal structure of carbon dioxide and the structures of low-temperature modifications of deuterium and nitrogen can be explained on the assumption of (6).

Although this approximation to the intermolecular potential is simple and useful, the accuracy of the core potential itself has not been determined, especially with respect to the van der Waals attraction, which corresponds to the second term of the core potential. Our aim is to derive a more accurate expression for the van der Waals attraction and to substitute this for the second term of the core-potential part in (6).

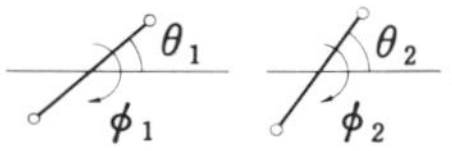

Fig. 2. Polar angles indicating the mutual orientations of two symmetric linear molecules.

III. THE VAN DER WAALS ATTRACTION

A. Potential of the London Dispersion Force

We consider two symmetric linear molecules of the same kind in their ground states lying at a great distance from each other, and we determine the energy of their interaction.

To solve this problem we apply perturbation theory, regarding the two isolated molecules as the unperturbed system and the potential energy of their interaction as the perturbation operator. This electrical interaction at a large distance r can be expanded in a series beginning with a dipole-dipole interaction proportional to r^{-3}, a dipole-quadrupole interaction proportional to r^{-4}, and so on.

In Section III.A we restrict ourselves to the leading term, the dipole-dipole interaction. In the ground states of the molecules under consideration, the mean value of the electric dipole moment is zero; hence the required energy of the interaction is zero in the first approximation of perturbation theory. In the second approximation, we obtain a nonvanishing result of the form $-\mu r^{-6}$, which is proportional to r^{-6}.

The potential energy J between two electric dipoles $\mathbf{p}(1)$ and $\mathbf{p}(2)$ is given by

$$J = r^{-3}\left[p_x(1)p_x(2) + p_y(1)p_y(2) - 2p_z(1)p_z(2)\right]$$

where the z-axis is taken along the line connecting the two molecules.

We express the (x,y,z)-components of the dipole in terms of the components p_1, p_2, p_3 with respect to the molecular axis. Here p_3 is the component parallel to the molecular axis, and p_2 is the component perpendicular to both the molecular and the z-axes. When the angles $\theta_1, \phi_1, \theta_2, \phi_2$ in Fig. 2 are used, we have

$$p_x(1) = p_1(1)\cos\theta_1\cos\phi_1 - p_2(1)\sin\phi_1 + p_3(1)\sin\theta_1\cos\phi_1$$

$$p_y(1) = p_1(1)\cos\theta_1\sin\phi_1 + p_2(1)\cos\phi_1 + p_3(1)\sin\theta_1\sin\phi_1$$

$$p_z(1) = -p_1(1)\sin\theta_1 \qquad\qquad\qquad + p_3(1)\cos\theta_1$$

with similar expressions for $\mathbf{p}(2)$.

The interaction J then takes the form

$$J = r^{-3}\sum_{i=1}^{3}\sum_{k=1}^{3} p_i(1)T_{ik}P_k(2) \tag{8}$$

where

$$T_{ik} = \begin{pmatrix} \cos\theta_1\cos\phi_1 & \cos\theta_1\sin\phi_1 & -\sin\theta_1 \\ -\sin\phi_1 & \cos\phi_1 & 0 \\ \sin\theta_1\cos\phi_1 & \sin\theta_1\sin\phi_1 & \cos\theta_1 \end{pmatrix}\begin{pmatrix} 1 & 0 & 0 \\ 0 & 1 & 0 \\ 0 & 0 & -2 \end{pmatrix}$$

$$\times \begin{pmatrix} \cos\theta_2\cos\phi_2 & -\sin\phi_2 & \sin\theta_2\cos\phi_2 \\ \cos\theta_2\sin\phi_2 & \cos\phi_2 & \sin\theta_2\sin\phi_2 \\ -\sin\theta_2 & 0 & \cos\theta_2 \end{pmatrix}$$

Let us denote by 0 and ρ the ground and excited states, respectively, of molecule 1, and by $\hbar\omega_\rho$ the energy difference between these two states. The corresponding quantities for molecule 2 are denoted by σ and $\hbar\omega_\sigma$. Then, the energy of interaction up to the second-order perturbation is given by

$$-\mu r^{-6} = \sum_{\rho\neq 0}\sum_{\sigma\neq 0}\frac{\langle 00|J|\rho\sigma\rangle\langle\rho\sigma|J|00\rangle}{\hbar\omega_\rho + \hbar\omega_\sigma} \tag{9}$$

where $\langle 00|J|\rho\sigma\rangle$ indicates the matrix element between the ground state $(0,0)$ and an excited state (ρ,σ). When (8) is used, the coefficient μ is calculated to be

$$\mu = \sum_i\sum_k (T_{ik})^2\sum_\rho\sum_{\sigma\neq 0}\frac{\langle 0|p_i(1)|\rho\rangle\langle\rho|p_i(1)|0\rangle\langle 0|p_k(2)|\sigma\rangle\langle\sigma|p_k(2)|0\rangle}{\hbar\omega_\rho + \hbar\omega_\sigma}$$

Following Dalgarno and Davison,[7] we use the identity

$$\frac{1}{a+b} = \frac{2}{\pi} \int_0^\infty \frac{ab}{(a^2+u^2)(b^2+u^2)} \, du, \quad (a,b>0)$$

and transform (10) into

$$\mu = \sum_i \sum_k (T_{ik})^2 \frac{\hbar}{2\pi} \int_0^\infty \alpha_i^{(1)}(u) \alpha_k^{(2)}(u) \, du \tag{11}$$

where

$$\alpha_i^{(1)}(u) = \frac{2}{\hbar} \sum_{\rho \neq 0} \frac{\omega_\rho}{\omega_\rho + u^2} \langle 0|p_i(1)|\rho\rangle\langle\rho|p_i(1)|0\rangle \tag{12}$$

and $\alpha_i^{(2)}(u)$ is defined by a similar expression. In (11), μ is expressed in the form of a product of the dynamic polarizabilities integrated over imaginary frequencies.

Let $\alpha_\parallel(u)$ and $\alpha_\perp(u)$ denote the components of the dynamic polarizability tensor parallel and perpendicular to the molecular axis, respectively:

$$\alpha_1(u) = \alpha_2(u) = \alpha_\perp(u), \qquad \alpha_3(u) = \alpha_\parallel(u) \tag{13}$$

Then (11) is calculated to be

$$\mu = \frac{\hbar}{3\pi} \int_0^\infty \left[2\alpha_\perp^{(1)}(u) + \alpha_\parallel^{(1)}(u)\right]\left[2\alpha_\perp^{(2)}(u) + \alpha_\parallel^{(2)}(u)\right] du$$

$$+ (3\cos^2\theta_1 - 1)\frac{\hbar}{6\pi} \int_0^\infty \left[\alpha_\parallel^{(1)}(u) - \alpha_\perp^{(1)}(u)\right]\left[2\alpha_\perp^{(2)}(u) + \alpha_\parallel^{(2)}(u)\right] du$$

$$+ (3\cos^2\theta_2 - 1)\frac{\hbar}{6\pi} \int_0^\infty \left[2\alpha_\perp^{(1)}(u) + \alpha_\parallel^{(1)}(u)\right]\left[\alpha_\parallel^{(2)}(u) - \alpha_\perp^{(2)}(u)\right] du$$

$$+ \left[(\sin\theta_1 \sin\theta_2 \cos(\phi_1 - \phi_2) - 2\cos\theta_1 \cos\theta_2)^2 - \cos^2\theta_1 - \cos^2\theta_2\right]$$

$$\times \frac{\hbar}{2\pi} \int_0^\infty \left[\alpha_\parallel^{(1)}(u) - \alpha_\perp^{(1)}(u)\right]\left[\alpha_\parallel^{(2)}(u) - \alpha_\perp^{(2)}(u)\right] du \tag{14}$$

We now consider the case of the two molecules being of the same kind.

B. The Role of Electrostatic Polarizabilities

It is probable that the ratio

$$\int_0^\infty \alpha_\parallel(u)^2\,du : \int_0^\infty \alpha_\parallel(u)\alpha_\perp(u)\,du : \int_0^\infty \alpha_\perp(u)^2\,du$$

is close to the ratio

$$\alpha_\parallel^2 : \alpha_\parallel\alpha_\perp : \alpha_\perp^2$$

Here $\alpha_\parallel$ and $\alpha_\perp$ are parallel and perpendicular components of the static polarizability tensor, which are given in Table III. In fact, this is true for hydrogen. Making use of dynamic polarizability obtained by Victor and Dalgarno,[8] we have

$$\int_0^\infty \alpha_\parallel(u)^2\,du : \int_0^\infty \alpha_\parallel(u)\alpha_\perp(u)\,du : \int_0^\infty \alpha_\perp(u)^2\,du = 1 : 0.746 : 0.560$$

which is very close to

$$\alpha_\parallel^2 : \alpha_\parallel\alpha_\perp : \alpha_\perp^2 = 1 : 0.769 : 0.591$$

For the potential between two inert-gas atoms, the constant $2r_0^6 U_0$ in the Lennard-Jones potential, (1), is substantially larger than the coefficient of

TABLE III
Molecular Polarizabilities[a]

	$\alpha_\parallel(\text{Å}^3)$	$\alpha_\perp(\text{Å}^3)$
H_2	0.934	0.718
N_2	2.38	1.45
O_2	2.35	1.21
F_2	1.72^b	0.86^b
CO_2	4.05	1.95

[a]From Landolt-Börnstein, *Zahlenwerte und Funktionen*, Springer Verlag, Berlin, 1951, Vol. I, Part 3, p. 510.
[b]Calculated from $(\alpha_\parallel + 2\alpha_\perp)/3 = 1.15$ on the assumption that $\alpha_\parallel = 2\alpha_\perp$.

dipole-dipole interaction

$$\frac{6\hbar}{2\pi}\int_0^\infty \alpha(u)^2\,du$$

(The ratios are: 1.45 for Ne, 1.7 for Ar, 1.9 for Kr, and 2.1 for Xe.) This is because the effects of dipole-quadrupole interaction are tacitly included in the attractive part of the intermolecular potential. It is also probable that the dipole-quadrupole interaction energy is similar to the dipole-dipole interaction energy as far as the orientation dependence is concerned.

Thus, the potential of the van der Waals attraction, which is denoted by W, may be approximated by

$$W = -W_0 F(\theta_1, \theta_2, \phi_1 - \phi_2) r^{-6} \tag{15}$$

where

$$F(\theta_1, \theta_2, \phi_1 - \phi_2) = 2(2\alpha_\perp + \alpha_\parallel)^2$$

$$+ (3\cos^2\theta_1 + 3\cos^2\theta_2 - 2)(\alpha_\parallel - \alpha_\perp)(2\alpha_\perp + \alpha_\parallel)$$

$$+ 3\Big[(\sin\theta_1 \sin\theta_2 \cos(\phi_1 - \phi_2) - 2\cos\theta_1 \cos\theta_2)^2$$

$$- \cos^2\theta_1 - \cos^2\theta_2 \Big](\alpha_\parallel - \alpha_\perp)^2 \tag{16}$$

Here the factor W_0 is assumed to be independent of the orientations of the molecules.

C. Effects of Octopolar Induction

Although the van der Waals attraction derived in the foregoing does depend on the mutual orientations of the two molecules, an important part of true dependence is missing. In (8), the dipoles were assumed to be located at the molecular centers. This is not a good approximation at all unless the intermolecular distance r is very large.

Let us consider a homonuclear diatomic molecule in a uniform (static or slowly oscillating) electric field. At the center of the molecule, the field induces a dipole, an octopole, and so forth. The induced octopole is equivalent to the octopole produced by two half-strength dipoles located on the molecular axis. The locations are only slightly different for different directions of the electric field. (The locations are independent of the directions of the field in the case of a dielectric prolate spheroid and also in the case of a quantum mechanical anisotropic harmonic oscillator.)

Calculation for a hydrogen molecule shows that the locations of the two dipoles are not far from the positions of atomic nuclei.[9]

We therefore make the following assumptions: for D_2, N_2, O_2, and F_2, two half-strength dipoles are induced at the end points of the rodlike molecular core; for CO_2, three one-third-strength dipoles are induced at the mid- and end points of the core.

Thus far we have considered one molecule in a uniform electric field. But our replacement of one dipole at the molecular center by a system of two or three similar dipoles on the molecular axis can be applied to the interaction between two molecules, namely, to $p(1)$ and $p(2)$ in (8).

We first consider diatomic molecules with core length l. For the orientation $\theta_1 = \theta_2 = 0$, r^{-3} in (8) should be replaced by

$$\frac{1}{4}\left(\frac{2}{r^3} + \frac{1}{(r-l)^3} + \frac{1}{(r+l)^3} \right)$$

and r^{-6} should be replaced by the square of this expression, which is

$$\frac{1}{(r^2-2l^2)^3}\left[1 - \frac{7l^6}{r^6} + \cdots \right]$$

or $(r^2 - 2l^2)^{-3}$, in good approximation. Similarly, for the orientation $\theta_1 = \theta_2 = \pi/2$ with $\phi_1 - \phi_2 = \pi/2$, r^{-6} should be replaced by $(r^2 + l^2)^{-3}$. In general, the equation

$$W = -W_0 F(\theta_1,\theta_2,\phi_1-\phi_2)\left[r^2 - \frac{1}{2}l^2(3\cos^2\theta_1 + 3\cos^2\theta_2 - 2) \right]^{-3} \quad (17)$$

greatly improves on (15) as the potential for the van der Waals attraction between two diatomic molecules.[9] [In (17), small terms containing $(\alpha_{\parallel} - \alpha_{\perp})l^2/(\alpha_{\parallel}+\alpha_{\perp})r^2$ have been neglected.] We use this expression instead of the second term $2U_0(\rho_0/\rho)^6$ in (6).

For CO_2, we obtain

$$W = -W_0 F(\theta_1,\theta_2,\phi_1-\phi_2)r^{-2}\left[r^2 - \frac{1}{2}l^2(3\cos^2\theta_1 + 3\cos^2\theta_2 - 2) \right]^{-2} \quad (18)$$

In Section III.A, we considered only the dipole-dipole interaction J_{dd} as the perturbation operator J. The operator J can be expanded in a series

$$J = J_{dd} + J_{dq} + J_{do} + J_{qq} + \cdots$$

where J_{dq} is the dipole-quadrupole interaction, J_{do} is the dipole-octopole interaction, and so forth. Nonvanishing terms in the second approximation of perturbation theory are a product of J_{dd} and J_{dd} which is proportional to r^{-6}, a product of J_{dq} and J_{dq} which is proportional to r^{-8}, and a product of J_{dd} and J_{do} which is also proportional to r^{-8}. The difference between (15) and (17) or (18) is equivalent to inclusion of this product of J_{dd} and J_{do}.

D. Determination of the Constant W_0

The coefficient W_0 is to be determined in such a way that the second virial coefficient calculated on the basis of our intermolecular potential comes close to the observed values. When the intermolecular potential U is expressed as a function of the core-to-core distance ρ and the angles θ_1, θ_2, and $\phi_1 - \phi_2$ of the molecular axes, the second virial coefficient $B(T)$ is given by

$$B(T) = \int_0^\infty \left\langle 1 - \exp \frac{-U(\rho, \theta_1, \theta_2, \phi_1 - \phi_2)}{kT} \right\rangle_{Av}$$

$$\times \left(\frac{\pi}{4} l^2 + 2\pi l \rho + 2\pi \rho^2 \right) d\rho$$

where $\langle \ \rangle_{Av}$ indicates the average with respect to angular variables [cf. (3)].

It is easy to prove the following theorem: if a function $g(\theta, \phi)$ is unchanged by inversion

$$g(\theta, \phi) = g(\pi - \theta, \phi + \pi)$$

and is a linear combination of spherical harmonics of zeroth, second, and fourth orders, then the following identity holds:

$$\frac{1}{4\pi} \int_0^{2\pi} \int_0^\pi g(\theta, \phi) \sin \theta \, d\theta \, d\phi$$

$$= \frac{2}{5} \left[\frac{1}{3} g(0, \phi) + \frac{1}{3} g\left(\frac{\pi}{2}, 0 \right) + \frac{1}{3} g\left(\frac{\pi}{2}, \frac{\pi}{2} \right) \right]$$

$$+ \frac{3}{5} \left[\frac{1}{4} g\left(\theta_0, \frac{\pi}{4} \right) + \frac{1}{4} g\left(\theta_0, \frac{3\pi}{4} \right) + \frac{1}{4} g\left(\theta_0, \frac{5\pi}{4} \right) + \frac{1}{4} g\left(\theta_0, \frac{7\pi}{4} \right) \right]$$

where

$$\theta_0 = \cos^{-1} \left(\frac{1}{\sqrt{3}} \right)$$

For the mutual orientations of the two molecules, we define

$[zz]$ where $\theta_1 = \theta_2 = 0$

$[zx]$ where $\theta_1 = 0, \quad \theta_2 = \dfrac{\pi}{2}$

$[xx]$ where $\theta_1 = \theta_2 = \dfrac{\pi}{2}, \quad \phi_1 - \phi_2 = 0$

$[xy]$ where $\theta_1 = \theta_2 = \dfrac{\pi}{2}, \quad \phi_1 - \phi_2 = \dfrac{\pi}{2}$

$[zd]$ where $\theta_1 = 0, \quad \theta_2 = \cos^{-1}\left(\dfrac{1}{\sqrt{3}}\right)$

$[xd]$ where $\theta_1 = \dfrac{\pi}{2}, \quad \theta_2 = \cos^{-1}\left(\dfrac{1}{\sqrt{3}}\right), \quad \phi_1 - \phi_2 = \dfrac{\pi}{4}$

$[dd]$ where $\theta_1 = \theta_2 = \cos^{-1}\left(\dfrac{1}{\sqrt{3}}\right), \quad \phi_1 - \phi_2 = 0$

$[dd']$ where $\theta_1 = \theta_2 = \cos^{-1}\left(\dfrac{1}{\sqrt{3}}\right), \quad \phi_1 - \phi_2 = \dfrac{\pi}{2}$

$[dd'']$ where $\theta_1 = \theta_2 = \cos^{-1}\left(\dfrac{1}{\sqrt{3}}\right), \quad \phi_1 - \phi_2 = \pi$

Here d stands for diagonal (Fig. 3).

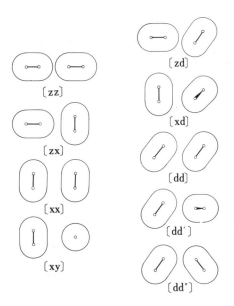

Fig. 3.   The nine typical orientations.

On the basis of the above-mentioned theorem, a good approximation is to substitute the average $\langle\ \rangle_{Av}$ in the integrand by a sum

$$\frac{4}{25}\left[\frac{1}{9}(zz)+\frac{4}{9}(zx)+\frac{2}{9}(xx)+\frac{2}{9}(xy)\right]$$

$$+\frac{12}{25}\left[\frac{1}{3}(zd)+\frac{2}{3}(xd)\right]$$

$$+\frac{9}{25}\left[\frac{1}{4}(dd)+\frac{2}{4}(dd')+\frac{1}{4}(dd'')\right]$$

Here parentheses indicate the values in the corresponding orientations, for which fundamental quantities are listed in Table IV. By this method, W_0 is determined to be

$$\frac{(2\alpha_\perp+\alpha_\parallel)^2 W_0}{(\rho_0+l/2)^6 U_0} = \qquad 0.90 \qquad \text{for} \quad D_2$$

$$0.88 \qquad \text{for} \quad N_2$$

$$0.84 \qquad \text{for} \quad O_2$$

$$0.76 \qquad \text{for} \quad F_2$$

$$0.63 \qquad \text{for} \quad CO_2$$

TABLE IV
Fundamental Quantities for Each Orientation [cf. (6), (16), (17), and (18)]

	$F(\theta_1,\theta_2,\phi_1-\phi_2)$	$r^2-\dfrac{l^2}{2}(3\cos^2\theta_1+3\cos^2\theta_2-2)$	Quadrupole interaction
$[zz]$	$3(4\alpha_\parallel^2+2\alpha_\perp^2)$	r^2-2l^2	$\dfrac{6Q^2}{r^5}$
$[zx]$	$3(5\alpha_\parallel\alpha_\perp+\alpha_\perp^2)$	$r^2-\dfrac{l^2}{2}$	$\dfrac{-3Q^2}{r^5}$
$[xx]$	$3(\alpha_\parallel^2+5\alpha_\perp^2)$	r^2+l^2	$\dfrac{9Q^2}{4r^5}$
$[xy]$	$3(2\alpha_\parallel\alpha_\perp+4\alpha_\perp^2)$	r^2+l^2	$\dfrac{3Q^2}{4r^5}$
$[zd]$	$(2\alpha_\perp+\alpha_\parallel)(4\alpha_\parallel+2\alpha_\perp)$	r^2-l^2	0
$[xd]$	$(2\alpha_\perp+\alpha_\parallel)(\alpha_\parallel+5\alpha_\perp)$	$r^2+\dfrac{l^2}{2}$	0
$[dd]$	$6(2\alpha_\parallel\alpha_\perp+\alpha_\perp^2)$	r^2	$\dfrac{-7Q^2}{3r^5}$
$[dd']$	$\dfrac{2(2\alpha_\parallel^2+14\alpha_\parallel\alpha_\perp+11\alpha_\perp^2)}{3}$	r^2	$\dfrac{-Q^2}{3r^5}$
$[dd'']$	$\dfrac{2(8\alpha_\parallel^2+2\alpha_\parallel\alpha_\perp+17\alpha_\perp^2)}{3}$	r^2	$\dfrac{3Q^2}{r^5}$

IV. INTERMOLECULAR POTENTIALS AND CRYSTAL STRUCTURES

A. Carbon Dioxide

By making use of (18) instead of the second term in (6), we finally obtain the intermolecular potential of carbon dioxide. In Fig. 4 this potential appears as functions of the center-to-center distances for the typical orientations mentioned previously. The well of the intermolecular potential for CO_2 is deepest in or near orientation [dd].

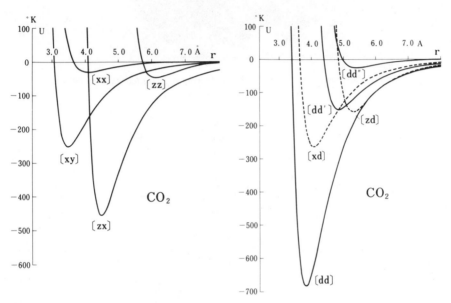

Fig. 4. The intermolecular potential for CO_2 in typical orientations; r is the center-to-center distance.

The characteristic features of this orientation dependence of the intermolecular potential can be represented by Kihara's[6] quadrupolar molecular models, made of two ferrite magnets with plastic or wooden spherical caps (see Fig. 5).

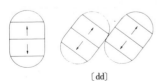

Fig. 5. Magnetic model of the carbon dioxide molecule.

Fig. 6. The cubic *Pa*3 structure of solid carbon dioxide represented by magnetic molecular models.

TABLE V

Crystal Structures

Phase	Temperature range (°K)	Lattice	Space group	Molecules per unit cell
CO_2	0 –194.6	Cubic	*Pa*3	4
β-D_2	2.8– 18.7	Hexagonal	(*hcp*)	2
α-D_2[a]	0 – 2.5	Cubic	*Pa*3	4
β-N_2	35.6– 63.1	Hexagonal	$P6_3/mmc$	2
α-N_2[b]	0 – 35.6	Cubic	$P2_13$	4
γ-O_2[c]	43.7– 54.3	Cubic	$Pm3n$	8
β-O_2[d]	23.9– 43.7	Rhombohedral	$R\bar{3}m$	1
α-O_2[e]	0 – 23.9	Monoclinic	$C2/m$	2
β-F_2[f]	45.6– 53.5	Cubic	$Pm3n$	8
α-F_2[g]	0 – 45.6	Monoclinic	$C2/m$ or $C2/c$	4

[a] Ref. 11.
[b] Ref. 13.
[c] T. H. Jordan, W. E. Streib, H. W. Smith, and W. N. Lipscomb, *Acta Crystallogr.*, **17**, 777 (1964).
[d] Ref. 14.
[e] Ref. 15.
[f] T. H. Jordan, W. E. Streib, and W. N. Lipscomb, *J. Chem. Phys.*, **41**, 760 (1964).
[g] Ref. 16.

An assembly of such molecular models simulates the crystal structure of carbon dioxide (see Fig. 6).

The crystal structure of carbon dioxide belongs to the cubic system. The carbon atoms in a crystal of carbon dioxide form a face-centered-cubic lattice, which is composed of four primitive cubic lattices. On each primitive cubic lattice, all axes of the CO_2 molecules are parallel to a particular body diagonal; the direction of the respective axes is different for each of the four primitive lattices. The space group is $Pa3$.

The structures of molecular crystals under consideration are summarized in Table V.

B. Deuterium and Nitrogen

By using (17) instead of the second term in (6), we obtain the intermolecular potential of deuterium, which is shown in Fig. 7. The potential energy takes its minimum value in orientation [zx], as Kolchanski[10] first pointed out.

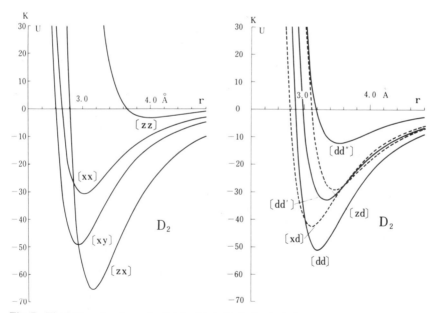

Fig. 7. The intermolecular potential for D_2 in typical orientations.

The characteristic features of the orientation dependence of the intermolecular potential can be represented by magnetic models of molecules (see Fig. 8).

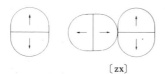

[zx]   **Fig. 8.** Magnetic model of the D_2 or N_2 molecule.

The structure in the low-temperature phase of solid deuterium is cubic $Pa3$ with four molecules per unit cell.[11] This crystal structure is similar to that of carbon dioxide and is simulated by an assembly of our magnetic models.[6]

The core potential with electrostatic quadrupole interaction, (6), is a useful approximation for CO_2 and D_2, since it also takes its minimum energy near orientation [dd] for CO_2 and in orientation [zx] for D_2.

The calculated intermolecular potential for nitrogen is plotted in Fig. 9. The minimum potential energies in orientations [zx] and [xy] are deeper than those in the others. In this case also the predicted crystal structure at low temperatures is cubic $Pa3$, with four molecules per unit cell.

This structure is the same as that first proposed for α-nitrogen,[12] which is stable below 35.6°K. According to more recent single-crystal diffraction studies by Jordan et al.,[13] however, molecular centers in α-nitrogen are replaced by 0.1 to 0.2 Å from the centrosymmetric positions in $Pa3$ along

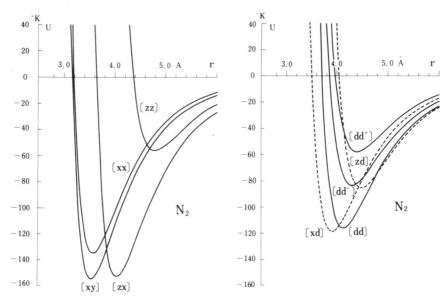

Fig. 9. Intermolecular potential for N_2 in typical orientations.

the [111] directions, thus the symmetry is reduced to $P2_13$. The nature of this replacement is not known.

C. Oxygen and Fluorine

The intermolecular potentials for oxygen and for fluorine, appearing in Figs. 10 and 11, respectively, take their minimum values in orientation [xx], in which the two molecules are parallel and side by side. The electrostatic quadrupolar interaction has been neglected for oxygen and fluorine.

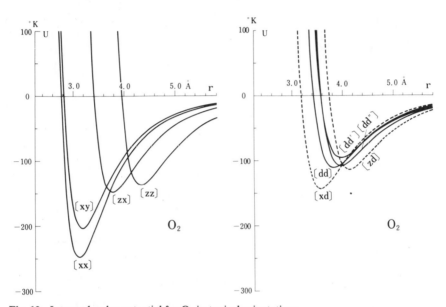

Fig. 10. Intermolecular potential for O_2 in typical orientations.

The structure of solid β-oxygen (stable between 23.8 and 43.8°K) is rhombohedral $R\bar{3}m$ with $a = 4.21$ Å and $\alpha = 46°16'$; the molecules are parallel to the crystal axis.[14] This structure corresponds to the minimum-energy assembly of O_2 molecules with the intermolecular potential derived earlier. The calculated values are $a = 4.1$ Å and $\alpha = 46°$.

The crystal structure of antiferromagnetic α-oxygen (stable below 23.8°K) is monoclinic $C2/m$ with two molecules per unit cell.[15] This structure cannot be explained on the basis of our intermolecular potential alone. It will be necessary to take account of antiferromagnetic interaction —namely, attraction between antiparallel spins.

The crystal structure of α-fluorine (stable below 45.6°K) is monoclinic

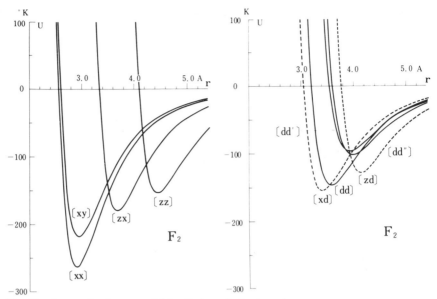

Fig. 11.   Intermolecular potential for F_2 in typical orientations.

$C2/m$ or $C2/c$ with four molecules per unit cell.[16] This is one of the closest packing structures of F_2 molecules, of which the electron-density contour[4] shows a "waist" between the two atoms. This "waist" is not taken into consideration by our intermolecular potential. As a minimum-energy configuration, our potential would give rhombohedral $R\bar{3}m$, which is very close to the real structure of α-fluorine.

V. CONCLUSION

The intermolecular potentials for D_2, N_2, O_2, F_2, and CO_2 have been determined on the basis of the second virial coefficients, the polarizabilities parallel and perpendicular to the molecular axes, and the electric quadrupole moments. The repulsive parts of the potentials are taken from the corresponding rod–core potentials. The effects of octopolar induction are taken into consideration in a unique way.

The potential U depends on the relative orientations of the two molecules as well as the distance r between the molecular centers. Let U_{min} denote the potential depth where $\partial U / \partial r = 0$, and let $\langle (U_{min} - \langle U_{min} \rangle_{Av})^2 \rangle_{Av}$ be the squared deviation from its mean value averaged over the molecular orientations. Then the quantity

$$\frac{\left[\left\langle \left(U_{min} - \langle U_{min} \rangle_{Av} \right)^2 \right\rangle_{Av} \right]^{1/2}}{\langle U_{min} \rangle_{Av}}$$

which is a measure of anisotropy of the potential depth, is 0.36 for D_2, 0.26 for N_2, 0.25 for O_2, 0.27 for F_2, and 0.72 for CO_2. The remarkable anisotropy for CO_2 and D_2 is due to strong electrostatic quadrupole interactions for these molecules.

The dependence of the potential depth on the molecular orientations governs the structure of the molecular crystals. The intermolecular potentials obtained in this chapter are consistent with the crystal structures.

References

1. J. E. Lennard-Jones, *Proc. Roy. Soc. (London) A*, **106**, 463 (1924); *Proc. Phys. Soc. London*, **43**, 461 (1931); *Physica*, **4**, 941 (1937).

2. T. Kihara and M. Kotani, *Proc. Phys.-Math. Soc. Japan*, **25**, 602 (1943). J. O. Hirschfelder, R. B. Bird, and E. L. Spotz, *J. Chem. Phys.*, **16**, 968 (1948); **17**, 1343 (1949).

3. T. Kihara, *Revs. Modern Phys.*, **25**, 831 (1953); *Advances in Chemical Physics*, Vol. 5, Wiley-Interscience, New York, 1963, p. 147.

4. R. F. W. Bader, W. H. Henneker, and P. E. Cade, *J. Chem. Phys.*, **46**, 3341 (1967). See also S. C. Wang, *Phys. Rev.*, **31** 579 (1928).

5. J. H. Dymond and E. B. Smith, *The Virial Coefficients of Gases*, Clarendon Press, Oxford, 1969.

6. T. Kihara, *Acta Crystallogr.*, **16**, 1119 (1963); *Acta Crystallogr.*, **21**, 877 (1966); *Advances in Chemical Physics*, Vol. 20, Wiley-Interscience, New York, 1971, p. 1. See also *Acta Crystallogr. A.*, to be published.

7. A. Dalgarno and W. D. Davison, *Advances in Atomic and Molecular Physics* Vol. 3, Academic Press, New York, 1966, p. 1.

8. G. A. Victor and A. Dalgarno, *J. Chem. Phys.*, **53**, 1316 (1970).

9. A. Koide and T. Kihara, *Chem. Phys.*, **5**, 34 (1974).

10. E. Kochanski, *J. Chem. Phys.*, **58**, 5823 (1973).

11. K. F. Mucker, P. M. Harris, D. White, and R. A. Erickson, *J. Chem. Phys.*, **49**, 1922 (1968).

12. L. H. Bolz, M. E. Boyd, F. A. Mauer, and H. S. Peiser, *Acta Crystallogr.*, **12**, 247 (1959).

13. T. H. Jordan, H. W. Smith, W. E. Streib, and W. N. Lipscomb, *J. Chem. Phys.*, **41**, 756 (1964). A. F. Schuch and R. L. Mills, *J. Chem. Phys.*, **52**, 6000 (1970).

14. E. M. Hörl, *Acta Crystallogr.*, **15**, 845 (1962).

15. C. S. Barrett, L. Meyer, and J. Wasserman, *J. Chem. Phys.*, **47**, 592 (1967).

16. L. Meyer, C. S. Barrett, and S. C. Greer, *J. Chem. Phys.*, **49**, 1902 (1968).

COOPERATIVE CONFORMATIONAL KINETICS OF SYNTHETIC AND BIOLOGICAL CHAIN MOLECULES

Laboratoire d'Acoustique Moléculaire, *
Université Louis Pasteur
Strasbourg, France*

CONTENTS

I. INTRODUCTION

It is well known that various biological functions of nucleic acids and proteins are controlled by their conformations, which are most sensitive to minor external effects. Such high sensitivity is the result of the cooperative nature of the elementary steps of conformational changes.

Equally well known is the importance of the kinetic aspects of conformational changes—for example, in the transfer of genetic information or in the process of enzymatic reactions. Thus kinetic studies founded on simple models that exhibit cooperativity are of particular interest.

*Research group associated with the Centre National de la Recherche Scientifique.

Linear systems, commonly found among biopolymers, furnish good examples that readily lend themselves to both theoretical and experimental studies. Perhaps one of the most thoroughly investigated conformational changes is the helix–coil transition of poly-α-aminoacids composed of like units (i.e., those polypeptides in which each amino acid residue–CO–CHR–NH–bears the same side group R). In the α-helical conformation, the peptide nitrogen atom is hydrogen bonded to the peptide (carbonyl) oxygen atom of the fourth residue down the chain.

When dissolved in an appropriate solvent, a helical polypeptide can undergo a sharp transition to a randomly coiled chain as the temperature, or the pH, is varied through a certain range. In this context, the term "random coiled chain" represents a large collection of disordered conformations.

Whereas the amido group is a rigid planar structure, there is some freedom of rotation about the bonds to the adjacent carbon atoms;[1] the randomly coiled conformation results if the hydrogen bonds are broken, leading to intramolecular rotations.

Hence each residue of the chain is considered to present essentially two conformations: the helical conformation, as exhibited in the α-helix, and the coiled conformation, which corresponds to a collection of conformations. With this simplified picture, we can represent the chain molecule as a linear Ising lattice (i.e., as a linear array of spins, each of which is pointing in one of two opposite directions). The elementary step of a conformational change then consists in a jump involving spin reversal, as shown in Fig. 1. A chain composed of a large number N of residues exhibits 2^N states, when end effects are neglected.

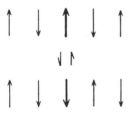

Fig. 1. Elementary step of the conformational change of a poly-α-aminoacid, as represented by the linear Ising model.

The process considered presents the typical features of cooperativity, since the elementary step of converting a residue (or flipping a spin) is affected by the properties of other parts of the chain. In the present case, immediate neighbors of like conformation are favored (positive cooperativity).

The nucleation parameter $\sigma \ll 1$ of Zimm and Bragg[2] permits us to assign a low statistical weight to the interruption of a coiled region (i.e., to the formation of the first turn of the helix). This step is rendered difficult by a large reduction of entropy; once formed, however, this turn acts as a nucleus to which further turns can be added by hydrogen bonding.

The reciprocal σ^{-1} of the nucleation parameter is a measure of the cooperativity; the smaller the value of σ, the longer the uninterrupted sequences of residues of like conformation. For example, for a very long chain, the uninterrupted helical sequences are composed on the average of $\sigma^{-1/2}$ residues (cooperativity length[2]).

A similar problem is encountered in synthetic chain molecules such as vinyl-polymers $(CH_2–CHR)_N$ which, again, exhibit a very large number of conformations produced by rotational isomerism. As indicated in Fig. 2, each C–C bond must form a well-defined angle θ close to $109°$ with the preceding bond. The set of different conformations results from partial freedom of rotation of the bonds—for example, of bond j in Fig. 2 around bond $j-1$. The various positions on the valency cone, however, are not equally probable: as a result of various attractive and repulsive forces arising from other elements of the chain, particular conformations are especially stable (e.g., the *trans* position (*t*) and the two *gauche* (*g*) positions shown in Fig. 2). Conformational changes thus require that a potential barrier be overcome. Such a chain exhibits 3^N states, when end effects are neglected.

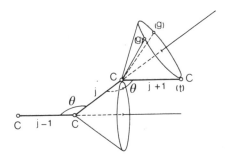

Fig. 2. Carbon skeleton of a vinyl chain.

If only bond j in Fig. 2 were to jump to another position on the cone, the two moieties of the chain (the one ending at bond $j-1$, and the one starting at bond j) would have to rotate as a whole with respect to each other. Since each of these moieties may comprise thousands of atoms, certain conformational changes are likely to consist of a cooperative local reorganization of the chain, involving several bonds, to keep other parts of the chain as unperturbed as possible.

Many authors[3-9] have attempted to describe local movements on cubic and tetrahedral lattices. It still seems to be difficult to assign an exact probability to each of the many possible movements that involve, say, 3 to 10 bonds, especially when considering such factors as inhibition of double occupancy, nearest-neighbor contact, obstructing effects of neighboring parts of the chain, and distortions of bond angles and bond lengths in the crossing of potential barriers.

No exact description of these local motions is necessary here, since we wish to discuss only their cooperative character and certain properties of these motions that might be reflected in the slow modes of deformation of the chain.

An important conclusion of the studies just mentioned based on simple lattice models is the migrational character of many local conformational changes in vinyl-polymers. This is illustrated by Fig. 3, and the symbolism is the same used in Fig. 1 for describing conformational changes of polypeptides. In this oversimplified picture, the bonds (the spins) move two at a time. Thus two spins essentially exchange their directions while jumping. The couples that are jumping are the heavy lines in Fig. 3. No jump occurs along the second line of the reaction scheme of Fig. 3. The two conformations appearing on this line are identical, the jumping couple, however, has moved one step to the right. Two jumps are shown, and the net result is the migration of the circled spin.

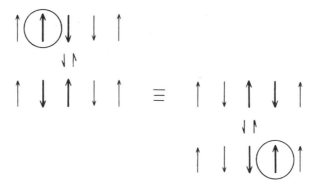

Fig. 3. Migrational process in a simplified picture of conformational changes of vinyl-polymers.

The importance of migrational processes in polymers is evident because they generate the well-known Rouse modes,[10] exhibited by the widely used bead-and-spring model. This was first shown analytically by Orwoll and Stockmayer,[11] using a freely jointed chain model in which two adjoining bond vectors are exchanged in space at each jump. If L_j is the three-dimensional vector drawn from an arbitrary origin to the jth skeletal atom, the resulting displacement vector $\Delta \mathbf{L}_j$ of atom j, when jumping, is equal to

$$\Delta \mathbf{L}_j = \mathbf{L}_{j+1} - 2\mathbf{L}_j + \mathbf{L}_{j-1} \tag{1}$$

This leads directly to relaxation times τ_p that are proportional to those of Rouse; that is, for a chain of N units, they are proportional to λ_p^{-1} with

$$\lambda_p = 4\sin^2\left(\frac{p\pi}{2(N+1)}\right) \tag{2}$$

where p is an integer ($p = 1, 2, \ldots, N$). Moreover, the τ_p's are proportional to the reciprocal of the jump probability.[11]

Not all the local motions are migrational, since certain short stereoregular sequences, for example, must exhibit transitions that occur back and forth at the same position. Thus a first distinction between migrational and transitional motions must be established. The four-bond motion considered by Monnerie and Geny[6] provides another example of a transitional motion. In general, mixed motions exhibiting both migrational and transitional character may also exist.

In the stochastic approach used to describe the conformational transitions of polypeptides and vinyl-polymers examined previously, local motions are assumed to be Markovian. Perhaps one should keep in mind that whereas the processes are fundamentally Markovian, a simplified picture based on the reduction of the number of variables can lead to non-Markovian behavior. Such reductions have been used in the models described, since only the most stable rotational isomers have been retained. This procedure was advocated long ago by Volkenstein.[12]

Rather general conditions under which a cascade of local motions can be described by a bead-and-spring diffusion equation, as far as the lowest modes are concerned, have been given by Iwata,[13] who used a coarse-graining procedure. Iwata's slow relaxation times are proportional to the inverse of a mean probability $P_{\bar{v}}$ of local movements.

It is often considered[14] that the inverse jump probabilities are proportional to the solvent viscosity η_0. This corresponds to diffusional behavior of local movements when treated as the escape over a potential barrier of a particle embedded in a viscous medium. The complete treatment of this problem by Kramers,[15] however, leads to a general expression for P, of which the diffusion limit is but a particular case and corresponds to large viscous friction.

If the phenomenological concept of internal viscosity as introduced by Kuhn and Kuhn[16] is applicable, the corresponding effects would arise from the nondiffusional behavior of local movements (cf. also Ref. 17). The expression of P given by Kramers permits[18] (for slow modes) the reduction of Iwata's coarse-grained diffusion equation to the diffusion equation for the bead-and-spring model with internal viscosity[19] (cf. Section III.A). In addition to the family of relaxation times of the Rouse[10] (and Zimm[20]) models, of which the subset τ_p enters into the hydrodynamic quantities, the above-mentioned diffusion equation leads to a second family:[19]

$$\tau_p' = \tau_p(1 + \rho_p v_p) \tag{3}$$

Each τ_p' is associated with one of Rouse's modes and contains an additional internal viscosity term $\tau_p \rho_p v_p$, independent of η_0; in (3), $\rho_p = \Phi_p / (g\zeta)$, where the η_0-independent Φ_p's are the elements of the diagonal matrix Φ of internal viscosity;[19] the v_p's differ from unity when the hydrodynamic interactions between the beads are taken into account; g is the number of monomeric units per subchain of the bead-and-spring model, and ζ is the effective coefficient of friction of the monomeric unit.

Equation (3) shows a linear dependence of the τ_p''s on η_0, similar to that obtained for the relaxation time that characterizes simple models (elastic sphere,[21] dumbbell[22]). Information can clearly be gained by using solvents of various viscosities, and we discuss results obtained in this way by several techniques.

We shall consider further the possibility[23] of the coexistence of local movements that are either diffusional or nondiffusional, in connection with a discussion of pending problems in polymer dynamics.

Section II of this chapter deals with the kinetics of the linear Ising lattice. In principle, the stochastic process under consideration can be described if the $2^N \times 2^N$ transition rate matrix $\mathbf{Q}$ can be written down and diagonalized. The nondiagonal elements of $\mathbf{Q}$, which are nonnegative, are the transition rates from one state of the lattice to another state; the values of the diagonal elements follow from the condition that the sum by row of the elements be equal to zero.

If the probabilities of the 2^N states are ranged into a row vector $\mathbf{p}(t)$, the rate equation reads in matrix form

$$\frac{d}{dt}\mathbf{p}(t) = \mathbf{p}(t)\mathbf{Q}(t) \tag{4}$$

Steps to the solution for a special form of the elements of the matrix $\mathbf{Q}$ have been given by Glauber.[24]

In many instances it is helpful to focus attention on certain particular sequences of chain elements rather than to look for the probability of each state of the chain. Let X_j and Y_i be sequences comprising j and i consecutive elements, respectively, for given states of an infinitely long chain. If, for example, we choose $j = 5$, the sequence X_j might be ABBAA, where A and B now denote the two possible states of each element. The probability $p(X_j)$ for the sequence X_j to show up obeys the following linear differential equation (Silberberg and Simha[25]):

$$\frac{d}{dt}p(X_j) = \sum_{i=1}^{j+1} C_i p(Y_i) \tag{5}$$

where the C_i's are linear combinations of the elements of the matrix $\mathbf{Q}$. Thus the differential equation for $p(X_j)$ contains probabilities of sequences of $j+1$ elements; for an infinitely long chain, this obviously leads to an infinite set of equations. For a chain of finite length, the problem is even more complicated because $p(X_j)$ depends on where the sequence X_j is located along the chain.

Equations (5) have been written for an infinitely long chain up to $j=6$, as well as for short chains containing up to seven elements.[26] Approximate methods developed by Gô,[27] by Silberberg and Simha,[25,26] and by Schwarz,[28] permit reduction of the equations to $4N-5$ for a finite chain, and to 4 for an infinitely long chain. Certain important quantities had already been obtained by simple calculations—for example, the initial rate of change after a sudden perturbation of the fraction of chain elements that are in one given state.[29]

Section II.A is an account of the calculation of Schwarz,[28,30] leading in a straightforward manner to analytical expressions of the four relaxation times for an infinitely long, highly cooperative chain. A generalization of Schwarz's calculation, of which the results have been given recently,[31] is presented in some detail in Section II.B. Essentially, to calculate the relaxation amplitudes, we take into account the effect of fluctuations of the number of uninterrupted sequences of like elements (A or B). It may perhaps be anticipated that this effect will increase when the molecular weight is lowered. The result will be compared with measurements of the ultrasonic absorption in solutions of polypeptides, as recently reported.[32]

An interesting case, which is examined in Section II.C, arises when the process is sequential (i.e., when it exhibits a well-defined sequence of steps). Processes that involve short chains often lead to such sequences. For example, if we consider a polypeptide that is too short to present two helical sequences with a nonnegligible probability, we may wish to consider the following set of states:[33] the coiled chain, one nucleus formed, one, two, etc.,. helical elements added to the nucleus. If the nucleus can form anywhere in the chain, each of the preceding states represents a collection of conformations.

An especially interesting example of a sequential process is found in the base-pairing in oligonucleotides, as studied by Eigen and Pörschke.[34-36]

Their experiments led to a detailed description of the nucleation step, whence conclusions could be drawn about the kinetics of transferring the genetic information. A short account of Eigen and Pörschke's work (included in Section II.C), is our only incursion into the realm of nucleic acids.

Sequential processes lead to kinetic problems that can be resolved with the aid of (4). The transition rate matrix $\mathbf{Q}$ is then tridiagonal, and we can in general calculate its eigenvalues and other useful related quantities fairly easily (examples are given in Section II.C).

The reversible unfolding and refolding of globular proteins (not necessarily short ones) has also been treated as a sequential process,[37-40] and is discussed in Section II.C.

In the Appendix to Section II we develop the most rudimentary sequential model—the two-state model. There are two reasons for including some of the aspects of this very simple model; first, protein unfolding and refolding has sometimes been considered cooperative enough to be amenable to treatment with a two-state model; second, results for the two-state model are useful in discussing aspects of polymer dynamics already mentioned, which are considered in Section III.

II. KINETICS OF THE LINEAR ISING LATTICE; SELECTED APPLICATIONS TO BIOPOLYMER DYNAMICS

A. Infinite Chain; The Four Relaxation Times

To reduce their infinite set of equations already cited [Eqs. (5)] to a finite number of equations, Silberberg and Simha[25] explored various possibilities of introducing closure conditions.

It should first be recalled that at equilibrium, closure occurs at the doublets. Indeed, equilibrium properties are determined by the free energy distribution among the possible states which, in turn, yields the partition function of the system. For the linear Ising lattice, the free energy of a state of the chain is given by the sum of free energies of the various elements plus the free energy contributions arising from the interactions between nearest neighbors. One may, for example, include formally the interaction term in the contribution of one of the elements involved—say, always the right-hand one.[28] Accordingly, the overall free energy could be composed of terms, with each assigned to one element of the lattice, and depending on the state of that element as well as on the state of the left neighbor.

It follows that the probability of any sequence at equilibrium can be expressed entirely in terms of probabilities of singlets and of doublets.

Consider, for example, the sequence $ZYXU,\ldots$, where each letter stands for one of the states, A or B, of the corresponding element.

Once the doublet ZY is assumed to exist, the nature of the second element is known. The probability that the sequence ZYX obtains thus clearly equals the probability of ZY, multiplied by the conditional probability of YX, when the first element of the latter doublet is known to be Y. Calling these various probabilities $\bar{p}$ in an infinite chain, with the overbar designating the equilibrium state, one thus has to multiply $\bar{p}(ZY)$ by $\bar{p}(YX)/\bar{p}(Y)$. Analogous reasoning applies to further doublets, yielding

$$\bar{p}(ZYXV\cdots)=\frac{\bar{p}(ZY)\bar{p}(YX)\bar{p}(XV)}{\bar{p}(Y)\bar{p}(X)}\cdots \tag{6}$$

Next the pertinent probabilities for singlets and doublets are introduced, namely;

$$\bar{p}(A)=1-\bar{p}(B) \tag{7a}$$

$$\bar{p}(AB)=\bar{p}(BA) \tag{7b}$$

$$\bar{p}(AA)=1-\bar{p}(B)-\bar{p}(AB) \tag{7c}$$

$$\bar{p}(BB)=\bar{p}(B)-\bar{p}(AB) \tag{7d}$$

As shown by Schwarz,[28] (6) and (7) constitute a basis for the description. of equilibrium properties in terms of two probabilities—for example, $\bar{p}(B)$ and $\bar{p}(AB)$. The chemical equilibrium constants of $AAB \rightleftharpoons ABB$ and $AAA \rightleftharpoons ABA$, considered by Zimm and Bragg,[2] may be introduced as s and σs, respectively. Then, by application of (6) and (7), the following equations are obtained:[28]

$$\frac{\bar{p}(ABB)}{\bar{p}(AAB)}=\frac{\bar{p}(BB)}{\bar{p}(AA)}\cdot\frac{\bar{p}(A)}{\bar{p}(B)}=s \tag{8a}$$

$$\frac{\bar{p}(ABA)}{\bar{p}(AAA)}=\left(\frac{\bar{p}(AB)}{\bar{p}(AA)}\right)^{2}\cdot\frac{\bar{p}(A)}{\bar{p}(B)}=\sigma s \tag{8b}$$

Equations (7) and (8) lead to the classical result obtained by the matrix method[2] for $\bar{p}(B)$; $\bar{p}(B)$ will also be denoted Θ (helicity), when B represents the helical conformation of a unit in a polypeptide; the preceding equations lead further to

$$\bar{p}(AB) = \frac{(\sigma s)^{1/2}(\Theta(1-\Theta))^{1/2}}{\lambda_0} \tag{9}$$

with

$$\lambda_0 = 1 + (\sigma s)^{1/2}\left(\frac{\Theta}{(1-\Theta)}\right)^{1/2} $$

Finally, Θ satisfies the relationship[41]

$$\frac{\sigma s}{(1-s)^2} = \frac{\Theta(1-\Theta)}{(1-2\Theta)^2} \tag{10}$$

If we return now to the kinetic problem, it is found that closures at different levels may be introduced. The triplet closure, for instance, permits one to express in a nonequilibrium state the probability of any sequence in terms of probabilities of triplets and doublets, as follows:

$$P(ZYXVU\cdots) = \frac{p(ZYX)p(YXV)p(XVU)}{p(YX)p(XV)p(VU)}\cdots \tag{11}$$

Closures at the triplets and quadruplets have been compared by Rabinowitz, Silberberg, Simha, and Loftus.[26] Numerical solutions of the resulting nonlinear differential equations have been tested by variation of the magnitudes of the rate parameters over a wide range. Singlet, doublet, and triplet probabilities as well as number- and weight-average sequence lengths have been examined as the system passes from one equilibrium state toward a second. No significant differences between the two sets of approximations have been observed.

Thus although assertion of rigorous closure at the triplet level is erroneous (as shown by Silberberg and Simha[42]), triplet closure seems to provide a numerically adequate approximation. It has been elegantly applied by Schwarz[28,30] to deduce the relaxation times of the "infinitely long" linear Ising lattice. Although Schwarz has also written the kinetic equations for a finite chain (which are $4N - 5$ in number, for a chain of N elements), we report here only the results for asymptotically large values of N.

In the kinetic treatment, four independent variables are needed, which can be chosen as $p(B)$, $p(AB)$, $p(ABA)$, $p(BAB)$. The remaining triplet probabilities are

$$p(AAB) = p(BAA) = p(AB) - p(BAB) \tag{12a}$$

$$p(ABB) = p(BBA) = p(AB) - p(ABA) \tag{12b}$$

$$p(AAA) = 1 - p(B) - 2p(AB) + p(BAB) \tag{12c}$$

$$p(BBB) = p(B) - 2p(AB) + p(ABA) \tag{12d}$$

According to Schwarz[29] transitions between states of the system are described by considering elementary rate processes. The growth process of an uninterrupted sequence of like units is associated with two kinetic coefficients k_F and k_D:

$$AAB \underset{k_D}{\overset{k_F}{\rightleftharpoons}} ABB$$

which are related by the condition $k_F/k_D = s$.

Next, the two nucleation processes are considered:

$$AAA \underset{f_A k_D}{\overset{\sigma f_A k_F}{\rightleftharpoons}} ABA$$

$$BBB \underset{f_B k_F}{\overset{\sigma f_B k_D}{\rightleftharpoons}} BAB$$

Clearly there exists a maximum of three independent kinetic coefficients, which may be taken as k_F, f_A, f_B, and σ is again Zimm and Bragg's[2] statistical nucleation parameter. (In the particular case in which the interactions between neighboring elements are only of dipolar origin, the kinetic coefficients take the more special form proposed by Glauber.[24])

To formulate the rate equation for any chosen variable, one must sum up rate contributions due to all the elementary processes that affect the corresponding sequence of A's and/or B's. The time dependence of $p(B)$ and $p(AB)$ can be directly expressed in terms of triplets, whereas in the case of $p(ABA)$ and $p(BAB)$, quadruplets have to be introduced when transitions of the outer states are considered. After applying triplet closure to any p-function, the set of rate equations (14) obtains[28] in the limit of $N \rightarrow \infty$. These equations should be applicable when N is large as compared with the cooperativity length—that is, when

$$N \gg \sigma^{-1/2} \tag{13}$$

$$\frac{dp(B)}{dt} = 2k_F p(AAB) - 2k_D p(ABB) + f_A(\sigma k_F p(AAA) - k_D p(ABA))$$

$$+ f_B(k_F p(BAB) - \sigma k_D p(BBB)) \tag{14a}$$

$$\frac{dp(AB)}{dt} = f_A(\sigma k_F p(AAA) - k_D p(ABA)) - f_B(k_F p(BAB) - \sigma k_D p(BBB))$$

$$\tag{14b}$$

$$\frac{dp(ABA)}{dt} = 2k_D \frac{p^2(ABB)}{p(BB)} - 2k_F \frac{p(ABA)p(AAB)}{p(AB)}$$

$$+ f_A(\sigma k_F p(AAA) - k_D p(ABA))$$

$$- 2f_B\left(k_F \frac{p(ABA)p(BAB)}{p(AB)} - \sigma k_D \frac{p(ABB)p(BBB)}{p(BB)}\right) \tag{14c}$$

$$\frac{dp(BAB)}{dt} = 2k_F \frac{p^2(AAB)}{p(AA)} - 2k_D \frac{p(ABB)p(BAB)}{p(AB)}$$

$$+ 2f_A\left(\sigma k_F \frac{p(AAB)p(AAA)}{p(AA)} - k_D \frac{p(ABA)p(BAB)}{p(AB)}\right)$$

$$- f_B(k_F p(BAB) - \sigma k_D p(BBB)) \tag{14d}$$

In combination with (7) and (12), these equations represent a closed system of rate equations for the kinetic analysis of the infinite linear Ising lattice.

The conclusion that only four equations are needed in the case of an infinitely long lattice had already been reached by Gô,[27] who calculated the relaxation times by computer.

Equations (14) can be linearized by the usual procedures[43,44] if applied to chemical relaxation after slight perturbation of equilibrium. For this purpose, we introduce the variables

$$x_1 = p_1 - p_1^\circ = p(B) - p^\circ(B) \tag{15a}$$

$$x_2 = p_2 - p_2^\circ = p(AB) - p^\circ(AB) \tag{15b}$$

$$x_3 = p_3 - p_3^\circ = p(ABA) - p^\circ(ABA) \tag{15c}$$

$$x_4 = p_4 - p_4^\circ = p(BAB) - p^\circ(BAB) \tag{15d}$$

which represent the deviations of the p variables from their values p° in an appropriate time-independent state. The instantaneous equilibrium values of the p_i's are denoted $\bar{p}_i$, and those of the x_i's are denoted $\bar{x}_i$ ($i = 1, 2, 3, 4$). Introducing the column-vectors $\mathbf{x} = (x_1, x_2, x_3, x_4)$ and $\bar{\mathbf{x}} = (\bar{x}_1, \bar{x}_2, \bar{x}_3, \bar{x}_4)$, the linearized rate equations read, in matrix notation

$$\frac{d\mathbf{x}}{dt} = \mathbf{A}(\mathbf{x} - \bar{\mathbf{x}}) \tag{16}$$

When unimolecular reactions are considered, we generally use row vectors (as in (4)), to ensure that the matrix $\mathbf{Q}$ is of the form that is familiar in the theory of Markov processes. In (16), on the other hand, we have held to the tradition of using column vectors in studies of chemical relaxation. It follows that when (16) is applied to unimolecular reactions, as we do in Section II.C, the $\mathbf{A}$ matrix is the transpose of the corresponding $\mathbf{Q}$ matrix.

The complete expressions of the 16 elements of the $\mathbf{A}$ matrix have been given by Schwarz,[30] who also used a simplified form [Eq. (18)] that is valid for high cooperativity ($\sigma^{1/2} \ll 1$), and when the additional conditions (17) are fulfilled:[30]

$$\sigma^{1/2} \ll f_A, \qquad f_B \ll \sigma^{-1/2} \tag{17}$$

Conditions (17) mean that the thermodynamic instability of an A- or B-nucleus is essentially caused by its slow rate of formation; thus the rate constant for the destruction of a nucleus becomes comparable with that of a growth reaction. With these assumptions, and putting $p(B) = \Theta$, the $\mathbf{A}$

matrix can be simplified to[30]

$$
k_D^{-1}\mathbf{A} = \begin{vmatrix}
-\sigma(f_A+f_B) & 2\sigma^{1/2}\dfrac{2\Theta-1}{(\Theta(1-\Theta))^{1/2}} \\[2em]
-\sigma(f_A-f_B) & -2\sigma(f_A+f_B) \\[1em]
-\sigma\left(2\dfrac{1-\Theta}{\Theta}+f_A\right) & 4\sigma^{1/2}\left(\dfrac{1-\Theta}{\Theta}\right)^{1/2} \\[1.5em]
\sigma\left(2\dfrac{\Theta}{1-\Theta}+f_B\right) & 4\sigma^{1/2}\left(\dfrac{\Theta}{1-\Theta}\right)^{1/2}
\end{vmatrix}
$$

$$
\begin{matrix}
2-f_A & -(2-f_B) \\[1.5em]
-f_A & -f_B \\[1em]
-(2+f_A) & 2\sigma^{1/2}(1-f_B)\left(\dfrac{1-\Theta}{\Theta}\right)^{1/2} \\[1.5em]
2\sigma^{1/2}(1-f_A)\left(\dfrac{\Theta}{1-\Theta}\right)^{1/2} & -(2+f_B)
\end{matrix}
\Bigg| \quad (18)
$$

The relaxation times must satisfy the secular equation

$$
|\mathbf{A}+\tau^{-1}\mathbf{I}|=0 \tag{19}
$$

where $\mathbf{I}$ is the unit matrix. Employing Viète's rule, which relates the roots of (19) to its coefficients, Schwarz deduced within the scope of the foregoing approximation the following analytical expressions of the four relaxation times:[30]

$$
\tau_1^{-1}=\frac{\sigma}{\Theta(1-\Theta)}k_D \tag{20a}
$$

$$
\tau_2^{-1}=4\left(\frac{\sigma}{\Theta(1-\Theta)}\right)^{1/2}\left(\frac{f_A}{2+f_A}(1-\Theta)+\frac{f_B}{2+f_B}\Theta\right)k_D \tag{20b}
$$

$$
\tau_3^{-1}=(2+f_B)k_D \tag{20c}
$$

$$
\tau_4^{-1}=(2+f_A)k_D \tag{20d}
$$

The problem of calculating relaxation amplitudes is considered in Section II.B. If a small stepwise perturbation is applied at $t=0$, and one measures directly $p(B)=\Theta$ for $t>0$, the relaxation behavior is readily seen to be dominated by the longest time τ_1.[45] It follows from (20a) that τ_1

exhibits a maximum at midtransition between the full A and full B forms ($\Theta = 0.5$), which is sharp with respect to the variation of the equilibrium constant s when the transition itself is sharp (i.e., when the process is highly cooperative). Here we have a particular example of the slowing down of processes by cooperativity.

Schwarz has shown[29] that the initial rate $(d\Theta/dt)_{t=0}$ is related to a mean relaxation time τ^* by

$$(\tau^*)^{-1} = (\Delta\Theta)^{-1} \left(\frac{d\Theta}{dt} \right)_{t=0} \tag{21}$$

where $\Delta\Theta$ is the difference between the final and initial values of Θ. When a small perturbation is applied, τ^* turns out to be close to τ_1^{30} (and equal to it at midtransition). Craig and Crothers[46] have shown that τ^* can be calculated even when the perturbation is not small and that the measurement of the initial rate $(d\Theta/dt)_{t=0}$ permits evaluation of rate constants from kinetic curves under rather general conditions.

B. Relaxation Amplitudes; Fluctuations of the Number of Uninterrupted Sequences

This section presents information on the kinetics of the linear Ising lattice obtained from relaxation experiments based on the application of a weak sinusoidal perturbation. The periodic displacement of the "chemical" equilibrium (here the equilibrium between the various states of the system) produces a dissipation of energy.

Typically, the required information is obtained by measuring the ultrasonic absorption as a function of the frequency, for various values of the equilibrium constant s. In his early work, Schwarz[29] treated the problem of ultrasonic absorption as if the system exhibited single-relaxation behavior, and he identified the corresponding relaxation time with the mean relaxation time τ^* defined by (21).

In aqueous solutions, the relaxation amplitude was furthermore found to be proportional to[47]

$$(\partial V)^2 \left(\frac{\partial \Theta}{\partial \ln s} \right) \tag{22a}$$

hence to

$$2\sigma^{-1/2}(\delta V)^2 (\Theta(1-\Theta))^{3/2} \tag{22b}$$

where δV is the difference in partial molar volumes of the chain unit when it affects the B and A forms, respectively (e.g., the helix and the coil forms, respectively).

We may interpret this result in the following manner: in aqueous solutions, perturbation is due only to the periodic change of pressure, since the wave produces no temperature variation. Thus the energy dissipated is proportional to

$$-\left(\frac{\partial V}{\partial P}\right)_S \int P\,dP \tag{23}$$

where the integration extends to one period of the applied field; V denotes the volume of the solute, and P the pressure. The derivative $(\partial V/\partial P)_S$ is taken at constant entropy. If the only contribution to volume changes arises from the conformational changes A$\rightleftharpoons$B of the chain units, $(\partial V/\partial P)_S$ is proportional to $\delta V(\partial\Theta/\partial P)_S$ for a given concentration of the solute. In aqueous solutions, the partial derivatives can be taken at will at constant entropy or at constant temperature T. On the other hand, neglecting $\partial\sigma/\partial P$ terms, we can write

$$\left(\frac{\partial\Theta}{\partial P}\right)_T \# \left(\frac{\partial\Theta}{\partial s}\right)\left(\frac{\partial s}{\partial P}\right)_T$$

Finally, since

$$\left(\frac{\partial \ln s}{\partial P}\right)_T = -\frac{\delta V}{RT} \tag{24}$$

where R is the gas constant, (23) is indeed seen to be proportional to (22).

We have followed in the preceding derivation a more general argument,[48] developed to account for experimental results[32] that appear to deviate from the behavior predicted by Schwarz. The detailed discussion of these experiments is temporarily postponed. We need only remark for the moment that if an average molar volume change $\overline{\Delta V}$ is associated with the loss of one uninterrupted A-sequence (i.e., one disordered sequence) per molecule, there will result at least at relatively high values of Θ, a contribution to the ultrasonic absorption proportional to[48]

$$-N^{-1}\delta V\overline{\Delta V}\left(\frac{\partial v}{\partial \ln s}\right) \tag{25}$$

where v is the average number of uninterrupted B-sequences (i.e., ordered, generally helical sequences) per molecule, and N is the number of A and B units. When the value of v calculated by Lifson and Roig[49] is used, it is easily seen that such a term would indeed account[48] for deviations from the behavior predicted by (22).

It is clear that the calculation of the ultrasonic absorption involves rather complex mechanisms. General expressions for the ultrasonic absorption associated with chemical relaxation have already been derived.[43,44] Explicit expressions of the measurable quantities for stepwise and for sinusoidal perturbation [(26) and (28) below] can be given[31] in terms of the cofactors A_r^{li} and $\tilde{A}_r^{mi}$ of the determinants $|\mathbf{A}_r| = |\mathbf{A} + \tau_r^{-1}\mathbf{I}|$, $(r = 1, 2, 3, 4)$, and of their transposes. It follows, indeed, from the thermodynamics of irreversible processes that under quite general conditions, the $\mathbf{A}$ matrix can be made symmetrical, thus can be diagonalized; the right and left eigenvectors therefore can be biorthonormalized.[44] The required cofactors turn out to have simple expressions for the infinite linear Ising lattice.

1. Stepwise Perturbation at $t = 0$.

The constant stationary state to be approached after the perturbation may be chosen as the reference state; thus the equilibrium values $\bar{p}_i$ of Section II.A are equal to the reference values p_i°, and the equilibrium values $\bar{x}_i$ are equal to zero. Let Z be a given measurable quantity, and δZ its deviation from the stationary value $\bar{Z}$. The time dependence of δZ involves the initial values $\overset{\circ}{x}_i$ at $t = 0$, and the derivatives $(\partial Z / \partial x_i)|_0$ taken at zero value of the x_i's (i.e., the $(\partial Z / \partial p_i)|_0$ taken at $p_i = \bar{p}_i$). For δZ, at $t > 0$ we can write

$$\delta Z = \sum_r \left(\sum_i A_r^{li} \frac{\partial Z}{\partial x_i}\Big|_0 \right) \left(\sum_j \tilde{A}_r^{mj} \overset{\circ}{x}_j \right) \left(\sum_k A_r^{lk} \tilde{A}_r^{mk} \right)^{-1} \exp - \frac{t}{\tau_r} \quad (26)$$

The integers l and m denote arbitrary rows. An adequate choice of l and m can effectively simplify the calculations.

2. Sinusoidal Perturbation.

Again, let us take as an example an aqueous solution, calculating the absorption of ultrasonic longitudinal waves. Assume that infinitesimal changes of the x_i's at constant pressure and entropy produce the infinitesimal volume change of the unit volume

$$dV = c \sum_i \delta V_i dx_i \quad (27)$$

where c is the concentration in moles of monomers per unit volume. The expression of $\alpha\lambda / \pi$ can be written

$$\frac{\alpha\lambda}{\pi} = -c\rho v^2 \sum_r \left(\sum_i A_r^{li} \delta V_i \right) \left(\sum_j \tilde{A}_r^{mj} \frac{\partial \bar{x}_j}{\partial P}\Big|_S \right) \left(\sum_k A_r^{lk} \tilde{A}_r^{mk} \right)^{-1} \frac{\omega \tau_r}{1 + \omega^2 \tau_r^2}$$

$$(28)$$

where ω is the circular frequency, $\alpha\lambda$ the absorption coefficient per wavelength, ρ the specific weight of the medium, and v the sound velocity.

Since the p_j°'s have constant values, $(\partial\bar{x}_j/\partial P)_S$ in (28) can be replaced by $(\partial\bar{p}_j/\partial P)_S$. These derivatives contain an oscillating contribution at the frequency ω. The amplitude of ultrasonic waves is quite small, however; therefore, $(\partial\bar{p}_j/\partial P)_S$ is practically the pressure derivate of the equilibrium value of p_j at zero applied field.

Turning again to the infinite linear Ising lattice, we assume in accordance with a previous remark that volume changes result not only from the A⇌B conformational changes, but also from fluctuations of the number of uninterrupted A or B sequences per molecule. Therefore (27) is written in the form

$$dV = c\left(\delta V\,dp\,(\mathrm{B}) + \Delta V\,dp\,(\mathrm{AB}) + \Delta V\,dp\,(\mathrm{BA})\right) \tag{29}$$

With the foregoing meaning of $\overline{\Delta V}$ (i.e., the average molar volume change associated with the loss of one disordered sequence per molecule), and noting that $p(\mathrm{AB}) = p(\mathrm{BA})$, we have $\overline{\Delta V} = -2\Delta V$; thus (29) can also be written

$$dV = c\left(\delta V\,dp_1 - \overline{\Delta V}\,dp_2\right) \tag{30}$$

To calculate the ultrasonic absorption associated with the longest time τ_1, all the cofactors A and $\tilde{A}$ with a subscript 1 are required *a priori*. Some of the terms in (28), however, make negligible contributions and are not retained. If we set

$$\frac{\alpha\lambda}{\pi} = -c\rho v^2 \sum A'_r \frac{\omega\tau_r}{1+\omega^2\tau_r^2} \tag{31}$$

the amplitude coefficient A'_1 is given to sufficient approximation by

$$A'_1 = \frac{A_1^{11}\tilde{A}_1^{11}}{A_1^{11}\tilde{A}_1^{11} + A_1^{12}\tilde{A}_1^{12}}\left(\delta V - \frac{A_1^{12}}{A_1^{11}}\overline{\Delta V}\right)\left(\frac{\partial\bar{p}_1}{\partial\ln s}\frac{\partial\ln s}{\partial P} + \frac{\partial\bar{p}_1}{\partial\ln\sigma}\frac{\partial\ln\sigma}{\partial P}\right.$$

$$\left. + \frac{\tilde{A}_1^{12}}{\tilde{A}_1^{11}}\left(\frac{\partial\bar{p}_2}{\partial\ln s}\frac{\partial\ln s}{\partial P} + \frac{\partial\bar{p}_2}{\partial\ln\sigma}\frac{\partial\ln\sigma}{\partial P}\right)\right)$$

$$\tag{32}$$

Since the nucleation parameter σ of Zimm and Bragg may be described as the equilibrium constant for the formation of an interruption in a sequence of B-units by a process that maintains a constant number of B-units (i.e., of helical units),[50] we can introduce the corresponding volume change by the following relation:

$$\frac{\partial \ln \sigma}{\partial P} = -\frac{\Delta V_\sigma}{RT} \tag{33}$$

At infinite chain length, $-\Delta V_\sigma$ should not differ greatly from $\overline{\Delta V}$, and at the present stage we make no distinction between these two volume changes. Putting $\bar{p}_1 = \Theta$, the required cofactors are found to be

$$k_D^{-1} A_1^{11} = k_D^{-1} \tilde{A}_1^{11} = -4\sigma^{1/2}(\Theta(1-\Theta))^{-1/2}(f_B(2+f_A)\Theta + f_A(2+f_B)(1-\Theta))$$

$$\tag{34a}$$

$$k_D^{-1} A_1^{12} = 2\sigma(2\Theta-1)(\Theta(1-\Theta))^{-1}(f_B(2+f_A)\Theta + f_A(2+f_B)(1-\Theta)) \tag{34b}$$

$$k_D^{-1} \tilde{A}_1^{12} = 2\sigma^{1/2}(\Theta(1-\Theta))^{-1/2}((2\Theta-1)((2-f_A)(2-f_B) - 4f_A f_B) + 4(f_A - f_B)) \tag{34c}$$

Since we are interested mainly in values of Θ that are significantly larger than 0.5, we assume that $1-2\Theta$ is not close to zero. If we assume furthermore that

$$|f_A - f_B| \ll f_A, f_B \tag{35}$$

and take into account (34), the last term in the second set of parentheses of (32) is neglibible, and A_1' reduces to

$$A_1' = -\frac{A_1^{11} \tilde{A}_1^{11}}{A_1^{11} \tilde{A}_1^{11} + A_1^{12} \tilde{A}_1^{12}} \frac{\partial \Theta}{\partial \ln s}\left(\delta V - \frac{A_1^{12}}{A_1^{11}} \overline{\Delta V}\right)\left(\frac{\delta V}{RT} - \frac{\partial \Theta / \partial \ln \sigma}{\partial \Theta / \partial \ln s} \frac{\overline{\Delta V}}{RT}\right)$$

$$\approx -2\sigma^{-1/2} \frac{(\delta V)^2}{RT}(\Theta(1-\Theta))^{3/2}\left(1 - \frac{\sigma^{1/2}}{2} \frac{\overline{\Delta V}}{\delta V} \frac{1-2\Theta}{(\Theta(1-\Theta))^{1/2}}\right)^2 \tag{36}$$

Thus the contribution to the ultrasonic absorption per wavelength of the slowest process is

$$\frac{(\alpha\lambda)_1}{\pi} \cong 2c\rho v^2 \sigma^{-1/2} \frac{(\delta V)^2}{RT} (\Theta(1-\Theta))^{3/2}$$

$$\times \left(1 - \frac{\sigma^{1/2}}{2} \frac{\overline{\Delta V}}{\delta V} \frac{1-2\Theta}{(\Theta(1-\Theta))^{1/2}}\right)^2 \frac{\omega\tau_1}{1+\omega^2\tau_1^2} \tag{37}$$

For $\overline{\Delta V} = 0$ the result of Schwarz[47] is recovered. On the other hand, if only the two terms proportional to $(\delta V)^2$ and to $\delta V \overline{\Delta V}$ are retained in (37), we introduce the average number $v = N\bar{p}(\mathrm{AB})$ of uninterrupted helical sequences per molecule, to obtain

$$\frac{(\alpha\lambda)_1}{\pi} \cong \frac{c\rho v^2}{RT} \left((\delta V)^2 \frac{\partial\Theta}{\partial\ln s} - 2N^{-1}\delta V \overline{\Delta V} \frac{\partial v}{\partial\ln s} + \cdots\right) \frac{\omega\tau_1}{1+\omega^2\tau_1^2} \tag{38}$$

The expression within parentheses in (38) must be positive. If the second term in the parentheses is positive itself, which is the case for the Θ values of interest here, we may consider that $(\alpha\lambda)_1$ is the sum of the term of Schwarz proportional to $(\delta V)^2(\partial\Theta/\partial\ln s)$, and of a term due to fluctuations of v. As predicted[48] by the simple argument recalled at the beginning of the section, this term is proportional to $-N^{-1}\delta V \overline{\Delta V} (\partial v/\partial\ln s)$.

It is customary in ultrasonic studies to consider the maximum value $(\alpha\lambda)_M$ of $\alpha\lambda$, obtained for the frequency at which $\omega\tau = 1$. In the present case, we are interested in the dependence of $(\alpha\lambda)_M$ on Θ. However, since τ_1 itself, as given by (20a), depends on Θ, we would have to consider the value of $\alpha\lambda$ obtained at a frequency properly chosen for each value of Θ, and compare the experimental plot of $(\alpha\lambda)_M$ versus Θ with the theory. This may not be easy to carry out in practice, since in addition the measurements do not seem to support single relaxation behavior.[32]

Let us nevertheless examine how the term of Schwarz of (37) and (38)

$$2c\rho v^2 \sigma^{1/2} \frac{(\delta V)^2}{RT} (\Theta(1-\Theta))^{3/2} \frac{\omega\tau_1}{1+\omega^2\tau_1^2}$$

varies with Θ under different conditions. When (20a) is used, this term is seen to be proportional to

$$(\Theta(1-\Theta))^{5/2} \quad \text{if} \quad \omega\tau_1 \ll 1$$

$$(\Theta(1-\Theta))^{3/2} \quad \text{if} \quad \omega\tau_1 = 1$$

$$(\Theta(1-\Theta))^{1/2} \quad \text{if} \quad \omega\tau_1 \gg 1$$

In all these conditions, the contribution to $(\alpha\lambda)_1$ shows a maximum at a value Θ_m corresponding to midtransition ($\Theta_m = 0.5$).

The result is different if the complete equation (37) is used. It is easily seen that the value Θ_m at which Θ is a maximum is then given, for the corresponding approximation, by

$$\Theta_m = 0.5 + \sigma^{1/2} \frac{1+\omega^2\tau_1^2}{5+\omega^2\tau_1^2} \frac{(\overline{\Delta V})_m}{\delta V} \tag{39}$$

leading to

$$\Theta_m = 0.5 + \frac{\sigma^{1/2}}{5} \frac{(\overline{\Delta V})_m}{\delta V} \quad \text{if} \quad \omega\tau_1 \ll 1 \tag{39a}$$

$$\Theta_m = 0.5 + \frac{\sigma^{1/2}}{3} \frac{(\overline{\Delta V})_m}{\delta V} \quad \text{if} \quad \omega\tau_1 = 1 \tag{39b}$$

$$\Theta_m = 0.5 + \sigma^{1/2} \frac{(\overline{\Delta V})_m}{\delta V} \quad \text{if} \quad \omega\tau_1 \gg 1 \tag{39c}$$

In the derivation of (39), $\overline{\Delta V}$ has been considered as a constant, a condition that need not be rigorously exact, as the following discussion indicates. Therefore, $(\overline{\Delta V})_m$ has been used to denote the value of $\overline{\Delta V}$ at $\Theta = \Theta_m$. The derivation is seen to amount to neglecting the change of $\overline{\Delta V}$ with Θ in the vicinity of the value $\Theta = \Theta_m$.

Suppose now that $\overline{\Delta V}$ is of the same sign as δV. It is shown at the end of the section that this assumption is likely to be fulfilled in the case of poly-L-glutamic acid (PLGA); for results of ultrasonic measurements[32] of this substance, see Fig. 4.

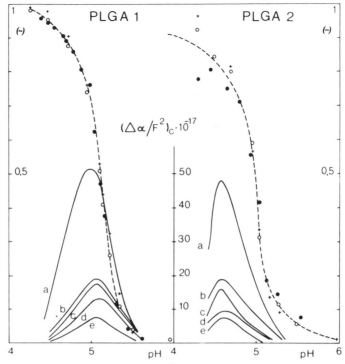

Fig. 4. Conformational contribution $(\Delta\alpha/F^2)_c$ to the ultrasonic absorption as a function of the pH at different frequencies, and transition curves for two samples of poly-L-glutamic acid of different molecular weights, in H_2O–0.2 M NaCl.[32] Transition curves are obtained by ultraviolet polarimetry (solid dots) and circular dichroism at 222 nm (open dots) and 208 nm (plus signs). Frequencies: (a) 1.04, (b) 2.82, (c) 3.55, (d) 5.04, (e) 7.96 MH$_z$.

As pointed out by Zana,[51] ultrasonic measurements in aqueous solutions of polypeptides have sometimes been interpreted erroneously as arising from conformational transitions. In the case considered here, proton-transfer reactions occurring between the carboxylic groups and the solvent, as well as site binding of counterions, are known to contribute to the observed ultrasonic absorption.[52,53] These effects must therefore be separated from the conformational contribution to the ultrasonic absorption. In Fig. 4 the purely conformational contribution $(\Delta\alpha/F^2)_c$ to $\Delta\alpha/F^2$ (where the subscript c means "conformational") is considered. The $\Delta\alpha$ represents the excess absorption $\alpha - \alpha_0$ of the solution over the solvent, and F the frequency. Results obtained for two samples of PLGA (1 and 2), of respective molecular weights 90,000 and 12,000, dissolved in H_2O–0.2 M NaCl, at the concentration $2 \times 10^{-2} g/cm^3$, are plotted as a function of the pH at different frequencies. The effect of site binding of counterions has been eliminated[54] by employing a method already applied by Tondre and Zana[53] to other substances (i.e., the use of a voluminous counterion). The existence of a conformational contribution to the absorption has been further confirmed by comparison of the results with those of similar measurements for a sample of poly-DL-glutamic acid.[54]

Figure 4 also shows, for both samples of PLGA, the value of the helicity Θ as a function of the pH, as determined by ultraviolet polarimetry and circular dichroism.[54] When allowance is made for the various sources of error in the measurement of the ultrasonic absorption α and the helicity Θ, it appears that for PLGA 1 the maximum of $(\Delta\alpha/F^2)_c$ occurs for a value of Θ lying between 0.5 and 0.7, which is therefore compatible at the limit with the value found by other authors.[55,56]

For PLGA 2 the maximum of $(\Delta\alpha/F^2)_c$ occurs at a value of Θ lying between 0.7 and 0.85. It is of course possible that the displacement of the absorption peak for PLGA 2 is connected with end effects in a chain of finite length.

However, PLGA 2 contains about 94 monomer units, whereas the cooperativity length $\sigma^{-1/2}$ is about 14, if the value $\sigma = 5 \times 10^{-3}$ is used.[50] The requirement imposed by (13) can therefore be considered fulfilled, and it may seem meaningful to attempt to apply the theory that is valid at infinite length. In this case, however, we must not neglect to recognize that the loss of one disordered sequence per molecule may produce a larger volume change for a smaller molecule. Illustratively, note that for the two samples considered here, the larger one (PLGA 1) comprises on the average about 20 uninterrupted helical sequences, whereas the smaller one (PLGA 2) comprises on the average only 2.5. Thus when a molecule of PLGA 1 loses one disordered sequence at constant θ, about 16 elements redistribute among 20 disordered sequences, whereas for PLGA 2 about 10 elements redistribute among about 3 sequences. Therefore, one may well expect a larger value of $\overline{\Delta V}$ for PLGA 2 than for PLGA 1.

Thus let us consider that to a certain extent we can account for the molecular weight dependence of Θ_m by assuming $\overline{\Delta V}$ in the first set of parentheses of (32) to be larger for PLGA 2 than for PLGA 1. However, apart from introducing the preceding molecular-weight-dependent volume change $\overline{\Delta V}$, we continue to use the results for an infinitely long chain. Therefore, the corresponding nucleation coefficient σ, as given by (33), appears in the second set of parentheses in (32).

The molecular-weight-dependent $\overline{\Delta V}$ is then written

$$\overline{\Delta V} = -\gamma^{-1}\Delta V_\sigma \tag{40}$$

$$-\epsilon \leqslant \gamma < 0 \quad \text{or} \quad 0 < \gamma \leqslant 1 \tag{41a}$$

$$0 < \epsilon \ll 1 \tag{41b}$$

According to (41a) and (41b), we do not exclude the possibility that $-\Delta V_\sigma$ can be of sign opposite to that of $\overline{\Delta V}$, but, in this case, much smaller in absolute value than $\overline{\Delta V}$. In a previous paper, owing to an unfortunate choice of notation, a coefficient α had been introduced in (40), leading to possible confusion with the absorption coefficient α. Equations (38) and (39) are now transformed into

$$\frac{(\alpha\lambda)_1}{\pi} \simeq \frac{c\rho v^2}{RT}\left((\delta V)^2\frac{\partial\Theta}{\partial\ln s} - (1+\gamma)N^{-1}\delta V\,\overline{\Delta V}\frac{\partial v}{\partial\ln s} + \cdots\right)\frac{\omega\tau_1}{1+\omega^2\tau_1^2}$$

$$\tag{42}$$

and

$$\Theta_m = 0.5 + \frac{1+\gamma}{2}\sigma^{1/2}\frac{1+\omega^2\tau_1^2}{5+\omega^2\tau_1^2}\frac{(\overline{\Delta V})_m}{\delta V} \tag{43}$$

or

$$\Theta_m = 0.5 + \frac{1+\gamma}{10}\sigma^{1/2}\frac{(\overline{\Delta V})_m}{\delta V} \qquad \text{if} \quad \omega\tau_1 \ll 1 \tag{43a}$$

$$\Theta_m = 0.5 + \frac{1+\gamma}{6}\sigma^{1/2}\frac{(\overline{\Delta V})_m}{\delta V} \qquad \text{if} \quad \omega\tau_1 = 1 \tag{43b}$$

$$\Theta_m = 0.5 + \frac{1+\gamma}{2}\sigma^{1/2}\frac{(\overline{\Delta V})_m}{\delta V} \qquad \text{if} \quad \omega\tau_1 \gg 1 \tag{43c}$$

It may be noticed that the pressure derivative of the helicity is given by

$$\frac{d\Theta}{dP} = -2\sigma^{1/2}\frac{\delta V}{RT}(\Theta(1-\Theta))^{3/2}\left(1 - \frac{\gamma}{2}\sigma^{1/2}\frac{1-2\Theta}{(\Theta(1-\Theta))^{1/2}}\frac{\overline{\Delta V}}{\delta V}\right) \tag{44}$$

Thus $-d\Theta/dP$, when plotted against Θ, shows a maximum for $\Theta = \Theta'_m$:

$$\Theta'_m = 0.5 + \frac{\gamma}{6}\sigma^{1/2}\frac{(\overline{\Delta V})'_m}{\delta V} \tag{45}$$

where $(\overline{\Delta V})'_m$ denotes the value of $\overline{\Delta V}$ for $\Theta = \Theta'_m$. This gives an independent equation containing the molecular-weight-dependent factor γ.

According to the ultrasonic measurements of Barksdale and Stuehr,[56] the lowest relaxation frequency is of the order of 200 Hz for PLGA. Therefore, it appears probable that our experiments[32] have been carried out under the condition $\omega\tau_1 \gg 1$, and Θ_m should be expressed by (43c). On the other hand, δV is known to be positive and of the order of 1 cm^3/mole for PLGA.[57]

Let us assume for PLGA 2 a value of $N^{-1}\overline{\Delta V}$ (the volume change referred to one monomer unit) that is significantly smaller than δV—say, such that $N^{-1}\overline{\Delta V}/\delta V \cong 0.06$; if we again take $\sigma = 5\times10^{-3}$ and neglect γ as compared to 1, the value $\Theta = 0.7$ is obtained. Thus even a small value of $N^{-1}\overline{\Delta V}/\delta V$ may lead to a large displacement of the ultrasonic absorption peak towards Θ values greater than 0.5, provided $\overline{\Delta V}$ is of the same sign as δV (i.e., provided $\overline{\Delta V}$ is positive in the present case).

When a molecule loses one uninterrupted disordered sequence at constant helicity, the segments that were previously a part of the disordered sequence redistribute among the other disordered sequences, whose average length and degree of folding therefore increase. This conformational change may produce a positive volume change $\overline{\Delta V}$ in two ways.

1. In favoring in each disordered sequence the formation of hydrogen bonds linking γ-carboxylic groups to other γ-carboxylic groups, or to groups of the main chain.

2. In favoring hydrophobic interactions that involve nonpolar groups of the lateral chains.

The first mechanism increases the volume by removing ionized γ-carboxylic groups that were previously embedded in water, thus diminishing the electrostrictive effect. The second effect is also known to lead to an increase in volume.[58] Since the corresponding molar changes may amount to several cubic centimeters, the value of $\overline{\Delta V}$ assumed previously does not appear unreasonable.

Thus the fluctuations of the number of uninterrupted sequences of like units do furnish a possible explanation for the displacement of the absorption peak. The finiteness of the chain does not provide an explanation in itself, if only Schwarz's term in the expression of $\alpha\lambda_1$ is used [cf. (22a) and (38)], since for a finite chain the maximum of $\partial\Theta/\partial\ln s$ occurs at $\Theta < 0.5$. This follows from the results for finite chains of Lifson and Roig[49] (cf. also Fig. 10 of Ref. 48).

It is worth noting, on the other hand, that our measurements[32] do not exhibit single relaxation behavior for either sample 1 or sample 2. This may result from interactions between elements that are more remote than adjacent neighbors, or it may indicate that the second relaxation time τ_2 [in (20b)] furnishes a nonnegligible contribution to the ultrasonic absorption. We are presently investigating the latter possibility, and we intend to carry out further measurements.

Forces responsible for the stability and conformational transitions of proteins are involved in the molecular mechanisms which have been considered. Thus, at least under the conditions of the experiments referred to, ultrasonic measurements seem to permit a dynamic approach to the study of these forces.

C. Sequential Processes; Oligopeptides and Oligonucleotides; Protein Unfolding and Refolding

As a first example of a sequential process, let us consider the base pairing in oligonucleotides. According to Eigen and Pörschke,[34,36] the formation of an (A, U) double helix of oligoadenylic and oligouridylic acids at neutral pH proceeds in accordance with the following scheme:

$$
\begin{array}{l}
\text{A} \quad \text{U} \qquad \text{A} - \text{U} \qquad \text{A} - \text{U} \qquad \text{A} - \text{U} \\
\;| \quad\;\; | \qquad\;\; | \qquad | \qquad\; | \qquad | \qquad\; | \qquad | \\
\text{A} \quad \text{U} \qquad \text{A} \quad\; \text{U} \qquad \text{A} - \text{U} \qquad \text{A} - \text{U} \\
\;| \quad\;\; | \qquad\;\; | \qquad | \qquad\; | \qquad | \qquad\; | \qquad | \\
\text{A} \;\underset{+}{} \text{U} \xrightleftharpoons[k_{10}]{k_{01}} \text{A} \quad\; \text{U} \xrightleftharpoons[k_{21}]{k_{12}} \text{A} \quad\; \text{U} \xrightleftharpoons[k_{32}]{k_{23}} \text{A} - \text{U} \xrightleftharpoons[k_{43}]{k_{34}} \quad (46)\\
\;| \qquad | \qquad\;\; | \qquad | \qquad\; | \qquad | \qquad\; | \qquad | \\
\text{A} \quad \text{U} \qquad \text{A} \quad\; \text{U} \qquad \text{A} \quad\; \text{U} \qquad \text{A} \quad\; \text{U} \\
\;| \quad\;\; | \qquad\;\; | \qquad | \qquad\; | \qquad | \qquad\; | \qquad | \\
\text{A} \quad \text{U} \qquad \text{A} \quad\; \text{U} \qquad \text{A} \quad\; \text{U} \qquad \text{A} \quad\; \text{U} \\
\;| \quad\;\; | \qquad\;\; | \qquad | \qquad\; | \qquad | \qquad\; | \qquad |
\end{array}
$$

First a nucleus is formed which consists of three base pairs. This bimolecular process is followed by the pairing of the fourth, fifth, etc., base pairs. In the experimental conditions considered, the monomolecular growth process was about 10^4 times faster than the initial nucleation step.[35,36]

That a well-defined sequence of states may be considered in this case can be explained with the aid of a population analysis giving the relative fractions of the helix species present at equilibrium.[59] In the typical example of a $A(pA)_8 + U(pU)_8$ solution ($A : U = 1 : 1$), many of the possible helix species can be neglected.[35,36] The species to be considered are the unpaired single strands, the completely paired double helix, and, to a lesser degree, helices that have one or two base pairs less than the maximum number possible. Pörschke and Eigen[35,36] show that there results a simplified picture of the reactions proceeding in solution. For example, in a temperature jump from 0 to 3.4°C, about 70% of the total change of paired bases is due to the transition of the completely paired double helix to the single strands. Although partially paired helices appear as intermediates, their concentration cannot be detected. Thus the formation of helices proceeds as a second-order reaction according to the following all-or-none scheme:

$$A_N + U_N \underset{k_D}{\overset{k_R}{\rightleftharpoons}} (A+U)_N \tag{47}$$

Here k_R and k_D are, respectively, the rate constants of recombination and of dissociation.

Apart from deviations due to a fast effect, the kinetic experiments performed in 0.05-M sodium cacodylate solution at pH 6.9 could be represented by one relaxation process, with the time τ corresponding to reaction (47), that is,

$$\tau^{-1} = k_R(c_A + c_U) + k_D \tag{48}$$

In (48), c_A and c_U are the free concentrations of oligoadenylic and oligouridylic acids, respectively. The rate constants are obtained from plots of the reciprocal relaxation time *versus* the sum of free concentrations.

The kinetics of the helix–coil transitions have been measured for the chain lengths 8, 9, 10, 11, 14, and 18. The rate constants of dissociation show a strong temperature dependence. Increase of temperature leads to increased rates of dissociation; the longer the nucleotide chain, the higher the corresponding activation enthalpies. When the probability of dissociating a base pair is low compared with the probability of forming it, the rate of dissociation of long helices is small (of the order of 1 sec for $N=15$), although the elementary times of closing and opening of individual base pairs are very short (between 10^{-7} and 10^{-6} sec).

On the other hand, the rate constants of recombination are of the same order of magnitude for all the chains. This follows from the relative difficulty of forming a nucleus. The formation of a nucleus is therefore rate-determining, and it is irrelevant how many base pairs are formed subsequently as long as the cooperativity length is not exceeded.

The rate constants of recombination decrease with increasing temperature, and the corresponding activation enthalpies are negative. Pörschke and Eigen[35,36] conclude that the rate-determining step of recombination consists in three elementary steps. Thus k_R is given by

$$k_R = \frac{k_{01}}{k_{10}} \cdot \frac{k_{12}}{k_{21}} \cdot k_{23} \qquad (49)$$

where the meaning of the kinetic coefficients of the right-hand member of (49) is to be taken from reaction scheme (46).

The thermodynamic study of Pörschke[35] shows that the equilibrium constants k_{01}/k_{10} and k_{12}/k_{21} associated with the formation of the first and second base pairs have reaction enthalpies of about -4 and -5.5 kcal/mole, respectively. The enthalpy of formation of the third base pair is estimated to be 1 kcal/mole, leading to a total activation enthalpy of -8.5 kcal/mole, which is close to the experimental value -9 kcal/mole.

As pointed out by Eigen,[34] the result that three base pairs form a stable nucleus is of interest in studies of the mechanism of genetic information transfer. Indeed, it may be recalled that the codon–anticodon interaction at the ribosome proceeds by way of triplets (i.e., by way of a number of base pairs corresponding to the nucleation length given previously). The transfer of information by way of triplets of base pairs is of course the most economical way to code for 20 aminoacids. Eigen[34] concludes that besides this alphabetic advantage, there must also have been some mechanistic advantage; otherwise this particular way of coding would not have occurred. Two properties are required for an optimal code system: first, the transfer of information must be accurate and reliable; that is, a long codon would be the best. However, a second requirement is important: the process of information transfer must be fast; in this respect, the complementarity between long nucleotide sequences would take too much time to become established. For these reasons, the interaction between triplets appears to be optimal because it permits an opening and closing of the code unit in times as short as those of the order of 10^{-7} sec.

In the preceding experiments, the formation of a double helix follows a sequential pattern that can be described using the all-or-none approximation (47), as long as short chains are considered. As pointed out by Eigen and Pörschke,[35,36] deviations from the all-or-none behavior appear for chains containing 18 base pairs or more, since partially paired nucleotides can no longer be neglected.

Another example of a transition that both exhibits a well-defined sequence of states and lends itself to all-or-none treatment, is the transition between the two helical conformations of short poly-*l*-proline molecules (Winklmaier, Engel, and Ganser[60]). The transition between the two helices (i.e., the right-handed I-helix with all peptide bonds in *cis* conformation, and the left-handed II-helix with all peptide bonds in *trans* conformation) is an example of an order $\rightleftharpoons$ order transition. This process is quite slow, as illustrated by the values of the relaxation times, which are of the order of minutes or hours, according to the chain length. For further details, and for a description of the kinetic behavior of longer molecules of poly-*l*-proline, the reader is referred to the original article,[60] as well as to a review by Schwarz and Engel.[45]

There is increasing evidence, on the other hand, that in polymodal melting of natural DNA, well-defined segments melt in accordance with sequential processes that follow quite precise rules.[61] The complicated features exhibited by long DNA molecules are, however, outside the scope of this chapter. Attention is focused, rather, on processes that belong to an intermediate category that fulfills the following two conditions: (*a*) the concentrations of intermediates are not all negligible (thus the process is not all-or-none); (*b*) the processes do not present the complications inherent in nonnegligible probability of more than one boundary (e.g., between an ordered and a disordered region). Such sequential processes can consist in the conformational changes of polypeptides, or in the double-helix formation of polynucleotides when the chain lengths are of the order of the cooperativity length. In the first case the reaction is monomolecular; in the second case there is an initial bimolecular step.

Models leading to such sequential processes have been considered by many authors[62–68] in the steady-state approximation, which postulates that the net flow of molecules across each kinetic step in the reaction scheme is the same. This assumption is equivalent to the requirement that the concentrations of all intermediates be constant; thus the rate of disappearance of the initial form equals the rate of appearance of the final form. This simplification results in the replacement of N relaxation times by only one.

It may now be noticed that the solution of the kinetic equations for the most general monomolecular reaction (possibly with transitions among every couple of states), takes a general and simple form for a finite number of states and for given initial conditions. The rate equations have the form (4), with appropriate time-independent values of the elements of the transition rate matrix $\mathbf{Q}$. The solution of (4) is given by the following matrix equation:

$$\mathbf{p}(t) = \mathring{\mathbf{p}} \exp(\mathbf{Q}t) \qquad (50)$$

where $\mathring{\mathbf{p}}$ is the row vector of the initial values $\mathring{p}_j$ of the occupancies of the states at time $t = 0$; the matrix $\exp(\mathbf{Q}t)$ is obtained from the element-by-element sum of the exponential series expansion:

$$\mathbf{I} + \mathbf{Q}t + \mathbf{Q}^2 \frac{t^2}{2!} + \cdots$$

where $\mathbf{I}$ denotes the unit matrix.

Let us call $\mathbf{B}$ the diagonal matrix of the equilibrium values $\bar{p}_j$ of the p_j's. The symmetric matrix $\mathbf{B}^{1/2}\mathbf{Q}\mathbf{B}^{-1/2}$ can be transformed to diagonal form:

$$\mathbf{G}^{-1}\mathbf{B}^{1/2}\mathbf{Q}\mathbf{B}^{-1/2}\mathbf{G} = \Lambda = \mathrm{diag}(\lambda_r) \qquad (51)$$

where $\mathrm{diag}(\lambda_r)$ is the diagonal matrix of the eigenvalues λ_r of the $(N+1) \times (N+1)$ matrix $\mathbf{Q}$; among these eigenvalues, N are nonzero and negative. The relaxation times are $\tau_r = -(\lambda_r)^{-1}$.

For the row vector $\mathbf{p}(t)$ of the occupancies, we write

$$\mathbf{p}(t) = \mathring{\mathbf{p}}\mathbf{B}^{-1/2}\mathbf{G}\,\mathrm{diag}(\exp\lambda_r t)\mathbf{G}^{-1}\mathbf{B}^{1/2} \qquad (52)$$

Equation (53) below, for the deviations $\delta p_j(t) = p_j(t) - \bar{p}_j$ from stationary values of the elements of the row vector $\mathbf{p}(t)$, can be derived from (52). For small deviations, (53) can also be obtained from (26); the δZ value we are seeking in this case is just one of the $\delta p_j(t)$'s, which we call $\delta p_s(t)$. Thus in Eq. (26) we have the condition $\partial Z/\partial x_i = 0$, except for $Z = p_s(t)$ and $x = x_s$, in which case $\partial Z/\partial x_s = 1$. Hence

$$\delta p_s(t) = \sum_r A_r^{ls}\left(\sum_j \tilde{A}_r^{mj}\delta\mathring{p}_j\right)\left(\sum_k A_r^{lk}\tilde{A}_r^{mk}\right)^{-1}\exp-\frac{t}{\tau_r} \qquad (53)$$

where $\delta\mathring{p}_j = \mathring{p}_j - \bar{p}_j$.

As we stated previously in connection with (16), the cofactors that appear in (53) are those associated with the matrix $\mathbf{A} = \tilde{\mathbf{Q}}$. These cofactors can be replaced by any convenient form of the components of the right and left eigenvectors of the matrix $\mathbf{A}$. On the other hand, from (52) the expression (53) is not restricted to small perturbations.

Leaving now the general monomolecular reaction mechanism, we devote the last part of this section to a few sequential processes involving problems like those just discussed. In a sequential process, the transition rate matrix is tridiagonal (i.e., its only nonzero elements are those of the main diagonal and those of the two adjacent diagonals). For such tridiagonal matrices, some general properties can be found in treatises of mathematics (e.g., Ref. 69).

Some of the sequential processes we deal with present one slow nucleation step located at one end of the reaction path, or two slow steps, one at each end of the reaction path. Since it makes little difference in the present treatment whether one of these exceptional steps is bimolecular (pairing of two chains), or monomolecular (conformational change of a chain), only the second case is considered.

1.   Oligopeptide Conformational Changes and Oligonucleotide Double-Helix Formation

The example of a sequence of reversible steps in which all but the first have the same forward rate constant, whereas all reverse steps are mutually equivalent, has often been considered and is mentioned briefly. The reaction scheme is

$$0 \underset{q'}{\overset{q_0}{\rightleftharpoons}} 1 \underset{q'}{\overset{q}{\rightleftharpoons}} 2 \cdots \underset{q'}{\overset{q}{\rightleftharpoons}} j \underset{q'}{\overset{q}{\rightleftharpoons}} \cdots \underset{q'}{\overset{q}{\rightleftharpoons}} N \qquad (54)$$

in which q_0 is generally assumed to be much smaller than q. The characteristic equation

$$|\mathbf{Q} - \lambda \mathbf{I}| = 0 \qquad (55)$$

can be transformed using the auxiliary variables ϕ_r defined by

$$-\lambda_r = q + q' - 2(qq')^{1/2} \cos \phi_r \qquad (56)$$

The nonzero λ_r's are obtained from the following equation for the ϕ_r's[33, 70] $(0 < \phi < \pi)$:

$$\sin(N+1)\phi + D \sin(N\phi) = 0 \qquad (57a)$$

$$D = q_0(qq')^{-1/2} - \left(\frac{q}{q'}\right)^{1/2} \qquad (57b)$$

When it is true that

$$c^{1/2} = \left(\frac{q'}{q}\right)^{1/2} < \left(1 - \frac{q_0}{q}\right) \Big/ \left(1 + \frac{1}{N}\right) \qquad (58)$$

which is the case for short chains throughout the transition region, (57a) has one imaginary solution to which corresponds the longest relaxation time τ_1, given to a first approximation by[33,71]

$$\tau_1^{-1} \# q(1-c)\left(\frac{q_0}{q} + (1-c)c^N\right) \tag{59}$$

It is easily seen[33] that the longest time τ_1 is approximately equal to the relaxation time associated with the complete conformational change $0 \rightleftharpoons N$. The approximate value of τ_1 given by (59) coincides with the one derived by Baldwin[68] from the steady-state assumption. Exact values of all the relaxation times can be obtained by solving (57a) numerically. Condition (58) furthermore requires that a series of inequalities be fulfilled[33] by the times $\tau_2 > \tau_3 > \cdots > \tau_N$. These inequalities show that unless $c = q'/q$ is close to unity, all the times $\tau_2 \cdots \tau_N$ are of the same order of magnitude.

For the approximation of (59), the shortest time τ_N and the longest time τ_1 are related by the following equation;[71]

$$\tau_1^{-1} \# \tau_N^{-1}\left(\frac{q_0}{q} + (1-c)c^N\right)(1 - c^{1/2})/(1 + c^{1/2}) \tag{60}$$

which for $c \ll 1$ reduces to

$$\tau_1 \# \left(\frac{q}{q_0}\right)\tau_N \tag{61}$$

In the case of the *cis-trans* isomerization of the poly-*l*-proline molecule, it is permissible to apply directly the sequential model with nucleation steps at the ends [although the model will necessarily be slightly more complicated than (54)]. This follows because nucleations within the chain can be neglected, their rate of formation being much slower than the formation of a nucleus at an end.[60]

If, however, nucleation can occur with a nonnegligible probability in the interior of the molecule, the possibility of using the present model for describing the kinetics depends, as pointed out in Section I, on a proper definition of the sequence of states considered. This sequence consists of: the coiled chain (state zero), any chain that contains one nucleus (state 1), any chain that contains one nucleus plus one helical element (state 2), and so on.

It is then necessary to take into account that propagation can occur on both sides of an interior nucleus and that the value of the nucleation coefficient q_0/q must reflect the possibility that the nucleus is formed all along the chain.[33,71]

2. Protein Unfolding and Refolding

The aminoacid sequence of a globular protein is known to contain the information required for the formation of its three-dimensional biologically active structure.[72-74] Therefore, the folding of a polypeptide chain represents a significant step toward making the information of the genetic code available for the function of the corresponding protein. Furthermore, the elucidation of the mechanistic and thermodynamic principles of unfolding and refolding proteins may contribute significantly to the understanding of their dynamic behavior, therefore of their biological activity.

A major disruption of a structure that comprises many aminoacid residues is likely to be a highly cooperative process. This of course is why a quite simple model—namely, the all-or-none, or two-state model—has been successful in describing conformational changes of proteins.[75] Its use is based on the assumption that the various states of the molecule can be divided into two main categories, which are stable below and above a certain "transition temperature," as the case may be. In this instance, changing the temperature will induce a sharp conformational change between these "native" and "denatured" structures. Apart from the sharpness of the transition, cooperativity also becomes manifest by extraordinarily high activation energies. The existence of very high activation energies in the denaturation of proteins near their pH of maximum stability was known to Arrhenius.[76] It is of course highly instructive to measure separately the enthalpic and entropic contributions to the activation energy; this is achieved by determining denaturation and renaturation rates as a function of the temperature.

In contrast to the all-or-none behavior, simple inspection sometimes reveals that a transition occurs in two or more stages. Even when two-state behavior is apparently followed according to various criteria (as discussed by several authors[75,77]), one should expect to be able to detect intermediates in unfolding and refolding reactions. First, the complexity of the primary structure of the molecule is likely to offer several possibilities of stable (or metastable) tertiary structures. Second, the unfolding and refolding reactions, although highly cooperative, are likely to involve a rapid rate of reshuffling between different conformational isomers.

Kinetic methods have proved quite successful in this respect, and many examples are known today in which kinetic curves show several phases, not only in the unfolding but also in the refolding of the protein. In particular, rapid kinetic methods have now disclosed fast steps in these processes for certain proteins. Interpretation of the results generally necessitates extensive study in which, in particular, the effect of varying the environmental conditions on equilibrium and kinetic properties must be examined. A great deal of work has been done in these fields, and many aspects have been comprehensively reviewed.[77-80]

In beginning a discussion of some of the kinetic studies, an example should be mentioned in which the sole "abnormal" temperature dependence of the renaturation rate constant (as deduced from the application of the two-state model) has led to the consideration of an intermediate structure.[37]

The measurements referred to were performed by Phol[81–83] for a series of pancreatic proteins, especially trypsin, α-chymotrypsin, and ribonuclease A. The reversible unfolding of these proteins at pH values close to 2 (transition I), is usually described as a strongly cooperative all-or-none conformational change. The kinetic experiments carried out using the "slow temperature-jump" technique,[81] which permits measurements of relaxation times in the second to minute range, seemed to confirm this view. In particular, the large number of measurements for α-chymotrypsin showed that one single relaxation time τ is observed; moreover, the value of this time is independent of the wavelength of the ultraviolet radiation used to follow the process and is also unaltered if the optical rotation is measured.

Thus the following expressions for the all-or-none process were used to determine the unfolding and refolding rates k_U and k_R:

$$\tau^{-1} = k_U + k_R \tag{62}$$

$$K = \frac{k_U}{k_R} \tag{63}$$

where K is the equilibrium constant. The refolding rate constant k_R, however, showed anomalous temperature dependence in that a negative reaction enthalpy was obtained in the upper temperature range.[81,82]

It therefore seemed worthwhile to determine whether Pohl's results could be explained by a mechanism involving an intermediate state, and we investigated the following two mechanisms:[37,71]

$$N \rightleftharpoons U \rightleftharpoons U' \tag{64}$$

$$U \rightleftharpoons N \rightleftharpoons U' \tag{65}$$

where N denotes the native structure, and U and U' are two unfolded conformations. Of course N, U, and U' may each represent more than one microscopic state; in this event, measured properties are averaged over appropriate ensembles.

In the case of reaction scheme (64), unfolding proceeds along a two stage pathway in one direction. For reaction scheme (65), the two unfolded conformations would lie on two distinct pathways starting from the native structure.

To ascertain whether such two-step reaction mechanisms are compatible with the observation of a single relaxation time, we must calculate the amplitudes associated with each of the two times that characterize the system. This is readily done using (26), and generally valid expressions have been given for a moderately large temperature jump.[37,71]

The results are especially simple when the two steps are thermodynamically and kinetically identical (i.e., when they are characterized by the same enthalpies and entropies of reaction and of activation). In an even more particular case, the two optical coefficients associated with the unfolded conformations are equal. To emphasize the equality of the associated optical coefficients, the two unfolded conformations have been denoted in this case by the same symbol U. It was found that the simple model, which fulfills the preceding conditions, accounts quite well for Pohl's results for α-chymotrypsin.[37]

Only the longest relaxation time is observed, since either the two relaxation times associated with this three-state model are nearly equal, or the amplitude associated with the shortest relaxation time can be neglected. On the other hand, the curves that give the reciprocal observable relaxation time τ^{-1}, and those giving the two rate constants k_U and k_R as a function of the reciprocal absolute temperature T^{-1} in a semilogarithmic plot, must obey several precise geometrical conditions, which appear to be all fulfilled for α-chymotrypsin.[37] In fact, slightly more complicated quantities than just τ^{-1}, k_U, and k_R must be plotted if simple properties are to be obtained.

Somewhat less particular reaction schemes of type (64)—for which, however, the same restriction relating the optical constants is assumed—have also been examined.[84]

Each of the two steps in the proposed mechanism must reflect a highly cooperative conformational change, as demonstrated by the high values of the associated enthalpies and entropies of reaction and of activation.[37] Of course, more work is necessary to determine whether the equality of the optical constants of the conformations U and U′ actually holds with sufficient accuracy to support the proposed simplified model.

A refolding mechanism that initially involves a multistep nucleation process could lead to qualitatively similar results. Pohl[85] has pointed out that hydrophobic interactions may play a part in such a process, since they involve fast preequilibria that enter into the overall rate constants.

On the other hand, we should like to inquire whether the success of the simple three-state model might be connected to the existence[86] of two distinct structural regions in α-chymotrypsin. The implications of these regions, viewed as comprising a nucleus plus structure added by a growth process, has been discussed in detail by Wetlaufer.[87]

In contrast to the preceding example, the kinetic studies to be discussed now furnished biphasic curves under appropriate conditions, indicating the existence of at least two reaction steps. However, one or both of these steps may involve multiple elementary steps. There are indeed several mechanisms by which a complex reaction may exhibit single relaxation behavior. A first example was mentioned at the beginning of the section, in the discussion of all-or-none mechanisms involving both a nucleation step and rapid propagation steps. A second example has just been provided by a three-state system, in which two conformations are experimentally indistinguishable because their optical constants are equal.

The refolding of staphylococcal nuclease from the acidified form on neutralization, as measured by changes in tryptophan fluorescence in stopped-flow (pH-jump) experiments, was described by Schechter, Chen, and Anfinsen[88] as a sequence of two first-order processes with half-times of about 55 and 350 msec, respectively. Epstein et al.[38] consider that the most probable mechanism of refolding is a sequential reaction similar to (64) that involves an intermediate conformation X:

$$N \rightleftharpoons X \rightleftharpoons D \tag{66}$$

In the noncrosslinked single chain of this nuclease, the acid-induced change represents a major disruption. Thus the unfolded structure is denoted D ("denatured").

The measurements of the refolding kinetics over a temperature range of 13 to 38°C indicate a difference in the physical basis of the two processes. The faster process shows no significant dependence on temperature in the range investigated. On the other hand, the rate constant of the slower process is increased fourfold in passing from 13 to 38°C. On the basis of these observations, and of the three-dimensional model of the nuclease, the authors propose the following reaction scheme.[88] The entropically driven nucleation of two helices that largely form the "pocket" in which the indole ring of the staphylococcal nuclease lies, would occur first, and would be reflected in the rapid rate process. This first step of the refolding mechanism would be followed by the slower positioning of one or more of the hydrophobic side chains, or segments of side chains, around the indole ring.

Equilibrium and kinetic studies by Ikai, Fish, and Tanford[39,89] for the reversible denaturation of horse heart ferricytochrome c by guanidine hydrochloride (stopped-flow experiments) led the authors to suggest the existence of metastable intermediates that are not on the direct pathway between the native and the denatured structures. Although most of the kinetic data may be described in terms of two exponential decay terms, mechanisms involving three species only were found not to account fully for the results. Among the four-species reactions examined,[90] the most satisfactory proved to be the mechanism

$$N \rightleftharpoons X_1 \rightleftharpoons D \rightleftharpoons X_2 \qquad (67)$$

in which X_1 is an intermediate in the pathway between the native and the denatured structures, whereas X_2 is a relatively highly ordered structure on a dead-end pathway. This structure has been interpreted[39,89] as representing an incorrectly folded form of the polypeptide chain. The rate constants required to account for the data are such that the first step in the conversion of the denatured structure D to the native protein would be the rapid formation of X_2. The depletion of this incorrectly folded structure and the ultimate conversion to the native structure would occur subsequently, the rate-limiting step in the transition zone being the reaction $D \rightarrow X_1$.

Thus among all the possible folded structures produced from disordered conformations, only certain selected ones could constitute true intermediates along the pathway to the native structure.[89] Other initial attempts at folding the chain would bring together residues that are not in contact in the native structure: these would not survive and would have to be reversed before the native structure could be attained. According to Ikai, Fish, and Tanford,[89] pathways that include incorrectly folded structures during the refolding process might be a rather general feature for chains that are minimally constrained after denaturation (e.g., because they are devoid of disulfide bonds, or because disulfide bonds link only short streches of the chain, as in β-lactoglobulin). This view is consistent with the results of Tanford, Aune, and Ikai[91] for lysozyme: the guanidine "denatured" structures of lysozyme, although devoid of noncovalent bonds, are highly constrained by four disulfide bonds, and under most conditions they show no intermediate forms during renaturation.[91]

A fast initial phase in the millisecond time range has been found in the unfolding of ribonuclease A by stopped-flow pH-jump experiments (Tsong, Baldwin, and Mc Phie[40]) and by T-jump experiments (Tsong, Baldwin, and Elson[92]); a compound fast phase was detected by T-jump experiments in the same time range in the unfolding of chymotrypsinogen A (Tsong and Baldwin[93]). All these experiments were carried out over the whole transition region of the protein by varying the temperature, under conditions in which the disulfide bonds remain intact.

The work of Baldwin, Elson, et al.[40,94,95] has permitted comparison of the experimental results with the predictions of nucleation-dependent sequential reaction schemes. Both reaction scheme (54) and the following more general scheme

$$0 \underset{q_0'}{\overset{q_0}{\rightleftharpoons}} 1 \underset{q'}{\overset{q}{\rightleftharpoons}} 2 \underset{q'}{\overset{q}{\rightleftharpoons}} \cdots \underset{q'}{\overset{q}{\rightleftharpoons}} j \underset{q'}{\overset{q}{\rightleftharpoons}} \cdots \underset{q'}{\overset{q}{\rightleftharpoons}} N \qquad (68)$$

were examined. The sequential reaction (68) differs from (54) by the existence of a slow rate constant q_0' in the $1 \rightarrow 0$ step. For a detailed discussion, the complete relaxation spectrum of the model must be known (i.e., the relaxation amplitudes must be calculated). Several authors have made the required calculations for the sequential reaction (54)[70,96] and for both (54) and (68).[94]

Elson[94] concludes that under a variety of conditions it is possible to detect both a transient and a steady-state phase in the unfolding reaction. The steady-state approximation postulates that the net flow of molecules across each step in the reaction scheme is the same. As was pointed out earlier, this assumption is equivalent to the requirement that the concentrations of all intermediates be constant. Instead, Elson describes a quasi-steady state that becomes established when the molecules flow from the more ordered structures to the less ordered conformations.

Apart from the notations N and D already employed for the native and completely denatured structures, these conformations will also be denoted, when convenient, by their respective numbers in the sequences (54) or (68) —that is, by N and 0 respectively (cf. also Fig. 5).

During unfolding, the quasi-steady-state condition is established when enough of the more ordered structures attain concentrations whose ratios are close to the equilibrium value q'/q. However, this condition is still far from being fulfilled for the less ordered structures.[94] A quasi-steady state is established even when $q_0' = q'$—that is, when reaction (68) reduces to (54). In fact, Elson devotes the most detailed numerical discussion to reaction (54). The establishment of the quasi-steady state in this case is closely related to the condition (58). Strictly speaking, however, this condition is neither necessary nor sufficient for the establishment of a bimodal regime. Inequality (58), indeed, entails that the slowest relaxation time be well separated from all the others, a condition that certainly favors the establishment of such a regime. The amplitude of the fast mode could, however, be negligible. Yet even if condition (58) is not fulfilled, one relaxation other than the slowest one could appear with a particularly large amplitude, owing to special conditions relating the values of the optical coefficients that characterize the various states.

Qualitative reasoning permits visualization of how bimodal regimes may appear in both the refolding and unfolding processes represented by the simple sequential scheme (54). Figure 5 is a schematic diagram in which

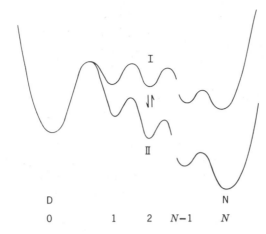

Fig. 5. Qualitative explanation of a bimodal regime in both the refolding and the unfolding of a protein along a sequential path (cf. text). This schematic diagram gives the Gibbs free energy of the states numbered 0, 1, 2,..., N. State N is the native structure N; state 0 is an unfolded state, or the denatured state D. Transition I→II is a refolding process; transition II→I is an unfolding process.

the Gibbs free energy of the states is plotted against an abscissa proportional to the number of the state. The lines represent two states I and II of the system, in which the structures N and D are differently populated.

Consider first the refolding process represented by transition I→II, in which the population of structure N will increase. In a first stage, there will be a rapid reequilibration of the populations among the more ordered structures, leading to the fast transient regime. However, the population of D must eventually adjust itself to its new equilibrium value; this process will be slower than the first one, owing to the high barrier 0→1. A quasisteady state will become established because of the slow flowing of molecules across the barrier 0→1, whereas the various states are progressively occupied at their exact level of population.

Consider now the transition II→I, an unfolding process in which the population of structure D will increase. The reasoning closely follows the preceding case, except that we cannot argue that the transition 1→0 is slow. However, since the molecules must pass a given sequence of gateways leading to D, there will again exist, after a transient phase, a slow quasisteady phase in the reequilibration process during which the required amount of molecules will attain, and then cross, the barrier 1→0.

The most detailed experimental testing of the model (68) has been carried out by Tsong, Baldwin, and Elson[97] for ribonuclease A. The measurements were performed under conditions comparable to those of

Tsong, Baldwin, and McPhie[40] referred to earlier.

The entire set of kinetic results could be reproduced semi quantitatively by assignment of values to the parameters q_0/q, q_0'/q', $c = q'/q$, and N in (68). As pointed out by the authors,[97] model (68) fails to account for the existence of a "midrange" kinetic phase observed at low temperature of unfolding and refolding, as had already been described and discussed in connection with intermediate states by Scott and Scheraga.[98]

Although the evidence for a nucleation step and a series of steps leading to the native structure in the refolding process appears conclusive, the description of the refolding process for the various proteins that have been considered is still approximate. Experiments that permit detection of movements of different well-positioned groups of the protein are promising in this respect.

Tsong and Baldwin[95] have already investigated whether for ribonuclease A the fast and slow processes are separate unfolding reactions involving different tyrosine groups (whose state of exposure is detected by absorbancy at 286.5 nm), or whether these processes are different kinetic phases of a general unfolding reaction. To answer this question, they studied 41-dinitrophenyl ribonuclease A. The kinetics of exposure of the single dinitrophenyl group, measured at 358 nm, are biphasic, and the rates of both the fast and slow processes agree closely with those measured by tyrosine groups at 286.5 nm. Tsong and Baldwin[95] conclude that their results are consistent with sequential unfolding, each step of which involves a far-reaching conformational change.

It is perhaps unnecessary to emphasize that the compatibility of a model with a series of experiments need not prove its validity. Tsong, Baldwin, and Elson[97] noted in this connection, that Summers and McPhie[99] had to insert values of the parameters somewhat different from the ones they used themselves, to reconcile model (68) with kinetic measurements on ribonuclease A. They further emphasize[97] that a more general model—that would, for example, involve alternative pathways and different rate constants at each step of unfolding and refolding—may be required to give a realistic description of the kinetics.

The kinetic methods can be used advantageously in conjunction with sensitive methods that permit detecting and characterizing conformational changes in proteins. The use of nuclear magnetic resonance (NMR) can only be mentioned in this chapter. Aside from the review of Roberts and Jardetzky,[100] we refer to three recent publications, which contain references to previous work.

Roberts and Benz,[101] using the proton Fourier transform method, have obtained information about the local environment of individual residues throughout the transition. Thus they were able to suggest a precise sequence for the unfolding of ribonuclease A by the action of urea, guanidine hydrochloride, or heat, at low pH.

Allerhand, Childers, and Oldfield[102] have shown that with the use of 20-mm probes it is possible to attain sufficient signal-to-noise ratios for detection of single-carbon resonances in proton-decoupled natural ^{13}C Fourier transform NMR spectra. Nonprotonated aromatic carbon atoms of a native protein yield narrow ^{13}C resonances, whereas protonated aromatic carbons yield broad resonances. The same authors have demonstrated that the folding of a protein into its native conformation may produce sufficient chemical shift nonequivalence to make possible the observation of numerous resolved single-carbon resonances in the aromatic region of the ^{13}C spectrum.[102]

Campbell, Dobson, Williams, and Xavier[103] have discussed the possibility of determining the structure of a protein in solution by methods based on paramagnetic ion-induced shifts, and broadening of resonances associated with particular nuclei, to place these nuclei in space in relation to the paramagnetic ion.

Leaving for an instant the physical techniques, we should note that Wetlaufer and Ristow[80] have recently stressed that a finding telling much about the renatured material, in contrast to results of physical "low-information methods," is the regaining of enzymatic or other specific biological activity in a controlled test of protein renaturation. In connection with Wetlaufer and Ristow's statement as well as with the preceding discussion of kinetic experiments, Garel and Baldwin[104] have reported that both the fast and slow reactions in the refolding of ribonuclease A have a common product—fully active ribonuclease A. This finding deserves further investigation in relation to models treated previously in this section, but its discussion is unfortunately beyond the scope of this chapter.

Whereas the treatment of models that involve up to four states (not necessarily in sequence) can be found in an article by Ikai and Tanford,[90] we return to a sequential kinetic model considered long ago by Flory[63] in a different context. It can be shown that it is possible to treat this model in a simple way by methods already mentioned [cf. Section II.B and the treatment of model (54)].

The sequential model to be considered involves N steps: all the forward steps have the same rate constant q, except the first step $0 \to 1$, which has a slower rate constant $q_0 < q$; all the reverse steps have the same rate constant q', except the first step $N \to N-1$, which has a slower rate constant $q_0' < q'$. In general, we even assume that $q_0 \ll q$ and that $q_0' \ll q'$. The reaction scheme is

$$0 \underset{q'}{\overset{q_0}{\rightleftharpoons}} 1 \underset{q'}{\overset{q}{\rightleftharpoons}} 2 \underset{q'}{\overset{q}{\rightleftharpoons}} \cdots \underset{q'}{\overset{q}{\rightleftharpoons}} j \underset{q'}{\overset{q}{\rightleftharpoons}} \cdots \underset{q'}{\overset{q}{\rightleftharpoons}} N-1 \underset{q_0'}{\overset{q}{\rightleftharpoons}} N \qquad (69)$$

The **Q** matrix for this reaction is of the form

$$
\mathbf{Q} =
\begin{vmatrix}
-q_0 & q_0 & & & & \\
q' & -(q+q') & q & & & \\
 & q' & -(q+q') & q & & \\
 & & q' & -(q+q') & q & \\
 & & & q'_0 & -q'_0 &
\end{vmatrix}
\tag{70}
$$

The elements not written are equal to zero.

If a schematic diagram similar to that of Fig. 5 is used, the potential profile giving the Gibbs free energy of the states $0, 1, 2, \ldots, N$ has the aspect shown in Fig. 6.

When the model is applied to proteins, N again represents the native structure N and the step $N \to N-1$ may represent a nucleation process connected with the difficulty of achieving the first step that leads to the destruction of the special structure N.

Use of the model does not require, however, that the step $N \to N-1$ be a nucleation step. One may wish to apply the model to express that the growth process leading from D to N becomes easier after a certain number of steps are accomplished, whereas the steps in the last rapid growth process are not distinguished (see Fig. 6). Under such conditions, of course, there is no clear-cut distinction between the step $N \to N-1$ and a nucleation step. The situation then becomes comparable to that encountered for oligonucleotides, as discussed at the beginning of the section.

When the auxiliary variable ϕ_r given by (56) is introduced, it is seen that the N nonzero eigenvalues of the matrix **Q** are obtained from the following

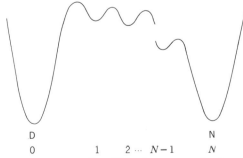

D   N
0   1   2 ⋯ N-1   N

Fig. 6. Schematic diagram of (69). The Gibbs free energy is plotted against the number $(0, 1, 2, \ldots, N)$ designating the state. As in Fig. 5, N is the native structure N, and 0 is an unfolded state, or the fully denatured state D.

equation for the ϕ_r's:

$$\sin(N+1)\phi+(D+D')\sin N\phi+DD'\sin(N-1)\phi=0 \tag{71a}$$

where

$$D=q_0(qq')^{-1/2}-\left(\frac{q}{q'}\right)^{1/2} \tag{71b}$$

$$D'=q_0'(qq')^{-1/2}-\left(\frac{q'}{q}\right)^{1/2} \tag{71c}$$

The following expressions of the cofactors $A_r^{N+1,i}$ and $\tilde{A}_r^{N+1,k}$ [cf. (53)] may be useful:

$$A_r^{N+1,i}=-q_0'(q')^{N-i}((q_0+\lambda_r)A_{i-2}+q_0q'A_{i-3}) \tag{72a}$$

$$\tilde{A}_r^{N+1,k}=-q^{N+1-k}((q_0+\lambda_r)A_{k-2}+q_0q'A_{k-3}) \tag{72b}$$

where

$$A_i=(-1)^i(qq')^{i/2}\frac{\sin(i+1)\phi}{\sin\phi} \tag{72c}$$

The particular case of reaction (54) is obtained by putting $q_0'=q'$, whence

$$D'=0$$

and (71a) reduces to (57a).

D. Appendix: A Two-State Non-Arrhenius Model

The following sections are devoted to synthetic polymers, in particular vinyl-polymers. We assume that the ultrasonic absorption can be ascribed primarily to segments that undergo a cooperative transitional change[105,48] (cf. Section I). It is further assumed that these segments are found among adjacent sequences each including an average of p chain units. Moreover, the conformational changes exhibited by these segments are assumed to be mutually independent. Thus the model is reduced to a set of two-state systems. For a chain molecule, the picture is obviously oversimplified.

Once these assumptions are made, the form of the relaxation time is known, and the relaxation amplitude follows in a straightforward manner from the classical result[106] for two-state systems. Ultrasonic absorption of polymers in solution has already been interpreted in terms of a two-state model.[107] However, the relaxation was assumed to result from a particular movement that involves more specially three bonds.[107]

No *a priori* assumption is made here about the number of relaxing systems per molecule, and we consider equations that are valid for systems that show cooperativity.[105,48] The equations relating to the relaxation time would apply to any two-state system like those discussed in the preceding section. The main concern, however, is to obtain information from the values of the relaxation time and of the ultrasonic absorption for polymers in solution.

Cooperativity is introduced in a two-state model by using Gibbs free energies of reaction and activation, which vary rapidly with the temperature. The entropic term in the Gibbs free energy of reaction may compensate the enthalpic term in the range of temperatures of interest, whence a transition temperature T_t is obtained.

We may recall in this connection that the first equilibrium theory of the helix–coil transition of polyaminoacids by Schellman[108] was based on the principle just outlined. Using the molar Gibbs free energy of unfolding ΔG_U (instead of the Helmholtz free energy), we may write, according to Schellman[108]

$$\Delta G_U = N(\Delta H_{\text{res}} - T\Delta S_{\text{res}}) + C \tag{73}$$

where N is the number of residues participating in the transition; ΔH_{res} and ΔS_{res} are the enthalpic and entropic contributions to ΔG_U per residue. For polyaminoacids, ΔH_{res} is approximately equal to the heat of dissociation per hydrogen bond, and ΔS_{res} to the configurational entropy produced by the freedom of rotation of the three bonds per residue in the unfolded chain; C is mainly related to end effects and is practically constant.

If the all-or-none assumption is made, the fraction of molecules existing in the ordered structure is given at any temperature by[108]

$$f = \left(1 + \exp\left(-\frac{\Delta G_U}{RT}\right)\right)^{-1} = \frac{1 + \tanh(\Delta G_U/2RT)}{2} \tag{74}$$

where R is the gas constant. Schellman defines the temperature at the midpoint of the curve giving f *versus* $\Delta G_U/2RT$ (i.e., the temperature for which $\Delta G_U = 0$) as the transition temperature T_t. The transition becomes sharper as N increases and would correspond to a real discontinuity of f—that is, to

$$f = 1 \quad \text{for} \quad T < T_T$$

$$f = 0 \quad \text{for} \quad T > T_T$$

in the limit of $N \to \infty$.

The preceding model was applied to kinetic studies in connection with the observation of a small temperature dependence of the ultrasonic relaxation times of vinyl-polymers in solution[109] (cf. Section III.B).

Let us insert in the expression of the relaxation time of a two-state system:

$$\tau^{-1} = k_{12} + k_{21} \tag{75}$$

the values of the forward and backward rates k_{12} and k_{21}, as deduced from the theory of absolute reaction rates[110] (e.g., for the transition $i \to j$):

$$k_{ij} = \left(\frac{kT}{h} \right) \exp \left(\frac{-\Delta G_{ij}}{RT} \right) \tag{76}$$

In (76), ΔG_{ij} is the Gibbs free energy of activation; k and h are Boltzmann's and Planck's constants. The "transmission coefficient" has been put equal to 1, since its deviation from unity could be included in the entropic term. Similarly, degeneracy of a state would conveniently be introduced[84] as an entropic contribution to the Gibbs free energy of reaction

$$\Delta G = \Delta G_{12} - \Delta G_{21} = \Delta H - T \Delta S$$

Since we are concerned with relaxation times that apparently vary only slightly with temperature (non-Arrhenius behavior), we may ask what condition is required in order that

$$\frac{d\tau}{dT} = 0 \tag{77a}$$

Since the process considered may be expected to exhibit cooperativity and large variations of the Gibbs free energies, it may not be necessary to take into account the temperature dependence of the preexponential term in (76). Rather, it is of interest to determine the conditions under which

$$\frac{d(T\tau)}{dT} = 0 \tag{77b}$$

For (77b) to hold, the following condition must be fulfilled:[105]

$$\exp \left(-\frac{\Delta G}{RT} \right) = -\frac{\Delta H_{21}}{\Delta H_{12}} \tag{78}$$

Condition (78) requires that the two activation enthalpies be of opposite sign, therefore that one of them be negative. The corresponding activation entropy also must be negative to permit the corresponding Gibbs free energy to be positive. Such a situation is depicted in Fig. 7, where a segment of the chain is shown in a disordered conformation (state 1) and in an ordered conformation (state 2). A negative activation enthalpy ΔH_{12} is obtained if the transition state $\mathfrak{T}$ is more ordered than the conformation 1.

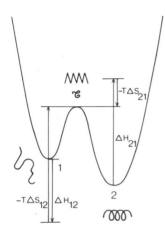

Fig. 7. A two-state non-Arrhenius model. One activation enthalpy and the corresponding activation entropy are negative [105] (cf. text).

One may, of course, question the significance of the abscissa in Fig. 7. As long as the transition $1 \rightleftharpoons 2$ necessitates, at some stage of the conformational change, that the state of the segment considered be close to a certain structure $\mathfrak{T}$, we do not require an exact description of the path followed. Actually, state 1 represents a collection of disordered states, and by introducing a negative activation entropy we express the fact that many paths lead from state 1 to the transition state $\mathfrak{T}$. The negative activation enthalpy may be connected with preequilibria, as in one of the examples of Section II.C; however, we do not exclude the possibility that the transitional conformation $\mathfrak{T}$ really has an enthalpy smaller than that of disordered conformations.

If the maximum of τ occurs at a temperature T_s which is not too different from T_t [i.e., according to (78) if $\Delta H_{12} \cong -\Delta H_{21}$], one has[105] for $T = T_s$

$$T\tau \frac{d^2}{dt^2} \frac{1}{T\tau} = -\frac{\Delta H_{12} \cdot \Delta H_{21}}{(RT^2)^2} \tag{79}$$

Thus for $\Delta H_{12} \cdot \Delta H_{21} < 0$, and $|\Delta H_{12} \cdot \Delta H_{21}|/(RT)^2 \gg 1$, the relaxation time τ would exhibit a sharp maximum as the temperature is varied.

Actually, what is observed for polymer solutions in the Megahertz range appears to be more closely represented by a small variation of τ as function of the temperature (cf. the discussion in Section III.B). However, we neither expect that the relaxing systems considered in this case will be highly cooperative nor that the oversimplified picture used will be quantitatively valid.

It is nevertheless useful to investigate the applicability of the model in a precise manner, and we need to determine all the parameters involved, using the classical equation for the relaxation amplitude, defined as the variation of $\Delta(\alpha/\omega^2)$ over the whole frequency range $(0<\omega<\infty)$; α represents the absorption coefficient and ω the circular frequency. This amplitude is proportional to the the the maximum value μ_M of $\alpha\lambda$, where λ is the wavelength: $\Delta(\alpha/\omega^2)=\tau\mu_M/(\pi v)$. The value of $\Delta(\alpha/\omega^2)$ is known to be

$$\Delta\left(\frac{\alpha}{\omega}\right)^2 = ac_{RS}\tau\left(\frac{\Delta H}{RT}\right)^2\left(1-\frac{\Delta V\cdot C_p}{\Delta H\cdot\beta V}\right)^2\frac{\exp(-\Delta G/RT)}{(1+\exp(-\Delta G/RT))^2} \quad (80)$$

In (80) ΔV is the molar volume of reaction, v the sound velocity, V the molar volume of the solution, C_p the specific heat at constant pressure, β the expansion coefficient, and c_{RS} the concentration of relaxing systems in moles per gram. Thus

$$c_{RS} = \frac{c}{\rho pm}$$

where c is the concentration of the solution in grams per cubic centimeter, ρ its specific weight, and m the molecular weight of the monomeric unit; p has already been defined as the mean number of chain units of the segment that contains one relaxing system. Furthermore,

$$a = \frac{R(\gamma-1)}{2vC_p}$$

where γ is the ratio of specific heats.

As is often done, the ΔV term in (80) is neglected. Thus the model is described by five parameters: ΔH, ΔS, ΔS_{12}, ΔS_{21}, and p. Under these conditions $(\Delta V\cong 0)$, the function (80) is sometimes said to be similar to a Schottky anomaly of specific heat. This is true, except when the system is fairly cooperative. For such a system, $\Delta(\alpha/\omega^2)$ exhibits a maximum that is produced by the exponential functions alone, with no contribution of the factor T^{-2} (nor of the factor a). The maximum is obtained in this case for

$\Delta G = 0$, thus for $T = T_t$. Moreover, if we set $\mathfrak{A} = a^{-1}\Delta(\alpha/\omega^2)$, for $T = T_t$ we have[105]

$$\frac{d}{dT^2} \ln \frac{T^2\mathfrak{A}}{\tau} = -\frac{1}{2}\left(\frac{\Delta H}{RT^2}\right)^2 \tag{81}$$

Also, for $T = T_s$ we write

$$c_{RS}^{-1} = \frac{T^3\tau^2}{\mathfrak{A}}\frac{d^2}{dT^2}\frac{1}{T\tau} \tag{82}$$

The measurements are often carried out in a narrow temperature range. In such cases, one merely determines the slopes of the curves giving $\mathfrak{A}$ and τ as a function of T.

The following relationships are then conveniently used:[105, 84]

$$\frac{\Delta H_{12}}{RT} + \frac{\Delta H_{21}}{RT} = B - 2\Gamma \tag{83}$$

$$\exp\left(-\frac{\Delta G}{RT}\right) = \left(\frac{\Delta H}{RT} - B\right)\left(\frac{\Delta H}{RT} + B\right)^{-1} = -\left(\frac{\Delta H_{21}}{RT} + \Gamma\right)\left(\frac{\Delta H_{12}}{RT} + \Gamma\right)^{-1} \tag{84}$$

where

$$B = \frac{d\ln(T^2\mathfrak{A}\tau^{-1})}{d\ln T}$$

$$\Gamma = \frac{d\ln(T\tau)}{d\ln T}$$

Other useful relationships may be derived,[111] and one may also wish simply to fit the data to (75), (76), and (80).[112] The limitations of the model appear in Section III.B, where comparison with experiments is made. However, it is not premature to note that: (a) even for small molecules (and no cooperativity) the two-state model may be far from quantitative,[113] and (b) the model's primary limitation for polymer molecules is likely to arise from the assumption that the relaxing systems are independent.

In most applications of the model, the two activation enthalpies were found to be of opposite sign in accordance with (78) (cf. Section III.B). However, it has also happened that the activation enthalpies were both negative,[114, 115] although in this case one of them was small in absolute value.

Suppose a two-state system with fairly high cooperativity has two negative activation enthalpies. These are necessarily significantly different from each other, to allow a transition between states 1 and 2. Furthermore, the inverse relaxation time will show a minimum only if the preexponential term of one of the rate constants contributes to the temperature dependence. This necessarily happens at a temperature markedly higher than the transition temperature T_t. In the corresponding range, one rate constant— say, k_{12}—greatly exceeds the other. It follows that the following condition is necessary if (77a) is to be fulfilled:

$$T = -\frac{\Delta H_{12}}{R} \tag{85}$$

This, indeed, corresponds to a fairly low absolute value (i.e., RT) of the smallest activation enthalpy, showing how far below zero the second negative activation enthalpy may go.

III. DYNAMICS OF LINEAR CHAIN MOLECULES: A DISCUSSION OF PENDING PROBLEMS

A. Statement of the Problems

In stating the problems to be examined, it is useful to start with the stochastic approach to polymer dynamics in Iwata's version.[13] Thus we take his coarse-grained diffusion equation, which can be reduced[13] to a form resembling that of the bead-and-spring model diffusion equation

$$\frac{\partial \psi(\mathbf{s},t)}{\partial t} = \nabla^{\mathrm{T}}\mathbf{D}(3l^{-2}\mathbf{M}\mathbf{s}+\nabla)\psi(\mathbf{s},t) \tag{86}$$

The transposed differential operator ∇^{T} is the divergence of the column vectors on which it operates. The formation of gaussian subchains of g bonds has led in (86) to an N'-dimensional normal coordinate $\mathbf{s}$

$$\mathbf{s} = (\mathbf{s}_1, \mathbf{s}_2, \ldots, \mathbf{s}_p, \ldots, \mathbf{s}_{N'}); \qquad N' \ll N \tag{87}$$

where each $\mathbf{s}_p$ is three-dimensional. The mean-square end-to-end distance of the subchains is $l^2 = gb^2$, where b is the "effective bond length." The zeroth normal coordinate, which represents the position of the center of resistance of the polymer chain, has been eliminated. The matrix $\mathbf{M}$ is diagonal and has the eigenvalues given by (2), with N' replacing N. In making this statement, particular aspects associated with the ring-shaped character of Iwata's chain model are disregarded.

If P_v denotes the probability that a conformational change of a segment of $v+1$ bonds occurs within unit time, and if $\bar{v}$ denotes the average value of v, the elements of the diagonal diffusion matrix **D** are given by:[13]

$$D_p = \frac{\delta_p P_{\bar{v}}}{g} \tag{88}$$

For slow modes, δ_p reduces to a constant value δ_0, which is of the order of Δ^2, where Δ^2 is the average of the mean squares of the displacement vectors of the atoms that take part in the local conformational change.[13] If, for these modes, one uses the approximation:

$$D_p \cong D_0 = \frac{kT}{g\zeta} \tag{89}$$

where ζ is the "effective friction coefficient" of the monomeric unit, as Iwata pointed out, (86) reduces to Rouse's equation. In (89), D_0 is inversely proportional to the solvent viscosity η_0, because ζ is proportional to η_0. The choice of the value of D_p as given by (89) is in the Kirkwood tradition;[116] however, it is not the only possible one. One may also[18] start with (88) and express $P_{\bar{v}}$ using the result of Kramers[15] for the probability that a particle embedded in a viscous medium crosses a potential barrier $\Delta w \gg kT$. The solution of the corresponding one-particle Fokker-Planck equation leads to the following expression[15] of P:

$$P = \frac{\omega_A}{2\pi\omega_C}\left(\frac{z}{2m}\left(1 + \frac{4m^2\omega_C^2}{z^2}\right)^{1/2} - \frac{z}{2m}\right)\exp\left(-\frac{\Delta w}{kT}\right) \tag{90}$$

where m and z are the mass and friction coefficients of the moving particle, ω_A^2 and ω_C^2 the curvatures at the bottom and at the top of the $\Delta w/m$ curve (Fig. 8), ω_A the harmonic oscillator frequency near A and m/z the time constant for the decay of the velocity autocorrelation of a Brownian particle. A quasistationary state is assumed, in which practically no particles have yet arrived at B.

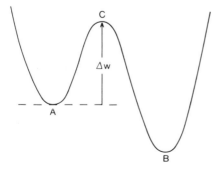

Fig. 8. Double minimum potential curve. The particle escapes from A to B.

In the case of high friction, $z/m \gg 2\omega_C$, the motion over the barrier is diffusional. The Fokker-Planck equation in phase space of the Brownian particle then reduces to a one-dimensional diffusion equation in configurational space (Smoluchowski's equation), leading to

$$P^{-1} = \frac{2\pi z}{\omega_A \omega_C m} \exp\left(\frac{\Delta w}{kT}\right) \tag{91}$$

On the other hand, for low friction, $2\omega_A \ll z/m \ll 2\omega_C$, we may use the following development of P^{-1}, which follows from (90):

$$P^{-1} = \frac{2\pi}{\omega_A}\left(1 + \frac{z}{2\omega_C m}\right)\exp\left(\frac{\Delta w}{kT}\right) \tag{92}$$

When the second term in (92) is neglected, the result of the transition-state method obtains.

Figure 9 is a plot[17] of P^{-1} as a function of the solvent viscosity η_0, when all other parameters are kept constant. Since z is proportional to η_0, the diffusion limit obtains at large η_0 values, where the P^{-1} curve is seen to approach a straight line going through the origin. At small η_0 values (i.e., when the behavior is nondiffusional), the slope of the P^{-1} curve (straight line d in Fig. 9) is only half of that of the preceding line—compare (91) and (92).

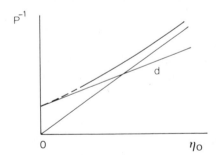

Fig. 9.   Inverse transition probability P^{-1} as a function of the solvent viscosity η_0 (cf. text).

In the range of p-values for which (89) holds, substitution of P^{-1} as given by (91) into (88) leads to[18]

$$\zeta = \frac{2\pi k T z}{\delta_0 \omega_A \omega_C m} \exp\left(\frac{\Delta w}{kT}\right) \tag{93}$$

However, if (92), which contains an η_0-independent term, is used instead of (91), the resulting diffusion equation is[18]

$$\frac{\partial \psi(\mathbf{s}, t)}{\partial t} = \nabla^{\mathrm{T}}\left(\mathbf{I} + (g\zeta')^{-1}\mathbf{\Phi}\right)^{-1} D(3l^{-2}\mathbf{M}\mathbf{s} + \nabla)\psi(\mathbf{s}, t) \tag{94}$$

where $\mathbf{I}$ is the unit matrix and $\mathbf{\Phi}$ a diagonal matrix; we can put D equal to $kT/(g\zeta')$, with ζ' given by

$$\zeta' = \frac{\pi k T z}{\delta_0 \omega_A \omega_C m} \exp\left(\frac{\Delta w}{kT}\right) \tag{95}$$

This choice of D requires that all the elements Φ_p of $\mathbf{\Phi}$ be equal to the same value Φ given by[18]

$$\frac{\Phi}{g} = \frac{2\pi k T}{\delta_0 \omega_A} \exp\left(\frac{\Delta w}{kT}\right) \tag{96}$$

Other possibilities are examined in Section IIIC.

Equation (94) exhibits the structure of the diffusion equation for the bead-and-spring model with internal viscosity,[19] when the external flow is set equal to zero in the latter equation. In earlier papers[19] the Φ_p's were considered as functions of the index p of the mode. The particular choice $\Phi_p = p\Phi_1$ was proposed for interpreting flow birefringence measurements of Leray.[117]

However, it was emphasized[19] that these experimental results were the only justification for such a choice of the Φ_p's. The effect on measurable quantities when mode-independent Φ's as given by (96) are used, is discussed in Section III.C.

It may be noticed that the introduction of an η_0-independent frictional term into (86), using (92), represents an attempt to account for nondiffusional behavior of local rearrangements, within the framework of a diffusion equation.

This, of course, will only be meaningful if (91) is not a universally valid approximation. Data that provide direct information on the behavior of P^{-1} in solvents of various viscosities are treated in Section III.B, where we conclude that both diffusional processes obeying (91) and nondiffusional ones obeying (92) are likely to exist.

As a consequence, the use of Iwata's $P_{\bar{v}}$, the probability for an "average local jump," may not be sufficient to develop a satisfactory dynamical theory of polymer chain molecules. For this, we require a detailed knowledge of the various processes and of the extent to which each kind of rearrangement participates in the renewal of conformations (i.e., whether a rearrangement does or does not migrate along the chain, according to the different possibilities considered in Section I).

Since such a complete theory is still lacking, we are obliged to examine various assumptions that can be made with respect to the type of process that dominates the dynamical behavior in a given range of η_0 values, discussing their respective consequences.

The stochastic theory has not been developed so far that the overall rotation of the molecule can be separated from deformational modes, as required for describing the behavior of fairly rigid and/or medium-sized molecules. For such molecules, Stockmayer[14] used an expression of the relaxation time taken from theories based on simple models. As an example, it may be recalled in this connection that the overall relaxation time of an elastic sphere is given by[118]

$$\tau^{-1} = \tau_{def}^{-1} + \tau_{or}^{-1} \tag{97}$$

where τ_{def} and τ_{or} are the deformational and orientational relaxation times.

We thus feel justified in basing the discussion of Section III,C on results obtained for the bead-and-spring model with internal viscosity, using the corresponding diffusion equation. This equation, in which we insert the expressions of ζ' and Φ given earlier, includes the rotational terms. In normal coordinate space it may be stated[19]

$$\frac{\partial \psi(s,t)}{\partial t} = \nabla^T \big(I + (g\zeta')^{-1}\Phi \big)^{-1} \Big(D(3l^{-2}Ms + \nabla) - \dot{s}^\circ - (g\zeta')^{-1}\Phi\, \dot{s}_\Omega \Big) \psi(s,t)$$

$$\tag{98}$$

In (98), $\dot{s}^\circ$ is the unperturbed velocity field and $\dot{s}_\Omega$ is the rotational velocity of the molecule. (A recent discussion of the value of $\dot{s}_\Omega$ can be found in Ref.[119]). The relaxation times τ_p and τ_p' associated with (98) were considered in Section I.

When the local conformational change under consideration affects the attached polymeric chains, "effective" values[120] of the parameters employed may have to be used in the expression of P. Finally, it should be noted that the effect of hydrodynamic interactions between the chain elements is omitted in the following discussion. In general, this approximation is known to affect only numerical coefficients when the dynamic properties in weak velocity fields (the only ones considered here) are expressed in terms of a reduced velocity gradient or of a reduced frequency. The reduced gradient, or frequency, is proportional to the intrinsic viscosity $[\eta]_0$ of the solution at zero velocity gradient and zero frequency. It may be recalled that (98) leads[19] to the Rouse-Zimm expression of the intrinsic viscosity $[\eta]_0$, which is a function of the τ_p's only.

As is well known, the τ_p's are η_0-proportional. Thus the nondiffusional character of the local movements would not be reflected in the value of

$[\eta]_0$. On the contrary, it may be reflected in the value of the initial slope of the curve giving the extinction angle χ of the flow birefringence as a function of the velocity gradient [cf. (105a)].

B. Some Characteristics of Local Movements.

Particular attention is devoted in this section to measurements of the ultrasonic absorption α of solutions of polymers and oligomers. It should be emphasized that many erroneous results have been obtained from analysis of measurements obtained in a narrow frequency range. Moreover, the necessity of measuring $\alpha - \alpha_0$, the difference between the absorption of a dilute solution and that of the solvent, always imposes severe limits on accuracy.

It was correctly concluded from early experiments[121] that local movements are involved; conformational changes due to rotational isomerism were thus postulated, and their cooperative nature was suggested. However, these measurements were carried out in the narrow frequency range 1 to 25 MHz. Under these conditions the frequency dependence of the absorption showed an abnormal behavior, which led us to discard an interpretation based on relaxational effects. Nevertheless, we are of the opinion that the results were in error, in spite of recent apparent confirmation.[122]

Hässler and Bauer,[107] who carried out measurements in polystyrene solutions in the range 1 to 60 MHz, correctly concluded that relaxational behavior occurs. However, the results were analyzed in terms of a single relaxation process. On the other hand, the measurements of Fünfschilling, Lemaréchal, and Cerf[109,114] showed that in solutions of polystyrene, polymethylmethacrylate, and polyvinylpyridine, at least two relaxations must exist in this range. A similar conclusion has been drawn by Ohsawa and Wada[123] for polystyrene, and recently by Nomura, Kato, and Miyahara[124,125] for polyvinylpropionate, polyvinylbutyrate, and polyvinylacetate.

In discussing relaxational behavior, we consider, whenever possible, results obtained over nearly three decades of frequency, and sometimes more. This, however, does not ensure lack of ambiguity in the analysis of spectra. It is known that relaxation processes exist beyond 100 MHz, as a later part of this section indicates. Therefore doubts always remain about whether a limiting value

$$\frac{\alpha_\infty}{F^2} = \lim_{F \to \infty} \frac{\alpha}{F^2} \tag{99}$$

can be obtained from the measurements; furthermore, α_∞ can be larger or smaller than the solvent absorption α_0. According to Lemaréchal,[126] this circumstance seems to have been insufficiently considered by Ludlow, Wyn-Jones, and Rassing.[127]

Measurement of the velocity of propagation of the waves is a promising technique for characterizing the high-frequency relaxational behavior of polymer solutions.[128] Such measurements, although providing the same information as those of the absorption, do not suffer from the loss of sensitivity exhibited by the latter ones at high frequency.

Before going into the ultrasonic relaxation behavior of polymeric chains, let us recall various results that suggest the local character and conformational nature of the mechanisms involved. For example, the above-defined difference $\alpha - \alpha_0$ is proportional to the polymer concentration c, up to at least $10^{-1} g/cm^3$ for polymethylmethacrylate,[109] and the value of $\alpha - \alpha_0$ shows comparable behavior for polyvinylpyridine and polystyrene.[114] Furthermore, $\alpha - \alpha_0$ is sensitive to stereoregularity, at least below a few tens of megahertz,[129] and to chain structure, as evidenced by the contrasting behavior of polystyrene and its derivative of greater stiffness, poly-ortho-bromostyrene,[115,130] at the lower-frequency end of the spectra. On the other hand, according to Bader and Cerf,[128] the molecular weight M has no influence on the observed relaxation frequency, down to $M_w = 4800$ for polystyrene. A similar conclusion has been reached recently by Cochran, Dunbar, North, and Pethrick,[131] who used the resonator method of Eggers[132] at the lower frequencies. Behavior of oligomers is discussed later in the section.

Figure 10 shows results for polyvinylpyridine in dioxane[133] at four temperatures ranging from 15 to 45°C and frequencies between 0.3 and 185 MHz. Results for two samples of molecular weight larger than 200,000 have been used to establish the curves. It has been verified that the two samples furnish identical results. Three different devices have been employed to cover the frequency range studied: the conventional interferometer and pulse devices at frequencies above 2 MHz, and a recently developed reverberation technique[134] for the range 0.3 to 1.2 MHz.

If it is assumed that relaxation processes that occur above 100 MHz are well separated from those studied here, or that they have a negligible influence on the measured values of α, it is of interest to consider the quantity

$$\delta\alpha = \alpha - \alpha_\infty = \Delta\alpha - (\Delta\alpha)_\infty \qquad (100)$$

In (100) $\Delta\alpha$ is the value of $\alpha - \alpha_0$ at the frequency F, and $\Delta\alpha_\infty$ that of $\alpha_\infty - \alpha_0$, where α_∞ is defined by (99). In Fig. 10, the incremental absorption per wavelength $\delta\alpha \cdot \lambda$ is plotted against λ in a double-logarithmic scale.

With this plot, a single relaxation is represented by a curve, whose slopes are equal to $+1$ and -1, at low and high frequencies, respectively (dotted

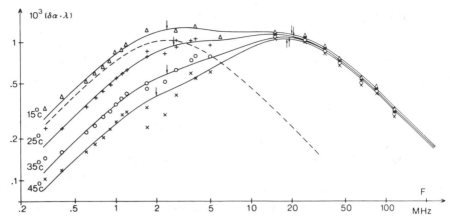

Fig. 10. Incremental absorption per wavelength plotted against the frequency for polyvinyl-pyridine in dioxane at four temperatures.[133] Arrows locate the relaxation frequencies deduced when a two-relaxation process is postulated. The concentration is 3×10^{-2} g/cm^3.

curve in Fig. 10). It is noteworthy that slopes equal to $+1$ are observed in Fig. 10 at the low-frequency end of the spectra, over a range that may approach one decade of frequency. A similar observation has been reported for polystyrene in a series of measurements carried out in solvents of various viscosities[135] (cf. also Fig. 12) and for polystyrene at various temperatures.[115,130]

On the other hand, in the high-frequency region (i.e., above 10 MHz), the shape of the experimental curves depends markedly on the validity of the earlier assumption—namely, that relaxations that may occur beyond 100 MHz are not detected in measurements carried out in the frequency range under consideration.

It is difficult to decide whether the results should be analyzed in terms of a discrete set of relaxations or whether a continuous relaxation spectrum extending to frequencies that may be greater than 100 MHz should be considered. We notice, however, that at 15°C, the spectrum approaches single-relaxation behavior up to 2 MHz (dotted curve in Fig. 10). If we assume that the spectra can be decomposed into two single-relaxation curves, we have the relaxation frequencies $F_r = (2\pi\tau)^{-1}$ and the corresponding amplitudes μ_M (i.e., the maximum values of $\delta\alpha \cdot \lambda$) given in Table I. The extreme values of the parametric compatible with the measurements appear in the Table for the two relaxations, which are denoted 1 and 2, respectively.

From the preceding analysis, the relaxation times appear to exhibit a small temperature dependence (non-Arrhenius behavior). The recent measurements for polystyrene solutions ($M_n = 200,000$) referred to pre-

TABLE I

Relaxation Frequencies and Amplitudes as Deduced from the Spectra of
Fig. 10, Analyzed in Terms of Two Single-Relaxation Processes

$T(°C)$	$(\mu_M)_1 \times 10^3$			$(F_r)_1$ (MHz)			$(\mu_M)_2 \times 10^3$			$(F_r)_2$ (MHz)		
15	0.9	to	1.15	2.0	to	2.8	0.9	to	1.0	18	to	22
25	0.55	–	0.8	2.0	–	2.9	0.9	–	1.0	18	–	22
35	0.3	–	0.4	1.8	–	3.0	0.9	–	1.1	18	–	22
45	0.15	–	0.25	1.5	–	2.4	0.9	–	1.1	18	–	22

viously,[115,130] have furnished ultrasonic spectra quite comparable to those of Fig. 10 for polyvinylpyridine, although the value of the absorption is somewhat smaller for polystyrene.

It must be emphasized, however, that a different temperature dependence of the relaxation times may be derived if a specific continuous relaxation spectrum is postulated. Such an example is provided by the above-mentioned work on polystyrene,[131] from which Arrhenius behavior of a "mean-relaxation time" was obtained at high molecular weight. Nevertheless, even in this work[131] a small temperature dependence of the relaxation time was obtained for a sample of molecular weight $M = 4000$.

Assuming that the observed ultrasonic absorption reflects transitions between rotational isomeric states, we derive molecular parameters from the data of Table I, using the two-state non-Arrhenius model considered in Section II,D. In view of the uncertainties relating to the high-frequency region of the spectra, the analysis is carried out for relaxation 1 only.

For the sake of simplicity, volume changes between the states are not taken into account. This leads to the values given in Table II for the enthalpies (kcal/mole) and entropies [cal/(mole)(deg)] of reaction and of activation, as well as to the values indicated for the transition temperature $T_t = \Delta H / \Delta S$. It has been assumed that there is one relaxing unit every $p = 20$ monomer units. Smaller values of p have been reported[114] for relaxation 2. With the exception of T_t, the various parameters depend only slightly on the value of the number p—which was defined in Section II,D and should not be confused either with the index of a Rouse mode or with the probability $p(ZY\ldots)$ of a state, in the theory of the linear Ising lattice. The extreme values of the parameters compatible with the measurements are given in Table II.

Application of the two-state model described amounts to assuming that each relaxation process is produced by one kind of relaxing system, or

TABLE II

Enthalpies and Entropies of Reaction and Activation, as well as Transition Temperature T_t, for the Lower-Frequency Relaxation in Fig. 10, Using a non-Arrhenius Two-State Model.

ΔH_{12}	ΔH_{21}	$\Delta H = \Delta H_{12} - \Delta H_{21}$	ΔS_{12}	ΔS_{21}	$\Delta S = \Delta S_{12} - \Delta S_{21}$	T_t (°C)
−5.6 to −14.2	1.4 to −3.9	−7.0 to −10.3	−49 to −79	−22 to −39	−27 to −40	−14 to −16

unit, and that the different units (i.e., segments of different lengths) are excited independently from one another.[48] This assumption is reasonable if all the relaxing units are transitional, in the sense of Section I. Even in this case, however, two relaxation processes can originate from a three-state transitional system.

Actually, there exists some evidence for interactions between different movements; thus at least a partially migrational character of the processes under consideration is probable.

One may first notice that if the different movements are not independent of one another, those leading to the relaxation frequency $(F_r)_1$ could split progressively when the temperature T is increased into movements that would involve shorter segments. In this case, the effective change of $(\mu_M)_1$ with T would be smaller than that deduced from Table II. According to (83), this would lead to a value of $\Delta H_{12} + \Delta H_{21}$, which would be less negative than that given in Table II, therefore perhaps preferable.

Moreover, although polystyrene solutions have furnished ultrasonic spectra which are comparable to those of Fig. 10 for polyvinylpyridine, a smaller temperature dependence of $(\mu_M)_1$ has been observed,[115,130] as One may speculate that in this case, the movements leading to relaxation 1 decompose less easily into smaller ones when T is increased, owing to the greater stiffness of the chain.

From all the results quoted, ultrasonic measurements are seen to reveal a strong "high frequency" absorption band. Furthermore, if conformational changes are responsible for the observed effects, their temperature dependence suggests that relaxation 1 is produced by a cooperative process.

Oshawa and Wada[123] have shown that the value of the absorption α lies much higher than would be expected from the dynamic shear viscosity. In this work, a resonance reverberation technique[136,137] was used, in the range 10 to about 200 kHz. The results of Ono, Shintani, Yano, and Wada[138] show that in the lower frequency range the relaxation is molecular-weight dependent (cf. Fig. 11). We should like to suggest[133] that this relaxation may result from a cascade of movements that absorb in the "high-frequency" band, in much the same way that the Rouse modes, evidenced by measuring the dynamic shear viscosity, are generated by local movements (Section I).

Figure 11 indicates that an oligostyrene $(M_n = 2100)$ exhibits a smaller absorption than the other samples. Oligostyrenes $(M_n = 2100, M_n = 600)$ have been studied in toluene at different temperatures by Froelich, Noël, Lewiner, and Monnerie.[112] A decrease in absorption was also observed, and both relaxation frequencies $(F_r)_1$ and $(F_r)_2$ were found to increase

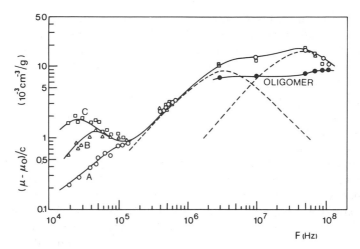

Fig. 11. Ultrasonic absorption of polystyrene in xylene.[138] The excess absorption per wavelength divided by the frequency is plotted against the frequency for three samples, of average molecular weights 2.1×10^5 (A), 3.7×10^5 (B), 10^6 (C). Dashed lines are single-relaxation curves.

when the molecular weight decreases. These observations, which have been reported[112,115] to be of value in determining p, may indicate that the movements close to a free end of the oligomer molecule can occur via a milder reorganization process than in the polymer.

Cochran, Dunbar, North, and Pethrick[131] have observed a similar decrease in absorption and have also obtained higher relaxation frequencies for oligomers than for polymers. However, the decrease in absorption was less pronounced than that found by the previous authors,[112,115] perhaps because a mean relaxation frequency was considered.

The preceding observations appear to support the suggestion already mentioned that there exist local movements in a long chain which require a cooperative reorganization of segments comprising several monomeric units.

Studies involving polymers of various structures, as well as oligomers, should be fruitful. Variation of temperature, in a wider range than has been considered so far, should provide a test of the validity and limitation of the non-Arrhenius two-state model that was applied. In particular, this should permit a check on whether the quantity $T^2\mathfrak{A}/\tau$ goes through a maximum at a value of T that can be interpreted as a transition temperature (see Section II,D). At the same time, more insight could be obtained with respect to the suggested cooperative nature of local conformational changes.

We discuss next measurements of the ultrasonic absorption carried out in solvents of various viscosities; later in this section, we mention fluorescence depolarization and nuclear magnetic resonance measurements. The measurement of the high frequency shear viscosity is considered in Section III,C.

Figure 12 shows results of ultrasonic measurements for polystyrene. A fraction prepared by radical polymerization ($M_w = 630,000$) was studied at 20°C in four solvents[135] (methylethylketone, decalin, diethylphthalate, and dibutylphthalate). The viscosities ranged from 0.44 to 20.7 centipoise (cP).

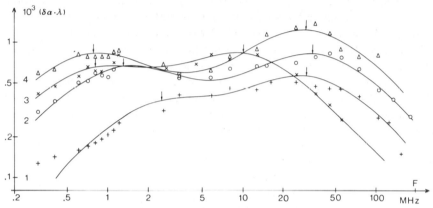

Fig. 12. Incremental absorption per wavelength plotted against the frequency for polystyrene solutions: (1) in methylethylketone, (2) in decalin, (3) in diethylphthalate, (4) in dibutylphthalate.[135] Arrows locate the relaxation frequencies deduced when a two-relaxation process is postulated. The concentration is 2.5×10^{-2} g/cm³. Reproduced by permission of North-Holland Publishing Company, Amsterdam.

The ultrasonic absorption of each solvent and solution was measured between 0.3 and 185 MHz. The absorption excess $\Delta\alpha = \alpha - \alpha_0$ divided by the square of the frequency was plotted against the frequency F. It appeared that $\Delta\alpha / F^2$ approaches a (nonzero) limiting value at high frequency; thus the process described earlier—see (100)—was applied to deduce $\delta\alpha \cdot \lambda$, as shown in Fig. 12. In methylethylketone, deviation from the two-relaxation behavior appears at the low-frequency end of the spectrum. Theoretical curves for two-relaxation processes were fitted to the experimental points.

If we consider only the low-frequency relaxation, Fig. 12 indicates that the relaxation frequency $(F_r)_1$ decreases significantly when η_0 is increased from 0.44 cP (methylethylketone) to 2.4 cP (decalin). On the other hand, $(F_r)_1$ decreases at most by a factor of 3 over the whole range of viscosities when η_0 is increased by a factor of 50. The preceding results point to the existence of nondiffusional processes. Indeed, to take into account the effect of these processes, we must insert values of the k_{12} and k_{21} (which are of the form of (90)), into the expression of the relaxation time as given by (75). We are thus led to a qualitative agreement with the observed behavior of $(F_r)_1$ when η_0 is varied.

In the limit of small η_0's the resulting expression of F_r reduces to the η_0-independent value given by the transition state limit. In fact, this is the value that has been considered hitherto in all studies of the thermal relaxation by ultrasonic methods.

Although the observed behavior of $(F_r)_1$ is typical of nondiffusional processes, it should be kept in mind that diffusional behavior (i.e., an η_0-proportional relaxation time) must be restored in the limit of very large η_0's (cf. Fig. 9). It is therefore advisable that measurements be carried out in a large number of solvents of different η_0's.

Arguments in favor of the existence of nondiffusional processes have also been drawn[23] from the analysis of measurements of the fluorescence depolarization and the nuclear magnetic resonance. Figure 13 shows experimental curves obtained by Biddle and Nordström[139] for polystyrene containing a small amount of fluorescent monomer. The rotational relaxation time τ was deduced from Perrin's[140] equation:

$$\frac{1}{p} - \frac{1}{3} = \left(\frac{1}{p_0} - \frac{1}{3}\right)\left(1 + \frac{3\theta}{\tau}\right) \tag{101}$$

in which p is the degree of polarization and p_0 the value p attains in the absence of rotation; θ and τ are the time constants for fluorescent decay and rotational diffusion, respectively.

A recent attempt to relate p to the local jump probability[141] P led to

$$\frac{1}{p} - \frac{1}{3} = \left(\frac{1}{p_0} - \frac{1}{3}\right)\left(1 + \epsilon(P\theta)^{1/2}\right) \tag{102}$$

in which ϵ is a conformational parameter of the order of 0.5 to 2.5. The assumption that local movements occur on the diamond lattice appears critical[141] for the validity of (102).

However, according to Weill,[142] (101) must be formally valid, as follows from the linear dependence of $1/p - 1/3$ on θ when the latter time is varied by adding various amounts of a quenching substance. We therefore admit that the values of τ given by Biddle and Nordström represent a local correlation time, with the necessary reservations that arise from the use of (101). In Fig. 13 the τ versus η_0 curves are concave downward at the lower temperatures. A similar observation had already been made by Wahl,

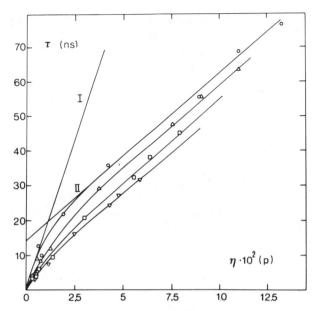

Fig. 13. Relaxation time deduced from fluorescence depolarization measurements[139] (cf. text) as a function of solvent viscosity. Data for polystyrene of degree of polymerization 650. Straight lines I and II have been added. Curves for results obtained at four temperatures as follows: 30°C (O), 35°C (Δ), 45°C (□), and 55°C (∇).

Meyer, and Parrod.[143] It may be noted that the correlation time deduced by Hermann[144] from NMR measurements in solutions of polyoxyethylene exhibits a comparable behavior when plotted as a function of η_0. In the case of polystyrene, the correlation time τ, as deduced from measurements of both the nuclear magnetic resonance and the fluorescence depolarization, obeys the inequality[144]

$$10 < 10^8 \frac{\tau}{\eta_0} < 20$$

The time τ is expressed here in seconds and η_0 in poise. In low-viscosity solvents (i.e., for $\eta_0 < 1$ cP), this corresponds to a relaxation frequency higher than 100 MHz, hence higher than the ultrasonic relaxation frequency $(F_r)_2$, in accord with a previous remark.

Let us take the example of the fluorescence depolarization measurements obtained at 30°C. The τ *versus* η_0 curve is seen to approach two straight lines I and II (Fig. 13) at low and high values of η_0, respectively.

In one possible[23] interpretation of these results, we assume that the highly diversified local movements of linear polymeric chains belong mainly to two quasi-independent classes I and II that are characterized by times τ_I and τ_{II}, which are represented by the above-mentioned straight lines I and II, respectively. It may be noticed that the τ *versus* η_0 law represented by line I corresponds to diffusional behavior, whereas the law represented by line II corresponds to nondiffusional behavior (cf. Section I).

With this assumption, the relevant value of the local relaxation time τ for each value of η_0 follows simply because the renewal of conformations is controlled by the fast process. For small values of η_0, we have $\tau_I < \tau_{II}$ (cf. respective positions of straight lines I and II in Fig. 13); thus $\tau \cong \tau_I$. However, for large values of η_0 we have $\tau_{II} < \tau_I$; thus $\tau \cong \tau_{II}$. The curves are then concave downward, provided the slope of straight line I is greater than that of straight line II—a situation that should obtain if the diffusional processes are produced by segmental rearrangements which are characterized by relatively large viscous friction coefficients.

On the other hand, according to Weill, Hermann, and Bentz,[145,146] deviations from purely exponential behavior of the autocorrelation function of local orientation may be reflected in the τ *versus* η_0 law.

Fast local processes involving low potential barriers are expected to obey the diffusion-limit in any case, and it may be that the corresponding viscous friction coefficients are smaller than those of the supposedly coexisting slower non-diffusional local processes. Nonetheless, the latter processes may still become manifest as a consequence of their greater importance for the long range conformational changes. Also, a situation may then result in which the curve giving the local relaxation time τ as a function of η_0 would exhibit a sigmoidal shape (i.e., it would resemble the curve represented in a different context in Fig. 14b).

Keeping in mind the reservations already made, we consider in the next section various possible implications of nondiffusional character of certain local movements for the slow modes and for the corresponding measurable quantities.

C. Consequences for Slow Modes; Comparison with Experiments

We now consider the question: to what extent do local rearrangements of either type—diffusional and nondiffusional—affect the values of the longest τ_p''s? In other words, to what extent do they enter into the matrix Φ of internal viscosity (Section III,A)?

The behavior of the τ_p''s can be predicted in several cases that may arise. For a given chain molecule, comparison with the experimental results will permit drawing conclusions about the category to which the molecule belongs. The relation of nondiffusionality to cooperativity is discussed at the end of the section.

1. Diffusional Processes Dominate in the Whole Range of η_0 Values

The bead-and-spring model relaxation times are given by the Rouse[10] and Zimm[20] expressions; that is, the τ_p''s are not distinguished from the τ_p's. When hydrodynamic interactions are neglected, the τ_p's are given by

$$\tau_p = \frac{6}{\pi^2} \frac{M[\eta]_0 \eta_0}{RT} p^{-2} \tag{103}$$

where M is the molecular weight. On the other hand, when hydrodynamic interactions are taken into account, the form of the τ_p's is conserved, whereas their dependence on p is different.[20] Thus at high p values, τ_p is proportional to $p^{-3/2}$.

The point of interest in the present discussion is that here τ_p is proportional to the solvent viscosity η_0 in the whole range of η_0 values. Hence the initial slope at infinite dilution $(\tan \alpha)_0$ of the curve giving the extinction angle χ of the flow birefringence[20]

$$(\tan \alpha)_0 = \frac{1}{2} \frac{\sum\limits_{p=1}^{N'} \tau_p^2}{\sum\limits_{p=1}^{N'} \tau_p} \tag{104a}$$

is itself proportional to η_0. The summation in (104a) is extended to N' subchains. The result, however, does not depend on the value chosen for N', and one has[20]

$$(\tan \alpha)_0 = 0.2 \frac{M[\eta]_0 \eta_0}{RT} \tag{104b}$$

or

$$(\tan \alpha)_0 = 0.1 \frac{M[\eta]_0 \eta_0}{RT} \tag{104c}$$

according to whether the hydrodynamic interactions are weak or strong, respectively.

2. Nondiffusional Processes Dominate in the Whole Range of η_0 Values

The τ_p''s are given by (3), and (104a) is replaced by[19]

$$(\tan \alpha)_0 = \frac{1}{2} \frac{\sum\limits_{p=1}^{N'} \tau_p \tau_p'}{\sum\limits_{p=1}^{N'} \tau_p} \tag{105a}$$

This expression can also be written

$$(\tan\alpha)_0 = \frac{1}{2}\frac{\sum\limits_{p=1}^{N'} \tau_p^2}{\sum\limits_{p=1}^{N'} \tau_p} + \frac{1}{2}\frac{\sum\limits_{p=1}^{N'} \tau_p(\tau_p'-\tau_p)}{\sum\limits_{p=1}^{N'} \tau_p} \qquad (105b)$$

Accordingly, $(\tan\alpha)_0$ contains an η_0-independent term [the internal viscosity term, i.e., the second term in the right-hand member of (105b)]. This term is responsible for the finite intercept of the straight line D ("deformation") in Fig. 14a and 14b. The form (105b) of $(\tan\alpha)_0$ is similar to results for simple models.[21,22]

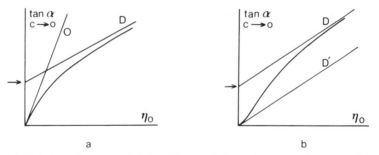

Fig. 14. Initial slope $(\tan\alpha)_0$ at infinite dilution of the extinction angle curve of the flow birefringence. Theoretical predictions: (a) when nondiffusional processes dominate; (b) when diffusional processes dominate at low η_0 values, whereas nondiffusional processes dominate at large η_0 values (cf. text). Intercepts designated by an arrow indicate the η_0-independent internal viscosity term.

As pointed out by Kuhn and Kuhn,[16] the relaxation time for rotational diffusion of the molecules can be predominant in the present case at low η_0 values. Thus at small η_0's and small shear rates, the chain molecules may behave as if they were "rigidified," whereupon they present the orientational effect that is known for rigid particles subjected to a transverse velocity field.[147] The variation of $(\tan\alpha)_0$ as a function of η_0 shows passage from orientational to deformational effect; the corresponding curve can be calculated for an elastic sphere,[118] and it exhibits the shape illustrated in Fig. 14a. In the present context, the orientational effect may be ascribed to a very large internal viscosity.

The value of $(\tan\alpha)_0$ has also been calculated for the bead-and-spring model in the limit when the orientational effect dominates (low η_0's). If the axes of the molecules are assumed to lie in the plane of flow, we obtain[19,148]

$$(\tan \alpha)_0 = 0.9 \frac{M[\eta]_0 \eta_0}{RT} \qquad (106a)$$

or

$$(\tan \alpha)_0 = 0.7 \frac{M[\eta]_0 \eta_0}{RT} \qquad (106b)$$

according to whether hydrodynamic interactions are weak or strong, respectively. When the distribution function of the axes is three-dimensional, somewhat smaller numerical coefficients[149] are obtained in (106a) and (106b): 0.67 and 0.5, respectively. In the limit where the orientational effect prevails, the $(\tan \alpha)_0$ *versus* η_0 curve approaches the straight line O ("orientation") in Fig. 14a.

3. Diffusional Processes Dominate at Low η_0's, Whereas Nondiffusional Processes Dominate at Large η_0's

Our consideration is suggested by the results of methods that permit characterizing local movements (preceding section). If we assume that nondiffusional processes may take precedence beyond a certain range of η_0 values, we must keep in mind that even for these processes and in the limit of very large η_0's diffusional behavior is recovered (cf. Fig. 9 and the discussion of Section III,A).

In the present case (104) and (105) must hold at low and high η_0's, respectively. Thus the $(\tan \alpha)_0$ *versus* η_0 curve exhibits[23] the behavior shown in Fig. 14b.

4. Comparison with Experiments; Discussion

The measurement of $(\tan \alpha)_0$ has been carried out in solvents of various viscosities for polystyrene,[150] for desoxyribonucleic acid (DNA),[151] and for a variety of other chain molecules. The examples of polystyrene and DNA will enable us to outline pending problems.

Figure 15 shows values of $(\tan \alpha)_0$ obtained for DNA at various η_0's by adding glycerol to water.[151] The curve is seen to follow the pattern of Fig. 14a. As pointed out by Harrington,[152] these measurements, as well as those of Leray,[117] were carried out under conditions for which the intrinsic viscosity $[\eta]_0$ (at a given molecular weight) depended on the sample. However, when η_0 was varied in repeated experiments[117] the $(\tan \alpha)_0$ *versus* η_0 curve also showed the trend of Fig. 15, thus confirming the result of Ref. 151.

Whereas the bead-and-spring model need not be quantitatively valid for DNA, the experiments just mentioned indicate orientational behavior of

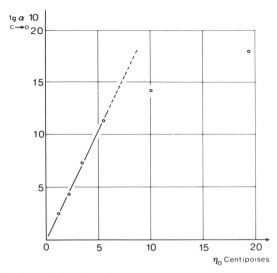

Fig. 15. Measured values of the slope $(\tan\alpha)_0$ plotted against the solvent viscosity η_0 for DNA.[151] The viscosity was changed by adding various amounts of glycerol to the solvent. The measurements exhibit the trend of Fig. 14a. From *Helvetica Chimica Acta*.

DNA at low η_0's. The existence of an orientational effect has been reported on the basis of dielectric measurements[153,14] for polysulfones with repeating units of the structure $-CH_2CR_1R_2SO_2-$ in chains of molecular weights up to as high as 10^6. A similar effect has been observed[154] for halostyrene polymers having molecular weights below about 2×10^4.

For polystyrene, the initial slope $(\tan\alpha)_0$ has been obtained[150] for various values of η_0, using cyclohexanone as a solvent, in the range of temperature 0 to 60°C. As emphasized by Janeschitz-Kriegl,[155] the values of $(\tan\alpha)_0$ obtained in these measurements, as well as in those of Leray,[117] are too high when compared with his own results.[155]

The discrepancy is smaller with the first series of measurements[150] when the true value of the molecular weight, as given in a footnote to Ref. 156, is taken into consideration. Furthermore, when Janeschitz-Kriegl's results for low value of the velocity gradient are considered, the discrepancy appears even smaller (cf. Fig. 3 of Ref. 155).

Figure 16, in which $(\tan\alpha)_0$ is plotted against $M[\eta]_0\eta_0/RT$, shows recent results of Budtov and Pen'kov[157] for several fractions of polystyrene of different molecular weights and for different values of η_0. The curve has the form of Fig. 14a. However, according to Stockmayer,[14] orientational effect at low η_0's is unlikely for polystyrene except at low molecular weight. On the other hand, for the high molecular weight fraction[157] the curve might be similar to that of Fig. 14b. In this case, the limiting slope at high

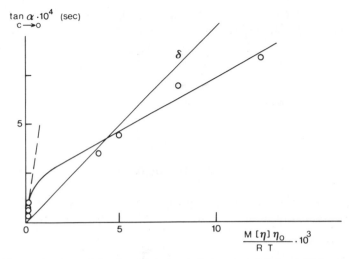

Fig. 16. Measured values of the slope $(\tan\alpha)_0$ plotted against $M[\eta]_0\eta_0/RT$ for fractions of polystyrene in various solvents and at different temperatures.[157] The straight line δ, which has been added, is the theoretically predicted behavior when diffusional processes dominate. Reprinted from Ref. 157 with permission from Pergamon Press Ltd.

values of the abscissa would be smaller than the value given by (104c), which is represented by the straight line δ in Fig. 16.

The case of polymethylmethacrylate,[158,159] for which orientation of the molecules at low η_0's has been reported, raises doubts similar to those encountered in this respect for polystyrene. On the other hand, for cellulose derivatives internal viscosity effects appear to exist from two independent series of measurements.[160,161] (cf. also end of this section).

In short, from the experiments just discussed, it seems that an influence of nondiffusional local processes on the longest relaxation times τ_p' may exist for a variety of chain molecules. In fact, such an influence appears to be particularly important for certain less flexible chains.

Here the usual distinction between conformational and kinetic flexibility should be made. High kinetic flexibility requires, for the most part, that low barriers be overcome in transitions between isomeric states, thus leading to high conformational flexibility. On the other hand, low conformational flexibility usually entails low kinetic flexibility, in accordance with the observations just reported.

Very useful information is also provided by the measurement of the viscoelastic behavior of polymeric solutions in high-frequency shear fields, especially by the measurement of the high-frequency limit η_∞' of the real part η' of the viscosity. Such a limit follows from the theory for both rigid[162,163] and deformable particles.

For an elastic sphere, the high-frequency limit $[\eta']_\infty$ of the real part of the intrinsic viscosity is given by[164]

$$\frac{[\eta']_\infty}{[\eta]_0} = 1 - \frac{2.5\eta_0}{\mu\tau} \tag{107}$$

where τ is the relaxation time of the sphere embedded in the liquid[21,164,165]

$$\tau = \frac{1.5\eta_0 + \eta_{sph}}{\mu} \tag{108}$$

In (108), η_{sph} is the viscosity of the sphere and μ its Lamé elastic shear coefficient.

On the other hand, when there is only one relaxation time τ, its value is directly obtained from the measurement of $(\tan\alpha)_0$, since

$$(\tan\alpha)_0 = \frac{\tau}{2} \tag{109}$$

This equation was first obtained for rigid ellipsoids oriented in a transverse velocity field.[147] However, it has more general significance,[19] and it is valid, for example, when applied to a deformable sphere.[118] If the latter model is employed for a polymeric chain molecule, it may be useful to define an internal viscosity

$$\eta_{int} = \eta_{sph} - \eta_0 \tag{110}$$

in which case the following simple relation between $[\eta']_\infty$ and the internal viscosity term $(\tan\alpha)_{0,int}$ in the expression of $(\tan\alpha)_0$ is obtained[166]

$$\frac{[\eta']_\infty}{[\eta]_0} = \frac{(\tan\alpha)_{0,int}}{(\tan\alpha)_0} = \frac{\eta_{int}}{2.5\eta_0 + \eta_{int}} \tag{111}$$

For the present model, two consequences follow from (111), as well as from the original form of the results:[21,164]

1. The ratio $[\eta']_\infty/[\eta]_0$ must be comparable to the experimental value of $(\tan\alpha)_{0,int}/(\tan\alpha)_0$.

2. The ratio $[\eta']_\infty/[\eta]_0$ must exhibit the same dependence on molecular weight and on solvent viscosity as the ratio $(\tan\alpha)_{0,int}/(\tan\alpha)_0$.

On the other hand, $[\eta']_\infty$ was known to be small compared with $[\eta]_0$.[167–169] It was therefore suggested[19] that (98) might not apply for a high-frequency shear gradient. Moreover, results of Massa, Schrag, and Ferry[170] for polystyrene show that the high-frequency limit of η' does not depend on the solvent viscosity η_0 when η_0 is varied by a factor of 30. The excess viscosity due to the polymer is represented in Fig. 17 in a semilogarithmic plot *versus* the reduced frequency ωa_T (a_T is the reduction factor). These results can be regarded from two different standpoints.

1. It can be assumed that (98), the bead-and-spring diffusion equation with internal viscosity, is applicable at high frequency. Under these conditions, it follows that the corresponding limiting value $[\eta']_\infty$ is given by[171]

$$\frac{[\eta']_\infty}{[\eta]_0} = \frac{\displaystyle\sum_{p=1}^{N'} \tau_p (\tau_p' - \tau_p)/\tau_p'}{\displaystyle\sum_{p=1}^{N'} \tau_p} \tag{112}$$

and exhibits a dependence on η_0 that is contradicted by the results of Fig. 17. From this point of view, it follows that the η_0-independent internal viscosity term does not exist for polystyrene and that the internal viscosity must be proportional to the solvent viscosity.[170]

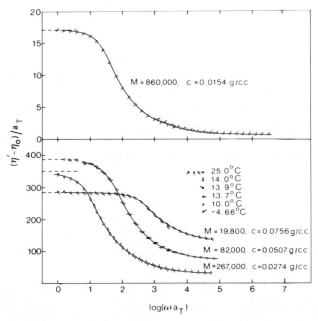

Fig. 17. Real part of reduced contribution of polymer to dynamic viscosity *versus* the reduced frequency in semilogarithmic scale.[170] Results for various fractions of polystyrene in a wide range of η_0 values. Reprinted with permission from *Macromolecules*, **4**, 210 (1971), Fig. 7.

It has been proposed that lateral displacements involved in the crossing of potential barriers are not reflected in the movements of the beads.[175,176] Therefore, if the bead-and-spring model is used, it would be necessary to introduce an η_0-proportional internal viscosity.

The quantitative validity found[170,172] for the expressions of the real and imaginary parts of $[\eta]$ as deduced[171] from (98) must be examined critically, however, even if accord with the birefringence measurements[173] in an oscillatory shear gradient is obtained. Not only does one use several adjustable parameters (including the number N' of subchains, as first proposed by Thurston and Schrag[174]), but also Φ_p's are employed that are proportional to the index p of the mode (cf. the discussion of Section III,A).

2. On the other hand, one may envisage[166] the internal viscosity coefficients as being frequency dependent. Recall that the vanishing (or near vanishing) of $[\eta']_\infty$ was successfully attributed by Rouse[10] to the ability of each part of the molecule to move with the velocity of the surrounding liquid in a high-frequency shear field. It may be, however, that the response of the chain to the rapid establishment of a small-amplitude velocity gradient G first involves relative displacement of fairly large groups of monomers. In this picture, local deformations would first occur mainly in the segments connecting these groups. If G were maintained at its new value, the number of rearrangements produced per unit time would increase and finally would reach a stationary value. If, instead, G is decreased to zero within a time of the order of the shortest relaxation times of the bead-and-spring model, only relatively few local rearrangements would have been produced, and internal viscosity would play a smaller role.

In other words, one may wonder whether at high frequency η_0-independent internal viscosity effects should not be omitted, especially for the slower modes.

In describing the response of the chain to a stepwise perturbation we have considered non-Rousian deformational modes. The existence of such modes is rendered possible by the large number of degrees of freedom of the chain. This number is of the order of 3^N (cf. Section I), as compared with the number of Rouse modes, which is smaller than N.

Since each of the foregoing groups of monomers is not fully adapted to the velocity field, the mechanism described would in itself lead to an η_0-independent contribution to $[\eta']_\infty$—that is, to an η_0-proportional internal viscosity.

The possibility that there exists an η_0-proportional internal viscosity leads us to consider reductions of Iwata's coarse-grained diffusion equation to a form more general than the one considered in Section III,A. If an

η_0-proportional internal viscosity is introduced in (94), more general relationships than (93), (95), and (96) are in fact obtained.[166]

When diffusional processes (class I processes) are assumed to dominate the dynamic behavior, internal viscosity coefficients ρ can be introduced by writing $\zeta^I(1+\rho_p^I)$ in place of ζ in the left-hand member of (93). (The superscript I indicates that class I processes are considered.) The reduction of (86) to the form (94) leads to the following relationship between ζ^I and ρ_1^I:

$$\frac{2\pi kT z^I}{\delta_0^I \omega_A^I \omega_C^I m^I} \exp\left(\frac{w^I}{kT}\right) = \zeta^I(1+\rho_1^I) \tag{113a}$$

Similarly, for ρ_p^I we have

$$\frac{1+\rho_p^I}{1+\rho_1^I} = \frac{\delta_0^I}{\delta_p^I} \tag{113b}$$

From Iwata's mode dependence of the δ_p's we may set[166]

$$\left(\delta_p^I\right)^{-1} \cong \left(\delta_0^I\right)^{-1}\left(1+\epsilon^I(p-1)^2\right) \cong \left(\delta_0^I\right)^{-1}\left(1+\epsilon^I p^2\right) \tag{114}$$

as long as δ_p is not too small in comparison with δ_0 (and for p's not too close to unity). The coefficient ϵ^I may be deduced from Fig. 2 of Iwata's work.[9]

Thus in the range validity of (114) and for values of ρ that are small with respect to unity, the following approximation applies:

$$\rho_p^I \cong \rho_1^I + \epsilon^I p^2 \tag{115}$$

The corresponding elements $\Phi_p = g\zeta\rho_p$ of the internal viscosity matrix are η_0-proportional. This conclusion would hold even if $\rho_1^I = 0$, showing that the mode dependence of Iwata's δ_p's is sufficient to produce an η_0-proportional internal viscosity. In addition, introduction of an internal viscosity permits inclusion in the bead-and-spring model of modes higher than those considered in Iwata's reduction, for which $\delta_p \cong \delta_0$. In Fig. 14b, no contribution of an η_0-proportional internal viscosity has been taken into account.

When nondiffusional processes (class II processes) are assumed to take precedence, we use (95) and (96), with superscripts II. Furthermore, the left-hand members of these equations are now written as $\zeta^{II}(1+\rho_p^{II})$ and $\zeta^{II}\sigma_p^{II}$, respectively. The ρ_p^{II}- and σ_p^{II}-proportional terms are, respectively, η_0-proportional and η_0-independent contributions to internal viscosity.

In the limit of large η_0's, (93) with superscripts II holds. However, the use of (93) instead of (95) in this range of viscosities would simply lead to a difference in numerical coefficients in the expression of the effective friction coefficient ζ of the monomeric unit in terms of molecular parameters. This difference need not be significant and can be disregarded.[166]

If we neglect the η_0-proportional contribution to internal viscosity, $\sigma_p^{II} = n$ is mode independent and equal to the value of $\Phi/(g\zeta)$ as given by (96). This leads to

$$\tau_p' = \tau_p(1+n) \tag{116a}$$

in which

$$n = \frac{2\omega_c^{II}m^{II}}{z^{II}} \tag{116b}$$

Further, the insertion of (95) and (116b) into (92) leads to:

$$1+n = \frac{kT}{\delta_0^{II}\zeta^{II}}(P^{II})^{-1} \tag{117}$$

Equations (113) to (117) are merely the result of a formal reduction of Iwata's coarse-grained diffusion equation to the form of (94). However, several conclusions can be drawn from the preceding discussion.

1. If it is assumed that nondiffusional behavior of local movements is reflected in the slow relaxation times [i.e., if (116a) is inserted into (105b)], there results an η_0-independent internal viscosity term in $(\tan\alpha)_0$ which is proportional to $M[\eta]_0$.[18, 166]

Measurements of the extinction angle of the flow birefringence for nitrocellulose fractions by Tsvetkov and Shtennikova[160] can be interpreted in terms of orientation of stiff molecules, or internal viscosity of moderately flexible chains.* The measured dependence of the corresponding internal viscosity term on both M and η_0 would be in accord with the form of $(\tan\alpha)_0$ just mentioned.

2. The condition for nondiffusional behavior of local movements is that the value of n is not negligible as compared with unity. This is equivalent to the condition that the ratio $z/(2\omega_C m)$ in (92) is not considerably larger than unity. It was pointed out in Section III.B that nondiffusional

---

*For cellulose tricarbanilate fractions, extinction angles obtained at high values of the reduced velocity gradient compare well with those obtained for highly flexible chains.[177] These results are not necessarily incompatible with others, since even at high molecular weight they show deviation from the flexible chain behavior at the lower values of the reduced gradient.

character may be found for low-friction local movements. Reorganizations of this type would maintain other parts of the chain unperturbed. The reorganizations probably would also require substantial cooperativity, hence the crossing of a high-potential barrier, leading to rather low relaxation frequency. If we assume that the one-particle scheme considered in Section III.A remains valid, this would also require a high value of ω_C.* The latter condition, in turn, favors nondiffusional character.

Thus as a rule we can expect cooperativity and nondiffusionality of the local movements in chain molecules to be concomitant properties. It would, of course, be helpful to have a numerical estimate of the value of n. However, it is still difficult to calculate this parameter for a chain molecule.

On the other hand, it is useful to study the ultrasonic relaxation of small molecules exhibiting rotational isomerism in solvents of widely varying viscosities. Experiments which we have recently carried out with Dr. Rogez indicate that small molecules would represent an extreme case of non-diffusional behavior $(n \gg 1)$, since the ultrasonic relaxation frequency appears to be practically independent of the solvent viscosity.

This is consistent with the results of ultrasonic measurements for chain molecules as described in Section III, B. In fact, these measurements show that nondiffusional local processes characterized by rather long relaxation times (of the order of 10^{-7} sec) are likely to play an important role (cf. Section III.B). It is possible that influence of these movements on slow viscoelastic modes, especially in moderately flexible chains, is responsible for internal viscosity effects.

Final proof that these local processes consist in cooperative rearrangements among rotational isomers is, however, still lacking.

One should also keep in mind that the non-diffusional character of segmental motions in chain molecules may be reduced by movements of the adjacent parts. Therefore, we still require further evidence for evaluating the influence of non-diffusional processes upon the slow modes, especially in the more flexible molecules. Evidence from non-newtonian flow of solutions will be discussed in a forthcoming publication.

IV. CONCLUDING REMARKS AND ACKNOWLEDGMENTS

The foregoing presentation has necessarily been incomplete. Even in the restricted area of cooperative conformational kinetics, which we have

---

*The τ_p's and τ_p'''s do not contain the potential barriers explicitly (see (103)). These barriers are nevertheless accounted for in the expression of the internal viscosity term (the second term in (116a)) owing to their effect on ω_c.

attempted to review, many contributions could not be quoted in the scope of this report. In Ref. 14 dielectric studies were analyzed in more detail than in this context. Aspects of the dynamics of helix–coil transitions in biopolymers have been presented recently.[178,179] Among monographs relevant to the subject, we should like to mention those of Birshtein and Ptitsyn[180] and Kinsinger.[181]

I am grateful to Drs. Caloin and Lemaréchal for valuable discussions.

Gauthier-Villars (Paris), the Journal de Physique (Paris), the Society of Polymer Science of Japan (Tokyo), North-Holland Publishing Company (Amsterdam), the Royal Swedish Academy of Science (Stockholm), the Swiss Chemical Society (Basel), Pergamon Press Ltd. (Oxford), and the American Chemical Society (Washington D. C.) have kindly given permission to reproduce figures from their journals.

References

1. L. Pauling, R. B. Corey, and H. R. Branson, *Proc. Nat. Acad. Sci. (U.S.)*, **37**, 205 (1951).
2. B. H. Zimm and J. K. Bragg, *J. Chem. Phys.*, **31**, 526 (1959).
3. P. H. Verdier and W. H. Stockmayer, *J. Chem. Phys.*, **36**, 227 (1962).
4. P. H. Verdier, *J. Chem. Phys.*, **45**, 2118, and 2122 (1966).
5. K. Iwata and M. Kurata, *J. Chem. Phys.*, **50**, 4008 (1969).
6. L. Monnerie and F. Gény, *J. Chim. Phys.*, **66**, 1691 (1969).
7. L. Monnerie, F. Gény, and J. Fouquet, *J. Chim. Phys.*, **66**, 1698 (1969).
8. F. Gény and L. Monnerie, *J. Chim. Phys.*, **66**, 1708 (1969).
9. K. Iwata, *J. Chem. Phys.*, **58**, 4184 (1973).
10. P. E. Rouse, *J. Chem. Phys.*, **21**, 1272 (1953).
11. R. A. Orwoll and W. H. Stockmayer, *Advances in Chemical Physics*, Vol. XV, Interscience, New York, 1969, p. 305.
12. M. V. Volkenstein, *Dokl. Akad. Nauk SSSR*, **78**, 879 (1951).
13. K. Iwata, *J. Chem. Phys.*, **54**, 12 (1971).
14. W. H. Stockmayer, *Pure Appl. Chem. Suppl., Macromol. Chem.*, **8**, 379 (1973).
15. H. A. Kramers, *Physica*, **7**, 284 (1940).
16. W. Kuhn and H. Kuhn, *Helv. Chim. Acta*, **28**, 1533 (1945); **29**, 71 (1946).
17. R. Cerf, *C. R. Acad. Sci. (Paris)*, **250**, 3599 (1960).
18. R. Cerf, *C. R. Acad. Sci. (Paris), ser. B*, **272**, 1143 (1971); *Chem. Phys. Lett.*, **16**, 42 (1972).
19. R. Cerf, *J. Phys. Rad.*, **19**, 122 (1958); *Advances in Polymer Science* Vol. 1, Springer-Verlag, Berlin, 1959, p. 382.
20. B. H. Zimm, *J. Chem. Phys.*, **24**, 269 (1956).
21. R. Cerf, *J. Chim. Phys.*, **48**, 59 (1951).
22. R. Cerf, *J. Polym. Sci.*, **20**, 216 (1956).
23. R. Cerf, *Chem. Phys. Lett.*, **22**, 613 (1973).
24. R. J. Glauber, *J. Math. Phys.*, **4**, 294 (1963).
25. A. Silberberg and R. Simha, *Biopolymers*, **6**, 479 (1968).
26. P. Rabinowitz, A. Silberberg, R. Simha, and E. Loftus, *Advances in Chemical Physics*, Interscience, New York, Vol. XV 1969, p. 281.
27. N. Gô, *J. Phys. Soc. Japan*, **22**, 416 (1967).
28. G. Schwarz, *Ber. Bunsenges. Phys. Chem*, **75**, 40 (1971).
29. G. Schwarz, *Ber. Bunsenges. Phys. Chem*, **68**, 843 (1964).

30. G. Schwarz, *J. Theoret. Biol.*, **36**, 569 (1972).
31. R. Cerf, *C. R. Acad. Sci. (Paris), ser. C*, **278**, 811 (1974).
32. B. Michels and R. Cerf, *C. R. Acad. Sci. (Paris), ser. D*, **274**, 1096 (1972).
33. R. Cerf, *C. R. Acad. Sci. (Paris), ser. D*, **269**, 2140 (1969).
34. M. Eigen, in *Fast Reactions and Primary Processes in Chemical Kinetics*, S. Claesson, Ed., Interscience, New York, 1967, p. 333.
35. D. Pörschke, thesis, Braunschweig, 1968.
36. D. Pörschke and M. Eigen, *J. Mol. Biol.*, **62**, 361 (1971).
37. R. Cerf, *C. R. Acad. Sci. (Paris), ser. D*, **271**, 2403 (1970); **272**, 747 (1971).
38. H. F. Epstein, A. N. Schechter, R. F. Chen, and C. B. Anfinsen, *J. Mol. Biol.*, **60**, 499 (1971).
39. A. Ikai and C. Tanford, *Nature*, **230**, 100 (1971).
40. T. Y. Tsong, R. L. Baldwin, and P. McPhie, *J. Mol. Biol.*, **63**, 453 (1972).
41. G. Schwarz, *Eur. J. Biochem.*, **12**, 442 (1970).
42. A. Silberberg and R. Simha, *Macromolecules*, **5**, 332 (1972).
43. M. Eigen and L. De Maeyer, in *Technique of Organic Chemistry*, Vol. 8, Part 2, A. Weissberger, Ed., Interscience, New York, 1963, p. 895.
44. G. Schwarz, *Rev. Mod. Phys.*, **40**, 206 (1968).
45. G. Schwarz and J. Engel, *Angew, Chem.*, **84**, 615 (1972).
46. M. E. Craig and D. M. Crothers, *Biopolymers*, **6**, 385 (1968).
47. G. Schwarz, *J. Mol. Biol.*, **11**, 64 (1965).
48. R. Cerf, *Suppl. J. Phys.*, **33C6**, 99 (1972).
49. S. Lifson and A. Roig, *J. Chem. Phys.*, **34**, 1963 (1961).
50. J. Applequist, *J. Chem. Phys.*, **38**, 934 (1963).
51. R. Zana, *Suppl. J. Phys.*, **33C6**, 108 (1972).
52. R. Zana and J. Lang, *J. Phys. Chem.*, **74**, 2734 (1970).
53. C. Tondre and R. Zana, *J. Phys. Chem.*, **75**, 3367 (1971).
54. B. Michels, thesis, Université Louis Pasteur, Strasbourg, 1972.
55. T. K. Saksena, B. Michels, and R. Zana, *J. Chim. Phys.*, **65**, 597 (1968).
56. A. B. Barksdale and J. E. Stuehr, *J. Am. Chem. Soc.*, **94**, 3334, (1972).
57. J. Rifkind and J. Applequist, *J. Phys. Chem.*, **86**, 4207 (1964).
58. G. Némethy and H. A. Scheraga, *J. Phys. Chem.*, **66**, 1773 (1962).
59. J. Applequist and V. Damle, *J. Am. Chem. Soc.*, **87**, 1450 (1965).
60. D. Winklmair, J. Engel, and V. Ganser, *Biopolymers*, **10**, 721 (1971).
61. C. Reiss, *J. Mol. Biol.*; to appear.
62. M. Saunders and P. D. Ross, *Biochem. Biophys. Res. Commun.* **3**, 314 (1960).
63. P. J. Flory, *J. Polym. Sci.*, **49**, 105 (1961).
64. N. R. Kallenbach, D. M. Crothers, and R. G. Mortimer, *Biochem. Biophys. Res. Commun.*, **11**, 213 (1963).
65. P. D. Ross and J. Sturtevant, *J. Am. Chem. Soc.*, **84**, 4503 (1962).
66. R. D. Blake, L. C. Klotz, and J. Fresco, *J. Am. Chem. Soc.*, **90**, 3556 (1968).
67. D. M. Crothers, N. Davidson, and N. R. Kallenbach, *J. Am. Chem. Soc.*, **90**, 3560 (1968).
68. R. L. Baldwin, in *Molecular Associations in Biology*, B. Pullman Ed., Academic Press, New York, p. 145.
69. N. R. Amundson, *Mathematical Methods in Chemical Engineering, Matrices and their Application*, Prentice-Hall, Englewood Cliffs, N. J., 1966.
70. B. Ninham, R. Nossal, and R. Zwanzig, *J. Chem. Phys.*, **51**, 5028 (1969).
71. R. Cerf, in *Dynamic Aspects of Conformational Changes in Biological Macromolecules*, C. Sadron Ed., Reidel, Dordrecht, 1972, p. 247.
72. M. Sela, F. H. White, Jr., and C. B. Anfinsen, *Science*, **125**, 691, (1957).

73. C. J. Epstein, R. F. Goldberger, and C. B. Anfinsen, *Cold Spring Harbor Symp. Quant. Biol.*, **28**, 439 (1963).
74. C. B. Anfinsen, *Harvey Lect.*, **61**, 95 (1967).
75. R. Lumry, R. Biltonen, and J. F. Brandts, *Biopolymers*, **4**, 917 (1966).
76. S. Arrhenius, *Immunochemistry*, Macmillan, New York, 1907.
77. C. Tanford, *Advances in Protein Chemistry, Vol. 23*, Academic Press, New York, 1968, p. 121.
78. R. Lumry and R. Biltonen, in *Structure and Stability of Biological Macromolecules*, Vol. 2, S. Timasheff and G. D. Fasman, Eds., Dekker, New York, 1969, p. 65.
79. F. Brandts, in *Structure and Stability of Biological Macromolecules*, Vol. 2, S. Timasheff and G. D. Fasman, Eds., Dekker, New York, 1969, p. 213.
80. D. B. Wetlaufer and S. Ristow, *Ann. Rev. Biochem.*, **42**, 135 (1973).
81. F. Pohl, *Eur. J. Biochem.*, **4**, 373 (1968).
82. F. Pohl, *Eur. J. Biochem.*, **7**, 146 (1969).
83. F. Pohl, *FEBS Lett.*, **3**, 60 (1969).
84. R. Cerf, *Suppl. J. Phys.*, **32C5a**, 275 (1971).
85. F. Pohl, *Angew. Chem., Int. Ed.*, **11**, 894 (1972).
86. D. Blow, in *The Enzymes*, P. D. Boyer, Ed., Vol. 3, Academic Press, New York, 1971, p. 185.
87. D. B. Wetlaufer, *Proc. Nat. Acad. Sci. (U. S.)*, **70**, 697 (1973).
88. A. N. Schechter, R. F. Chen, and C. B. Anfinsen, *Science*, **167**, 886 (1970).
89. A. Ikai, W. W. Fish, and C. Tanford, *J. Mol. Biol.*, **73**, 165 (1973).
90. A. Ikai and C. Tanford, *J. Mol. Biol.*, **73**, 145 (1973).
91. C. Tanford, K. C. Aune, and A. Ikai, *J. Mol. Biol.*, **73**, 185 (1973).
92. T. Y. Tsong, R. L. Baldwin, and E. L. Elson, *Proc. Nat. Acad. Sci. (U. S.)*, **68**, 2712 (1971).
93. T. Y. Tsong and R. L. Baldwin, *J. Mol. Biol.*, **69**, 145 (1972).
94. E. L. Elson, *Biopolymers*, **11**, 1499 (1972).
95. T. Y. Tsong and R. L. Baldwin, *J. Mol. Biol.*, **69**, 149 (1972).
96. J. A. Ferretti, B. W. Ninham, and V. A. Parsegian, *Macromolecules*, **3**, 34 (1970).
97. T. Y. Tsong, R. L. Baldwin, and E. L. Elson, *Proc. Nat. Acad. Sci. (U. S.)*, **69**, 1809 (1972).
98. R. A. Scott and H. A. Scheraga, *J. Am. Chem. Soc.*, **85**, 3866, (1963).
99. M. R. Summers and P. McPhie, *Biochem. Biophys. Res. Commun.*, **47**, 831 (1972).
100. G. C. K. Roberts and O. Jardetsky, *Advances in Protein Chemistry*, Vol. 24, Academic Press, New York, 1970, p. 447.
101. G. C. K. Roberts and F. W. Benz, *Ann. N. Y. Acad. Sci.*, **222**, 130 (1973).
102. A. Allerhand, R. F. Childers, and E. Oldfield, *Ann. N. Y. Acad. Sci.*, **222**, 764 (1973).
103. I. D. Campbell, C. M. Dobson, R. J. P. Williams, and A. V. Xavier, *Ann. N. Y. Acad. Sci.*, **222**, 163 (1973).
104. J. R. Garel and R. L. Baldwin, *Proc. Nat. Acad. Sci. (U. S.)*, **70**, 3347 (1973).
105. R. Cerf, *C. R. Acad. Sci. (Paris), ser. C*, **270**, 1075 (1970).
106. J. Lamb, in *Dispersion and Absorption of Sound by Molecular Processes*, D. Sette, Ed., Academic Press, New York, 1963.
107. H. Hässler and H. J. Bauer, *Disc. Faraday Soc.*, **49**, 238 (1970).
108. J. A. Schellmann, *C. R. Lab. Carlsberg, ser. Chim.*, **29**, 230 (1955).
109. O. Fünfschilling, P. Lemaréchal, and R. Cerf, *C. R. Acad. Sci. Paris, ser. C*, **270**, 659 (1970).
110. S. Glasstone, K. J. Laidler, and H. E. Eyring, *The Theory of Rate Processes*, McGraw-Hill, New York, 1941.

111. P. Lemaréchal, *Acustica*, **24**, 232 (1971).
112. B. Froelich, C. Noël, J. Lewiner, and L. Monnerie, *C. R. Acad. Sci. (Paris), ser. C*, **277**, 1089 (1973).
113. E. Wyn-Jones and W. J. Orwille-Thomas, *Advan. Mol. Relax. Proc.*, **2**, 201 (1972).
114. O. Fünfschilling, P. Lemaréchal, and R. Cerf, *Chem. Phys. Lett.*, **12**, 365 (1971).
115. B. Froelich, thesis, third cycle, Université de Paris VI, 1974.
116. J. G. Kirkwood, *J. Polym. Sci.*, **12**, 1 (1954).
117. J. Leray, *J. Polym. Sci.*, **23**, 167 (1957); *J. Chim. Phys.*, **37**, 323 (1960).
118. R. Cerf, *J. Polym. Sci.*, **12**, 35 (1954).
119. E. R. Bazúa and M. C. Williams, *J. Chem. Phys.*, **59**, 2858 (1973).
120. E. Helfand, *J. Chem. Phys.*, **54**, 4651 (1971).
121. R. Cerf, J. Lang, and S. Candau, *J. Polym. Sci., C*, **7**, 163 (1965).
122. K. Fritzsche, P. Hauptmann, and H. Jung, *Acustica*, **26**, 153 (1972).
123. T. Ohsawa and Y. Wada, *Polym. J.*, **1**, 465 (1970).
124. H. Nomura, S. Kato, and Y. Miyahara, *J. Soc. Mater. Sci. Japan*, **21**, 476 (1972).
125. H. Nomura, S. Kato, and Y. Miyahara, *Nippon Kagaku Kaishi*, **8**, 1191 (1972).
126. P. Lemaréchal, *Chem. Phys. Lett.*, **16**, 495 (1972).
127. N. Ludlow, E. Wyn-Jones, and J. Rassing, *Chem. Phys. Lett.*, **13**, 477 (1972).
128. M. Bader and R. Cerf, *Acustica*, **23**, 31 (1970).
129. C. Tondre and R. Cerf, *J. Chim. Phys.*, **65**, 1105 (1968).
130. B. Froelich, H. Ott, C. Noël, L. Monnerie, and R. Cerf, *in Eur. Polym. J.*, **11**, 15 (1975).
131. M. A. Cochran, J. H. Dunbar, A. M. North, and R. A. Pethrick, *J. C. S. Faraday II*, **70**, 215 (1974).
132. F. Eggers, *Acustica*, **19**, 323 (1967/68).
133. H. Ott and R. Cerf, *C. R. Aca. Sci. (Paris), ser. C*, **278**, 1173, (1974).
134. H. Ott, *Acustica*, **27**, 353 (1972).
135. H. Ott, R. Cerf, B. Michels, and P. Lemaréchal, *Chem. Phys. Lett.*, **24**, 323 (1974).
136. T. Ohsawa and Y. Wada, *Japan J. Appl. Phys.*, **6**, 1351 (1967).
137. T. Ohsawa and Y. Wada, *Japan J. Appl. Phys.*, **8**, 411 (1969).
138. K. Ono, H. Shintani, O. Yano, and Y. Wada, *Polym. J.*, **5**, 164 (1973).
139. D. Biddle and T. Nordström, *Ark. Kem*, **32**, 359 (1970).
140. F. Perrin, *J. Phys. Rad.*, **7**, 390 (1926); *Ann. Phys. (Paris)*, **10**, 169 (1929).
141. B. Valeur, L. Monnerie, and J. P. Jary, *C. R. Acad. Sci. (Paris), ser. C*, **278**, 589 (1974).
142. G. Weill, *C. R. Acad. Sci. (Paris), ser. B*, **272**, 116 (1971).
143. P. Wahl, G. Meyer, and J. Parrod, *Eur. Polym. J.*, **J.6**, 585 (1970).
144. G. Hermann, thesis, Université Louis Pasteur, Strasbourg, 1973.
145. G Hermann, personnal communication.
146. G. Weill and J. P. Bentz, personnal communication.
147. A. Peterlin and H. A. Stuart, *Z. Phys.*, **112**, 1 (1939).
148. The first paper of Ref. 19 contains an error in the calculation of the distribution function for the orientational effect. This error does not affect the values obtained for $(\tan \alpha)_0$ and has been corrected by R. Koyama and R. Cerf, *J. Phys. Rad.*, **21**, 503 (1960).
149. C. E. Chaffey, *J. Chim. Phys.*, **10**, 1379 (1966).
150. R. Cerf, *C. R. Acad. Sci. (Paris)*, **230**, 81 (1950); *J. Chim. Phys.*, **48**, 85 (1951).
151. H. Schwander and R. Cerf, *Helv. Chim. Acta*, **34**, 436 (1951).
152. E. Harrington, in *Polymer Encyclopedia*, Vol. 7 Interscience, New York, 1967, p. 100.
153. T. W. Bates, K. J. Ivin, and G. Williams, *Trans. Faraday Soc.*, **63**, 1964 (1967).
154. W. H. Stockmayer and K. Matsuo, *Macromolecules*, **5**, 766 (1972).
155. H. Janeschitz-Kriegl, *Koll. Z., Z. Polym.*, **203**, 119 (1965).

156. C. Wolff, *J. Chim. Phys.*, **57**, 712 (1960).
157. V. P. Budtov and S. N. Pen'kov, *Polym. Sci. USSR*, **15**, 45 (1973).
158. V. N. Tsvetkov and V. P. Budtov, *Polym. Sci. USSR*, **6**, 17 (1964).
159. V. N. Tsvetkov and V. P. Budtov, *Polym. Sci. USSR*, **6**, 1332 (1964).
160. V. N. Tsvetkov and I. N. Shtennikova, *Polym. Sci. USSR*, **6**, 349, and 1146 (1964).
161. E. Penzel, F. Debeauvais, P. Gramain, and H. Benoit, *J. Chim. Phys.*, **67**, 471 (1970).
162. J. G. Kirkwood and P. L. Auer, *J. Chem. Phys.*, **19**, 281 (1951).
163. R. Cerf, *J. Phys. Rad.*, **13**, 458 (1952).
164. R. Cerf, *J. Chem. Phys.*, **20**, 395 (1952).
165. R. Roscoe, *J. Fluid Mech.*, **28**, 273 (1967).
166. R. Cerf, *Chem. Phys. Lett.*, **24**, 317 (1974).
167. P. E. Rouse and K. Sittel, *J. Appl. Phys.*, **24**, 690 (1953).
168. G. Harrison, J. Lamb, and A. J. Matheson, *J. Phys. Chem.*, **68**, 1072 (1964).
169. J. D. Ferry, L. A. Holmes, J. Lamb, and A. J. Matheson, *J. Phys. Chem.*, **70**, 1685 (1966).
170. D. J. Massa, J. L. Schrag, and J. D. Ferry, *Macromolecules*, **4**, 210 (1971).
171. A. Peterlin, *J. Polym. Sci.*, *A*-2, **5**, 179 (1967).
172. K. Osaki and J. L. Schrag, *Polym. J.*, **2**, 541 (1971).
173. G. B. Thurston and J. L. Schrag, *J. Polym. Sci.*, *A*-2, **6**, 1331 (1968).
174. G. B. Thurston and J. L. Schrag, *J. Chem. Phys.*, **45**, 3373 (1966).
175. A. Peterlin, *Polymer Lett.*, **10**, 101 (1972).
176. A. Peterlin, *J. Polym. Sci.*, Symposium 43, 187 (1973).
177. H. Janeschitz-Kriegl and W. Burchard, *J. Polym. Sci., A-2* **6**, 1953 (1968).
178. T. Tanaka, K. Soda, and A. Wada, *J. Chem. Phys.*, **58**, 5707 (1973).
179. T. Tanaka and M. Suzuki, *J. Chem. Phys.*, **59**, 3795 (1973).
180. T. M. Birshtein and O. B. Ptitsyn, *Conformations of Macromolecules*, Interscience, New York, 1966.
181. J. B. Kinsinger, in *Markov Chains and Monte Carlo Calculations in Polymer Science*, G. G. Lowry, Eds. Dekker, New York, 1970.

THE ELECTRODYNAMICS
OF ATOMS AND MOLECULES

Trinity Hall
Cambridge CB2 1TJ, England

CONTENTS

I. INTRODUCTION

The quantum theory of the interaction of electromagnetic radiation with matter has always been closely associated with the behavior of atoms and molecules in an electromagnetic field. It is notable that the early papers of Dirac (which contain the foundations of quantum field theory and the second quantization formalism) were concerned with the interaction of the

153

electromagnetic field with an atom rather than with a single charged particle, and they showed how one could use the quantum theory to calculate the Einstein A and B coefficients and the Kramers-Heisenberg dispersion formula without recourse to Bohr's correspondence principle and classical electromagnetism.[1,2] Nevertheless it is evident that the inability to set up the theory in accordance with the principle of special relativity was recognized as a serious defect, and Dirac's theory was viewed as a first step toward a relativistic theory of radiation phenomena.[3] Accordingly, after the discovery of the relativistic wave equation, quantum electrodynamics was largely concerned with the development of the covariant electrodynamics of the electron-positron field, although some work continued on the original nonrelativistic theory and its application to atoms and molecules.[4,5] Largely because of the development of the laser, however, in the postwar years there has been renewed interest in low-energy radiation phenomena, and it is the nonrelativistic formalism appropriate to the electrodynamics of atoms and molecules that we discuss here. An extensive account of the application of this formalism to the atom-laser system has been given recently by Stenholm.[6]

The existence of different hamiltonian operators that are unitarily equivalent is one of the most important features of this formalism. In his first paper Dirac[1] had used the electric dipole approximation to write the operator describing the coupling of the atom and the field in terms of the atomic dipole moment

$$\mathcal{H} = H^{\text{atom}} + H^{\text{field}} - \mathbf{d} \cdot \mathbf{E}(0)^{\perp} \qquad (1\text{-}1)$$

$$\mathbf{d} = \sum_{i=1}^{N} e_i \mathbf{R}_i$$

but in the subsequent paper[2] he reverted to the usual classical expression based on the particle momentum $\mathbf{p}_i$ and the vector potential of the electromagnetic field $\mathbf{A}(\mathbf{x})$, which was reinterpreted as an operator

$$H = H^{\text{atom}} + H^{\text{field}} - \frac{1}{2} \sum_{i=1}^{N} \left(\frac{e_i}{m_i} \right) \{ \mathbf{p}_i \cdot \mathbf{A}(\mathbf{R}_i)$$

$$+ \mathbf{A}(\mathbf{R}_i) \cdot \mathbf{p}_i \} + \frac{1}{2} \sum_{i=1}^{N} \left(\frac{e_i^2}{m_i} \right) \mathbf{A}(\mathbf{R}_i) \cdot \mathbf{A}(\mathbf{R}_i) \qquad (1\text{-}2)$$

The elimination of the longitudinal electric field in favor of the instantaneous Coulomb interaction between the charges is implied by the use of the atomic hamiltonian H^{atom}, but this does not carry the implication that the vector potential $\mathbf{A}(\mathbf{x})$ is purely transverse, and the particle momentum $\mathbf{p}_i$ and $\mathbf{A}(\mathbf{R}_i)$ do not necessarily commute. Dirac, however, employed the Coulomb gauge condition $\text{Div}\,\mathbf{A}(\mathbf{x}) = 0$ as a further restriction on the vector potential, in which case the ordering of $\mathbf{p}_i$ and $\mathbf{A}(\mathbf{R}_i)$ in (1-2) is immaterial. Some years later Göppert-Mayer derived both these expressions starting from a classical lagrangian and the principle of least action, and she demonstrated that in the electric dipole approximation the two hamiltonians were connected by a canonical transformation.[7]

This argument, which had been based on semiclassical radiation theory, was rediscovered by Richards[8] but was first explored in the context of molecular quantum electrodynamics by Power and co-workers,[9,10] who clarified many of the details of the previous work. They showed that if the Coulomb gauge condition is imposed as a subsidiary condition on the vector potential, it is possible to relate the conventional hamiltonian H and the "multipole" hamiltonian, $\mathcal{H}$ by a unitary transformation

$$\mathcal{H} = \Lambda H \Lambda^{-1} \qquad |\Psi\rangle' = \Lambda |\Psi\rangle$$

$$\Lambda = \exp\left(-\frac{iS}{\hbar}\right) \qquad S = \int d^3\mathbf{x}\,\mathbf{P}(\mathbf{x}) \cdot \mathbf{A}(\mathbf{x})$$

$$\text{Div}\,\mathbf{A}(\mathbf{x}) = 0 \tag{1-3}$$

where $\mathbf{P}(\mathbf{x})$, the electric polarization field for the atom, is related to the atomic charge density $\rho(\mathbf{x})$,

$$\rho(\mathbf{x}) = -\boldsymbol{\nabla} \cdot \mathbf{P}(\mathbf{x}) \tag{1-4}$$

Power and Zienau[10] used a multipole series representation of the polarization field $\mathbf{P}(\mathbf{x})$. Although they only retained terms in the hamiltonian $\mathcal{H}$ up to the electric quadrupole and the magnetic dipole contributions in a multipole expansion, their work showed how one might dispense with a truncated multipole expansion, and so formulate without approximation a theory equivalent to the conventional hamiltonian H, which includes retardation factors. This step was taken by Atkins and Woolley[11] who, following Power and Zienau, expressed the electric and magnetic polarization fields as infinite multipole series, to generalize directly the lagrangian approach of Göppert-Mayer.[7] It was recognized subsequently that the multipole series could be summed up exactly into an integral representation and that the transformation operator Λ could be expressed in terms of a path integral over the transverse vector potential

$$\Lambda = \exp\left\{ -\left(\frac{i}{\hbar}\right) \sum_{i=1}^{N} e_i \int_{\mathbf{r}}^{\mathbf{R}_i} d\mathbf{z} \cdot \mathbf{A}(\mathbf{z}) \right\} \tag{1-5}$$

In this expression **r** is an arbitrary point that may conveniently be chosen to coincide with the center of mass of the atomic system, and each integral in the sum is taken over the straight line path joining **r** and $\mathbf{R}_i$, the position coordinate of particle i.

An essentially equivalent approach is to write the interaction term in the lagrangian using the path integral representation of the electric and magnetic polarization fields; the interaction term can then be expressed as the integral of the Lorentz force taken over the path just described, which is a natural generalization of a well-known result in electrostatics and is valid even though the Lorentz force is not a conservative force.[12] The lagrangian thus becomes intrinsically path dependent, but as we learn later, different choices of path are equivalent to within a canonical (or unitary) transformation. If there were no path dependence, there would be no interaction between the charge and the electromagnetic field. As shown by Woolley,[13,14] an important consequence of the use in the unitary transformation of the exact path dependent operator, (1-5), is that it permits the making of a complete identification of all the contributing terms in the resulting hamiltonian $\mathcal{H}$. In the earlier treatments based on the use of a multipole series, the correct identification of the Coulomb energies and certain self-energies was not achieved.

It is noteworthy that phase factors of the form displayed in (1-5) have had a long history in quantum electrodynamics. A relativistic generalization appropriate to a single charge e may be written as

$$\exp\left\{ \frac{ie}{\hbar} \int_{x_1}^{x_2} dx_\mu A(x)^\mu \right\} \tag{1-6}$$

where x_1 and x_2 are two space-time points, and $A(x)^\mu$ is the four-potential of the electromagnetic field $A(x)^\mu = \mathbf{A}(\mathbf{x}, t), \phi(\mathbf{x}, t)$. This expression was used in the early subtraction (renormalization) procedures of Dirac,[15] Peierls,[16] and Fock,[17] and in more recent discussions of the electrodynamics of the Dirac equation due to Schwinger[18] and Valatin,[19] since it was found that apart from this phase factor, the formalism could be expressed in terms of integrals over the electromagnetic field tensor $f_{\mu\nu} = (\partial_\nu A^\mu - \partial_\mu A^\nu)$, allowing the establishment of a gauge-invariant calculus. More recently this phase factor has been discussed in terms of the arbitrariness of the phase of the wavefunction of charged particles in the presence of an electromagnetic field that manifests itself in, for example, the Bohm-Aharonov effect.[20–23] It also provides a natural way of setting up a gauge-invariant electrodynamics of sources of finite size[24] and of introducing an explicit nonlocality into field theories.[25,26] Recalling that the nonrelativistic hamiltonian $\mathcal{H}$ can be expressed in terms of integrals over the field variables,[13] we see at once the close formal connection between the diverse relativistic field theories just mentioned and the transformed non-relativistic molecular theory. The similarity is obscured by the use of a truncated multipole expansion of the polarization field $\mathbf{P}(\mathbf{x})$, but when this is avoided and the transformation operator Λ is expressed as in (1-5), the formal analogy becomes transparently clear.

The exact nonrelativistic hamiltonian $\mathcal{H}$ for a single atom interacting with the radiation field can be written as

$$\mathcal{H} = H^{\text{atom}} + H^{\text{field}} - \frac{1}{2} \sum_{i=1}^{N} \left(\frac{e_i}{m_i} \right) \left\{ \mathbf{p}_i \cdot \tilde{\mathbf{A}}(\mathbf{R}_i) + \tilde{\mathbf{A}}(\mathbf{R}_i) \cdot \mathbf{p}_i \right\}$$

$$+ \frac{1}{2} \sum_{i=1}^{N} \left(\frac{e_i^2}{m_i} \right) \tilde{\mathbf{A}}(\mathbf{R}_i) \cdot \tilde{\mathbf{A}}(\mathbf{R}_i) + \int d^3\mathbf{x}\, \mathbf{P}(\mathbf{x})^{\perp} \cdot \mathbf{P}(\mathbf{x})^{\perp}$$

$$- \int d^3\mathbf{x}\, \mathbf{P}(\mathbf{x}) \cdot \mathbf{E}(\mathbf{x})^{\perp}$$

where $\tilde{\mathbf{A}}(\mathbf{R}_i)$ is the integral over the magnetic induction defined in (1-7) below and $\mathbf{P}(\mathbf{x})^{\perp}$ is the transverse component of the atomic polarization field operator.[13]

The path integral representation of Λ should be adopted as the fundamental form irrespective of whether a multipole expansion (Taylor series expansion) of the integrand is possible, as implied in the original derivations;[11,13] this means that the transformation may be taken to apply to unbound particle states (continuum states) as well as to bound atomic and molecular states. It is important to recognize a fundamental distinction between the interpretations of the operator Λ in semiclassical electrodynamics and quantum electrodynamics because of the possibility of *virtual photon* excitations in the quantum field theory. Whereas Λ is obviously unitary in the case of an external electromagnetic field (i.e., only real photon processes are admitted), this is not strictly true in quantum electrodynamics, since in gauges in which the vector potential is *not* orthogonal to the atomic polarization field $\mathbf{P}(\mathbf{x})$, the contributions of the virtual photons cause S to be unbounded and therefore not hermitian.[24,26] The two procedures should therefore be thought of as being equivalent only to the extent that they both lead to a multipolar representation of the interaction terms. It may also be remarked that a phase factor of a similar character is used in quantum chemistry in the definition of so-called gauge-invariant atomic orbitals,[27] and in the "momentum translation" approximation, which Reiss introduced recently to discuss multiphoton processes in atoms.[28] In these applications, however, the phase factor is not path dependent.

If we regard S of (1-3) as the generating function of a gauge transformation

$$\tilde{\mathbf{A}}(\mathbf{R}_i) = \mathbf{A}(\mathbf{R}_i) + \nabla_i S$$

from some arbitrary gauge, we find that the new vector potential may be written as an integral over the magnetic induction[19,29]

$$\tilde{\mathbf{A}}(\mathbf{R}_i) = \int_0^1 ds\, s\, \mathbf{B}(\mathbf{r} + s\mathbf{D}_i) \times \mathbf{D}_i \qquad (1\text{-}7)$$

where $\mathbf{D}_i = \mathbf{R}_i - \mathbf{r}$. It is readily verified that $\tilde{\mathbf{A}}(\mathbf{R}_i)$ may be regarded as a vector potential, since it is easy to see that the usual relation $\mathbf{B} = \text{Curl}\,\mathbf{A}$ is satisfied with $\mathbf{A}$ given by (1-7).[19] From the definition we may also infer the simple result

$$\mathbf{D}_i \cdot \mathbf{A}(\mathbf{R}_i) = 0 \qquad (1\text{-}8)$$

The vector potential $\tilde{\mathbf{A}}(\mathbf{R}_i)$ can therefore be considered to be a potential at the point $\mathbf{R}_i$ such that its component along the straight line connecting $\mathbf{R}_i$ to another fixed point in space $\mathbf{r}$ vanishes by (1-8). As remarked by Dirac and quoted in the paper by Valatin just cited, the last condition suffices to fix the gauge of the potential, and $\tilde{\mathbf{A}}(\mathbf{R}_i)$ can be characterized in this way by this condition. It is of course optional whether (1-7) is regarded as a vector potential or instead the magnetic induction is used as the fundamental dynamical variable. The manifest nonlocality of (1-7) is an important feature, since this nonlocality enables us to describe the Bohm-Aharonov effect[20] without explicit reference to the field potentials.[12] In the case of a homogeneous field, (1-7) reduces at once to $\mathbf{A} \propto \mathbf{B} \times \mathbf{R}_i$, which is a well-known result, and gauge transformations of this kind for homogeneous, static fields are of some importance in the simplification of molecular hamiltonians.[30–32]

Recently the method of obtaining the multipole interaction hamiltonian for charges interacting with the electromagnetic field has been reconsidered in the framework of semiclassical electrodynamics in which the (external) field is assumed to be determined by the Maxwell equations.[33] It was argued that with an appropriate choice of gauge one may pass directly to a multipolar representation, thus eliminating the need to perform a unitary transformation of the usual Coulomb gauge theory to achieve this result. With the aid of the remarks of the previous paragraph, a simple interpretation of this suggestion is possible, since it is easy to verify that the vector potential proposed by Barron and Gray consists of the leading terms of a multipole expansion about $\mathbf{r}$ of the potential $\tilde{\mathbf{A}}(\mathbf{R}_i)$ in (1-7).[29,34] Furthermore, in view of (1-8) we see that each of the phase integrals involved in the unitary transformation operator Λ, (1-5), vanishes separately

$$\int_{\mathbf{r}}^{\mathbf{R}_i} d\mathbf{z} \cdot \mathbf{A}(\mathbf{z}) = 0 \tag{1-9}$$

if we put $\mathbf{A}(\mathbf{z}) \equiv \tilde{\mathbf{A}}(\mathbf{z})$. Thus the proposal of Barron and Gray[33] amounts to choosing the gauge of the vector potential in such a way that the operator S vanishes, reducing Λ to the identity operator.[34] It has been shown that (1-9) can be used as a supplementary condition that leads essentially to the hamiltonian $\mathcal{H}$ without recourse to the usual Coulomb gauge hamiltonian and the unitary transformation.[29] This is not as simple a matter as in semiclassical electrodynamics, since in the canonical formalism the field potentials are to be regarded as dynamical variables. Therefore one must obtain the Poisson-bracket relations of these potentials with the other dynamical variables, and since their Poisson brackets are not gauge invariant, a method is required for finding the Poisson brackets that are appropriate to the chosen gauge. This in turn implies the use of hamiltonian methods more sophisticated (but not more difficult!) than are commonly described in elementary accounts of radiation theory.

Since these hamiltonian schemes only differ, in effect, by the choice of gauge, one would expect them to lead to the same results when employed in calculations. The conventional Coulomb gauge formalism, however, contains several rather displeasing features. For example, although there is apparently a clear separation between the static atomic binding energies arising from the longitudinal electric field and the radiation, which is always transverse and propagates at the speed of light, the propagation function for the Coulomb gauge vector potential is not properly retarded. Many years ago Kennard[35] showed that the solution of the hamiltonian equation of motion for the Coulomb gauge vector potential contains nonretarded, static terms in addition to the usual Lienart-Wiechert expression (this calculation is discussed in Section V). When the hamiltonian H in (1-2) is generalized to describe several atomic systems interacting with radiation, one must not forget to include the static intermolecular Coulomb interaction potential V_{ab}^{Cou},

$$H = \sum_{a} H_a^{\mathrm{atom}} + H^{\mathrm{field}} + \sum_{a \neq b} V_{ab}^{\mathrm{Cou}}$$

$$- \sum_{a,i=1}^{N} \left(\frac{e_{ia}}{m_{ia}} \right) \mathbf{p}_{ia} \cdot \mathbf{A}(\mathbf{R}_{ia}) + \frac{1}{2} \sum_{a,i=1}^{N} \left(\frac{e_{ia}^2}{m_{ia}} \right) A(\mathbf{R}_{ia})^2$$

$$\mathrm{Div}\,\mathbf{A}(\mathbf{x}) = 0 \tag{1-10}$$

Using quantum electrodynamics to calculate the shift in energy between two atoms due to their mutual interaction via the electromagnetic field, we find that the contribution from the static intermolecular potential V_{ab}^{Cou} is canceled identically by a static contribution arising from the transverse vector potential; thus an overall retarded result is eventually obtained.[36] On the other hand, it is known that if the hamiltonian $\mathcal{H}$ in (1-1) is used for the same purpose, it takes the form (see Section VI)

$$\mathcal{H} = \sum_a H_a^{\text{atom}} + H^{\text{field}} - \sum_a \mathbf{d}(a) \cdot \mathbf{E}(a)^\perp$$

$$- \sum_a \mathbf{m}(a) \cdot \mathbf{B}(a) + \sum_a \underline{\chi}(a) : \mathbf{B}(a)\mathbf{B}(a) + \cdots \qquad (1\text{-}11)$$

($\mathbf{m}$ is the magnetic dipole operator and $\underline{\chi}$ is the diamagnetic susceptibility) in which there are no static intermolecular Coulomb energies, and a retarded result is obtained immediately.[37] In the limit of small separations this result naturally reduces to the usual interaction energy computed using only the Coulomb potential.

A more serious difficulty arises when we consider processes off the energy shell—for example, the radiative decay of excited atomic states. Power and Zienau[10] have pointed out the existence of spurious static contributions when studying the emission of light from an excited state by following the time evolution of the system in the Schrödinger representation using the Coulomb gauge formalism with the hamiltonian (1-2). These difficulties are avoided entirely by using instead the gauge defined in (1-9), since the hamiltonian $\mathcal{H}$ leads directly to the expected physical field development. As a rule, only the lower-order multipole contributions are required in calculations in molecular physics; thus we believe that the hamiltonian $\mathcal{H}$ should be used in preference to the conventional result, (1-2), for both practical and formal reasons.

The aim of this chapter is to give a comprehensive account of the formalism of quantum electrodynamics as it applies to atoms and molecules. We start from the classical Maxwell-Lorentz theory in the nonrelativistic approximation (Section II) and neglect the spin interactions, since spin seems to be so intimately connected with both quantum mechanics and special relativity that it is not possible to give more than an ad hoc account in a nonrelativistic framework. After describing the hamiltonian mechanics of dynamical fields and the problem of lagrangian degeneracy (Section III), we illustrate in Section IV the degenerate canonical formalism in a discussion of the electromagnetic field in the absence of charges (an essentially perturbation theoretic point of view!). Section V is given over to a detailed account of the classical and quantum electrodynamics of a closed system of a single charge and the electromagnetic field in the nonrelativistic approximation. Although it may be argued that this should really be done in a fully covariant manner, we believe that this nonrelativistic theory is of interest because it provides the basis for the many-particle theory, appropriate to atoms and molecules, which is discussed in Section VI, and for which there is no known covariant theory.

To keep the discussion as general as possible, we make only isolated references to the molecular multipoles; in any event, the (exact) integral representations of the polarization fields afford a genuinely more compact notation than the rather clumsy infinite multipole series. We also feel that it is desirable to describe the unsatisfactory features of the theory as well as the successes, and this leads us to consider the question of self-interactions in some detail. Even in covariant electrodynamics the divergences in perturbation theory cannot be avoided, and they appear to signal an inconsistency in the formulation of electrodynamics so fundamental that only a radical revision of our ideas could lead to a theory without infinities. A finite theory of the electrodynamics of atoms and molecules seems to be a very distant prospect; equally, there seems to be little scope for modification of the existing formalism within the general framework described here, and this perhaps is the main reason for writing this account now.

II. CLASSICAL ELECTROMAGNETISM: THE MAXWELL–LORENTZ THEORY

II.A. The Maxwell Equations

The essential conceptual framework of the classical theory of charged particles is due to Lorentz,[38] who developed a microscopic theory of electrodynamics in which the interaction between the charges is mediated by the electromagnetic field. Although we do not pursue the arguments here, it is important that suitable statistical averaging procedures can be developed which if applied to the microscopic Lorentz theory enable us to recover the familiar laws of macroscopic (bulk) electrodynamics—for example, the Maxwell equations.[39,40] A striking feature of the theory is the important role of the electromagnetic *field*, and although Lorentz found difficulties with notions such as the structure of charged particles, his theory is the precursor of all modern classical and quantum theories of electrodynamics that are not based on action at a distance.[41] We therefore begin with an account of the essentials of the Maxwell-Lorentz theory.

Since our ultimate aim is to describe the interaction of electromagnetic radiation with molecular systems, we confine ourselves to a nonrelativistic approximation throughout; that is, the theory is appropriate for particles with velocities of magnitudes v that satisfy the inequality $(v/c)^2 \ll 1$, where c is the velocity of light. This approximation is a necessary consequence of the unavailability of any known covariant extension of the Dirac equation to molecular systems; it appears that attempts to construct relativistic quantum theories of N-particle systems ($N > 1$ and finite) usually founder on the absence of conservation of particle numbers in the relativistic regime.[42] Because the velocity of light is involved explicitly in the Newtonian inequality, the Maxwell equations have no nonrelativistic reduction, and a theory of charged particles in interaction with the electromagnetic field can have a nonrelativistic approximation only insofar as the particles are concerned. The resulting theory of interacting charges and radiation possesses neither the complete symmetry of the Galilean group characteristic of the Newtonian dynamics of point particles, nor that of the relativistic Lorentz group, which is the symmetry group of the Maxwell equations. We make no attempt here to take account of the spin interaction terms that cannot be derived naturally in a nonrelativistic theory and must be grafted onto the theory in an ad hoc (but plausible) way.[43]

The electromagnetic field is described by the electric field intensity vector $\mathbf{E}(\mathbf{x}, t)$ and the magnetic induction vector $\mathbf{B}(\mathbf{x}, t)$. In the (unobservable) idealized state characterized by the absence of charged particles, the dynamical behavior of the electromagnetic field is governed by the following set of partial differential equations involving the field variables $\mathbf{E}(\mathbf{x}, t)$ and $\mathbf{B}(\mathbf{x}, t)$:

$$\nabla \cdot \mathbf{B}(\mathbf{x}, t) = 0 \tag{2-1a}$$

$$\nabla \cdot \mathbf{E}(\mathbf{x}, t) = 0 \tag{2-1b}$$

$$\nabla \times \mathbf{E}(\mathbf{x}, t) = -\frac{\partial \mathbf{B}(\mathbf{x}, t)}{\partial t} \tag{2-1c}$$

$$\nabla \times \mathbf{B}(\mathbf{x}, t) = c^{-2} \frac{\partial \mathbf{E}(\mathbf{x}, t)}{\partial t} \tag{2-1d}$$

It is easily verified that the solutions of these equations are two wave fields, propagating with velocity c, which are orthogonal to each other and to the propagation direction, that is, they are purely transverse waves in accordance with the conditions expressed by (2-1a) and (2-1b). The six components of the two vector fields $\mathbf{E}(\mathbf{x}, t)$ and $\mathbf{B}(\mathbf{x}, t)$ are not all independent, however, because of relations between them implied by (2-1). As a first step in the reduction of this functional redundancy, we introduce the so-called field potentials $\phi(\mathbf{x}, t)$ and $\mathbf{A}(\mathbf{x}, t)$ through the relations

$$\mathbf{B}(\mathbf{x}, t) = \nabla \times \mathbf{A}(\mathbf{x}, t) \tag{2-2a}$$

$$\mathbf{E}(\mathbf{x}, t) = -\frac{\partial \mathbf{A}(\mathbf{x}, t)}{\partial t} - \nabla \phi(\mathbf{x}, t) \tag{2-2b}$$

which with the aid of several well-known vector identities are seen to be entirely consistent with (2-1). The $\phi(\mathbf{x}, t)$ and $\mathbf{A}(\mathbf{x}, t)$ are known conventionally as the scalar and vector potentials. A more revealing representation of the relationship between the fields $(\mathbf{E}, \mathbf{B})$ and the potentials $(\phi, \mathbf{A})$ is achieved by writing down the integral solutions of (2-2); if we define $T(\mathbf{x}, t) = \nabla \cdot \mathbf{A}(\mathbf{x}, t)$, we may write the general solutions for $\mathbf{A}(\mathbf{x}, t)$ and $\phi(\mathbf{x}, t)$ in the form[44]

$$\mathbf{A}(\mathbf{x}, t) = \mathbf{a}(\mathbf{x}, t) + \nabla K(\mathbf{x}, t) \tag{2-3a}$$

$$\phi(\mathbf{x}, t) = \nabla(\mathbf{x}, t) - \frac{\partial K(\mathbf{x}, t)}{\partial t} \tag{2-3b}$$

where

$$\mathbf{a}(\mathbf{x}, t) = \frac{\int d^3z [\nabla \times \mathbf{B}(\mathbf{z}, t)]}{(4\pi |\mathbf{z} - \mathbf{x}|)} \tag{2-4a}$$

$$V(\mathbf{x}, t) = \frac{\int d^3z [\nabla \cdot \mathbf{E}(\mathbf{z}, t)]}{(4\pi |\mathbf{z} - \mathbf{x}|)} \tag{2-4b}$$

$$K(\mathbf{x}, t) = -\frac{\int d^3z \, T(\mathbf{z}, t)}{(4\pi |\mathbf{z} - \mathbf{x}|)} \tag{2-4c}$$

Since $K(\mathbf{x}, t)$ is entirely arbitrary, we see that there are many equivalent choices of $\phi(\mathbf{x}, t)$ and $\mathbf{A}(\mathbf{x}, t)$ that satisfy (2-2) and therefore describe the same physical situation; the transformation described by (2-3) which connects one set of potentials to another is usually known as a guage transformation. Note that in the special case of the field in the absence of charged particles, (2-1b) implies that $V = 0$, and since $\nabla \cdot \mathbf{a}(\mathbf{x}, t)$ vanishes identically, we can reduce the free field potentials to the form

$$\phi(\mathbf{x}, t) = 0 \tag{2-5a}$$

$$\mathbf{A}(\mathbf{x}, t) \equiv \mathbf{a}(\mathbf{x}, t) \tag{2-5b}$$

That is, the free electromagnetic field may be characterized completely by the transverse vector field $\mathbf{a}(\mathbf{x}, t)$ defined by (2-4a).

To describe the interaction of the electromagnetic field with matter, we introduce the material charge and current densities $\rho(\mathbf{x}, t)$ and $\mathbf{j}(\mathbf{x}, t)$, respectively. The Maxwell equations for the field in interaction with a given charge-current distribution then take the form

$$\nabla \cdot \mathbf{B}(\mathbf{x}, t) = 0 \tag{2-6a}$$

$$\epsilon_0 \nabla \cdot \mathbf{E}(\mathbf{x}, t) = \rho(\mathbf{x}, t) \tag{2-6b}$$

$$\nabla \times \mathbf{E}(\mathbf{x}, t) = -\frac{\partial \mathbf{B}(\mathbf{x}, t)}{\partial t} \tag{2-6c}$$

$$\epsilon_0 c^2 \nabla \times \mathbf{B}(\mathbf{x}, t) = \frac{\epsilon_0 \partial \mathbf{E}(\mathbf{x}, t)}{\partial t} + \mathbf{j}(\mathbf{x}, t) \tag{2-6d}$$

If we decompose the electric field $\mathbf{E}(\mathbf{x}, t)$ into its transverse and longitudinal components $\mathbf{E}(\mathbf{x}, t) = \mathbf{E}(\mathbf{x}, t)^\perp + \mathbf{E}(\mathbf{x}, t)^\|$, where $\nabla \cdot \mathbf{E}(\mathbf{x}, t)^\perp = \nabla \times \mathbf{E}(\mathbf{x}, t)^\| \equiv 0$, it is evident from (2-6b) that the longitudinal component of the field arises from the charged particles. The implication of (2-3) and (2-6), however, is that the charges are *not* responsible for all electromagnetic phenomena, since the general solutions of the differential equations (2-6) will contain an arbitrary solution of the corresponding homogeneous equations (2-3), and it is this property which distinguishes Maxwell's theory from the action-at-a-distance philosophy. In the quantum theory these additional solutions are interpreted as arising from fluctuations of the vacuum state.[1,3] An important consequence of these equations follows from the recognition that spatial and time differentiations are commutative. This permits us to deduce a conservation law from the Maxwell equations involving $\rho(\mathbf{x}, t)$ and $\mathbf{j}(\mathbf{x}, t)$; since DivCurl vanishes identically, we deduce at once from (2-6b) and (2-6d) that

$$\nabla \cdot \mathbf{j}(\mathbf{x}, t) + \frac{\partial \rho(\mathbf{x}, t)}{\partial t} = 0 \tag{2-7}$$

which is the law of conservation of charge in differential form.

II.B. The Lorentz Force Law

An alternative procedure that is especially valuable when closed, non-conducting systems are considered is the introduction of two *polarization* fields in place of the charge and current densities, through the relations

$$\rho(\mathbf{x}, t) = -\nabla \cdot \mathbf{P}(\mathbf{x}, t) \tag{2-8a}$$

$$\mathbf{j}(\mathbf{x}, t) = \frac{d\mathbf{P}(\mathbf{x}, t)}{dt} + \nabla \times \mathbf{M}(\mathbf{x}, t) \tag{2-8b}$$

The $\mathbf{P}(\mathbf{x}, t)$ and $\mathbf{M}(\mathbf{x}, t)$ are usually called the electric and magnetic polarization fields, respectively. Although these quantities have previously been used only in the classical (macroscopic) theory of dielectrics, we see later that they are also useful auxiliary quantities in the theory of molecular quantum electrodynamics, provided always that we can construct explicit representations of them. The total time derivative in (2-8b) is intentional and is to be distinguished from the *partial* time derivative used in dielectric theory. It should be noted that (2-8) does not define $\mathbf{P}(\mathbf{x}, t)$ and $\mathbf{M}(\mathbf{x}, t)$ uniquely because new fields $\mathbf{P}(\mathbf{x}, t)'$ and $\mathbf{M}(\mathbf{x}, t)'$ defined by

$$\mathbf{P}(\mathbf{x}, t)' = \mathbf{P}(\mathbf{x}, t) + \nabla \times \mathbf{C}(\mathbf{x}, t) \tag{2-9a}$$

$$\mathbf{M}(\mathbf{x}, t)' = \mathbf{M}(\mathbf{x}, t) - \frac{d\mathbf{C}(\mathbf{x}, t)}{dt} \tag{2-9b}$$

where $\mathbf{C}(\mathbf{x}, t)$ is an arbitrary (differentiable) vector field, are equally acceptable; the transverse component of the electric polarization field $\mathbf{P}(\mathbf{x}, t)$ is therefore arbitrary. The freedom of choice in the polarization fields implied by (2-8) is completely analogous to the arbitrariness in the field potentials implied by (2-2), and it arises because of the conservation of electric charge.

We shall suppose that an elementary charge has no structure and may be taken to be a spinless point particle characterized solely by its charge e and its mass m. Since the charge and current densities are nonzero only at the position of the charge, we may express them in terms of the Dirac delta function;[45] for a collection of N charges we obtain

$$\rho(\mathbf{x}, t) = \sum_{i=1}^{N} e_i \delta^3(\mathbf{R}_i - \mathbf{x}) \tag{2-10}$$

$$\mathbf{j}(\mathbf{x}, t) = \sum_{i=1}^{N} e_i \dot{\mathbf{R}}_i \delta^3(\mathbf{R}_i - \mathbf{x}) \tag{2-11}$$

where $\mathbf{R}_i$ is the position vector of the ith particle and $\dot{\mathbf{R}}_i$ is its velocity. The corresponding expressions for the polarization fields are less familiar but may be obtained from (2-8), (2-10), and (2-11) to within an arbitrary transverse vector field. Historically it has been the practice to represent the polarization fields with the first few terms of their multipole expansions.[10] This procedure is not necessary, however, and in general investigations of the kind to be developed here, it is not desirable. It has been shown that the multipole series can be summed up exactly into an integral representation, which moreover is valid even when the multipole expansion is not appropriate; for example, one can define the electric and magnetic polarization fields for a single charged particle that has no multipole moments.[12, 13, 46, 47] In the case of a molecule, the most usual reference point for the multipole moments is the center of mass coordinate $\mathbf{r}$,

$$\mathbf{r} = \frac{\left\{ \sum_{i=1}^{N} m_i \mathbf{R}_i \right\}}{\left\{ \sum_{i=1}^{N} m_i \right\}} \tag{2-12}$$

This coordinate is also involved in the usual definition of the polarization fields for an overall neutral collection of charges

$$\mathbf{P}(\mathbf{x}, t) = \sum_{i=1}^{N} e_i \int_{\mathbf{r}}^{\mathbf{R}_i} d\mathbf{z} \, \delta^3(\mathbf{z} - \mathbf{x}) \tag{2-13}$$

$$\mathbf{M}(\mathbf{x}, t) = \sum_{i=1}^{N} e_i \int_{\mathbf{r}}^{\mathbf{R}_i} (d\mathbf{z} \times \dot{\mathbf{z}}) \, \delta^3(\mathbf{z} - \mathbf{x}) \tag{2-14}$$

In these expressions $d\mathbf{z}$ is an element of path on the straight line joining the vectors $\mathbf{r}$ and $\mathbf{R}_i$, and $\dot{\mathbf{z}} \equiv (d\mathbf{z}/dt)$ is to be interpreted as $(\mathbf{R}_i \cdot \nabla_i)\mathbf{z}$. We emphasize that this choice of the reference point $\mathbf{r}$, and of the path, is entirely a matter of convenience, since different choices are unitarily equivalent;[12] the choice of the center of mass as the reference point carries with it no dynamical implications because the reference point is to be understood as a *fixed* point in space.

In the general case the paths $\mathbf{z}$ are obtained as follows: for any position $\mathbf{R}(t)$ of the particle at a given instant in time t, we choose a curve $\mathbf{z}$ with an arbitrary starting point $\mathbf{r}$ (discussed below) and ending at the point $\mathbf{R}$. The paths to be specified are purely curves in space and have no time sequence attached to them; thus to obtain the equations of motion from the usual action principle using the polarization fields, we must prescribe first the position of the particle as a function of time, and in addition, for each point a separate curve $\mathbf{z}$. The latter curves must not be confused with the actual path of the particle during its motion.

In explicit calculations involving the polarization fields it is convenient to work with a parametric representation of the curve $\mathbf{z}$: we therefore introduce a parametric form $\mathbf{z} = \mathbf{z}(s)$, with $\min s = s_1$ giving $\mathbf{z} = \mathbf{r}$, and $\max s = s_2$ giving $\mathbf{z} = \mathbf{R}$; other values of s give points on the rest of the curve. The derivative $(\partial z^n / \partial R^p)$—which, for example, is involved in the evaluation of $\dot{\mathbf{z}}$—may then be obtained at once. Since $\mathbf{z}$ is only defined by the set of values of s, there is no meaning to be associated with any variation of $\mathbf{z}$ (or $\mathbf{r}$, or $\mathbf{R}$) that does not keep it on the curve. It is clear therefore that the only acceptable variations in the path lead to the result

$$\left(\frac{\partial z^n}{\partial R^p} \right) = f(s)\delta_{np}, \qquad f(s_1) = 0, \qquad f(s_2) = 1, \qquad n, p = x, y, z \quad (2\text{-}15)$$

For example, it is evident that if we start at $\mathbf{z} = \mathbf{R}$, the only $d\mathbf{z}$ that we are allowed to introduce must be such that $d\mathbf{z}$ and $d\mathbf{R}$ lie in the same direction (the tangent direction to the curve at this point), thus $d\mathbf{z} \equiv d\mathbf{R}$ there. The function $f(s)$ need not be specified in any more detail except that we suppose it to have (at least) a continuous first derivative.

With these results we can see under what circumstances (2-13) and (2-14) are consistent with (2-8). We use the usual summation convention for the components of vectors and tensors and employ the antisymmetric unit tensor E_{lmn} which has the property of vanishing if any two of its subscripts are equal and alternates in value (± 1) as its subscripts are permuted. Then the polarization fields for a single charge e may be written in the form

$$P(\mathbf{x}, t)^r = e \int_{s_1}^{s_2} ds \left(\frac{\partial z^r}{\partial s} \right) \delta^3(\mathbf{z} - \mathbf{x}) \qquad (2\text{-}16)$$

$$M(\mathbf{x}, t)^r = e\mathsf{E}_{rus}\dot{R}^s \int_{s_1}^{s_2} ds \left(\frac{\partial z^u}{\partial s} \right) f(s)\delta^3(\mathbf{z} - \mathbf{x}) \qquad (2\text{-}17)$$

The total time derivative of $P(\mathbf{x}, t)$ is easily obtained, since the time variation is due to the implicit time dependence of $\mathbf{R}$, and after an integration by parts, it may be written

$$\frac{dP(\mathbf{x}, t)^r}{dt} = e\dot{R}^r f(s)\delta^3(\mathbf{z} - \mathbf{x}) \Big|_{s=s_1}^{s=s_2}$$

$$- e\dot{R}^s \mathsf{E}_{prt}\mathsf{E}_{pus} \left(\frac{\partial}{\partial x^t} \right) \int_{s_1}^{s_2} ds\, f(s) \left(\frac{\partial z^u}{\partial s} \right) \delta^3(\mathbf{z} - \mathbf{x}) \qquad (2\text{-}18)$$

Provided the magnetization is defined as in (2-17), this equation is identical to the current equation (2-8b).

The divergence of the electric polarization field (2-16) is

$$\nabla \cdot \mathbf{P}(\mathbf{x}, t) = -e\delta^3(\mathbf{z} - \mathbf{x})\Big|_{s=s_1}^{s=s_2}$$

$$= -\rho(\mathbf{x}, t) + e\delta^3(\mathbf{x} - \mathbf{r}) \tag{2-19}$$

The additional term $e\delta^3(\mathbf{x} - \mathbf{r})$ corresponds to an "image" charge fixed at the arbitrary point $\mathbf{r}$; since it is static, it does not couple with the radiation field and so may be neglected in radiation problems (this is clear because the term makes no appearance in the current equation). To eliminate $e\delta^3(\mathbf{x} - \mathbf{r})$ entirely, one could for example choose $\mathbf{r} =$ spatial infinity, which is a natural choice for a single charge and is a procedure analogous to imposing a gauge condition on the field potentials. If, however, we have a collection of charged particles that is overall electrically neutral ($\Sigma_i e_i = 0$), the corresponding term derived from (2-19)—namely, $\Sigma_i e_i \delta^3(\mathbf{x} - \mathbf{r})$— vanishes by virtue of the electrical neutrality. In this situation, which is the usual case in molecular physics, the choice of $\mathbf{r}$ is of no consequence and no restrictions need be imposed.

To complete the Lorentz theory we need the equation of motion for a charged particle in the presence of a specified electromagnetic field $\{\mathbf{E}(\mathbf{x}, t), \mathbf{B}(\mathbf{x}, t)\}$. The equation of motion of a particle with momentum $\mathbf{p}$ is given by Newton's second law with the Lorentz force[48]

$$\dot{\mathbf{p}} = e\{\mathbf{E}(\mathbf{R}) + \dot{\mathbf{R}} \times \mathbf{B}(\mathbf{R})\} \tag{2-20}$$

which in the nonrelativistic regime reduces to the familiar form

$$m\ddot{\mathbf{R}} = e\{\mathbf{E}(\mathbf{R}) + \dot{\mathbf{R}} \times \mathbf{B}(\mathbf{R})\} \tag{2-21}$$

Together, (2-6) and (2-20) constitute the basic equations of the nonrelativistic Maxwell-Lorentz theory. These equations are not in the most convenient form for obtaining a quantum theory, however, and we reformulate them in the framework of classical canonical mechanics, permitting the derivation of both equations from a single hamiltonian function. We then obtain a quantum theory in the Heisenberg representation by means of the conventional reinterpretation of the classical Poisson-bracket relations as operator-commutator relations.[49] It is important to realize that the Maxwell-Lorentz equations (2-6) and (2-20) do not give a complete theory: they are only valid for specified charge distributions and fields, respectively, whereas it is clear physically that the sources and fields must be driven through each other's motion. We shall see that the canonical formalism is more general and contains terms that correspond to radiation damping. Although the classical radiation damping term is $O(v/c)^3$ and is thus strictly in the relativistic regime, its presence is required to maintain energy and momentum conservation. The preservation of these two conservation laws is well worthwhile however, since the conservation of the total energy implies that time-independent perturbation methods can be used in the quantum theory, and this leads to considerable simplification in such calculations in molecular physics as light-scattering cross-sections.

III. DEGENERATE LAGRANGIANS AND THE CANONICAL FORMALISM

III.A. The Euler-Lagrange Equation

The basic ideas of the lagrangian formulation of classical mechanics may be summarized as follows.[48] We define a quantity S, called the action, by the integral relation

$$S = \int_{t_1}^{t_2} L(R_i, \dot{R}_i, t)\, dt, \qquad i = 1, 2, \ldots, n \tag{3-1}$$

and use the calculus of variations to find the extremum values of S for arbitrary variations of the coordinates and velocities R_i and $\dot{R}_i$. We now assert that with an appropriate choice of the quantity L (the "lagrangian"), the conditions ensuring that the variations δS vanish, lead to the equations of motion for the dynamical system. If we so restrict the variation procedure that the variations vanish at t_1 and t_2, we have

$$\delta S = \int_{t_1}^{t_2} \left\{ \frac{\partial L}{\partial R_i} \delta R_i + \frac{\partial L}{\partial \dot{R}_i} \delta \dot{R}_i \right\} dt, \qquad i = 1 \ldots n$$

$$= \int_{t_1}^{t_2} \left\{ \frac{\partial L}{\partial R_i} - \frac{d}{dt}\left(\frac{\partial L}{\partial \dot{R}_i} \right) \right\} \delta R_i\, dt, \qquad i = 1 \ldots n \tag{3-2}$$

after an integration by parts and the use of the boundary conditions on the variations. The condition $\delta S = 0$ therefore implies the system of equations

$$0 = \frac{\partial L}{\partial R_i} - \frac{d}{dt}\left(\frac{\partial L}{\partial \dot{R}_i} \right), \qquad i = 1 \ldots n \tag{3-3}$$

which are known as the Euler-Lagrange equations, or as the lagrangian equations of motion. It is not difficult to see that two lagrangian functions, which differ only by a total time derivative

$$L' = L - \left(\frac{dF}{dt} \right) \tag{3-4}$$

lead to the same Euler-Lagrange equation; thus there is no one unique lagrangian corresponding to a given physical system.

For systems with an infinite number of degrees of freedom such as a field, we modify the formalism somewhat and introduce a so-called lagrangian density $\mathcal{L} = \mathcal{L}(V, V'^\alpha)$ expressed in terms of the field variables $V(x)$ and their derivatives $V'^\alpha = (\partial V(x)/\partial x_\alpha)$, $(x_0 = t, x_1, x_2, x_3 = \mathbf{x})$. The action integral is now written

$$S = \int_{t_1}^{t_2} \mathcal{L}(V, V'^\alpha) \, d\Omega, \qquad d\Omega = d^3\mathbf{x} \, dt \tag{3-5}$$

and the Euler-Lagrange equation is

$$\frac{\partial \mathcal{L}}{\partial V} - \left(\frac{\partial \mathcal{L}}{\partial V'^\alpha} \right)^{'\alpha} = 0 \tag{3-6}$$

or in vector notation,

$$\left(\frac{\partial \mathcal{L}}{\partial V} \right) - \frac{\partial}{\partial t} \left(\frac{\partial \mathcal{L}}{\partial (\partial V/\partial t)} \right) - \frac{\partial}{\partial \mathbf{x}} \left(\frac{\partial \mathcal{L}}{\partial (\nabla V)} \right) = 0 \tag{3-7}$$

Although it is not an essential procedure, it is very useful to transform from the lagrangian description to a hamiltonian description which implies the elimination of the lagrangian velocities in favor of the canonical momenta $= \partial L/\partial \dot{R}$. The quantization of hamiltonian systems, which is quite well understood and is widely known, is of course the normal practice in the quantum mechanics of atoms and molecules. It is possible for one or more of the canonical momenta to vanish (i.e., to have a lagrangian L which is independent of the corresponding velocities), and this situation requires a more general mathematical discussion than is usually found in textbooks of classical mechanics. The condition, referred to as lagrangian degeneracy, is the usual one in field theory.

III.B. Lagrangian Degeneracy

The theory of degenerate lagrangians was first studied by Dirac,[50–52] and we discuss and employ his procedures because they afford the most direct method of calculation in electrodynamics. The reader who is interested in more rigorous and comprehensive discussions of the mathematics of degenerate lagrangians is referred to the specialist literature.[53-55] For the sake of simplicity we initially discuss dynamical systems with a finite number (n) of degrees of freedom that may be described by a lagrangian $L = L(R_i, \dot{R}_i, t)$; the extension $n \to \infty$ is straightforward.[56]

From the variation of the action integral, we deduce the Euler-Lagrange equations (3-3). In the familiar case of elementary dynamics, these equations may be solved for the accelerations $\ddot{R}_i$ in terms of the coordinates and velocities, and one is lead to the Hessian matrix $\mathbf{J}$:

$$\mathbf{J}\ddot{\mathbf{R}} = \mathbf{B} \tag{3-8}$$

where

$$\mathbf{J} = \left\| \frac{\partial^2 L}{\partial \dot{R}_i \partial \dot{R}_j} \right\| \tag{3-9}$$

$$B_i = \left(\frac{\partial L}{\partial R_i} \right) - \left(\frac{\partial^2 L}{\partial \dot{R}_i \, dt} \right) - \left(\frac{\partial^2 L}{\partial \dot{R}_i \partial R_j} \right) \dot{R}_j \tag{3-10}$$

and $\mathbf{B}$ and $\ddot{\mathbf{R}}$ are $n \times 1$ column matrices. It is possible for the Hessian matrix to have a rank less than n, in which case L is said to be *degenerate*. Degeneracy reduces the differential order of some of the Lagrange equations and introduces the possibility that some equations may no longer be independent of the others: the remaining second-order Lagrange equations can no longer be solved for all the accelerations. Whereas for physical systems of finite n the usual theory is sufficient, field theories almost invariably involve degenerate lagrangians and the concomitant degenerate variational problem.[52,53] The dynamical consequences of the degeneracy may most conveniently be examined in the canonical formalism; in this way, one also obtains a basis for quantization because the Poisson brackets of the canonical dynamical variables define a Lie algebra suited to the quantization procedure.

Initially the domain of the lagrangian is taken to be the space $\mathcal{R}$ of $(2n + 1)$ dimensions: $\mathcal{R} = (t, R_n, \dot{R}_n)$. The momentum conjugate to the coordinate R_i is defined in the normal manner by

$$p_i = \frac{\partial L}{\partial \dot{R}_i}, \qquad i = 1, 2, \ldots, n \tag{3-11}$$

and this expression may be regarded as a mapping of $\mathcal{R}$ into the $(2n + 1)$-dimensional phase space $\mathcal{P} = (t, R_n, p_n)$. In the usual case the Hessian matrix $\mathbf{J}$, which now may be written $\| \partial p_i / \partial \dot{R}_j \|$ has rank n and the mapping is one–one, at least locally. The lagrangian is then said to be normal.[55] In the degenerate case the rank of $\mathbf{J}$ is reduced from n to $n-r$, which means that there must exist r distinct relations of the form

$$\psi_v(t, R_n, p_n) = 0, \qquad v = 1, 2, \ldots, r \qquad (3\text{-}12)$$

In this case the mapping of $\mathcal{R}$ into $\mathcal{P}$ is many-one (degenerate) and the whole of $\mathcal{R}$ is mapped onto the points of a $(2n+1-r)$-dimensional hypersurface $\mathcal{S}$, embedded in $\mathcal{P}$. The equations $\psi_v = 0$ are the *constraint equations* of Dirac's theory,[50-52] and since they are only valid in $\mathcal{S}$, they are called *weak equations* and the equality sign $=$ is replaced by $\approx$ to imply this restricted validity; equations that are true irrespective of these conditions are *strong equations* and are written with the normal equality sign. Thus (3-12) is written

$$\psi_v(t, R_n, p_n) \approx 0 \qquad (3\text{-}13)$$

The hamiltonian H may still be defined by the usual relation

$$H(R_n, p_n) \approx p_n \dot{R}_n - L \qquad (3\text{-}14)$$

but the relation must be written as a weak equation because the coordinates and momenta are restricted by (3-13). In Dirac's theory we look for the additional relations that ensure that the ψ_v are true for all times: these conditions may further delimit the dimensions of the hypersurface to which the actual physical motion is confined; this new space we write $\mathcal{S}'$. Thus we are led to a set of coherence conditions $\psi_\mu (\mu = 1, 2, \ldots, s; \; n \geqslant s \geqslant r)$ that satisfies $\dot{\psi}_\mu \approx [\psi_\mu, H] \approx 0$ and enable us to define the hamiltonian in $\mathcal{S}'$ as a strong equation. The resulting hamiltonian may however contain a number (not greater than s) of arbitrary terms, which arise because not all the canonical dynamical variables are independent, thus cannot determine uniquely the state of the system.

A fundamental classification of the dynamical variables is into the sets *first class* and *second class*: a dynamical variable $V(R_n, p_n)$ is first class if its Poisson brackets with all the constraints vanish at least weakly; otherwise it is second class. It may be shown that constraints that are themselves first class are relations that generate invariance groups of the system; that is, their importance is that they may be used as generating functions of contact transformations that leave unaltered the physical state of the system. It is just these quantities that give rise to the arbitrary terms in the total hamiltonian, since the existence of invariant relations is closely related to the functional dependence of the dynamical variables. In electrodynamics, for example, we find one invariant relation, and this is responsible for the existence of the gauge transformations in the theory.[57]

Finally we note that a dynamical variable may only correspond to a physical observable if it belongs to the set of first-class variables. In the following section we discuss the hamiltonian description of the electromagnetic field in the absence of charges; this example serves to illustrate the ideas briefly discussed here before passing on to the hamiltonian formulation of the electrodynamics of charged particles.

IV. THE CANONICAL DESCRIPTION OF THE RADIATION FIELD

IV.A. The Hamiltonian

Since only first-order derivatives of the field variables $\{\mathbf{E}(\mathbf{x},t), \mathbf{B}(\mathbf{x},t)\}$ occur in the Maxwell equations, it is usual to use the scalar and vector potentials $\phi(\mathbf{x},t)$ and $\mathbf{A}(\mathbf{x},t)$ as the lagrangian "coordinates" for the electromagnetic field. This is because the least action formulation requires that the equations of motion be (at least) second-order differential equations in the lagrangian variables. In the absence of charges, the Euler-Lagrange equations (3-7) must coincide with the free-field Maxwell equations (2-1). It is readily verified that a suitable choice for the lagrangian density is[37]

$$\mathcal{L} = \frac{\epsilon_0}{2} \left\{ E(\mathbf{x},t)^2 - c^2 B(\mathbf{x},t)^2 \right\}$$

$$= \frac{\epsilon_0}{2} \left\{ \left(\boldsymbol{\nabla}\phi(\mathbf{x},t) + \frac{\partial \mathbf{A}(\mathbf{x},t)}{\partial t} \right)^2 \right.$$

$$\left. - c^2 (\boldsymbol{\nabla} \times \mathbf{A}(\mathbf{x},t))^2 \right\} \qquad (4\text{-}1)$$

in terms of the field potentials. Now for a finite-dimensional system the momentum conjugate to the coordinate Q is defined in the normal way to be $(\partial L / \partial \dot{Q})$; for an infinite-dimensional field described by coordinates $V(\mathbf{x},t)$, the momentum densities are the coefficients of $\delta(\partial V(x,t)/\partial t)$ in the variation of the lagrangian with respect to the "velocities"

$$\delta_{\text{vel}} L = \int d^3\mathbf{x}\, U(\mathbf{x},t)\delta\left(\frac{\partial V(\mathbf{x},t)}{\partial t} \right) \qquad (4\text{-}2)$$

It is immediately apparent that $\mathcal{L}$ in (4-1) is independent of $\partial\phi(\mathbf{x},t)/\partial t$; thus the momentum $\Pi_0(\mathbf{x},t)$ conjugate to the scalar potential formally vanishes and the lagrangian system is degenerate. This result must therefore be written with the weak equality sign introduced in Section III

$$\Pi_0(\mathbf{x},t) \approx 0 \qquad (4\text{-}3)$$

and the remaining nonzero momenta and the hamiltonian must be written initially as weak equations. A straightforward calculation shows that $\Pi(\mathbf{x}, t)$, the momentum conjugate to the vector potential $\mathbf{A}(\mathbf{x}, t)$, is given by

$$\Pi(\mathbf{x}, t) \approx \epsilon_0 \left\{ \nabla \phi(\mathbf{x}, t) + \frac{\partial \mathbf{A}(x, t)}{\partial t} \right\} \tag{4-4}$$

The field variables $\phi(\mathbf{x}, t)$, $\mathbf{A}(\mathbf{x}, t)$, and their conjugate momenta are assumed to have only the following nonvanishing equal time Poisson brackets:

$$\left[A(\mathbf{x})^r, \Pi(\mathbf{x}')^s \right] = \delta_{rs} \delta^3(\mathbf{x} - \mathbf{x}') \tag{4-5}$$

$$\left[\phi(\mathbf{x}), \Pi_0(\mathbf{x}') \right] = \delta^3(\mathbf{x} - \mathbf{x}') \tag{4-6}$$

According to the theory of the previous section, the hamiltonian may still be defined in the usual way as

$$H \approx \int d^3\mathbf{x} \Pi(\mathbf{x}) \cdot \frac{\partial \mathbf{A}(\mathbf{x})}{\partial t} + \int d^3\mathbf{x} \Pi_0(\mathbf{x}) \frac{\partial \phi(\mathbf{x})}{\partial t} - L$$

$$= \frac{\epsilon_0}{2} \int d^3\mathbf{x} \left\{ \epsilon_0^{-2} \Pi(\mathbf{x})^2 + c^2 (\nabla \times \mathbf{A}(\mathbf{x}))^2 \right\} \tag{4-7}$$

$$- \epsilon_0^{-1} \int d^3\mathbf{x} \, \Pi(\mathbf{x}) \cdot \nabla \phi(\mathbf{x}) \tag{4-8}$$

where we have already made use of (4-3) to drop the term in $\Pi_0(\mathbf{x})$. An integration by parts on the last term in (4-8) brings the hamiltonian to the form

$$H \approx \frac{\epsilon_0}{2} \int d^3\mathbf{x} \left\{ \epsilon_0^{-2} \Pi(\mathbf{x})^2 + c^2 (\nabla \times \mathbf{A}(\mathbf{x}))^2 \right\}$$

$$+ \epsilon_0^{-1} \int d^3\mathbf{x} \phi(\mathbf{x}) (\nabla \cdot \Pi(\mathbf{x})) \tag{4-9}$$

We must now determine the consistency equations imposed by (4-3); the first condition required is that which ensures that $\Pi_0(\mathbf{x}, t)$ vanishes at all times. Since the equation of motion of a dynamical variable Ω may be written $\dot{\Omega} \approx [\Omega, H]$, we deduce at once that

$$\dot{\Pi}_0(\mathbf{x}, t) \approx \epsilon_0^{-1} \nabla \cdot \Pi(\mathbf{x}) \tag{4-10}$$

and thus infer that the right-hand side of this equation must vanish at least weakly for consistency. It is easy to verify that

$$\mathbf{\nabla} \cdot \mathbf{\Pi}(\mathbf{x}) \approx 0 \qquad (4\text{-}11)$$

is consistent with the field equations obtained with the aid of the Euler-Lagrange equations. Since (4-11) must also be true for all times, the bracket $[\mathbf{\nabla} \cdot \mathbf{\Pi}(\mathbf{x}), H]$ must vanish; it is readily verified that with H given by (4-9) this condition reduces to $0 = 0$, and so we have determined all the conditions required to make the theory consistent. The equations of constraint are simply (4-3) and (4-11), and using the fundamental Poisson-bracket relations (4-5) and (4-6), we can show that all their mutual Poisson brackets vanish

$$[\Pi_0(\mathbf{x}), \Pi_0(\mathbf{x}')] = [\Pi_0(\mathbf{x}'), \mathbf{\nabla} \cdot \mathbf{\Pi}(\mathbf{x})] = [\mathbf{\nabla} \cdot \mathbf{\Pi}(\mathbf{x}), \mathbf{\nabla}' \cdot \mathbf{\Pi}(\mathbf{x}')] = 0$$

That is, the equations of constraint are *first class*.

According to the general theory, the total hamiltonian H' may be obtained by adding all the first-class constraints with arbitrary coefficients to the hamiltonian H (because the first-class constraints commute with all physical observables the motion generated by H' remains in the surface $\mathbb{S}'$). The new hamiltonian is therefore defined in $\mathbb{S}'$ by the strong equation

$$H' = H + \int d^3\mathbf{x}\, u(\mathbf{x}) \Pi_0(\mathbf{x}) + \int d^3\mathbf{x}\, v(\mathbf{x})(\mathbf{\nabla} \cdot \mathbf{\Pi}(\mathbf{x})) \qquad (4\text{-}12)$$

$u(\mathbf{x})$ and $v(\mathbf{x})$ being arbitrary functions of space and time. When this hamiltonian is used, the equation of motion of a dynamical variable may be written as the strong equation

$$\dot{\Omega} = [\Omega, H'] \qquad (4\text{-}13)$$

If Ω is chosen to be $\phi(\mathbf{x})$, the equation of motion (4-13) implies that the time variation of $\phi(\mathbf{x})$ is arbitrary, since

$$\dot{\phi}(\mathbf{x}) \equiv [\phi(\mathbf{x}), H'] = u(\mathbf{x}) \qquad (4\text{-}14)$$

and $u(\mathbf{x})$ is an arbitrary function. We have already seen that $\Pi_0(\mathbf{x}) \approx 0$ always; therefore neither $\phi(\mathbf{x})$ nor $\Pi_0(\mathbf{x})$ is of interest in the description of the system and may be discarded from the theory. If a new arbitrary coefficient $w(\mathbf{x}) = v(\mathbf{x}) - \phi(\mathbf{x})\epsilon_0^{-1}$ is introduced, the hamiltonian may be written in the form

$$H' = \frac{\epsilon_0}{2} \int d^3\mathbf{x} \left\{ \epsilon_0^{-2} \mathbf{\Pi}(\mathbf{x})^2 + c^2 (\mathbf{\nabla} \times \mathbf{A}(\mathbf{x}))^2 \right\}$$
$$+ \int d^3\mathbf{x}\, w(\mathbf{x})(\mathbf{\nabla} \cdot \mathbf{\Pi}(\mathbf{x})) \qquad (4\text{-}15)$$

The hamiltonian in (4-15) is sufficient to give the equations of motion for all the variables of physical interest; the variables $\phi(\mathbf{x})$ and $\Pi_0(\mathbf{x})$ no longer appear or play any role in the theory. Note however that although we may specify completely the initial values of the dynamical variables, their values at later times computed by way of the hamiltonian equations of motion may not be specified uniquely because of the arbitrary coefficient $w(\mathbf{x})$. The last term in the hamiltonian leads to an arbitrary change in the vector potential $\mathbf{A}(\mathbf{x})$ which may be written as[51]

$$\delta A(\mathbf{x})^r = \int d^3\mathbf{x}'[A(\mathbf{x})^r, \nabla' \cdot \mathbf{\Pi}(\mathbf{x}')]w(\mathbf{x}')$$

$$= (\nabla w(\mathbf{x}))^r \tag{4-16}$$

The magnetic induction $\mathbf{B}(\mathbf{x})$, which is the physical observable, is however unchanged, since $\mathrm{Curl}\,\mathrm{Grad}\,w(\mathbf{x})$ vanishes. The quantity $(\nabla \cdot \mathbf{\Pi}(\mathbf{x}))$ may therefore be identified as the generating function of an infinitesimal contact transformation that leads to a new description of the same physical state; from the form of (4-16) it is clear that the transformation is simply a gauge transformation. The condition that $[\nabla \cdot \mathbf{\Pi}(\mathbf{x}), H'] \approx 0$ reduces to $0 = 0$ is of course precisely the condition that guarantees the invariance of the hamiltonian description to such transformations and motivates the inclusion of the term in $\nabla \cdot \mathbf{\Pi}(\mathbf{x})$ in (4-15) to give the most general hamiltonian.

A quantum theory furnishes no rules for the construction of the function $w(\mathbf{x})$, which should be interpreted as an operator, but it may be eliminated by a procedure first given by Dirac[50-52] and rationalized in later theories.[53,57] We recall that the term involving $w(\mathbf{x})$ appears because the constraint equation (4-11), being first class, could be used as a generator of infinitesimal transformations in the surface $\mathcal{S}'$ that left the physical state unchanged. If a supplementary condition on the vector potential $\mathbf{A}(\mathbf{x})$ is added to the theory, it may be written as an additional constraint equation $\psi[\mathbf{A}(\mathbf{x})] \approx 0$; if ψ is chosen so that its Poisson bracket with $\nabla \cdot \mathbf{\Pi}(\mathbf{x})$ is not zero, both constraints are *second class*, and the term in $w(\mathbf{x})$ must be eliminated from the hamiltonian H' because $\nabla \cdot \mathbf{\Pi}(\mathbf{x})$ may now induce migrations away from the surface $\mathcal{S}'$. Thus the term in $w(\mathbf{x})$ is discarded and we are left with the problem of quantizing a canonical system involving two second-class constraints. This is the type of problem that Dirac was the first to solve: the essential point of his theory is the redefinition of the Poisson brackets of the dynamical variables so that the new Poisson brackets (the "Dirac" brackets) of a dynamical variable with the (now) second-class constraints, vanish. This enables him to write the second-class constraints as strong equations. The new Poisson brackets still lead to the expected equations of motion and preserve the Lie algebra of the dynamical variables.[57]

The usual supplementary condition to introduce into the theory is the Coulomb gauge condition $\nabla \cdot \mathbf{A}(\mathbf{x}) = 0$, and for the moment this is the only possibility we consider. Let us write the two second-class constraints as

$$\psi_1(\mathbf{x}) = \nabla \cdot \mathbf{\Pi}(\mathbf{x}) \approx 0$$

$$\psi_2(\mathbf{x}) = \nabla \cdot \mathbf{A}(\mathbf{x}) \approx 0$$

(4-17)

and so construct the elements of a matrix $\Psi(\mathbf{x}, \mathbf{x}'')$ from their Poisson brackets

$$\Psi(\mathbf{x}, \mathbf{x}'')_{ij} = [\psi_i(\mathbf{x}), \psi_j(\mathbf{x}'')], \qquad i, j = 1, 2$$

(4-18)

This matrix has nonzero elements of the form

$$[\psi_1(\mathbf{x}), \psi_2(\mathbf{x}'')] = \nabla^2 \delta^3(\mathbf{x} - \mathbf{x}'')$$

(4-19)

and since it is nonsingular, its inverse Ψ^{-1} exists in $\mathcal{S}'$ and is defined by

$$\int d^3x \, \Psi(\mathbf{x}'', \mathbf{x})_{ij} \Psi(\mathbf{x}, \mathbf{x}')_{jk}^{-1} = \delta_{ik} \delta^3(\mathbf{x}' - \mathbf{x}'')$$

(4-20)

From (4-19) and (4-20) we thus obtain the elements of the inverse matrix as

$$\Psi_{11}^{-1} = \Psi_{22}^{-1} = 0$$

$$\Psi(\mathbf{x}, \mathbf{x}')_{12}^{-1} = -\Psi(\mathbf{x}, \mathbf{x}')_{21}^{-1} = \frac{1}{(4\pi|\mathbf{x} - \mathbf{x}'|)}$$

(4-21)

The Dirac bracket of two dynamical variables g and h, which we write $[g, h]^*$ is defined as[50-52]

$$[g, h]^* = [g, h] - \int d^3x \int d^3x' [g, \psi_i(\mathbf{x})] \Psi(\mathbf{x}, \mathbf{x}')_{ij}^{-1} [\psi_j(\mathbf{x}'), h] \quad (4\text{-}22)$$

and according to its definition it is antisymmetric, $[g, h]^* = -[h, g]^*$, since the inverse matrix Ψ^{-1} has the symmetry property $\Psi(\mathbf{x}', \mathbf{x})_{ji}^{-1} = -\Psi(\mathbf{x}, \mathbf{x}')_{ij}^{-1}$. It may be shown that the Dirac brackets also satisfy the Jacobi identity and so have exactly the same formal algebraic properties as the usual Poisson brackets.[57] It is readily verified that with this new definition of the bracket relation, the Dirac brackets of the constraint equations with any dynamical variable vanish automatically, for example,

$$[g,\psi_1(\mathbf{z})]^* = [g,\psi_1(\mathbf{z})] - \int d^3\mathbf{x} \int d^3\mathbf{x}'[g,\psi_i(\mathbf{x})]\Psi(\mathbf{x},\mathbf{x}')_{ij}^{-1}[\psi_j(\mathbf{x}'),\psi_1(\mathbf{z})]$$

which may be written as

$$[g,\psi_1(\mathbf{z})]^* = [g,\psi_1(\mathbf{z})] - \int d^3\mathbf{x} \int d^3\mathbf{x}'[g,\psi_i(\mathbf{x})]\Psi(\mathbf{x},\mathbf{x}')_{ij}^{-1}\Psi(\mathbf{x}',\mathbf{z})_{j1}$$

The integration over $\mathbf{x}'$ may be performed with the aid of (4-20), giving

$$[g,\psi_1(\mathbf{z})]^* = [g,\psi_1(\mathbf{z})] - \int d^3\mathbf{x}[g,\psi_1(\mathbf{x})]\delta^3(\mathbf{x}-\mathbf{z})$$

$$\equiv 0. \quad \text{Q.E.D.} \tag{4-23}$$

Thus provided we always work with the new definition of the bracket operation, we may write the equations of constraint, (4-17), as strong equations

$$\nabla \cdot \mathbf{\Pi}(\mathbf{x}) \equiv 0$$

$$\nabla \cdot \mathbf{A}(\mathbf{x}) \equiv 0 \tag{4-24}$$

and reduce the hamiltonian H' to the form

$$H'' = \frac{\epsilon_0}{2}\int d^3\mathbf{x}\left\{\epsilon_0^{-2}\mathbf{\Pi}(\mathbf{x})\cdot\mathbf{\Pi}(\mathbf{x}) + c^2(\nabla\times\mathbf{A}(\mathbf{x}))^2\right\} \tag{4-25}$$

The only nonzero Dirac bracket is that between $\mathbf{A}(\mathbf{x})$ and its conjugate $\mathbf{\Pi}(\mathbf{x})$, and using (4-22) we find

$$[A(\mathbf{x})^r,\mathbf{\Pi}(\mathbf{x}')^s]^* = \delta_{rs}\delta^3(\mathbf{x}-\mathbf{x}') - \nabla_r\nabla'_s|\mathbf{x}-\mathbf{x}'|^{-1}$$

$$\equiv \delta_{rs}^{\perp}(\mathbf{x}-\mathbf{x}') \tag{4-26}$$

which is the well-known Coulomb gauge equal-time Poisson bracket for the vector potential and its conjugate.[37] The usual elementary discussion of the canonical formulation of electrodynamics starts from the assumption of the existence of gauge transformations and introduces at the outset a gauge condition (such as the Coulomb gauge condition), together with a postulated Poisson bracket for the vector potential and its conjugate, appropriate to the given gauge. Here, however, all these features have followed as a consequence of the choice of the original lagrangian density. Although we have introduced the idea of the Dirac bracket in a heuristic fashion, this would not be necessary in a more thorough and mathematical account of the lagrangian degeneracy,[53,57] and provided we can calculate the inverse matrix defined in (4-20), we have the means of finding the Poisson brackets appropriate to any gauge. Of course the choice of guage should not affect the results of calculations, but as suggested in Section I, we find when we come to discuss the theory of the radiation field in the presence of charges that not all gauges are equally convenient; in particular, the familiar Coulomb gauge formalism is known to possess several ugly features that do not arise in the gauge defined by (1-9).[10,29]

IV.B. The Equations of Motion

We now investigate the equations of motion for $\mathbf{A}(\mathbf{x})$ and $\mathbf{\Pi}(\mathbf{x})$ that follow from (4-25) and (4-26). The canonical scheme is now normal, and the equation of motion is written as a strong equation in terms of the reduced hamiltonian H''

$$\dot{\Omega} = [\Omega, H'']^* \qquad (4\text{-}27)$$

where the * indicates that the Dirac bracket (4-26) is to be used in evaluating all brackets. Explicit calculation shows

$$\dot{\mathbf{A}}(\mathbf{x}) \equiv [\mathbf{A}(\mathbf{x}), H'']^* = \epsilon_0^{-1} \mathbf{\Pi}(\mathbf{x})^\perp \qquad (4\text{-}28)$$

$$\dot{\mathbf{\Pi}}(\mathbf{x}) \equiv [\mathbf{\Pi}(\mathbf{x}), H'']^* = -\epsilon_0 c^2 \nabla \times (\nabla \times \mathbf{A}(\mathbf{x})) \qquad (4\text{-}29)$$

from which we derive the usual wave equation

$$\left\{ \frac{c^{-2}\partial^2}{\partial t^2} + \nabla \times (\nabla \times \right\} \mathbf{A}(\mathbf{x}) = 0 \qquad (4\text{-}30)$$

which must be solved subject to the transversality condition (4-24). If we make the identification $\mathbf{\Pi}(\mathbf{x}) = -\epsilon_0 \mathbf{E}(\mathbf{x})$, it is obvious that (4-28) and (4-29) are consistent with the dynamical Maxwell equations for the field in the absence of charges.

We can obtain a solution of this equation as a superposition of plane waves if we confine the system to a cubic box of side L, demand that $\mathbf{A}(\mathbf{x})$ be periodic on the surfaces, and later let $L \to \infty$. Then we can write

$$\mathbf{A}(\mathbf{x}, t) = \frac{1}{\sqrt{\epsilon_0 V}} \sum_{\mathbf{k}} \left\{ \mathbf{q}^c(\mathbf{k}) e^{i(\mathbf{k} \cdot \mathbf{x} - wt)} + \mathbf{q}^c(\mathbf{k})^* e^{-i(\mathbf{k} \cdot \mathbf{x} - wt)} \right\} \qquad (4\text{-}31)$$

where the frequency $w = |\mathbf{k}|c$. In the limit of an infinite system the sum over $\mathbf{k}$ is converted to an integration by the rule

$$\sum_{\mathbf{k}} \Rightarrow \frac{V}{(2\pi)^3} \int d^3\mathbf{k} \qquad (4\text{-}32)$$

The vector Fourier coefficients $\mathbf{q}^c$, which are constants of the motion, must evidently have the property

$$\mathbf{q}^c \cdot \mathbf{k} = \mathbf{q}^{c*} \cdot \mathbf{k} = 0 \qquad (4\text{-}33)$$

in order that (4-24) be satisfied.

A more useful representation may be obtained if we introduce new time-dependent coefficients $\mathbf{q}(\mathbf{k})$ by the relation

$$\mathbf{q}(\mathbf{k}) = \mathbf{q}^c(\mathbf{k})e^{-iwt} \qquad (4\text{-}34)$$

so that the solution (4-31) becomes

$$\mathbf{A}(\mathbf{x},t) = \frac{1}{\sqrt{\epsilon_0 V}} \sum_{\mathbf{k}} \{\mathbf{q}(\mathbf{k}) + \mathbf{q}^*(-\mathbf{k})\} e^{i\mathbf{k}\cdot\mathbf{x}} \qquad (4\text{-}35)$$

Then with the aid of the equation of motion (4-28) we find

$$\mathbf{\Pi}(\mathbf{x},t) = -i\sqrt{\frac{\epsilon_0}{V}} \sum_{\mathbf{k}} w\{\mathbf{q}(\mathbf{k}) - \mathbf{q}^*(-\mathbf{k})\} e^{i\mathbf{k}\cdot\mathbf{x}} \qquad (4\text{-}36)$$

Equations (4-35) and (4-36) may be inverted to express the Fourier variables $\mathbf{q}$, $\mathbf{q}^*$ in terms of the field variables

$$(\mathbf{q}(\mathbf{k}) + \mathbf{q}^*(-\mathbf{k})) = \sqrt{\frac{\epsilon_0}{V}} \int d^3\mathbf{x}\, \mathbf{A}(\mathbf{x},t) e^{-i\mathbf{k}\cdot\mathbf{x}}$$

$$\equiv \sqrt{\frac{\epsilon_0}{V}}\ \mathbf{A}(\mathbf{k},t) \qquad (4\text{-}37)$$

$$-iw\epsilon_0(\mathbf{q}(\mathbf{k}) - \mathbf{q}^*(\mathbf{k})) = \sqrt{\frac{\epsilon_0}{V}} \int d^3\mathbf{x}\, \mathbf{\Pi}(\mathbf{x},t) e^{-i\mathbf{k}\cdot\mathbf{x}}$$

$$\equiv \sqrt{\frac{\epsilon_0}{V}}\ \mathbf{\Pi}(\mathbf{k},t) \qquad (4\text{-}38)$$

The important feature of these equations expressing the Fourier variables in terms of the field variables is that they do not involve the time variable t explicitly. It follows that we may regard $\mathbf{q}(\mathbf{k})$ and $\mathbf{q}^*(\mathbf{k})$ as new dynamical variables satisfying the hamiltonian equations of motion

$$\dot{q}(k) = [q(k), H'']$$ (4-39a)

$$\dot{q}^*(k) = [q^*(k), H'']$$ (4-39b)

The relationship of the $q^c(k)$ and the $q(k)$ (and their complex conjugates) should be clearly understood. The $q(k)$ are the important quantities of the nature of hamiltonian dynamical variables. They and their conjugate complexes at any time t are defined in terms of the field quantities at time t by equations that do not involve t explicitly. If the radiation field is interacting with charges, the $q(k)$, $q^*(k)$ variables may still be used as hamiltonian dynamical variables. The $q^c(k)$, on the other hand, are integrals of the equations of motion, not working dynamical variables, and they cease to be well defined when interactions are introduced into the theory.[58]

If we solve (4-37) and (4-38) for these new variables, we obtain

$$q(k) = \frac{1}{2}\sqrt{\frac{\epsilon_0}{V}} \int d^3x\, A(x)e^{-ik\cdot x} + \frac{i}{2}\sqrt{\frac{1}{\epsilon_0 V w^2}} \int d^3x\, \Pi(x)e^{-ik\cdot x}$$ (4-40)

$$q^*(k) = (q(k))^+$$ (4-41)

A straightforward calculation now shows that the hamiltonian, (4-25), may be written in terms of these new variables in the following form:

$$H = 2\sum_k w^2 q^*(k)\cdot q(k)$$ (4-41)

The associated Poisson brackets are

$$[q(k)^r, q(k')^s] = [q^*(k)^r, q^*(k')^s] = 0$$

$$[q(k)^r, q^*(k')^s] = -\left(\frac{i}{2w}\right)\delta_{rs}\delta(k-k')$$ (4-43)

As a check, we note that the equation of motion (4-39) has the explicit realization

$$\dot{q}(k)^r = [q(k)^r, H] = -iwq(k)^r$$ (4-44)

which agrees with the definition of the $q(k)$ in terms of the integrals of the motion $q^c(k)$. In accordance with the usual definitions in quantum theory, we introduce new (*classical*) variables $\mathbf{a(k)}$ and $\mathbf{a^+(k)}$,

$$\mathbf{a(k)} = \sqrt{2w}\ \mathbf{q(k)} \tag{4-45}$$

$$\mathbf{a^+(k)} = \sqrt{2w}\ \mathbf{q^*(k)} \tag{4-46}$$

in terms of which (4-42) and (4-43) become

$$H = \sum_{\mathbf{k}} w \mathbf{a^+(k)} \cdot \mathbf{a(k)} \tag{4-47}$$

$$[a(\mathbf{k})^r, a(\mathbf{k}')^s] = [a^+(\mathbf{k})^r, a^+(\mathbf{k}')^s] = 0$$
$$[a(\mathbf{k})^r, a^+(\mathbf{k}')^s] = -i\delta_{rs}\delta(\mathbf{k}-\mathbf{k}') \tag{4-48}$$

Finally we can return to the original expansions of the field variables and write

$$\mathbf{A(x},t) = \sum_{\mathbf{k}} \sqrt{\frac{1}{2\epsilon_0 Vw}}\ \{\mathbf{a(k)}e^{i\mathbf{k}\cdot\mathbf{x}} + \mathbf{a^+(k)}e^{-i\mathbf{k}\cdot\mathbf{x}}\} \tag{4-49}$$

$$\mathbf{\Pi(x},t) = -i\sum_{\mathbf{k}} \sqrt{\frac{w\epsilon_0}{2V}}\ \{\mathbf{a(k)}e^{i\mathbf{k}\cdot\mathbf{x}} - \mathbf{a^+(k)}e^{-i\mathbf{k}\cdot\mathbf{x}}\} \tag{4-50}$$

The radiation field is often described as a collection of independent oscillators, infinite in number. If we define oscillator variables

$$\mathbf{Q(k)} = \sqrt{\frac{\epsilon_0}{V}}\ \mathbf{A(k)} \tag{4-51}$$

$$\mathbf{P(k)} = \sqrt{\frac{1}{\epsilon_0 V}}\ \mathbf{\Pi(k)} \tag{4-52}$$

it is easily seen that

$$\mathbf{a(k)} = \sqrt{\frac{w}{2}}\ \mathbf{Q(k)} + i\sqrt{\frac{1}{2w}}\ \mathbf{P(k)} \tag{4-53}$$

$$\mathbf{a^+(k)} = \sqrt{\frac{w}{2}}\ \mathbf{Q(k)} - i\sqrt{\frac{1}{2w}}\ \mathbf{P(k)} \tag{4-54}$$

and (4-47) becomes

$$H = \tfrac{1}{2} \sum_{\mathbf{k}} \left\{ P(\mathbf{k})^2 + w^2 Q(\mathbf{k})^2 \right\} \tag{4-55}$$

which is the hamiltonian of a collection of independent oscillators of frequency $w = kc$. The $Q(\mathbf{k})$ and $P(\mathbf{k})$ are canonical variables because

$$\left[Q(\mathbf{k})^r, P(\mathbf{k}')^s \right] = \delta_{rs} \delta(\mathbf{k} - \mathbf{k}') \tag{4-56}$$

IV.C. The Constants of the Motion

We now consider some other variables associated with the radiation field, especially those combinations of the field variables which are constants of the motion. In hamiltonian mechanics the energy coincides with the hamiltonian if H has no explicit time dependence. The energy density of the electromagnetic field $\mathcal{E}(x)$ is therefore given by

$$\mathcal{E}(x) = \frac{\epsilon_0}{2} \left\{ \epsilon_0^{-2} \Pi(x) \cdot \Pi(x) + c^2 B(x) \cdot B(x) \right\} \tag{4-57}$$

With the aid of the equations of motion we easily find

$$\dot{\mathcal{E}}(x) = -\epsilon_0 c^2 \nabla \cdot E(x) \times B(x) \tag{4-58}$$

which is of the form of a conservation law. The flux of the conserved energy density, which we denote by the vector $S = \epsilon_0 c^2 E(x) \times B(x)$, is usually known as the Poynting vector. The momentum density of the field $\mathcal{g}$ is related directly to S,[59]

$$\mathcal{g} = c^{-2} S = \epsilon_0 E \times B \tag{4-59}$$

and the total field momentum G is then given by

$$G = \epsilon_0 \int dx^3 E(x) \times B(x) \tag{4-60}$$

If we introduce the canonical variables into (4-60) and use (4-49) and (4-50), we may express the field momentum in terms of the Fourier variables $a(\mathbf{k})$, $a^+(\mathbf{k})$

$$G = \sum_{\mathbf{k}} \mathbf{k} a^+(\mathbf{k}) \cdot a(\mathbf{k}) \tag{4-61}$$

This form for the momentum shows clearly that G is a conserved quantity ($[G, H] = 0$), as expected.

To complete the description of the radiation field, we must specify the polarization characteristics of the light. In view of (4-33), if we choose the

z-axis along the wave propagation direction, we shall always have $a(\mathbf{k})_z = a^+(\mathbf{k})_z = 0$. The Fourier coefficients may therefore be completely described by their components in two directions orthogonal to each other, and orthogonal to $\mathbf{k}$. With a right-handed coordinate system, we have

$$\hat{\mathbf{E}}_x \times \hat{\mathbf{E}}_y = \hat{k}, \qquad \hat{k} = \frac{\mathbf{k}}{|\mathbf{k}|} \tag{4-62}$$

An arbitrary light beam is of course a superposition of waves of different wave vectors, but let us initially confine our attention to a single mode of wave vector $\mathbf{k}$ which corresponds physically to strictly monochromatic radiation. In simple radiation problems in which the polarization information is not important, it is convenient to regard the light as plane polarized and put

$$a(\mathbf{k})_x = \mathbf{a}(\mathbf{k}) \cdot \hat{\mathbf{E}}_x, \qquad a(\mathbf{k})_y = \mathbf{a}(\mathbf{k}) \cdot \hat{\mathbf{E}}_y$$

$$a^+(\mathbf{k})_x = \mathbf{a}^+(\mathbf{k}) \cdot \hat{\mathbf{E}}_x, \quad a^+(\mathbf{k})_y = \mathbf{a}^+(\mathbf{k}) \cdot \hat{\mathbf{E}}_y \tag{4-63}$$

in terms of the base vectors defined in (4-62). In more general cases when the polarization changes are being investigated, it is advantageous to work with complex polarization unit vectors. The most general pure polarization state is specified by an azimuth α and an ellipticity β (see Fig. 1), referred to the axes $\hat{\mathbf{E}}_x$ and $\hat{\mathbf{E}}_y$. The components of the vector Fourier coefficients $\mathbf{a}$ along the minor and major axes u and v are $\pm(\pi/2)$ out of phase; thus, a phase factor $\exp(\pm i\pi/2) = \pm i$ is associated with the v-axis. We want a positive angle β to specify a right-handed ellipticity, which we define conventionally as a clockwise rotation of the electric field $\mathbf{E}$ as viewed by an observer receiving the wave. This obtains if a phase factor of $-i$ is associated with the v-axis. If we write

$$\mathbf{a}(\mathbf{k}) = a(\mathbf{k})\tau, \qquad \mathbf{a}^+(\mathbf{k}) = a(\mathbf{k})\tau^* \tag{4-64}$$

where $\tau = \tau_x \hat{\mathbf{E}}_x + \tau_y \hat{\mathbf{E}}_y$, is a complex polarization vector of unit length, the complex components τ_x, τ_y are obtained as

$$\tau_x = \cos\beta \cos\alpha - i \sin\beta \sin\alpha \tag{4-65}$$

$$\tau_y = -\cos\beta \sin\alpha - i \sin\beta \cos\alpha \tag{4-66}$$

and the components of τ^* are the conjugate complexes of these.

We may introduce a hermitian polarization tensor with elements

$$\xi_{ij} = a(\mathbf{k})_i a^+(\mathbf{k})_j, \qquad i,j = x,y \tag{4-67}$$

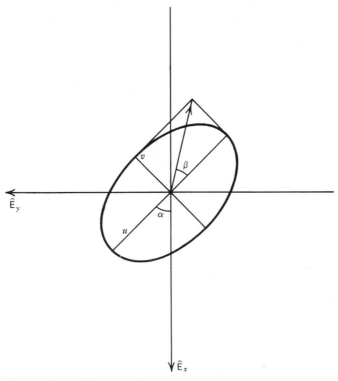

Fig. 1. The polarization ellipse.

It is clear from the equation of motion $\dot{\underline{\xi}} = [\underline{\xi}, H]$ that $\underline{\xi}$ is a conserved tensor quantity; in particular, each element of $\underline{\xi}$ is conserved separately so that linear combinations of these elements are also conserved. Since the tensor has only four independent components, it may also be represented in the following form:

$$\underline{\xi} = \boldsymbol{\eta} \cdot \boldsymbol{\sigma} + \eta_0 \mathbf{1} \tag{4-68}$$

where $\mathbf{1}$ is the unit matrix, the $\boldsymbol{\sigma}$ are the Pauli matrices, and the vector η_μ has components

$$\eta_0 = a(\mathbf{k})_x a^+(\mathbf{k})_x + a(\mathbf{k})_y a^+(\mathbf{k})_y$$

$$\eta_1 = a(\mathbf{k})_x a^+(\mathbf{k})_y + a(\mathbf{k})_y a^+(\mathbf{k})_x$$

$$\eta_2 = i\{a(\mathbf{k})_x a^+(\mathbf{k})_y - a(\mathbf{k})_y a^+(\mathbf{k})_x\} \tag{4-69}$$

$$\eta_3 = a(\mathbf{k})_x a^+(\mathbf{k})_x - a(\mathbf{k})_y a^+(\mathbf{k})_y$$

The $\eta_\mu(\mu=0,1,2,3)$ are conventionally known as the Stokes parameters.[60] One may either use the second-rank tensor $\underline{\xi}$ or the equivalent Stokes vector η_μ; both transform directly into the corresponding quantum mechanical operators on making the transition to quantum electrodynamics.[61,62] The representation of these quantities in terms of the angles α and β and the scalar Fourier coefficients $a(\mathbf{k})$, $a^+(\mathbf{k})$ is straightforward, since

$$\xi_{ij} = a(\mathbf{k})a^+(\mathbf{k})\tau_i\tau_j^* \tag{4-70}$$

Thus, for example,

$$\eta_2 \propto \tau_x\tau_y^* - \tau_y\tau_x^* = i\cos 2\beta \tag{4-71}$$

To conclude this discussion of the radiation field we investigate briefly the relationship between the conserved quantity $\underline{\xi}$ (or equivalently the Stokes vector η_μ) and the field variables $\mathbf{E}(\mathbf{x})$ and $\mathbf{B}(\mathbf{x})$. It has only quite recently been recognized that there is another conserved quantity for the radiation field which is independent of the two classical quantities identified previously. The full specification of this conserved quantity requires the knowledge of the components of a third-rank tensor built up from combinations of the field vectors $\mathbf{E}(\mathbf{x})$ and $\mathbf{B}(\mathbf{x})$ and their derivatives. For simplicity we consider here just one part of this tensor and refer the reader to the literature for the full details.[63]

We define a scalar field density $\gamma(\mathbf{x})$ by the relation

$$\gamma(\mathbf{x}) = \mathbf{E}(\mathbf{x})\cdot\nabla\times\mathbf{E}(\mathbf{x}) + c^2\mathbf{B}(\mathbf{x})\cdot\nabla\times\mathbf{B}(\mathbf{x}) \tag{4-72}$$

and using the hamiltonian (4-25) compute its time derivative as

$$\dot{\gamma}(\mathbf{x}) = [\gamma(\mathbf{x}), H]$$
$$= c^2\nabla\cdot\{\mathbf{E}(\mathbf{x})\times(\nabla\times\mathbf{B}(\mathbf{x})) - \mathbf{B}(\mathbf{x})\times(\nabla\times\mathbf{E}(\mathbf{x}))\} \tag{4-73}$$

which again is of the form of a conservation law [cf. (4-58)] in which the flux density $\phi(\mathbf{x})$ is

$$\phi(\mathbf{x}) = c^2\{\mathbf{E}(\mathbf{x})\times(\nabla\times\mathbf{B}(\mathbf{x})) - \mathbf{B}(\mathbf{x})\times(\nabla\times\mathbf{E}(\mathbf{x}))\} \tag{4-74}$$

An integration over all space leads to the total flux $\mathbf{\Phi}$ and the corresponding scalar Γ

$$\Gamma = \int d^3\mathbf{x}\,\gamma(\mathbf{x}) \tag{4-75}$$

$$\mathbf{\Phi} = \int d^3\mathbf{x}\phi(\mathbf{x}) \tag{4-76}$$

With the aid of the expansions (4-49) and (4-50), we express these quantities in terms of the Fourier variables. A simple calculation leads to the result

$$\Gamma = \left(\frac{2i}{\epsilon_0 c^2}\right) \sum_{\mathbf{k}} w^2 \hat{k} \cdot (\mathbf{a}(\mathbf{k}) \times \mathbf{a}^+(\mathbf{k})) \qquad (4\text{-}77)$$

$$\Phi = -\left(\frac{2i}{\epsilon_0}\right) \sum_{\mathbf{k}} w^2(\mathbf{a}(\mathbf{k}) \times \mathbf{a}^+(\mathbf{k})) \qquad (4\text{-}78)$$

It is obvious that the only nonzero element in the vector product is the z-component $= a(\mathbf{k})_x a^+(\mathbf{k})_y - a(\mathbf{k})_y a^+(\mathbf{k})_x$, and according to (4-71) this is proportional to the Stokes parameter $\eta_2 = i\cos(2\beta)$. It may be shown that the other elements of this conserved third-rank tensor are related to the other Stokes parameters, and therefore it is a conserved quantity describing the polarization characteristics of the radiation field.

An interesting representation for Γ may be obtained if we express the Fourier variables $\mathbf{a}(\mathbf{k})$ and $\mathbf{a}^+(\mathbf{k})$ in terms of the canonical oscillator variables $\mathbf{Q}(\mathbf{k})$ and $\mathbf{P}(\mathbf{k})$. We find at once that

$$\Gamma = -\left(\frac{1}{\epsilon_0 c^2}\right) \sum_{\mathbf{k}} w^2 \hat{k} \cdot (\mathbf{Q}(\mathbf{k}) \times \mathbf{P}(\mathbf{k})) \qquad (4\text{-}79)$$

$$= -\left(\frac{1}{\epsilon_0 c^2}\right) \sum_{\mathbf{k}} w^2 L(\mathbf{k})_z \qquad (4\text{-}80)$$

where we have introduced the *angular momentum* vector $\mathbf{L}(\mathbf{k}) = \mathbf{Q}(\mathbf{k}) \times \mathbf{P}(\mathbf{k})$ for the oscillator corresponding to the wave vector $\mathbf{k}$. Thus Γ can be interpreted in terms of an *intrinsic* angular momentum associated with the radiation field. This representation makes it perfectly clear that the new conserved quantity is independent of the energy and momentum of the field.

V. THE ELECTRODYNAMICS OF CHARGED PARTICLES

V.A. The Hamiltonian

We have summarized the Maxwell-Lorentz theory of charged particles (Section II) and described the essential techniques of the canonical formalism (Section III). As a first step toward our goal of obtaining a hamiltonian theory of charged particles, we next discussed the application of these techniques to the radiation field (Section IV). Now we extend the

discussion by reformulating the Maxwell-Lorentz theory in a classical
canonical framework; the subsequent quantization of this system of equations should provide the basis of an electrodynamics of atoms and molecules. We begin with an account of the conventional Coulomb gauge
hamiltonian formalism for a single charged particle; a straightforward
generalization then yields the hamiltonian for a collection of charged
particles.

The lagrangian for a *closed* system consisting of a charged particle and
the electromagnetic field may be written as a sum of three parts:[64]

$$L = L^{\text{particle}} + L^{\text{int}} + L^{\text{field}} \tag{5-1}$$

L^{particle} is a lagrangian for the particle in the absence of all interactions and
must thus describe the free motion of the particle. It is obvious that if we
choose

$$L^{\text{particle}} = \tfrac{1}{2} m \dot{R}^2 \tag{5-2}$$

the Euler-Lagrange equation (3-3) leads to the required equation of motion
$m\ddot{R} = 0$. Similarly L^{field} is a lagrangian for the electromagnetic field in the
absence of charged particles and may be taken to be

$$L^{\text{field}} = \int d^3x\, \mathscr{L}(\mathbf{x}, t) \tag{5-3}$$

where $\mathscr{L}(\mathbf{x}, t)$ is defined by (4-1). Finally L^{int} is the interaction lagrangian
describing the coupling between the charge and the electromagnetic field.
A possible choice for L^{int} expresses this coupling through the charge and
current densities, (2-10), (2-11), and the electromagnetic field potentials
$\phi(\mathbf{x}, t)$ and $\mathbf{A}(\mathbf{x}, t)$:

$$L^{\text{int}} = \int d^3x\, \mathbf{j}(\mathbf{x}, t) \cdot \mathbf{A}(\mathbf{x}, t) - \int d^3x\, \rho(\mathbf{x}, t)\phi(\mathbf{x}, t)$$

$$\equiv e\dot{\mathbf{R}} \cdot \mathbf{A}(\mathbf{R}) - e\phi(\mathbf{R}) \tag{5-4}$$

for a single charge. It is a simple matter to show that the combination of
(5-1) to (5-4) with the Euler-Lagrange conditions (3-3) and (3-7) leads to
the Maxwell equations (2-6) and the Lorentz force law (2-21), provided the
variations of the field and particle variables are made independently.[37] The
lagrangian variables are the coordinate $\mathbf{R}$ and the velocity $\dot{\mathbf{R}}$ for the
particle, and the potentials $\phi(\mathbf{x}, t)$ and $\mathbf{A}(\mathbf{x}, t)$ and their partial time derivatives for the electromagnetic field.

To pass from the lagrangian to the hamiltonian, we first define the
momenta conjugate to the lagrangian "coordinates" in the usual way—as
the partial derivatives of L with respect to the "velocities." We encounter

immediately the familiar difficulty that the derivative $\dot{\phi}(\mathbf{x},t)$ does not appear in (5-1); thus $\Pi_0(\mathbf{x})$, the momentum conjugate to $\phi(\mathbf{x},t)$, vanishes formally and L is therefore degenerate. Thus the momenta are to be written as weak equations;

$$\Pi_0(\mathbf{x})\approx 0 \tag{5-5}$$

$$\Pi(\mathbf{x})\approx\varepsilon_0\bigl(\nabla\phi(\mathbf{x},t)+\dot{\mathbf{A}}(\mathbf{x},t)\bigr) \tag{5-6}$$

$$\mathbf{p}\approx m\dot{\mathbf{R}}+e\mathbf{A}(\mathbf{R}) \tag{5-7}$$

The nonzero fundamental Poisson brackets are assumed to be canonical

$$[R^s,p^t]=\delta_{st} \tag{5-8}$$

$$[A(\mathbf{x})^s,\Pi(\mathbf{x}')^t]=\delta_{st}\delta(\mathbf{x}-\mathbf{x}') \tag{5-9}$$

$$[\phi(\mathbf{x}),\Pi_0(\mathbf{x}')]=\delta(\mathbf{x}-\mathbf{x}') \tag{5-10}$$

All Poisson brackets between particle and field variables are assumed to vanish: this condition corresponds to our earlier assumption that the particle and field variables can be varied independently when using the principle of least action to generate the equations of motion from (5-1). The hamiltonian is defined in the usual way:

$$H\approx\mathbf{p}\cdot\dot{\mathbf{R}}+\int d^3\mathbf{x}\Pi(\mathbf{x})\cdot\dot{\mathbf{A}}(\mathbf{x})+\int d^3\mathbf{x}\Pi_0(\mathbf{x})\dot{\phi}(\mathbf{x})-L \tag{5-11}$$

and on eliminating the velocities in favor of the conjugate momenta, we obtain, after simple manipulations (cf. Section IV)

$$H\approx\frac{1}{2m}(\mathbf{p}-e\mathbf{A}(\mathbf{R}))^2+\frac{\varepsilon_0}{2}\int d^3\mathbf{x}\bigl\{\varepsilon_0^{-2}\Pi(\mathbf{x})^2+c^2B(\mathbf{x})^2\bigr\}$$

$$+\int d^3\mathbf{x}\phi(\mathbf{x})\{\nabla\cdot\Pi(\mathbf{x})+\rho(\mathbf{x})\} \tag{5-12}$$

where $\mathbf{B}(\mathbf{x})=\mathrm{Curl}\,\mathbf{A}(\mathbf{x})$.

The consistency condition implied by (5-5) is that it should be true for all times; thus using the equation of motion $\dot{\Omega}\approx[\Omega,H]$, we obtain the condition

$$\psi_1(\mathbf{x})=\rho(\mathbf{x})+\nabla\cdot\Pi(\mathbf{x})\approx 0 \tag{5-13}$$

which must also be true for all time. Using (2-10) and (5-7), and the fundamental Poisson-bracket relations, it is easy to show that $[\psi_1(x),H]\approx 0$

reduces to $0 = 0$; thus (5-5) and (5-13) are the only constraint equations in the theory. We note that these constraints are first class, since

$$[\Pi_0(\mathbf{x}), \Pi_0(\mathbf{x}')] = [\Pi_0(\mathbf{x}), \psi_1(\mathbf{x}')] = [\psi_1(\mathbf{x}), \psi_1(\mathbf{x}')] = 0 \qquad (5\text{-}14)$$

As before, we obtain the total hamiltonian by adding all the first class constraints with arbitrary coefficients to the hamiltonian H; the presence of terms describing the particle-field coupling does not affect the demonstration that $\dot{\phi}(\mathbf{x})$ is an arbitrary quantity, Section IV, and we discard $\Pi_0(\mathbf{x})$ and $\phi(\mathbf{x})$ and write the hamiltonian as the strong equation

$$H' = \frac{1}{2m}(\mathbf{p} - e\mathbf{A}(\mathbf{R}))^2 + \frac{\varepsilon_0}{2}\int d^3\mathbf{x}\left\{\varepsilon_0^{-2}\Pi(\mathbf{x})^2 + c^2 B(\mathbf{x})^2\right\}$$
$$+ \int d^3\mathbf{x}\, w(\mathbf{x})\{\rho(x) + \nabla \cdot \Pi(\mathbf{x})\} \qquad (5\text{-}15)$$

The last term in the hamiltonian may be used as the generating function of gauge transformations. Under such an infinitesimal contact transformation, the vector potential changes as in (4-16), and the particle momentum $\mathbf{p}$ changes by an amount

$$\delta\mathbf{p} = e\nabla w(\mathbf{R}) \qquad (5\text{-}16)$$

leaving the kinetic momentum $= (\mathbf{p} - e\mathbf{A}(\mathbf{R}))$ unchanged by the transformation.

We must now eliminate the term in $w(\mathbf{x})$ by introducing a supplementary condition that has a nonvanishing Poisson bracket with $\psi_1(\mathbf{x})$. As before, we employ the Coulomb gauge condition and write

$$\psi_2(\mathbf{x}) = \nabla \cdot \mathbf{A}(\mathbf{x}) \approx 0 \qquad (5\text{-}17)$$

Since $[\rho(\mathbf{x}'), \nabla \cdot \mathbf{A}(\mathbf{x})] = 0$, the Ψ matrix obtained in Section IV is applicable here, and we can obtain the Dirac brackets at once using (4-21) and (4-22). The nonzero brackets are found to be

$$[R^s, p']^* = \delta_{st} \qquad (5\text{-}18)$$

$$[A(\mathbf{x})^s, \Pi(\mathbf{x}')^t]^* = \delta_{st}^{\perp}(\mathbf{x} - \mathbf{x}') \qquad (5\text{-}19)$$

$$[p^s, \Pi(\mathbf{x})^t]^* = ep^s \nabla_x^t \left\{\frac{1}{(4\pi|\mathbf{R} - \mathbf{x}|)}\right\} \qquad (5\text{-}20)$$

Since we may now put $\psi_i(\mathbf{x}) \equiv 0$, consideration of the brackets (5-19) and (5-20) suggests that we identify the canonical momentum $\Pi(\mathbf{x})$ with the total electric field strength to within a multiplicative factor $\Pi(x) = -\varepsilon_0 E(x)$ in SI units. Thus the Coulomb gauge hamiltonian for the charged particle–electromagnetic field system may be summarized as

$$H'' = \frac{1}{2m}(\mathbf{p} - e\mathbf{A}(\mathbf{R}))^2 + \frac{\varepsilon_0}{2}\int d^3x\{\mathbf{E}(x)\cdot\mathbf{E}(x) + c^2\mathbf{B}(x)\cdot\mathbf{B}(x)\} \quad (5\text{-}21)$$

$$\nabla\cdot\mathbf{A}(x) = 0 \quad (5\text{-}22)$$

$$\varepsilon_0\nabla\cdot\mathbf{E}(x) = \rho(x) \quad (5\text{-}23)$$

together with the Dirac brackets (5-18) to (5-20). [Hereafter we no longer explicitly include the * superscript, and we assume that the bracket operation corresponds to (5-18)-(5-20).]

V.B. The Equations of Motion

It is convenient to separate the electric field vector $\mathbf{E}(x)$ into its longitudinal and transverse parts

$$\mathbf{E}(x) = \mathbf{E}(x)^{\parallel} + \mathbf{E}(x)^{\perp} \quad (5\text{-}24)$$

since it is evident that the bracket (5-20) is nonzero because of the *longitudinal* component of $\Pi(x)$ [or $\mathbf{E}(x)$] which is associated with the (static) Coulomb field of the charge. For a single point charge, $\mathbf{E}(x)^{\parallel}$ gives an infinite contribution to the energy in (5-21); since

$$\mathbf{E}(x)^{\parallel} = -e\nabla_x\left\{\frac{1}{(4\pi\varepsilon_0|\mathbf{R} - x|)}\right\} \quad (5\text{-}25)$$

we have

$$\frac{\varepsilon_0}{2}\int d^3x\,\mathbf{E}(x)^{\parallel}\cdot\mathbf{E}(x)^{\parallel} = \frac{\varepsilon_0}{2}e^2\int d^3x\nabla\left(\frac{1}{4\pi\varepsilon_0|\mathbf{R} - x|}\right)\cdot\nabla\left(\frac{1}{4\pi\varepsilon_0|\mathbf{R} - x|}\right)$$

$$(5\text{-}26)$$

With the aid of the vector identity

$$\nabla\cdot(f\nabla g) = (\nabla f)\cdot(\nabla g) + f\nabla^2 g \quad (5\text{-}27)$$

and Gauss' theorem, (5-26) may be written as

$$-\frac{\varepsilon_0}{2}e^2\int d^3x\left(\frac{1}{4\pi\varepsilon_0|\mathbf{R} - x|}\right)\nabla^2\left(\frac{1}{4\pi\varepsilon_0|\mathbf{R} - x|}\right)$$

$$= \frac{\varepsilon_0}{2}e^2\int d^3x\left(\frac{1}{4\pi\varepsilon_0|\mathbf{R} - x|}\right)\delta^3(\mathbf{R} - x) = \infty \quad (5\text{-}28)$$

Here we have the first indication of a serious flaw in our model of a charged particle. However, since this infinity is independent of the dynamical variables (i.e., it is an infinite constant), we can at least say that it has no effect on the *equations of motion* obtained from the hamiltonian; hopefully, then, it may be discarded without further consideration. In the many-particle theory, (5-25) is generalized in an obvious way, and $\int d^3x E(x)^\parallel \cdot E(x)^\parallel$ contains the Coulomb interaction between the charges in addition to the infinite constant just obtained. The transverse electromagnetic field variables may be expressed in terms of the Fourier variables $a^+(k)$, $a(k)$, which, as we emphasized in Section IV, are useful working variables even when interactions with charged particles are admitted. In the following we make extensive use of the Fourier expansions developed in Section IV.

Since the hamiltonian H'' has no explicit time dependence, it follows that the total energy of the system, which is equal to H'', is a conserved quantity. This is merely another way of saying that the dynamical system of field plus charged particle is closed. The individual terms in H'' are not conserved separately, however, permitting us to say that the field and the charged particle can exchange energy between themselves while maintaining the same total energy. Similarly, whereas the (linear) field momentum G in (4-61) was found to be a conserved quantity for the radiation field, this is no longer true when interaction with the charge is allowed. Explicitly we calculate

$$\dot{G}_t = [G_t, H'']$$

$$= -e\dot{R}_s \sum_{k,\lambda} k_t [a_\lambda^+(k) a_\lambda(k), A(R)_s] \qquad (5\text{-}29)$$

which with the aid of the Fourier expansion of the vector potential (4-49) may be reduced to

$$\dot{G}_t = -ie\dot{R}_s \sum_k k_t \frac{1}{(2Vkc\varepsilon_0)^{1/2}} \{a_s(k) e^{ik\cdot R} - a_s^+(k) e^{-ik\cdot R}\}$$

$$= -e\dot{R}_s \nabla_t \sum_k \frac{1}{(2Vkc\varepsilon_0)^{1/2}} \{a_s(k) e^{ik\cdot R} + a_s^+(k) e^{-ik\cdot R}\}$$

$$= -e\nabla_t (\dot{R} \cdot A(R)) \qquad (5\text{-}30)$$

Thus momentum can also be exchanged between the field and the charge. We would expect, however, to find that the *total* momentum $(p + G)$ is a conserved quantity, since the system is assumed to be closed. A simple calculation shows that

$$\dot{p}_t = [p_t, H'']$$

$$= e\nabla_t(\dot{\mathbf{R}} \cdot \mathbf{A}(\mathbf{R})) \tag{5-31}$$

and this is equal to $-\dot{G}_t$ as required.

We now consider the equations of motion of the other canonical variables. Since

$$\dot{\mathbf{R}} = [\mathbf{R}, H''] = \frac{\mathbf{p} - e\mathbf{A}(\mathbf{R})}{m} \tag{5-32}$$

we obtain

$$m\dot{\mathbf{R}} = [(\mathbf{p} - e\mathbf{A}(\mathbf{R})), H''] \tag{5-33}$$

now $(\mathbf{p} - e\mathbf{A}(\mathbf{R}))$ does not "commute" with its square, which means that there is a contribution of the form

$$\left[(\mathbf{p} - e\mathbf{A}(\mathbf{R}))_t, \frac{(\mathbf{p} - e\mathbf{A}(\mathbf{R}))^2}{2m} \right]$$

$$= \left(\frac{e}{m} \right) (\mathbf{p} - e\mathbf{A}(\mathbf{R}))_s \left\{ \frac{\partial A(\mathbf{R})_t}{\partial R_s} - \frac{\partial A(\mathbf{R})_s}{\partial R_t} \right\}$$

$$= e(\dot{\mathbf{R}} \times \mathbf{B}(\mathbf{R}))_t \tag{5-34}$$

using (5-32) and the identification $\mathbf{p} \Rightarrow \partial/\partial\mathbf{R}$ in the evaluation of the bracket operation [c.f. (5-18)]. Similarly it is easy to see that

$$\left[(\mathbf{p} - e\mathbf{A}(\mathbf{R}))_t, \frac{\varepsilon_0}{2} \int d^3x \{ \mathbf{E}(\mathbf{x}) \cdot \mathbf{E}(\mathbf{x}) + c^2\mathbf{B}(\mathbf{x}) \cdot \mathbf{B}(\mathbf{x}) \} \right] = e E(\mathbf{R})_t$$

$$\tag{5-35}$$

and therefore

$$m\ddot{\mathbf{R}} = e\{ \mathbf{E}(\mathbf{R}) + \dot{\mathbf{R}} \times \mathbf{B}(\mathbf{R}) \} \tag{5-36}$$

which is the Lorentz force law. Next we show that the hamiltonian leads to the Maxwell equations for the field variables. We always have Div $\mathbf{B}(\mathbf{x})$ $= 0$, since we are using $\mathbf{B}(\mathbf{x}) = \text{Curl } \mathbf{A}(\mathbf{x})$; furthermore, the Maxwell equation (5-32) involving the charge density has been obtained as an identity in the development of the hamiltonian theory. The equation of motion for $\mathbf{A}(\mathbf{x})$ is the same as for the free radiation field [(4-28)], allowing us to write $\dot{\mathbf{B}}(\mathbf{x}) = -\text{Curl } \mathbf{E}(\mathbf{x})$ at once (but these are *interacting* fields of course). Finally in the equation of motion for $\mathbf{\Pi}(\mathbf{x})$ we have the contribution from the radiation field terms [(4-29)], plus an additional term due to the coupling with the charge

$$\delta \dot{\Pi}_s(\mathbf{x}) = \left[\Pi_s(\mathbf{x}), \frac{(\mathbf{p} - e\mathbf{A}(\mathbf{R}))^2}{2m} \right]$$

$$= e\dot{R}_t [\Pi_s(\mathbf{x}), (\mathbf{p} - e\mathbf{A}(\mathbf{R}))_t]$$

$$= e\dot{R}_s \delta(\mathbf{x} - \mathbf{R}) \qquad (5\text{-}37)$$

and therefore with the aid of (2-11), we recover the Maxwell equation (2-6d)

$$\varepsilon_0 \dot{\mathbf{E}}(\mathbf{x}) + \mathbf{j}(\mathbf{x}) = \varepsilon_0 c^2 \nabla \times \mathbf{B}(\mathbf{x})$$

Thus the canonical scheme displayed in (5-18) to (5-23) appears to be equivalent to the Maxwell-Lorentz theory described in Section II.

V.C. The Field Due to a Moving Charge

Further questions about the consistency of our hamiltonian formalism relate to the properties of the solutions of the coupled equations of motion.[64] To guarantee a consistent canonical formalism, we have to deal satisfactorily with two related problems. First we must identify a complete set of dynamical variables. By completeness we mean that the dynamical variables form the minimal set of conjugate pairs needed to determine the canonical equations of motion. We have assumed that the choice of the canonical variables $\{\mathbf{R}, \mathbf{p}\}$ and $\{\mathbf{A}, \mathbf{\Pi}\}$ is adequate, but the supposition must be investigated further. Having identified a complete set of dynamical variables, we are faced with the second problem. Suppose that the solutions of the inhomogeneous equations of motion are given as some functions of the solutions of the uncoupled equations. Since the identification of dynamical variables must be made for all times, the Poisson brackets for the uncoupled solutions will have to be specified for all times. In particular, if we impose conditions such that the solution for the particle variables reduces to the solution of the uncoupled equation at some initial time, the commutation rules for the uncoupled particle variables must then be fixed. The solution for the electromagnetic field variables, however, will contain a mixture of uncoupled field and particle variables in such a way that at the initial time we are left with a mixture of uncoupled field and uncoupled particle variables. One may then argue that the commutation rules for the uncoupled field variables should be determined by the requirement that they guarantee that the full set of dynamical variables satisfy the proper Poisson brackets. Suppose we require that the free-field variables satisfy free-field Poisson brackets. Will it then be consistent with the canonical Poisson brackets for the dynamical variables if the free-field variables are assumed to satisfy free-field Poisson brackets at all times?

Let us consider the following inhomogeneous equation of motion for the Coulomb gauge vector potential, which is easily derived from Hamilton's equations of motion for the field variables:

$$\left\{ c^{-2}\frac{\partial^2}{\partial t^2} - \nabla^2 \right\} \mathbf{A}(\mathbf{x},t) = (\varepsilon_0 c^2)^{-1}\mathbf{j}(\mathbf{x},t) \tag{5-38}$$

Let us define a Green function for this equation $\mathcal{D}(\mathbf{x},t)$, which incorporates the condition that $\mathbf{A}(\mathbf{x},t)$ must be divergenceless and must satisfy the given boundary conditions

$$\left(c^{-2}\frac{\partial^2}{\partial t^2} - \nabla^2 \right) \mathcal{D}(\mathbf{x},t) = \delta(t)\delta^{\perp}(\mathbf{x}) \tag{5-39}$$

Then the general solution of (5-38) may be written as

$$\mathbf{A}(\mathbf{x},t) = \mathbf{A}_h(\mathbf{x},t) + (\varepsilon_0 c^2)^{-1}\int d^3x' \int dt' \, \mathcal{D}(\mathbf{x}-\mathbf{x}';\, t-t') \cdot \mathbf{j}(\mathbf{x}',t')$$

$$= \mathbf{A}_h(\mathbf{x},t) + \frac{e}{\varepsilon_0 c^2}\int dt' \, \mathcal{D}(\mathbf{x}-\mathbf{R};\, t-t') \cdot \dot{\mathbf{R}}(t') \tag{5-40}$$

where $\mathbf{A}_h(\mathbf{x},t)$ is an arbitrary solution of the homogeneous wave equation. The appearance of the velocity $\dot{\mathbf{R}}(t')$ in (5-40) raises the possibility that the assumption that $\mathbf{A}_h$ satisfies free-field Poisson brackets is inconsistent with our earlier assumption that $[\mathbf{R}, \mathbf{A}(\mathbf{x})]$ vanishes. Although we could develop the solutions of the combined equations using the traditional methods of handling the Maxwell equations,[59] for our purposes it is more instructive to work directly with Hamilton's equations of motion. This is because we are going to base the quantum theory on the hamiltonian formalism, and there is a close correspondence between the classical hamiltonian formalism and the quantum theory in the Heisenberg representation. The following simple calculation of the field due to the charge will also yield a Fourier representation of the Green function dyadic $\mathcal{D}$ just introduced.

Since we are not interested in the polarization characteristics of the radiation due to the charge, it is simplest to use the Fourier expansion of the vector potential $\mathbf{A}(\mathbf{x})$ based on the polarization unit vectors $\hat{\mathbf{E}}_\lambda(\mathbf{k})$,

$$\mathbf{A}(\mathbf{x}) = \sum_{\mathbf{k},\lambda} \frac{1}{(2Vkc\varepsilon_0)^{1/2}} \hat{\mathbf{E}}_\lambda(\mathbf{k}) \{ a_\lambda(\mathbf{k})e^{i\mathbf{k}\cdot\mathbf{x}} + a_\lambda^+(\mathbf{k})e^{-i\mathbf{k}\cdot\mathbf{x}} \} \quad (5\text{-}41)$$

If we discard the infinite constant arising from $\mathbf{E}(\mathbf{x})^\parallel$, the hamiltonian may be written as

$$H = \sum_{\mathbf{k},\lambda} w a_\lambda^+(\mathbf{k})a_\lambda(\mathbf{k}) + \frac{1}{2m}(\mathbf{p} - e\mathbf{A}(\mathbf{R}))^2 \quad (5\text{-}42)$$

with $\mathbf{A}(\mathbf{x})$ given by (5-41). Then using Hamilton's equation of motion, $\dot{\Omega} = [\Omega, H]$, and (5-32), we find

$$\dot{a}_\lambda(\mathbf{k}) = -iw a_\lambda(\mathbf{k}) + \frac{ie}{(2Vkc\varepsilon_0)^{1/2}} \dot{\mathbf{R}} \cdot \hat{\mathbf{E}}_\lambda(\mathbf{k})e^{-i\mathbf{k}\cdot\mathbf{R}} \quad (5\text{-}43)$$

$$\dot{a}_\lambda^+(\mathbf{k}) = iw a_\lambda^+(\mathbf{k}) - \frac{ie}{(2Vkc\varepsilon_0)^{1/2}} \dot{\mathbf{R}} \cdot \hat{\mathbf{E}}_\lambda(\mathbf{k})e^{i\mathbf{k}\cdot\mathbf{R}} \quad (5\text{-}44)$$

For simplicity let us assume that the field vanishes until $t = 0$, when the charge starts to move; these equations can then be integrated formally to yield

$$a_\lambda(\mathbf{k},t) = a_\lambda(\mathbf{k},0) - \frac{ie}{(2Vkc\varepsilon_0)^{1/2}} \int_0^t dt' \dot{\mathbf{R}}(t') \cdot \hat{\mathbf{E}}_\lambda(\mathbf{k})e^{-i(\mathbf{k}\cdot\mathbf{R} + w(t-t'))} \quad (5\text{-}45)$$

$$a_\lambda^+(\mathbf{k},t) = a_\lambda^+(\mathbf{k},0) + \frac{ie}{(2Vkc\varepsilon_0)^{1/2}} \int_0^t dt' \dot{\mathbf{R}}(t') \cdot \hat{\mathbf{E}}_\lambda(\mathbf{k})e^{i(\mathbf{k}\cdot\mathbf{R} + w(t-t'))} \quad (5\text{-}46)$$

where the $a_\lambda(\mathbf{k},0)$ and $a_\lambda^+(\mathbf{k},0)$ are integration constants. If we combine (5-41), (5-45), and (5-46), we obtain the vector potential for the field due to the motion of the charge as [cf. (5-40)]

$$\mathbf{A}(\mathbf{x},t) = \mathbf{A}_h(\mathbf{x},0) + \mathbf{A}^*(\mathbf{x},t) \quad (5\text{-}47)$$

where $\mathbf{A}_h(\mathbf{x},0)$ is an arbitrary solution of the homogeneous wave equation at $t = 0$ [(4-30)], and $\mathbf{A}^*(\mathbf{x}, t)$ is given by

$$\mathbf{A}^*(\mathbf{x}, t) = \frac{e}{Vc\varepsilon_0} \sum_{\mathbf{k}, \lambda} \frac{\hat{\mathbf{E}}_\lambda(\mathbf{k})}{k} \int_0^t dt' \dot{\mathbf{R}}(t') \cdot \hat{\mathbf{E}}_\lambda(\mathbf{k}) \sin[\mathbf{k} \cdot (\mathbf{x} - \mathbf{R}) - kc(t - t')]$$

(5-48)

This expression may be further simplified if we evaluate the sum over the polarization states using the relation[37]

$$\sum_\lambda \hat{\mathbf{E}}_\lambda(\mathbf{k}) \hat{\mathbf{E}}_\lambda(\mathbf{k}) = (1 - \hat{k}\hat{k})$$

(5-49)

which is a simple consequence of the orthonormality of the $\hat{\mathbf{E}}_\lambda(\mathbf{k})$ and $\hat{k}$.* Equation (5-48) can now be evaluated explicitly in the following way. We convert the summation over $\mathbf{k}$ into an integration using (4-32) and interchange the order of the integrations over $\mathbf{k}$ and t'. Then $\mathbf{A}^*(x, t)$ can be written as

$$A_s^*(\mathbf{x}, t) = \frac{e}{(8\pi^3 c\varepsilon_0)} \int_0^t dt' \dot{R}_p(t') \int d^3k (1 - \hat{k}\hat{k})_{sp} \sin(\mathbf{k} \cdot \mathbf{r} - kc(t - t')) k^{-1}$$

$$= \frac{e}{(8\pi^3 c\varepsilon_0)} \operatorname{Im} \int_0^t dt' \dot{R}_p(t') \int_0^\infty dk\, k\, e^{-ikc(t - t')} \int d\Omega(\mathbf{k})(1 - \hat{k}\hat{k})_{sp} e^{i\mathbf{k}\cdot\mathbf{r}}$$

(5-50)

where $\mathbf{r} = \mathbf{x} - \mathbf{R}$, and Im means "take the imaginary part of." We may express the angular integration in terms of two second-rank tensors U and V, which are defined as

$$\mathsf{U} = (1 - \hat{r}\hat{r}), \qquad \mathsf{V} = (1 - 3\hat{r}\hat{r})$$

(5-51)

$$\frac{1}{4\pi} \int d\Omega(\mathbf{k})(1 - \hat{k}\hat{k}) e^{i\mathbf{k}\cdot\mathbf{r}} = \mathsf{U} \frac{\sin(kr)}{(kr)} + \mathsf{V} \left(\frac{\cos(kr)}{(kr)^2} - \frac{\sin(kr)}{(kr)^3} \right)$$

(5-52)

---

*It is easy to verify that the dyadic $\mathcal{D}$

$$\mathcal{D}(\mathbf{x} - \mathbf{x}'; t - t') = V^{-1} \sum_{\mathbf{k}} (1 - \hat{k}\hat{k}) \sin[\mathbf{k} \cdot (\mathbf{x} - \mathbf{x}') - kc(t - t')](kc)^{-1}$$

satisfies (5-40); we can write the solution (5-47), (5-48) in the form

$$\mathbf{A}(\mathbf{x}, t) = \mathbf{A}_h(\mathbf{x}, 0) + \int d^3x' \int_0^t dt' \mathcal{D}(\mathbf{x} - \mathbf{x}'; t - t') \cdot \mathbf{j}(\mathbf{x}', t')$$

so that (5-50) becomes

$$A_s^*(\mathbf{x},t) = -\frac{e}{(2\pi^2 c \varepsilon_0)} \int_0^t dt' \frac{\dot{R}_p(t')}{r^2} \int_0^\infty dx\, x \sin(Tx) \left\{ U_{sp} \frac{\sin(x)}{x} \right.$$

$$\left. + V_{sp} \left(\frac{\cos(x)}{x^2} - \frac{\sin(x)}{x^3} \right) \right\} \quad (5\text{-}53)$$

where $T = (t - t')c/r$ is a dimensionless parameter. Now

$$\int_0^\infty dx \sin(Tx) \sin(x) = \frac{\pi}{2} \{ \delta(1+T) - \delta(1-T) \} \quad (5\text{-}54)$$

$$\int_0^\infty dx \sin(Tx) \left(\frac{\cos(x)}{x} - \frac{\sin(x)}{x^2} \right) = \begin{cases} \pi T/2, & T \leq 1 \\ 0, & T > 1 \end{cases} \quad (5\text{-}55)$$

Thus if we define the dyadic of a static dipole field

$$\tau(t') = \frac{(1 - 3\hat{r}\hat{r})}{(4\pi\varepsilon_0 r^3)} \quad (5\text{-}56)$$

which depends on t' because of the implicit time dependence of $\mathbf{r}(t') = \mathbf{x} - \mathbf{R}(t')$, we finally obtain

$$\mathbf{A}^*(\mathbf{x},t) = \begin{cases} e \int_0^t dt'\, \tau(t') \cdot \dot{\mathbf{R}}(t')(t - t') & 0 < t < r/c \\[2mm] e \int_{t-r/c}^t dt'\, \tau(t') \cdot \dot{\mathbf{R}}(t')(t - t') & \\[2mm] \quad + \left(\frac{e}{c^2} \right) \left[\frac{(1 - \hat{r}\hat{r}) \cdot \dot{\mathbf{R}}(t')}{(4\pi\varepsilon_0 r(t'))} \right]_{t' = t - r/c} & t > r/c \end{cases} \quad (5\text{-}57)$$

This result, which is essentially the same as that obtained by Kennard many years ago,[35] is rather surprising because there is a false static term for $t < r/c$, and even for $t > r/c$ the vector potential is not properly retarded. The result expressed by (5-57), which should be recognized as a purely *classical* result, is peculiar to the Coulomb gauge theory and may be viewed as the origin of the difficulties in the quantum theory discussed by

Power and Zienau.[10] If we evaluate the field at the position of the charge [i.e., put $x = R$ in (5-50)], we obtain

$$A^*(R,t) = -\left(\frac{4e}{3c^2}\right)\left(\frac{1}{4\pi\varepsilon_0}\right)\ddot{R}(t) \qquad (5\text{-}58)$$

Since the system of charged particle and field is closed, this field must be the field due to the charge; thus (5-58) describes how the "self-field" acts back on the charge.

V.D. Self-Interactions and Renormalization: Classical and Quantum Theories

Let us now reconsider Hamilton's equations of motion for the particle, (5-31) and (5-32), since according to (5-40) and (5-58) the vector potential to be used in these equations may now be written explicitly as

$$A(R,t) = A_h(R,t) - \left(\frac{4e}{3c^2}\right)\left(\frac{1}{4\pi\varepsilon_0}\right)\ddot{R}(t) \qquad (5\text{-}59)$$

The important part of this equation is the term that depends on the particle's accleration; since it is independent of the coordinate R, it makes no contribution to the time derivative of the particle momentum. It will be convenient to ignore for the moment the contribution of $A_h(R,t)$ since its inclusion does not affect substantially the result we wish to derive. Thus with the neglect of $A_h(R,t)$, (5-31) and (5-32) become

$$m\dot{R}(t) = p + \left(\frac{4e^2}{3c^2}\right)\left(\frac{1}{4\pi\varepsilon_0}\right)\ddot{R}(t) \qquad (5\text{-}60)$$

$$\dot{p} = 0 \qquad (5\text{-}61)$$

so that in this approximation the momentum is a constant of the motion $p = p_0$. If we define a constant $w_0 = (3\pi\varepsilon_0 mc^2/e^2)$, the equation of motion for the particle, (5-60), may be written as

$$\ddot{R}(t) - w_0\dot{R}(t) = -\left(\frac{w_0}{m}\right)p_0 \qquad (5\text{-}62)$$

The general solution of the homogeneous differential equation

$$\left(\frac{d^2}{dt^2} - w_0\frac{d}{dt}\right)R_0 = 0 \qquad (5\text{-}63)$$

is

$$R_0 = R_1 + R_2\exp(w_0t) \qquad (5\text{-}64)$$

where $\mathbf{R}_1$ and $\mathbf{R}_2$ are arbitrary, constant vectors that must be chosen to satisfy the boundary conditions for a given problem. A particular integral of (5-62) is easily found—namely, $\mathbf{p}_0 t / m$, so that a complete solution of (5-62) may be written as

$$\mathbf{R}(t) = \mathbf{R}_1 + \mathbf{R}_2 \exp(w_0 t) + \frac{\mathbf{p}_0 t}{m} \qquad (5\text{-}65)$$

The particular integral depends of course on the boundary condition chosen; (5-65) is appropriate to *retarded* boundary conditions.

The significance of (5-65) is most clearly seen if we compute the velocity $\dot{\mathbf{R}}$

$$\dot{\mathbf{R}}(t) = w_0 \mathbf{R}_2 \exp(w_0 t) + \frac{\mathbf{p}_0}{m} \qquad (5\text{-}66)$$

We now have $\dot{\mathbf{R}}(-\infty) = \mathbf{p}_0 / m$ as expected, but it is not possible to have $\dot{\mathbf{R}}(+\infty) = \mathbf{p}_0 / m$ unless we choose $\mathbf{R}_2 \equiv 0$, because of the runaway exponential. [If the homogeneous solution $\mathbf{A}_h(\mathbf{R}, t)$ is included in the equation of motion, there is an additional term in the velocity $\dot{\mathbf{R}}$ of the form $\int_t^\infty dt' \mathbf{A}_h(\mathbf{R}, t') \exp(w_0(t - t'))$—i.e., retarded boundary conditions—which vanishes at $t = +\infty$; the runaway exponential derives from the homogeneous equation (5-62) and is therefore obtained regardless of whether we neglect $\mathbf{A}_h(\mathbf{R}, t)$ in (5-60).] Furthermore, the exponential term has the undesirable feature of an essential singularity at $e = 0(w_0 = \infty)$; one would hope to be able to construct solutions of the interacting charge and field problem that pass over smoothly into the solutions of the noninteracting problem as $e \rightarrow 0$.[65,66] The appearance of the runaway exponential in (5-65) is a manifestation of the failure of the canonical variables $\{\mathbf{R}, \mathbf{p}\}$ and $\{\mathbf{A}, \mathbf{\Pi}\}$ to form a complete set of dynamical variables; that is, strictly speaking one cannot base a hamiltonian formalism on these variables alone. Since Hamilton's equations of motion are first order, the solutions must be determined by specifying data at a single time. However it is clear from (5-65) that the requirement that $\dot{\mathbf{R}}(t)$ be finite at all times requires that $\mathbf{R}_2 = 0$, which in turn requires specifying $\dot{\mathbf{R}}(t)$ at some finite time as well as at $t = -\infty$. It may also be shown that when one includes the contribution from the homogeneous field $\mathbf{A}_h(\mathbf{R}, t)$, the Poisson bracket $[\mathbf{R}, \mathbf{A}]$ does not vanish, thereby contradicting the independence of the particle and field variables we assumed at the outset. If we continue to insist that $\mathbf{A}_h$ has the properties of a freefield, we are led to conclude that constraints as well as additional variables must be involved. These conclusions are not materially altered if we impose alternative boundary conditions on the equations of motion.[64] Equation (5-59) shows clearly that there is an acceleration dependence in the lagrangian (5-4) arising from the self-field of the charge; if this field is to be retained, we ought to regard the velocity $\dot{\mathbf{R}}$ as an additional lagrangian "coordinate" ($\ddot{\mathbf{R}}$ would then be its corresponding "velocity") in which case the relation $m\dot{\mathbf{R}} = \mathbf{p} - e\mathbf{A}(\mathbf{R})$, (5-7) would become an equation of constraint, since it could then be interpreted as an equation involving only canonical coordinates and momenta.

At this point it is convenient to interrupt the discussion and take stock of what has been found. We set out to reformulate the Maxwell-Lorentz theory of classical electrodynamics (Section II) in a hamiltonian framework as a first step toward a quantum theory of the interaction of radiation with atoms and molecules. We wrote down a lagrangian for a single charge interacting with the electromagnetic field, (5-4), from which we could recover the Maxwell equations and the Lorentz force law, and so passed to the hamiltonian (5-21) using the general canonical techniques described in Section III. When we described the Maxwell-Lorentz theory we pointed out that Maxwell's field equations refer to a specified charge-current distribution, whereas conversely the Lorentz force law describes the motion of a charged particle in a specified electromagnetic field. It was for this reason that we emphasized the independence of the particle and field dynamical variables in both the lagrangian and hamiltonian formulations. On solving Hamilton's equations of motion for the dynamical variables $\{\mathbf{R}, \mathbf{p}\}$ and $\{a_\lambda(\mathbf{k}), a_\lambda^+(\mathbf{k})\}$, however, we found that the charge was accelerated infinitely by the action of its *own* field. Moreover, because of the contribution of the self-field to the Poisson bracket $[\mathbf{R}, \mathbf{A}]$ the independence of the particle and field variables assumed initially cannot be justified [recall also (5-28), which shows that the self-Coulomb field of the particle gives an infinite contribution to the energy]. In summary, the central difficulty in the formulation of an electrodynamics of charged particles is the problem of how to handle the self-field of a charge.

The self-interactions cannot be simply ignored; as early as 1930 Oppenheimer raised the question of whether the radiative corrections to the differences in energy levels of an atom could be finite even though the self-energy of each level diverges.[67] The beautiful and extensive experiments of Lamb and co-workers[68-70] established very accurate values for the splitting between the $^2S_{1/2}$ and $^2P_{1/2}$ levels in atomic hydrogen ("the Lamb shift"), which is now accepted as the classic example of the phenomenon that Oppenheimer drew attention to. And the successful calculation of this correction using quantum electrodynamics and the ideas of renormalization theory was a major achievement of postwar theoretical

physics. The origins of renormalization theory can be found in Kramers's*
pioneering "structure-independent" approach to the electrodynamics of a
nonrelativistic charged particle interacting with its "proper" (self)
electromagnetic field in addition to external radiation and potential
fields.[71,72] Kramers showed that the interaction of the charge with its own
field could be incorporated into the theory by a renormalization of the
particle's mass, provided the appropriate choice is made for the self-field.
The resulting equations refer only to the *observed* mass and charge of the
charged particle (Kramers calls this the *structure-independent* aspect of the
theory), and the dynamical equations for the particle contain interaction
terms involving only the external fields. Kramers placed his structure-
independent equations in canonical form, and as a consequence of the
mass renormalization his resulting hamiltonian could be used to predict
the nonrelativistic Lamb shift.[74] Kramers's investigations lead one to the
conclusion that the lagrangian in (5-1) to (5-4) does *not* describe charged
particles in terms of the experimentally available quantities, such as the
mass and the charge of the particle. Instead the parameters e and m that
appear in the lagrangian must be interpreted as formal parameters rather
than observed quantities. One might have suspected this conclusion from a
consideration of the way the lagrangian was put together in the first place.
There is an obvious obscurity in the reference in (5-1) to the particle in the
absence of the electromagnetic field, since a charged particle cannot be
meaningfully separated from its *own* field; the particle that is observed is
always in interaction with its self-field. There cannot be self-interactions in
theories that use the observed particle parameters, since they are already
taken account of. It should be emphasized that the renormalization program is independent of the infinities in the theory in the sense that even a
convergent theory would have to be expressed in terms of the observed
experimental masses and other constants.[37] Electrodynamics, and all point
interaction theories, give infinite values for these renormalizations, and the
actual renormalization procedure is not well defined mathematically. The
mass renormalization referred to is the only renormalization required in
nonrelativistic electrodynamics; further renormalizations are called for
only in relativistic theories.[58]

Kramers's method unfortunately suffers from an ambiguous specifica-

*The following remark of Uhlenbeck[3,73] describes Kramers's feelings: "I can still hear
Kramers lament...he just could not understand why Dirac's radiation theory was so good,
even though it made insufficient distinction between the proper (self) field of the electron and
the external field, even though it did not give an account for the infinite electromagnetic
mass; and even though the correspondence with the classical theory of the electron was
entirely unclear.... He was aware earlier and more deeply than most of his contemporaries of
the imperfections in the theory."

tion of the canonical variables of the hamiltonian theory; we have already suggested that the existence of the self-interaction in the structure-dependent theory implies the presence of constraints and additional variables. The ambiguities in Kramers's theory arise because he has not formally incorporated the constraints (the relations among the canonical variables) into his hamiltonian formalism. Schiller and Schwartz[75] have reformulated Kramers's method to take account of the constraints, using essentially the generalized hamiltonian mechanics developed by Dirac[50,51] and discussed in Section III. They show that the mass renormalization techniques employed by Kramers, and the alternative methods of van Kampen[76] and Steinwedel,[77] can all be put into a consistent framework; all these structure-independent theories may be generated by canonical transformations from the original structure-dependent theory based on the lagrangian (5-1) to (5-4), provided the structure-dependent theory is modified to include the velocity $\dot{\mathbf{R}}$ and its conjugate momentum $\mathbf{P}_{\dot{\mathbf{R}}}(= \partial L / \partial \dot{\mathbf{R}})$ as canonical variables. It should be noted, however, that this program has been carried out only in the electric dipole approximation in which all retardation effects are to be ignored in evaluating the electron's charge distribution. As a result of this assumption, one neglects the magnetic forces acting on the particle, as well as particle recoil. The reader interested in further details of these workers' elegant formulation is referred to the original papers.[75]

To make the transition to a quantum theory, the conventional steps may be taken: the variables $\{\mathbf{R}, \mathbf{p}\}$ and $\{\mathbf{A}(\mathbf{x}), \mathbf{\Pi}(\mathbf{x})\}$ (or equivalently, $\{a_\lambda(\mathbf{k}), a_\lambda^+(\mathbf{k})\}$) are reinterpreted as operators, and the nonzero Dirac brackets of the classical theory are replaced by their quantum counterparts, the commutators:

$$[R^s, p^t] = i\hbar\delta_{st} \tag{5-67}$$

$$\left[A(\mathbf{x})^s, \Pi(\mathbf{x}')^t\right] = i\hbar\delta_{st}^{\perp}(\mathbf{x} - \mathbf{x}') \tag{5-68}$$

$$\left[p^s, \Pi(\mathbf{x})^t\right] = i\hbar e p^s \nabla_x^t\left(\frac{1}{(4\pi|\mathbf{R} - \mathbf{x}|)}\right) \tag{5-69}$$

$$[a_\lambda(\mathbf{k}), a_{\lambda'}^+(\mathbf{k}')] = \hbar\delta_{\lambda\lambda'}\delta(\mathbf{k} - \mathbf{k}') \tag{5-70}$$

The description of the electromagnetic field in the quantum theory in terms of photons is well known and is assumed here; thus the operators $a_\lambda(\mathbf{k})$ and $a_\lambda^+(\mathbf{k})$ are the usual photon annihilation and creation operators. The hamiltonian (5-21), which is reinterpreted as an operator, is not explicitly time dependent, and it is often convenient to transform to the Schrödinger representation. We can then write down the fundamental *eigenvalue* equation

$$(H - E)|\Psi\rangle = 0 \tag{5-71}$$

where $|\Psi\rangle$ is the Schrödinger picture state-vector for the combined system of charge and electromagnetic field, E is the total energy of the system, and

$$H = \left(\frac{1}{2m}\right)(\mathbf{p} - e\mathbf{A}(\mathbf{R}))^2 + \frac{\varepsilon_0}{2}\int d^3\mathbf{x}\{\mathbf{E}(\mathbf{x})\cdot\mathbf{E}(\mathbf{x}) + c^2\mathbf{B}(\mathbf{x})\cdot\mathbf{B}(\mathbf{x})\} \tag{5-72}$$

This transformation to the Schrödinger representation leads us to expect that the methods of solution of the dynamical problem will be very different from the classical calculation described earlier. The reader interested in equations of motion techniques in quantum electrodynamics should consult Dirac's masterly exposition.[58]

In practice, (5-71) is approached from the point of view of perturbation theory because we are interested in the exchange of energy between the different parts of the total system. The "self-interaction" of the charge with its own field is viewed as a process in which the charge emits one or more photons and subsequently reabsorbs them; such processes are called "virtual photon interactions." When this picture is combined with perturbation theory, it leads naturally to the idea of obtaining the approximate solutions of (5-71) as a power series in the fine structure constant $\alpha = (e^2/\hbar c)$, since each power of α can be linked with the emission and absorption of one photon. The mass renormalization in the quantum theory is thus usually performed with the aid of an (improper) unitary transformation of the structure-dependent hamiltonian operator (5-72); the question of the existence of constraints and additional variables in the theory is not faced, and instead one only aims to eliminate the self-interactions to a given order in α. The transformation that generates the mass renormalization to order α is called the Schwinger transformation[78] and is performed with an operator Σ[37]

$$\Sigma = \exp\left\{-\left(\frac{ie}{\hbar mc^2}\right)\mathbf{p}\cdot\mathbf{Z}(\mathbf{R})\right\} \tag{5-73}$$

In this expression m is the unrenormalized mass of the charge [as in (5-72)], and $\mathbf{Z}(\mathbf{R})$ is the Hertz operator for the electromagnetic field, which may be expresses in terms of the following integral over the transverse component of the canonical momentum $\mathbf{\Pi}(\mathbf{x})$:

$$Z(\mathbf{x}) = - \int d^3x' \frac{\mathbf{\Pi}^\perp(\mathbf{x}')}{(4\pi\varepsilon_0|\mathbf{x}-\mathbf{x}'|)} \tag{5-74}$$

Since $Z(\mathbf{x})$ is transverse, the order of the operators in (5-73) is immaterial. If we introduce the operator Fourier expansion of the canonical momentum $\mathbf{\Pi}(\mathbf{x})$ [cf. (5-40)] into (5-74), we obtain a more useful representation for the Hertz field operator

$$Z(\mathbf{x}) = i \sum_{\mathbf{k},\lambda} \left(\frac{c}{2V\varepsilon_0 k^3}\right)^{1/2} \hat{\mathbf{E}}_\lambda(\mathbf{k})\{a_\lambda(\mathbf{k})e^{i\mathbf{k}\cdot\mathbf{x}} - a_\lambda^+(\mathbf{k})e^{-i\mathbf{k}\cdot\mathbf{x}}\} \tag{5-75}$$

The effect of the virtual photon processes in Σ can be isolated by arranging the terms in the exponential such way that the annihilation operators act first.[24] This may be achieved with the aid of the Baker-Hausdorff formula

$$\exp(A+B) = \exp(A)\exp(B)\exp\left(\frac{[B,A]}{2}\right) \tag{5-76}$$

which is an identity provided that the commutator $[B,A]$ commutes with A and B. We put

$$A = \left(\frac{e}{\hbar mc^2}\right)\left(\frac{c}{2V\varepsilon_0}\right)^{1/2} \sum_{\mathbf{k},\lambda} \frac{\mathbf{p}\cdot\hat{\mathbf{E}}_\lambda(\mathbf{k})}{k\sqrt{k}} a_\lambda(\mathbf{k})e^{i\mathbf{k}\cdot\mathbf{R}} \tag{5-77}$$

$$B = \left(\frac{e}{\hbar mc^2}\right)\left(\frac{c}{2V\varepsilon_0}\right)^{1/2} \sum_{\mathbf{k},\lambda} \frac{\mathbf{p}\cdot\hat{\mathbf{E}}_\lambda(\mathbf{k})}{k\sqrt{k}} a_\lambda^+(\mathbf{k})e^{-i\mathbf{k}\cdot\mathbf{R}} \tag{5-78}$$

and evaluate the commutator $[B,A]$ with the aid of (5-49) and (5-70):

$$[B,A] = \left(\frac{e}{\hbar mc^2}\right)^2\left(\frac{c}{2V\varepsilon_0}\right) \sum_{\mathbf{k},\lambda}\sum_{\mathbf{k}',\lambda'} \frac{\mathbf{p}\cdot\hat{\mathbf{E}}_\lambda(\mathbf{k})\mathbf{p}\cdot\hat{\mathbf{E}}_{\lambda'}(\mathbf{k}')}{k\sqrt{k}\ k'\sqrt{k'}} e^{i(\mathbf{k}-\mathbf{k}')\cdot\mathbf{R}}$$
$$\times [a_{\lambda'}^+(\mathbf{k}'), a_\lambda(\mathbf{k})]$$
$$= -\left(\frac{e^2}{2\hbar m^2c^3\varepsilon_0}\right)\frac{1}{V} \sum_{\mathbf{k}} \frac{\mathbf{pp}:(1-\hat{\mathbf{k}}\hat{\mathbf{k}})}{k^3} \tag{5-79}$$

On converting the summation over $\mathbf{k}$ to an integration, (5-79) simplifies to

$$[B,A] = -\left(\frac{1}{6\pi^2\varepsilon_0}\right)\left(\frac{\alpha p^2}{m^2 c^2}\right)\int_0^\infty dk\, k^{-1} \qquad (5\text{-}80)$$

which involves a divergent integral. Consequently the transformation $\Sigma^{-1}H\Sigma$ cannot, strictly speaking, be regarded as a unitary transformation; it has become customary to call such transformations "improper (or unbounded) unitary transformations." These improper transformations occur very frequently in field theories. For example, whereas in the nonrelativistic quantum *mechanics* of point particles the coordinate and momentum representations are related by a Fourier transform, so that the interaction part and the unperturbed part ($=$ kinetic energy) of the hamiltonian can be separately diagonalized in representations that are related by a unitary transformation, only a weaker result holds in quantum field theory. Although it is usually possible to find different representations that diagonalize the interaction and unperturbed parts of the hamiltonian, such representations can be connected only by an improper unitary transformation.[79]

We confine our discussion of the Schwinger transformation to the electric dipole approximation; that is, we neglect the effects of retardation over the electron's charge distribution. The new variables are formally related to the old variables by the usual equation

$$\Omega^{\text{new}} = \Sigma^{-1}\Omega^{\text{old}}\Sigma \qquad (5\text{-}81)$$

To carry out the transformation; we use the following formula, which holds for two noncommuting quantities.[37]

$$\exp(A)B\exp(-A) = B + [A,B] + \tfrac{1}{2}[A,[A,B]] + \cdots \qquad (5\text{-}82)$$

This equation is easily derived if one uses the fact that exponential operators are defined by their power series expansions. In the electric dipole approximation, the particle momentum commutes with Σ, as does the canonical momentum of the field $\mathbf{\Pi}(\mathbf{x})$; thus we have to consider only the canonical "coordinates" $\mathbf{R}$ and $\mathbf{A}(\mathbf{x})$. The effect of Σ on the position coordinate $\mathbf{R}$ is simply to generate a displacement

$$\mathbf{R}^{\text{new}} = \mathbf{R}^{\text{old}} - \left(\frac{e}{mc^2}\right)\mathbf{Z}(\mathbf{R}) \qquad (5\text{-}83)$$

The transformation of the vector potential $\mathbf{A}(\mathbf{x})$ is most easily approached through the transformation of the Fourier operators $a_\lambda(\mathbf{k})$ and $a_\lambda^+(\mathbf{k})$; since the exponent in Σ is linear in the Fourier operators, the series in (5-82) terminates after the first commutator, and we find

$$a_\lambda(\mathbf{k})^{\text{new}} = a_\lambda(\mathbf{k})^{\text{old}} + \left(\frac{e}{mc^2}\right)\frac{c}{(2Vk^3\varepsilon_0)}\mathbf{p}\cdot\hat{\mathbf{E}}_\lambda(\mathbf{k})e^{-i\mathbf{k}\cdot\mathbf{R}} \qquad (5\text{-}84)$$

$$a_\lambda^+(\mathbf{k})^{\text{new}} = a_\lambda^+(\mathbf{k})^{\text{old}} + \left(\frac{e}{mc^2}\right)\frac{c}{(2Vk^3\varepsilon_0)}\mathbf{p}\cdot\hat{\mathbf{E}}_\lambda(\mathbf{k})e^{i\mathbf{k}\cdot\mathbf{R}} \qquad (5\text{-}85)$$

Substitution of these two equations into the quantum version of (5-41) leads to the transformed vector potential

$$\mathbf{A}(\mathbf{x})^{\text{new}} = \mathbf{A}(\mathbf{x})^{\text{old}} + \left(\frac{e}{mc^2\varepsilon_0}\right)(2\pi)^{-3}\int d^3\mathbf{k}\,\frac{\mathbf{p}\cdot(1-\hat{k}\hat{k})e^{i\mathbf{k}\cdot\mathbf{x}-\mathbf{R}}}{k^2} \qquad (5\text{-}86)$$

The angular integration can be carried out at once using (5-51) and (5-52); the remaining integral is then quite straightforward, and we obtain

$$\int d^3k\,\frac{(1-\hat{k}\hat{k})e^{i\mathbf{k}\cdot\mathbf{r}}}{k^2} = \left(\frac{\pi^2}{r}\right)(1+\hat{r}\hat{r}) \qquad (5\text{-}87)$$

Thus (5-86) becomes

$$\mathbf{A}(\mathbf{x})^{\text{new}} = \mathbf{A}(\mathbf{x})^{\text{old}} + \left(\frac{e}{2mc^2}\right)\frac{\mathbf{p}\cdot(1+\hat{r}\hat{r})}{4\pi\varepsilon_0 r} \qquad (5\text{-}88)$$

where $\mathbf{r} = \mathbf{x} - \mathbf{R}$. In the limit $\mathbf{x}\to\mathbf{R}$ we have

$$\mathbf{A}(\mathbf{R})^{\text{new}} = \mathbf{A}(\mathbf{R})^{\text{old}} + \left(\frac{2}{3\pi}\right)(e/mc^2)\frac{\mathbf{p}}{(4\pi\varepsilon_0)}\int_0^\infty dk \qquad (5\text{-}89)$$

Equations (5-88) and (5-89) should be compared with the related classical results obtained previously, (5-57) and (5-58), remembering that to the accuracy required, $\mathbf{p} = m\dot{\mathbf{R}}$. In the formalism of Kramers, (5-88) can be represented as

$$\mathbf{A}^{\text{external}} = \mathbf{A}^{\text{total}} - \mathbf{A}^{\text{self}} \quad \text{(to order } e) \qquad (5\text{-}90)$$

That is, the Schwinger transformation is a device for effecting, in an approximate way, the separation of the external and the proper fields due to the charge.

With these results we may obtain the transformed hamiltonian H^{new} $=\Sigma^{-1}H\Sigma$; we have

$$\Sigma^{-1}(\mathbf{p}-e\mathbf{A}(\mathbf{R}))\Sigma$$

$$=\mathbf{p}\left\{1-\left(\frac{2}{3\pi}\right)\left(\frac{e^2}{mc^2}\right)\left(\frac{1}{4\pi\varepsilon_0}\right)\int_0^\infty dk\right\}-e\mathbf{A}(\mathbf{R}) \qquad (5\text{-}91)$$

and

$$\Sigma^{-1}\left(\frac{\varepsilon_0}{2}\int d^3\mathbf{x}\{\mathbf{E}(\mathbf{x})\cdot\mathbf{E}(\mathbf{x})+c^2\mathbf{B}(\mathbf{x})\cdot\mathbf{B}(\mathbf{x})\}\right)\Sigma$$

$$=\frac{\varepsilon_0}{2}\int d^3\mathbf{x}\{\mathbf{E}(\mathbf{x})\cdot\mathbf{E}(\mathbf{x})+c^2\mathbf{B}(\mathbf{x})\cdot\mathbf{B}(\mathbf{x})\}+\left(\frac{e}{m}\right)\mathbf{p}\cdot\mathbf{A}(\mathbf{R})$$

$$+\left(\frac{2}{3\pi}\right)\left(\frac{e}{mc}\right)^2\frac{p^2}{(4\pi\varepsilon_0)}\int_0^\infty dk \qquad (5\text{-}92)$$

Thus if we collect terms and introduce the fine structure constant α, we obtain

$$H^{\text{new}}=\frac{p^2}{(2m)}\left\{1+\left(\frac{1}{12\pi^2\varepsilon_0}\right)\left(\frac{\alpha\hbar}{mc}\right)\int_0^\infty dk\right\}+\left(\frac{e^2}{2m}\right)A(\mathbf{R})^2$$

$$+\frac{\varepsilon_0}{2}\int d^3\mathbf{x}\{\mathbf{E}(\mathbf{x})\cdot\mathbf{E}(\mathbf{x})+c^2\mathbf{B}(\mathbf{x})\cdot\mathbf{B}(\mathbf{x})\}+0(e^3) \qquad (5\text{-}93)$$

Finally, let us define the renormalized mass $\tilde{m}$,

$$\frac{1}{\tilde{m}}=\frac{1}{m}\left\{1+\left(\frac{1}{12\pi^2\varepsilon_0}\right)\left(\frac{\alpha\hbar}{mc}\right)\int_0^\infty dk\right\}+0(\alpha^2) \qquad (5\text{-}94)$$

The structure-independent hamiltonian correct to order α is now

$$H^{\text{new}}=\frac{p^2}{(2\tilde{m})}+\left(\frac{e^2}{2\tilde{m}}\right)A(\mathbf{R})^2$$

$$+\frac{\varepsilon_0}{2}\int d^3\mathbf{x}\{\mathbf{E}(\mathbf{x})\cdot\mathbf{E}(\mathbf{x})+c^2\mathbf{B}(\mathbf{x})\cdot\mathbf{B}(\mathbf{x})\} \qquad (5\text{-}95)$$

in which there are no terms that give rise to self-interactions (to order α).

This transformation is not without possible practical consequences in the many-particle case. The Coulomb energies arising from $\frac{1}{2}\varepsilon_0 \int d^3\mathbf{x}\mathbf{E}(\mathbf{x})^{\|} \cdot \mathbf{E}(\mathbf{x})^{\|}$ are affected by the transformation, since according to (5-83) the individual position operators $\mathbf{R}_i$ are displaced by an amount proportional to $e_i\mathbf{Z}(\mathbf{R}_i)$, and the Coulomb energies involve $\mathbf{R}_i - \mathbf{R}_j$

$$\sum_{i,j=1}^{N} \frac{e_i e_j}{|\mathbf{R}_{ij}|} \overset{\Sigma}{\Longrightarrow} \sum_{i,j=1}^{N} \frac{e_i e_j}{|\mathbf{R}_{ij} - \mathbf{J}(\mathbf{R}_i,\mathbf{R}_j)|}$$

$$\mathbf{J}(\mathbf{R}_i,\mathbf{R}_j) = \frac{\left\{ \left(\frac{e_i}{m_i}\right)\mathbf{Z}(\mathbf{R}_i) - \left(\frac{e_j}{m_j}\right)\mathbf{Z}(\mathbf{R}_j) \right\}}{c^2}$$

One can then write a new atom-field hamiltonian as

$$H^{\text{new}} = H^{\text{atom}} + H^{\text{field}} + \sum_{i=1}^{N} (e_i^2/2m_i)A(\mathbf{R}_i)^2 - V$$

$$V = \sum_{i,j=1}^{N} e_i e_j \left\{ \frac{1}{|\mathbf{R}_{ij}|} - \frac{1}{|\mathbf{R}_{ij} - \mathbf{J}(\mathbf{R}_i,\mathbf{R}_j)|} \right\}$$

Although this hamiltonian has been obtained under the assumption of the electric dipole approximation, there is no essential change in the next approximation, that is, when we put $\exp(\pm i\mathbf{k} \cdot \mathbf{R}) \sim 1 \pm i\mathbf{k} \cdot \mathbf{R}$ in the Fourier expansion of the Hertz vector $\mathbf{Z}(\mathbf{R})$. In the usual theory this approximation corresponds to retaining magnetic dipole and electric quadrupole transition operators. If higher-order terms are needed, it is probably preferable to use the closed polarization field expressions discussed in Section VI B of the next chapter. In a perturbation theory treatment, the term proportional to $A(\mathbf{R}_i)^2$ can usually be dropped (the circumstances in which it can contribute in scattering processes are well known); and since V is no longer linear in the fine structure constant, the possibility exists of constructing a perturbation theory of *multiphoton* processes. In particular, it might be possible to base on H^{new} a theory of nonlinear light scattering from atoms and molecules for which the usual perturbation theory rapidly becomes unmanageable; nothing has been done on these lines, however, and this remains a topic for future investigation [cf. F. H. M. Faisal, *J. Phys. B*, **6**, L89-L92 (1973)].

In summary then, we have worked out in detail the hamiltonian theory of a nonrelativistic charge interacting with radiation, using the Coulomb gauge condition as a subsidiary condition. We have discovered a fundamental difficulty due to the self-interaction of the charge with its own field; this can be handled using renormalization theory, which we discuss further in the context of the self-interaction of atoms and molecules. We note at this state, however, that renormalization involves some cavalier handling of infinite quantities. The result of the renormalization is that the interaction terms refer only to *external* fields, and the theory is expressed in terms of the experimentally observable mass and charge of the particle. The other important feature of our calculations that may give rise to practical, as opposed to formal or logical, difficulties, is the appearance of apparently causality-violating static terms in the Coulomb gauge vector potential, (5-57), which derive from the Coulomb gauge propagation function, the dyadic $\mathcal{D}$ $(\mathbf{x}, \tau)$ (see footnote on p. 197). In the next section we meet similar problems in the context of the nonrelativistic electrodynamics of collections of bound charges.

VI. THE ELECTRODYNAMICS OF ATOMS AND MOLECULES

VI.A. The Coulomb Gauge Theory

The hamiltonian for a closed system of charged particles interacting with the electromagnetic field may be written down at once from the results obtained previously. If we introduce a subscript i on the particle variables, the hamiltonian equations (5-21) to (5-23) become

$$H = \tfrac{1}{2} \sum_{i=1}^{N} \frac{1}{m_i} (\mathbf{p}_i - e_i \mathbf{A}(\mathbf{R}_i))^2 + \frac{\varepsilon_0}{2} \int d^3\mathbf{x} \{ \mathbf{E}(\mathbf{x}) \cdot \mathbf{E}(\mathbf{x}) + c^2 \mathbf{B}(\mathbf{x}) \cdot \mathbf{B}(\mathbf{x}) \} \quad (6\text{-}1)$$

$$\nabla \cdot \mathbf{A}(\mathbf{x}) = 0 \quad (6\text{-}2)$$

$$\varepsilon_0 \nabla \cdot \mathbf{E}(\mathbf{x}) = \rho(\mathbf{x}) \quad (6\text{-}3)$$

where the charge density for the N-particle system is defined as in (2-10). The commutation relations generalize immediately if we remember that the operators referring to different particles commute. The generalization of (5-24) to (5-28) leads to the Coulomb interaction between the charges, together with the infinite constant which we shall ignore,

$$\frac{\varepsilon_0}{2} \int d^3\mathbf{x} \mathbf{E}(\mathbf{x})^{\parallel} \cdot \mathbf{E}(\mathbf{x})^{\parallel} = \frac{1}{(4\pi\varepsilon_0)} \sum_{i \neq j}^{N} \frac{e_i e_j}{|\mathbf{R}_i - \mathbf{R}_j|} \quad (6\text{-}4)$$

We have already remarked that the quantum theory is essentially committed to a perturbation theory point of view; thus it is convenient to regroup the terms in (6-1) into the form implied by (1-2):

$$H = H^{\text{atom}} + H^{\text{field}} + V \tag{6-5}$$

where

$$H^{\text{atom}} = \sum_{i=1}^{N} \frac{p_i^2}{2m_i} + \sum_{i \neq j}^{N} \frac{e_i e_j}{4\pi\varepsilon_0 |\mathbf{R}_i - \mathbf{R}_j|} \tag{6-6}$$

$$H^{\text{field}} = \frac{\varepsilon_0}{2} \int d^3\mathbf{x} \left\{ \mathbf{E}(\mathbf{x})^{\perp} \cdot \mathbf{E}(\mathbf{x})^{\perp} + c^2\mathbf{B}(\mathbf{x}) \cdot \mathbf{B}(\mathbf{x}) \right\} \tag{6-7}$$

$$V = -\sum_{i=1}^{N} \left(\frac{e_i}{m_i} \right) \mathbf{p}_i \cdot \mathbf{A}(\mathbf{R}_i) + \sum_{i=1}^{N} \left(\frac{e_i^2}{2m_i} \right) \mathbf{A}(\mathbf{R}_i) \cdot \mathbf{A}(\mathbf{R}_i) \tag{6-8}$$

and we have used the Coulomb gauge condition (6-2) to simply the perturbation operator V. These equations contain the unrenormalized parameters e_i and m_i, and consequently we must expect to find divergent expressions in the perturbation theory expansion. The significance of the infinities is now well understood however, and if we are prepared to set aside the fundamental question about the consistency of the theory, these infinite quantities can usually be separated out easily so as to leave physically significant results.

A common response to the dilemma caused by the appearance of a divergent perturbation theory expansion is to reject the philosophy of quantum electrodynamics and, instead, to work with a classical, external electromagnetic field using time-dependent perturbation theory. Such a "semiclassical" formalism was known even before Dirac's radiation theory but was regarded as inadequate because it could not account for the spontaneous emission of light by excited states. In recent years there has been renewed interest in low-energy (nonrelativistic) radiation phenomena, largely because of the discovery of the laser. One of the results of this resurgence of interest in nonrelativistic theory has been the attempt to revive the semiclassical method modified to account for spontaneous emission.[80,81] Some of the predictions of this new theory, which is often called the neoclassical theory (NCT) of electrodynamics are however not the same as those of quantum electrodynamics.

According to quantum electrodynamics, an excited quantum state should decay exponentially in time, in a manner independent of the way it was produced;[82] on the other hand, in NCT the decay behavior depends on the initial population of the excited state. Recent experiments on the radiative decay of the $2P$ level in atomic hydrogen produced by pulsed excitation from the $2S$ level, support the prediction of quantum electro-dynamics and are definitely not in agreement with the prediction of the neoclassical theory.[83] There is similar disagreement over the polarization fractions observed in cascade decays.[84] The experiment suggested origin-ally as a test of the two competing theories has also been recently carried out.[85] Quantum electrodynamics predicts that the incoherent resonance fluorescence from a two-level system is a maximum when only the excited state is occupied. A pure excited state can be produced in a sharp line absorber by a short coherent optical pulse if the pulse area Θ is equal to $(2n+1)\pi(n=0,1,2,\ldots)$. Here

$$\Theta = \left(\frac{2d}{\hbar}\right)\int_{-\infty}^{+\infty} dt\, E(t) \tag{6-9}$$

and d is the transition electric dipole moment, and $E(t)$ the slowly varying amplitude of the electric field. The fluorescence is zero when only the ground state is occupied, which occurs, for example, after illumination by a short pulse of area $2n\pi(n=0,1,2,\ldots)$. This oscillatory dependence of the fluorescence on input area has been demonstrated in rubidium vapor under conditions that approximate closely to an ideal two-level system. The neoclassical theory, however, predicts minima near $\Theta = n\pi(n = 1,2,\ldots)$, since in NCT a pure state is unable to radiate spontaneously. Gibbs also found that an exponential time decay was followed, whereas NCT predicts substantial deviation from a single exponential decay if the average effective pulse area exceeds $\pi/4$.[85]

Thus despite the internal inconsistencies in quantum electrodynamics discussed earlier, it does seem to be the best theory we have. Besides, the proposition that perturbation theory based on a quantized radiation field is more difficult to use and harder to understand than the time-dependent perturbation theory used in semiclassical theories appears to be doubtful at best, and there seems to be little if anything to be gained by giving up quantum electrodynamics as the customary vehicle for the description of radiation phenomena.

A critical discussion of the application of the Coulomb gauge equations (6-6) to (6-8), to the description of the interaction of atoms with the electromagnetic field has been given by Power and Zienau.[10] They con-clude that when the Coulomb gauge formalism is employed to describe radiative processes in atoms, special care is required to avoid the appearance of false static terms. These (unphysical) static terms, which take the form of precursor signals, appear in the propagation functions of the free-field variables even when the equations of motion are solved with retarded boundary conditions [cf. (5-57)]. This unexpected feature of the propagation function of the vector potential is actually necessary to guarantee causality, since in calculations involving the exchange of virtual photons between molecules[9, 10, 37] or molecular subsystems,[86, 87] the instantaneous Coulomb energies must not be forgotten. Furthermore, within the usual Wigner-Weisskopf model of the radiative decay of excited states, there are difficulties with false singularities in the development coefficients of the wavefunction for the atom + field system. To avoid all these difficulties, the authors propose a formalism in which instead of the vector potential $A(x)$, the working variables are the fields $E(x,t)$ and $B(x,t)$, and is obtained with the aid of a unitary transformation.[10] We discussed the history of this transformation in the Introduction, remarking that it may also be looked on as setting up the hamiltonian theory in another gauge. The detailed calculations required in these two procedures can be found in the literature[13,29] and are not repeated here; all that we wish to add to those discussions is a brief examination of the operator Λ, (1-5), in the case when the vector potential $A(x)$ is quantized and is not orthogonal to the atomic polarization field $P(x)$.

VI.B. Electrodynamics Without Potentials

We recall from Section I that the operator Λ may be written in the form

$$\Lambda = \exp\left(\frac{-iS}{\hbar}\right) \qquad (6\text{-}10)$$

$$S = \int d^3x\, P(x) \cdot A(x), \qquad \text{Div}\, A(x) = 0 \qquad (6\text{-}11)$$

If we combine the Fourier representation of the Coulomb gauge vector potential, (5-41), with (6-11), we obtain

$$S = \frac{1}{(2\varepsilon_0 Vc)^{1/2}} \int d^3x \sum_{k,\lambda} \frac{P(x) \cdot \hat{E}_\lambda(k)}{\sqrt{k}} \left\{ a_\lambda(k) e^{ik\cdot x} + a_\lambda^+(k) e^{-ik\cdot x} \right\} \qquad (6\text{-}12)$$

We now seek to isolate the effect of the virtual photon processes, which may be done as before (Section V) with the aid of the Baker-Haussdorf formula, (5-76). We set

$$A = \left(\frac{-i}{\hbar}\right) \frac{1}{(2\varepsilon_0 Vc)^{1/2}} \int d^3x \sum_{k,\lambda} \frac{P(x) \cdot \hat{E}_\lambda(k)}{\sqrt{k}} a_\lambda(k) e^{ik\cdot x} \qquad (6\text{-}13)$$

$$B = \left(\frac{-i}{\hbar}\right) \frac{1}{(2\varepsilon_0 Vc)^{1/2}} \int d^3x \sum_{k,\lambda} \frac{P(x) \cdot \hat{E}_\lambda(k)}{\sqrt{k}} a_\lambda^+(k) e^{-ik\cdot x} \qquad (6\text{-}14)$$

and with the aid of (5-70), we obtain the commutator $[B,A]$ as

$$[B,A] = (2\pi)^{-3} \frac{1}{(2\varepsilon_0 \hbar c)} \int d^3x \int d^3x' \int d^3k \, \mathbf{P}(x)\mathbf{P}(x') : (1 - \hat{k}\hat{k}) \frac{e^{i\mathbf{k}\cdot\mathbf{x}-\mathbf{x}'}}{k}$$

(6-15)

The integration over $\mathbf{k}$ is easily done if we use the results in (5-51) and (5-52):

$$\left(\frac{1}{4\pi}\right) \int d^3K (1 - \hat{\mathbf{K}}\hat{\mathbf{K}}) \frac{e^{i\mathbf{K}\cdot\mathbf{X}}}{K} = 2\frac{\hat{\mathbf{X}}\hat{\mathbf{X}}}{X^2}$$

(6-16)

where $\hat{\mathbf{X}}$ is the unit vector $\mathbf{X}/|\mathbf{X}|$; thus (6-15) reduces to

$$[B,A] = \frac{1}{(2\pi^2 \varepsilon_0 \hbar c)} \int d^3x \int d^3x' \mathbf{P}(x+x')\mathbf{P}(x') : \frac{\hat{\mathbf{x}}\hat{\mathbf{x}}}{x^2}$$

(6-17)

The easiest way to show that this integral is divergent is to consider the simple case of the hydrogen atom, for which we can write

$$\mathbf{P}(x) = e \int_{\mathbf{R}_1}^{\mathbf{R}_2} dz \, \delta^3(\mathbf{z} - \mathbf{x})$$

$$= e\mathbf{R} \int_0^1 ds \, \delta^3(\mathbf{R}_1 + s\mathbf{R} - \mathbf{x}), \qquad \mathbf{R} = \mathbf{R}_2 - \mathbf{R}_1$$

(6-18)

where $\mathbf{R}_1$ and $\mathbf{R}_2$ are the electron and proton coordinates, respectively.[13] Then

$$\int d^3x' \, \mathbf{P}(x+x')\mathbf{P}(x') = e^2\mathbf{R}\mathbf{R} \int_0^1 ds \int_0^1 ds' \delta^3[(s-s')\mathbf{R} - \mathbf{x}]$$

(6-19)

so that if we introduce the fine-structure constant $\alpha = e^2/\hbar c$, (6-17) reduces to

$$[B,A] = \frac{\alpha}{(2\pi^2 \varepsilon_0)} \int_0^1 ds \int_0^1 ds' |s-s'|^{-2} = -\infty$$

(6-20)

since the *direction* of the vector $(s-s')\mathbf{R}$ is independent of $(s-s')$. Thus just as in the case of the Schwinger transformation discussed in Section V, we are led to conclude that Λ is an improper unitary operator. Therefore it is not surprising to find that the transformation induced by Λ is a renormalization; the new hamiltonian contains a term that, for example, serves to subtract off the most divergent contribution to the atomic self-energy (the Lamb shift), computed using perturbation theory. It is not correct, however, to suppose that the new hamiltonian is restricted to the description of radiation processes in atoms (systems of *bound* charges); this misapprehension may arise if the polarization fields are represented as multipole series; but this is not a necessary step, and to emphasize this point we derive the hamiltonian for a single charge expressed in terms of its polarization fields and the electromagnetic field strengths $\mathbf{E}(\mathbf{x},t)$ and $\mathbf{B}(\mathbf{x},t)$ starting from the lagrangian formalism.[11,12]

The new lagrangian has the general structure shown in (5-1) and differs from the theory discussed in Section V only in the choice for the interaction part of the lagrangian

$$L^{\mathrm{int}} = \int d^3\mathbf{x}\mathbf{P}(\mathbf{x},t)\cdot\mathbf{E}(\mathbf{x},t) + \int d^3\mathbf{x}\mathbf{M}(\mathbf{x},t)\cdot\mathbf{B}(\mathbf{x},t) \qquad (6\text{-}21)$$

If we assume the usual connection between the field strengths and the field potentials, (2-2), and use the connection between the polarization fields and the charge-current density, it is easy to show that this lagrangian is related to (5-4) by the addition of a perfect derivative; therefore, as noted in Section III we expect to obtain the same equations of motion from the two lagrangians.[88] The explicit demonstration that this is true is not without interest, however, since it reveals a more intimate connection between the Lorentz force law and the Maxwell equations than could have been guessed from the conventional theory described in Section V. We shall not pursue this point here, merely noting that if the explicit form for the electric and magnetic polarization fields for a single charge, (2-16) and (2-17), are introduced into (6-21), we obtain at once

$$L^{\mathrm{int}} = \int_{\mathbf{r}}^{\mathbf{R}(t)} d\mathbf{z}\cdot\mathbf{F}(\mathbf{z},\dot{\mathbf{z}},t) \qquad (6\text{-}22)$$

where $\mathbf{F}$ is the Lorentz force defined in (2-21). Thus the interaction lagrangian is here written as the integral over the force acting on the charge evaluated along an arbitrary path $\mathbf{z}$ that ends at the instantaneous position $\mathbf{R}(t)$ of the charge. This is an immediate generalization of the well-known result in electrostatics, and it is valid even though the Lorentz force is *not conservative*. It is perfectly possible to adopt the point of view that the field strengths $\mathbf{E}(\mathbf{x},t)$ and $\mathbf{B}(\mathbf{x},t)$ are the fundamental quantities, whereupon the Maxwell equations (2-1a, 2-1c) need to be imposed as consistency equations to render (6-22) a suitable lagrangian.[12]

To recover the Maxwell equations from the combination of (5-2), (5-3), and (6-21) using the principle of least action, it is necessary to introduce the field potentials $\phi(\mathbf{x}, t)$ and $\mathbf{A}(\mathbf{x}, t)$ as lagrangian variables. This is because the least action formulation requires that the equations of motion be (at least) second-order differential equations in the lagrangian variables; this is true when the Maxwell equations are expressed in terms of ϕ and $\mathbf{A}$ but not when they are given in terms of $\mathbf{E}$ and $\mathbf{B}$. Since the starting equations of motion and the final hamiltonian are independent of ϕ and $\mathbf{A}$, we conjecture that there is a method of passing directly from one to the other without reference to ϕ and $\mathbf{A}$, but we have not been able to find it.

To obtain a hamiltonian description we use the momenta conjugate to the dynamical "coordinates" $\mathbf{R}, \phi(\mathbf{x}, t)$, and $\mathbf{A}(\mathbf{x}, t)$, in place of the "lagrangian velocities." On introducing the field potentials into the new lagrangian, we once again encounter the familiar difficulty that the momentum density $\Pi_0(\mathbf{x})$ conjugate to $\phi(\mathbf{x})$ vanishes formally. We have therefore a degenerate lagrangian system, and invariant relations may appear among the dynamical variables. The remaining nonzero momenta may be written

$$\mathbf{p} \approx m\dot{\mathbf{R}} + \mathbf{Q}(\mathbf{B}) \tag{6-23}$$

where

$$\mathbf{Q}(\mathbf{B}) = -e \int_{s_1}^{s_2} ds \, f(s) \mathbf{B}(\mathbf{z}) \times \left(\frac{\partial \mathbf{z}}{\partial s} \right) \tag{6-24}$$

and

$$\Pi(\mathbf{x}) \approx -(\varepsilon_0 \mathbf{E}(\mathbf{x}) + \mathbf{P}(\mathbf{x})) \tag{6-25}$$

We eliminate the velocities in the usual way and obtain the hamiltonian as a "weak" equation

$$\mathcal{H} \approx \left(\frac{1}{2m} \right) (\mathbf{p} + \mathbf{Q}(\mathbf{B}))^2 + \left(\frac{1}{2\varepsilon_0} \right) \int d^3\mathbf{x} \left\{ (\Pi(\mathbf{x}) + \mathbf{P}(\mathbf{x}))^2 \right.$$
$$\left. + \varepsilon_0^2 c^2 \mathbf{B}(\mathbf{x}) \cdot \mathbf{B}(\mathbf{x}) \right\} - \int d^3\mathbf{x} \phi(\mathbf{x}) (\nabla \cdot \Pi(\mathbf{x})) \tag{6-26}$$

As before, the Poisson brackets between the dynamical variables are assumed to have the canonical forms in (5-8) to (5-10).

The argument now follows exactly that given in Sections IV and V. The consistency condition ensuring that $\Pi_0(\mathbf{x}) \approx 0$ for all time is easily found to be

$$\nabla \cdot \Pi(\mathbf{x}) \approx 0 \tag{6-27}$$

and this is the invariant relation characterizing the new formalism [cf. (4-11) and (5-13)!]. To make (6-27) a "strong" (i.e., ordinary) equality, we introduce a supplementary condition on the vector potential—for example, the Coulomb gauge condition Div $\mathbf{A(x)} \approx 0$—and modify the definition of the Poisson brackets according to the general scheme discussed in Section IV. The final result of these manipulations is the following hamiltonian scheme in which the redundant variables $\phi(\mathbf{x})$ and $\Pi_0(\mathbf{x})$ no longer appear:

$$\mathcal{H}' = \left(\frac{1}{2m}\right)(\mathbf{p} + \mathbf{Q(B)})^2 + \left(\frac{1}{2\varepsilon_0}\right)\int d^3x \{\mathbf{\Pi(x)} \cdot \mathbf{\Pi(x)}$$

$$+ 2\mathbf{P(x)} \cdot \mathbf{\Pi(x)} + \mathbf{P(x)} \cdot \mathbf{P(x)} + \varepsilon_0^2 c^2 \mathbf{B(x)} \cdot \mathbf{B(x)}\} \qquad (6\text{-}28)$$

$$\nabla \cdot \mathbf{\Pi(x)} = 0 \qquad (6\text{-}29)$$

$$\nabla \cdot \mathbf{A(x)} = 0 \qquad (6\text{-}30)$$

$$[R^s, p']^* = \delta_{st} \qquad (6\text{-}31)$$

$$[A(\mathbf{x})^s, \Pi(\mathbf{x'})^t]^* = \delta_{st}^{\perp}(\mathbf{x} - \mathbf{x'}) \qquad (6\text{-}32)$$

A simple calculation shows that the Dirac bracket of the momentum $\mathbf{\Pi(x)}$ with the magnetic induction $\mathbf{B(x)}$ is

$$[\Pi(\mathbf{x})^s, B(\mathbf{x'})^t]^* = -\mathsf{E}_{ust}\left(\frac{\partial}{\partial x_u'}\right)\delta^3(\mathbf{x} - \mathbf{x'}) \qquad (6\text{-}33)$$

and therefore we may identify the momentum $\mathbf{\Pi(x)}$ with the electric field strength to within a constant factor

$$\mathbf{\Pi(x)} = -\varepsilon_0 \mathbf{E(x)} \qquad (6\text{-}34)$$

An important observation is that the gauge condition, (6-30), and the associated Dirace bracket (6-32) may be dispensed with, since the vector potential only occurs in the hamiltonian $\mathcal{H}'$ through its Curl (the magnetic induction); thus we may write the hamiltonian purely in terms of the canonical variables $\{\mathbf{R}, \mathbf{p}\}$ and the electromagnetic field variables $\{\mathbf{E(x)}, \mathbf{B(x)}\}$

$$\mathcal{H} = \left(\frac{1}{2m}\right)(\mathbf{p} + \mathbf{Q(B)})^2 - \int d^3x\, \mathbf{P(x)} \cdot \mathbf{E(x)}^{\perp} + \left(\frac{1}{2\varepsilon_0}\right)\int d^3x\, \mathbf{P(x)} \cdot \mathbf{P(x)}$$

$$+ \frac{\varepsilon_0}{2}\int d^3x \{\mathbf{E(x)}^{\perp} \cdot \mathbf{E(x)}^{\perp} + c^2 \mathbf{B(x)} \cdot \mathbf{B(x)}\} \qquad (6\text{-}35)$$

together with the following Poisson brackets:

$$[R^s, p'] = \delta_{st} \qquad (6\text{-}36)$$

$$[E(\mathbf{x})^s, B(\mathbf{x}')^t] = \varepsilon_0^{-1} \mathsf{E}_{ust}\left(\frac{\partial}{\partial x_u'}\right)\delta^3(\mathbf{x} - \mathbf{x}') \qquad (6\text{-}37)$$

The formal quantization of this canonical system is immediate and leads to a quantum theory in the Heisenberg representation. The hamiltonian in (6-35) is related to the conventional single particle hamiltonian, (5-21), by the improper unitary transformation induced by the operator Λ, in (6-10). Equation (6-35) therefore provides a *complete* description of a nonrelativistic spinless particle.

It may be remarked that the usual theory of the electrodynamics of a charged particle yields an invariant relation containing both field and particle variables— (5-13). As a result, the Poisson brackets of the particle momentum and the vector potential with the field momentum $\mathbf{\Pi}(\mathbf{x})$ are gauge dependent. Under a gauge transformation, however, these dynamical variables and their Poisson brackets change in such a way that the kinetic momentum $(\mathbf{p} - e\mathbf{A}(\mathbf{R}))$ is a gauge-invariant quantity. This behavior is to be contrasted with the formal simplicity of the present formulation.

When more than one charge is considered, the hamiltonian scheme in (6-35) to (6-37) may be generalized without difficulty. It suffices to give the particle variables an appropriate index and to introduce a summation over all particles; if desired, different arbitrary origins $\mathbf{r}_\alpha$ can be associated with subsets of the particles, making it possible to bring into the formalism the notion of well-defined atomic systems through the individual atomic polarization fields

$$\mathbf{P}(\mathbf{x})^{total} = \sum_\alpha^{\text{all atoms}} \mathbf{P}(\alpha : \mathbf{x}) \qquad (6\text{-}38)$$

$$\mathbf{P}(\alpha : \mathbf{x}) = \sum_{i=1}^{N} e_i \int_{\mathbf{r}_\alpha}^{\mathbf{R}_{\alpha i}} d\mathbf{z}\, \delta^3(\mathbf{z} - \mathbf{x}) \qquad (6\text{-}39)$$

It should be noted that this is a matter of convenience only, not an approximation. It is understood here that a separate curve $\mathbf{z}_i$ is to be defined for each particle.

We must now determine how we are going to employ the new hamiltonian in molecular physics using perturbation theory. Let us consider first the case of a single atom interacting with radiation. The obvious procedure is to try and separate terms with an explicit field dependence from those that depend on only the particle variables, for example,

$$\mathcal{H} = \mathcal{H}_1 + \mathcal{H}_2 + \mathcal{H}^{\text{field}} \tag{6-40}$$

where $\mathcal{H}^{\text{field}}$ is defined in (6-7) and

$$\mathcal{H}_1 = \sum_{i=1}^{N} \left(\frac{p_i^2}{2m_i} \right) + \left(\frac{1}{2\varepsilon_0} \right) \int d^3x \, \mathbf{P}(\mathbf{x}) \cdot \mathbf{P}(\mathbf{x}) \tag{6-41}$$

$$\mathcal{H}_2 = \mathcal{H} - (\mathcal{H}_1 + \mathcal{H}^{\text{field}}) \tag{6-42}$$

It is evident from (6-39) that the electric polarization $\mathbf{P}(\mathbf{x})$ depends on only the particle variables, and yet a relatively simple calculation shows that the integral over its square is infinite, therefore unable to provide the atomic binding energy.[13] The difficulty can be traced to the fact that the transverse component of $\mathbf{P}(\mathbf{x})$, which as pointed out in Section II is entirely arbitrary, is included in the integration. This remark, however, indicates how we should proceed; we must split the polarization field up into its longitudinal and transverse components and investigate the two contributions separately:

$$\int d^3x \, \mathbf{P}(\mathbf{x}) \cdot \mathbf{P}(\mathbf{x}) = \int d^3x \, \mathbf{P}(\mathbf{x})^{\|} \cdot \mathbf{P}(\mathbf{x})^{\|} + \int d^3x \, \mathbf{P}(\mathbf{x})^{\perp} \cdot \mathbf{P}(\mathbf{x})^{\perp} \tag{6-43}$$

The longitudinal component $\mathbf{P}(\mathbf{x})^{\|}$ is computed from $\mathbf{P}(\mathbf{x})$ with the aid of the longitudinal delta function,[37]

$$P(\mathbf{x})_r^{\|} = \int d^3x' \, P(\mathbf{x}')_s \delta_{rs}^{\|}(\mathbf{x} - \mathbf{x}'), \qquad r,s = x,y,z \tag{6-44}$$

and it is a simple matter to show that the result of this integration is [13]

$$\mathbf{P}(\mathbf{x})^{\|} = \sum_{i=1}^{N} e_i \nabla \left(\frac{1}{4\pi |\mathbf{R}_i - \mathbf{x}|} \right) \tag{6-45}$$

$$= -\varepsilon_0 \mathbf{E}(\mathbf{x})^{\|} \tag{6-46}$$

according to (5-25). Thus the integral over the longitudinal component provides the atomic binding energies; the integral over the transverse component is always infinite, however, and is treated formally as a perturbation term. We write therefore

$$\mathcal{H}^{\text{atom}} = \sum_{i=1}^{N} \left(\frac{p_i^2}{2m_i} \right) + \left(\frac{1}{2\varepsilon_0} \right) \int d^3\mathbf{x} \, \mathbf{P}(\mathbf{x})^{\|} \cdot \mathbf{P}(\mathbf{x})^{\|} \qquad (6\text{-}47)$$

and define the perturbation operator V as

$$V = \mathcal{H} - (\mathcal{H}^{\text{atom}} + \mathcal{H}^{\text{field}})$$

$$= - \int d^3\mathbf{x} \, \mathbf{P}(\mathbf{x}) \cdot \mathbf{E}(\mathbf{x})^{\perp} + \left(\frac{1}{2\varepsilon_0} \right) \int d^3\mathbf{x} \, \mathbf{P}(\mathbf{x})^{\perp} \cdot \mathbf{P}(\mathbf{x})^{\perp}$$

$$+ \text{ terms linear and quadratic in } \mathbf{B}(\mathbf{x}) \qquad (6\text{-}48)$$

The "multipole" form of the perturbation operator V is obtained by taking the path integral form of the polarization fields, and making a Taylor series expansions of their integrands about the arbitrary reference point $\mathbf{r}$. The integrals may then be done term by term, to lead to a multipolar representation.[11,12] The electric dipole approximation may be obtained at once, however, since it is appropriate when the field is spatially uniform; thus we may write

$$V_E' = - \int d^3\mathbf{x} \, \mathbf{P}(\mathbf{x}) \cdot \mathbf{E}(\mathbf{x})^{\perp} \sim - \mathbf{E}(0)^{\perp} \cdot \int d^3\mathbf{x} \, \mathbf{P}(\mathbf{x})$$

$$= - \mathbf{d} \cdot \mathbf{E}(0)^{\perp} \qquad (6\text{-}49)$$

where $\mathbf{d}$ is the usual electric dipole operator [see (1-1)]. It should be noted that the infinite self-energy arising from the Coulomb field of a single charge is contained in (6-47) and as usual is discarded. The infinity arising from the integral over $|\mathbf{P}(\mathbf{x})^{\perp}|^2$, however, is another kind of self-energy which we now proceed to investigate.

For our purposes it is sufficient to consider the many-particle case and to show that this term is a renormalization counter term serving to subtract off the most divergent contribution to the atomic self-energy computed in perturbation theory. We consider the process in which the atom emits a photon that is later reabsorbed by the same atom. Using the perturbation operator

$$V_E = - \int d^3\mathbf{x} \, \mathbf{P}(\mathbf{x}) \cdot \mathbf{E}(\mathbf{x})^{\perp} \qquad (6\text{-}50)$$

and the usual plane-wave expansion of $\mathbf{E}(\mathbf{x})^{\perp}$ [essentially, (4-50)], we obtain the energy shift from second-order perturbation theory as

$$\Delta E = -\left(\frac{1}{2\varepsilon_0 V}\right) \sum_{\mathbf{k},\lambda} \sum_{n} \frac{k}{(E_n + k)} \langle 0| \int d^3\mathbf{x}\, \mathbf{P}(\mathbf{x}) \cdot \hat{\mathbf{E}}_\lambda(\mathbf{k})|n\rangle$$

$$\times \langle n| \int d^3\mathbf{x}'\, \mathbf{P}(\mathbf{x}') \cdot \hat{\mathbf{E}}_\lambda(\mathbf{k})|0\rangle e^{i\mathbf{k}\cdot\mathbf{x}-\mathbf{x}'} \tag{6-51}$$

where the sum on n is taken over all excited atomic states and E_n is the transition energy from the ground state to the state $|n\rangle$. We have already integrated out the photon annihilation and creation operators. Now we can write

$$\frac{k}{E_n + k} = 1 - \frac{E_n}{E_n + k} \tag{6-52}$$

which means that (6-51) becomes

$$\Delta E = \left(\frac{1}{2\varepsilon_0 V}\right) \sum_{\mathbf{k},\lambda} \sum_{n} \frac{E_n}{E_n + k} \langle 0| \int d^3\mathbf{x}\, \mathbf{P}(\mathbf{x}) \cdot \hat{\mathbf{E}}_\lambda(\mathbf{k})|n\rangle$$

$$\times \langle n| \int d^3\mathbf{x}'\, \mathbf{P}(\mathbf{x}') \cdot \hat{\mathbf{E}}_\lambda(\mathbf{k})|0\rangle e^{i\mathbf{k}\cdot\mathbf{x}-\mathbf{x}'}$$

$$-\left(\frac{1}{2\varepsilon_0 V}\right) \sum_{\mathbf{k},\lambda} \sum_{n} \langle 0| \int d^3\mathbf{x}\, \mathbf{P}(\mathbf{x}) \cdot \hat{\mathbf{E}}_\lambda(\mathbf{k})|n\rangle \langle n| \int d^3\mathbf{x}'\, \mathbf{P}(\mathbf{x}') \cdot \hat{\mathbf{E}}_\lambda(\mathbf{k})|0\rangle$$

$$\times e^{i\mathbf{k}\cdot\mathbf{x}-\mathbf{x}'} \tag{6-53}$$

Now let us sum over the polarization states of the virtual photons, using closure on the atomic states $|n\rangle$ and the identity

$$V^{-1} \sum_{\mathbf{k}} (1 - \hat{k}\hat{k})_{rs} e^{i\mathbf{k}\cdot\mathbf{x}-\mathbf{x}'} = \delta_{rs}^{\perp}(\mathbf{x} - \mathbf{x}') \tag{6-54}$$

The second term in (6-53) is thus reduced to

$$-\left(\frac{1}{2\varepsilon_0}\right) \langle 0| \int d^3\mathbf{x}\, \mathbf{P}(\mathbf{x})^{\perp} \cdot \mathbf{P}(\mathbf{x})^{\perp}|0\rangle \tag{6-55}$$

which is exactly canceled by the matrix element of $\frac{1}{2}\varepsilon_0^{-1}\int d^3\mathbf{x}|\mathbf{P}(\mathbf{x})^{\perp}|^2$ in the hamiltonian, (6-48). Note that we have dropped the contribution of the form $\int d^3\mathbf{x}|\langle 0|\mathbf{P}(\mathbf{x})^{\perp}|0\rangle|^2$ which arises from the closure, since the transverse polarization is essentially off-diagonal with respect to the atomic states, therefore these matrix elements vanish identically.[47]

The physically significant energy change (a contribution to the nonrelativistic Lamb shift) therefore resides in the first part of (6-53). This expression was further analyzed by Power and Zienau[10] using the electric dipole approximation to the electric polarization field $\mathbf{P}(\mathbf{x}) \sim \mathbf{d}\delta^3(\mathbf{x})$. Strictly speaking, since all possible momenta of the virtual photons are included in the summation (integration) over $\mathbf{k}$, it is not legitimate to assume that the field is spatially uniform, and whereas Power and Zienau obtained a cubically divergent result that was reduced to Bethe's form[74] by other contributions in the hamiltonian, (6-53) is already only logarithmically divergent. To obtain this result, however, it is necessary to know the matrix elements of the polarization field, and for simplicity we restrict our discussion to the case of atomic hydrogen and the transition $1S \rightarrow 2P$.

We choose the polarization vectors along the x and y directions, thus restricting our attention to the states $2P_x$ and $2P_y$, which are degenerate. The basis functions are

$$|1S\rangle = (\pi)^{-1/2} \exp(-R) \tag{6-56}$$

$$|2P_x\rangle = \tfrac{1}{4}(2\pi)^{-1/2} R \exp(-\tfrac{1}{2}R) \sin\theta \sin\phi \tag{6-57}$$

$$|2P_y\rangle = \tfrac{1}{4}(2\pi)^{-1/2} R \exp(-\tfrac{1}{2}R) \sin\theta \cos\phi \tag{6-58}$$

Using (6-18) and the results in Appendix A, we easily find the matrix element of $\mathbf{P}(\mathbf{x})$ for a particle moving in the field of a center of force

$$\langle \psi_f | \mathbf{P}(\mathbf{x}) | \psi_i \rangle = \int_1^\infty ds\, s^2 \psi_f^*(s\mathbf{x})(e\mathbf{x})\psi_i(s\mathbf{x}) \tag{6-59}$$

so that

$$\langle 1S | \mathbf{P}(\mathbf{x}) | 2P_x \rangle = f(\mathbf{x})_{1S,2P} \hat{x} \sin\theta \sin\phi \tag{6-60}$$

$$\langle 1S | \mathbf{P}(\mathbf{x}) | 2P_y \rangle = f(\mathbf{x})_{1S,2P} \hat{x} \sin\theta \cos\phi \tag{6-61}$$

and

$$f(\mathbf{x})_{1S,2P} = \frac{e}{(4\pi\sqrt{2}\, x^2)} p^{-4} \left\{ 6 + 6(px) + 3(px)^2 + (px)^3 \right\} e^{-px}, \quad p = \frac{3}{2} \tag{6-62}$$

Now

$$\hat{x} \cdot \hat{\mathbf{E}}_x(\mathbf{k}) = \sin\theta \cos\phi, \qquad \hat{x} \cdot \hat{\mathbf{E}}_y(\mathbf{k}) = \sin\theta \sin\phi \tag{6-63}$$

and if we write $E_{2P} - E_{1S} = w$, we have

$$\Delta E_{1S,2P} = \left(\frac{w}{2\varepsilon_0 V}\right) \sum_k \frac{1}{(w+k)} \int d^3x \int d^3x' f(x)_{1S,2P} f(x')_{1S,2P}$$

$$\times e^{(i\mathbf{k}\cdot\mathbf{x}-\mathbf{x}')} \{ \sin^2\theta_x \cos^2\phi_x \sin^2\theta_{x'} \cos^2\phi_{x'}$$

$$+ \sin^2\theta_x \sin^2\phi_x \sin^2\theta_{x'} \sin^2\phi_{x'} \} \tag{6-64}$$

By lengthy but wholly straightforward manipulations, we can obtain the Fourier transforms of $f(x)_{1S,2P} \sin^2\theta_x \left\{ \begin{array}{c} \sin^2\phi_x \\ \cos^2\phi_x \end{array} \right\}$ (see Appendix A).

$$\int d^3x \sin^2\theta_x \left\{ \begin{array}{c} \sin^2\phi_x \\ \cos^2\phi_x \end{array} \right\} e^{i\mathbf{k}\cdot\mathbf{x}} e^{-px} \{ 6 + 6(px) + 3(px)^2 + (px)^3 \} x^{-2}$$

$$= 4\pi \left\{ \left(\frac{3}{k}\right) \tan^{-1}\left(\frac{k}{p}\right) + \frac{3p}{k^2+p^2} + \frac{2p^3}{(k^2+p^2)^2} \right\} \tag{6-65}$$

Finally, using the rule expressed in (4-32) to change the summation over $\mathbf{k}$ into an integration, and introducing the integration variable $t = k/p$, we obtain

$$\Delta E_{1S,2P} = \frac{we^2}{4\pi^2\varepsilon_0} p^{-7} \int_0^\infty dt (w+pt)^{-1} \left\{ 3\tan^{-1}(t) + \frac{3t}{1+t^2} + \frac{2t^3}{(1+t^2)^2} \right\}^2$$

$$\tag{6-66}$$

which is infinite, since

$$\int_0^\infty dt \frac{(\tan^{-1}(t))^2}{w+pt} \tag{6-67}$$

is logarithmically divergent, although the remaining integrals in (6-66) converge.

We must now see how the formalism can be extended to describe the interactions of several atoms with the electromagnetic field (and with each other by way of the electromagnetic field). To be able to identify individual atoms we must be able to assign the charged particles to different regions of space identified by local origins $\mathbf{r}_\alpha$; each atom can then be assigned its own polarization fields as in (6-38). As far as the *intraatomic* interactions are concerned, the generalization of the hamiltonian in (6-48) is straightforward and requires no further comment. To understand the description of the *interatomic* interactions in this formalism, we must consider further the integral over the square of the electric polarization field. Using (6-38), we may write

$$\int d^3\mathbf{x}\, \mathbf{P}(\mathbf{x}) \cdot \mathbf{P}(\mathbf{x}) = \int d^3\mathbf{x} \sum_\alpha \mathbf{P}(\alpha : \mathbf{x}) \cdot \mathbf{P}(\alpha : \mathbf{x}) + \int d^3\mathbf{x} \sum_{\alpha \neq \beta} \mathbf{P}(\alpha : \mathbf{x}) \cdot \mathbf{P}(\beta : \mathbf{x})$$

(6-68)

Because the first term on the right-hand side of (6-68) is decomposed into the atomic binding energies and self-energy contributions according to (6-43), it describes the intraatomic interactions. This decomposition is *not* extended to the interatomic contribution, however, and the entire integral is regarded as a contribution to the perturbation operator V. It then follows that since there are no explicit interatomic Coulomb energies in the hamiltonian, the interactions between different atoms must be described in terms of purely retarded transverse radiation transfers arising from $-\int d^3\mathbf{x}\, \mathbf{P}(\mathbf{x}) \cdot \mathbf{E}(\mathbf{x})^\perp$ and the related terms involving the magnetic induction $\mathbf{B}(x)$. This practice often leads to a form of perturbation theory that is simplified in comparison with the conventional Coulomb gauge theory—in which, as we noted earlier, static Coulomb terms may give rise to difficulties (see, e.g., Ref. 37). Of course in the limit of small separations, retardation effects are normally negligible, and we then recover the familiar static intermolecular interaction derived from Coulomb's law.

To complete this account we need the interaction terms that depend on the magnetic induction $\mathbf{B}(x)$. These terms arise from the expansion of the first part of (6-35):

$$\sum_{i=1}^{N} \left(\frac{1}{2m_i}\right)(\mathbf{p}_i + \mathbf{Q}_i(\mathbf{B}))^2 = \sum_{i=1}^{N} \left(\frac{1}{2m_i}\right)\left(p_i^2 + \mathbf{p}_i \cdot \mathbf{Q}_i(\mathbf{B}) + \mathbf{Q}_i(\mathbf{B}) \cdot \mathbf{p}_i + Q_i(\mathbf{B})^2\right)$$

(6-69)

where

$$\mathbf{Q}_i(\mathbf{B}) = e_i \int_{s_1}^{s_2} f(s)\, d\mathbf{z} \times \mathbf{B}(\mathbf{z})$$

(6-70)

and the path $\mathbf{z}$ is arbitrary. If we recall the discussion in Section I, we see that (6-70) is a more general form of (1-7), which may therefore be regarded either as a special form of vector potential or alternatively as a nonlocal functional of the magnetic induction. Since the paths $\mathbf{z}$ depend on the positions of the particles, the functional $Q_i(B)$ does not commute with the momentum p_i; thus the ordering of the factors in (6-69) is important. Classically the ordering does not matter, and we can reintroduce the magnetization field

$$M(x)' = \sum_{i=1}^{N} \left(\frac{e_i}{m_i} \right) \int_{s_1}^{s_2} f(s) \delta^3(z-x) dz \times p_i \qquad (6\text{-}71)$$

so that the interaction term, which is linear in the magnetic induction, can be written in the form

$$V_B = -\int d^3x \, M(x)' \cdot B(x) \qquad (6\text{-}72)$$

The prime on $M(x)'$ indicates that the magnetization field is expressed here in terms of the canonical variables $\{R_i, p_i\}$, rather than the lagrangian variables used in Section II.

Equation (6-71) cannot be taken over into the quantum theory directly because it is not hermitian; since there is no unique way of ensuring the hermiticity of the magnetization field, we require an additional condition to remove the arbitrariness in its definition. It seems natural to require that the Taylor series expansion of the magnetization field operator lead to a magnetic interaction energy, linear in the field, which can be displayed in the classical form

$$V_B = -m_\alpha B_\alpha - \frac{m_{\alpha\beta} B_{\alpha\beta}}{2} - \frac{m_{\alpha\beta\gamma} B_{\alpha\beta\gamma}}{6} - \cdots \qquad (6\text{-}73)$$

Here the tensors $m_\alpha, m_{\alpha\beta}, \ldots$ are the magnetic multipole operators and B_α and $B_{\alpha\beta}, \ldots$ are the magnetic induction and its derivatives, evaluated at the origin chosen for the expansion. It appears that this cannot be achieved with the usual symmetrized definition of the multipole tensors;[89] however, Raab[90] has demonstrated that it is possible to give a less symmetrical definition of the multipole operators which leads to an interaction energy of the form of (6-73). These multipole tensors are defined as

$$m_\alpha = \tfrac{1}{2} \sum_{i=1}^{N} \left(\frac{e_i}{m_i} \right) L_{i\alpha} \qquad (6\text{-}74a)$$

$$m_{\alpha\beta} = \tfrac{2}{3} \sum_{i=1}^{N} \left(\frac{e_i}{m_i} \right) \{ R_{i\beta} L_{i\alpha} - \tfrac{1}{2} [R_{i\beta}, L_{i\alpha}]_- \} \qquad (6\text{-}74b)$$

$$m_{\alpha\beta\gamma} = \tfrac{3}{4} \sum_{i=1}^{N} \left(\frac{e_i}{m_i} \right) \{ R_{i\beta} R_{i\gamma} L_{i\alpha} - \tfrac{1}{2} [R_{i\beta} R_{i\gamma}, L_{i\alpha}]_- \} \qquad (6\text{-}74c)$$

where $\mathbf{L}_i$ is the angular momentum operator $(\mathbf{R}_i \times \mathbf{p}_i)$. These tensors are symmetric in all their suffices except the first (α), and are manifestly hermitian.

The quantum-mechanical magnetization field is therefore defined as

$$\mathbf{M}(\mathbf{x})' = \sum_{i=1}^{N} \left(\frac{e_i}{m_i} \right) \int_{s_1}^{s_2} f(s) \left\{ \delta^3(\mathbf{z}-\mathbf{x}) d\mathbf{z} \times \mathbf{p}_i - \tfrac{1}{2} [\delta^3(\mathbf{z}-\mathbf{x}), (d\mathbf{z} \times \mathbf{p}_i)] \right\}$$

$$(6\text{-}75)$$

since if we choose the path $\mathbf{z} = s\mathbf{R}_i (s_1 = 0, s_2 = 1)$ and make a Taylor series expansion of the delta function about $\mathbf{x}$, we easily find

$$M(\mathbf{x})_\alpha' = \left\{ m_\alpha - \frac{m_{\alpha\beta} \nabla_\beta}{2} + \frac{m_{\alpha\beta\gamma} \nabla_\beta \nabla_\gamma}{3} - \cdots \right\} \delta^3(\mathbf{x}) \qquad (6\text{-}76)$$

as required (recall that an odd number of differentiations of the delta function lead to an overall minus sign when the integration over $\mathbf{x}$ is effected). The generalization of $\mathbf{M}(\mathbf{x})'$ to the many-atom case is exactly analogous to that given for the electric polarization field $\mathbf{P}(\mathbf{x})$, (6-38) and (6-39).

The contribution to the energy that is quadratic in the magnetic induction is a generalization of the familiar diamagnetic term. If we separate out the magnetic induction from $\mathbf{Q}_i(\mathbf{B})$ with the aid of an integration over a delta function

$$\mathbf{Q}_i(\mathbf{B}) = e_i \int d^3\mathbf{x} \int_{s_1}^{s_2} f(s) \delta^3(\mathbf{z}-\mathbf{x}) d\mathbf{z} \times \mathbf{B}(\mathbf{x}) \qquad (6\text{-}77)$$

we can write the interaction energy in the form

$$\sum_{i=1}^{N} \frac{Q_i(\mathbf{B})^2}{2m_i} = \int d^3\mathbf{x} \int d^3\mathbf{x}' X(\mathbf{x},\mathbf{x}')_{\alpha\beta} B(\mathbf{x})_\alpha B(\mathbf{x}')_\beta \qquad (6\text{-}78)$$

where the generalized susceptibility tensor field $\underline{X}$ is defined as

$$X(\mathbf{x},\mathbf{x}')_{\alpha\beta} = \sum_{i=1}^{N} \left(\frac{e_i^2}{2m_i} \right) \int_{s_1}^{s_2} f(s) \int_{s_1'}^{s_2'} f(s') \delta^3(\mathbf{z}-\mathbf{x}) \delta^3(\mathbf{z}'-\mathbf{x}')$$

$$\times (\delta_{\alpha\beta} d\mathbf{z} \cdot d\mathbf{z}' - dz_\alpha dz_\beta') \qquad (6\text{-}79)$$

If we now choose the linear paths $z = sR_i$, $z' = s'R_i$, (6-79) becomes

$$X(x,x')_{\alpha\beta} = \sum_{i=1}^{N} \left(\frac{e_i^2}{2m_i} \right) (R_i^2 \delta_{\alpha\beta} - R_{i\alpha} R_{i\beta}) \int_0^1 ds \, s \int_0^1 ds' \, s'$$

$$\times \delta^3(sR_i - x) \delta^3(s'R_i - x') \tag{6-80}$$

In the case of a homogeneous magnetic field, we can ignore the dependence of the delta function on s and s'; thus $\underline{X}$ reduces to

$$X(x,x')_{\alpha\beta} \sim \chi_{\alpha\beta} \delta^3(x) \delta^3(x') \tag{6-81}$$

where

$$\chi_{\alpha\beta} = \sum_{i=1}^{N} \left(\frac{e_i^2}{8m_i} \right) (R_i^2 \delta_{\alpha\beta} - R_{i\alpha} R_{i\beta}) \tag{6-82}$$

is the usual diamagnetic susceptibility tensor. The extension of the nonlocal tensor field $\underline{X}(x, x')$ to the many-atom case is again quite obvious and requires no further comment.

The full hamiltonian describing a closed system of atoms and the electromagnetic field can therefore be written in the form

$$\mathcal{H} = H^{\text{field}} + H^{\text{atom}} - \int d^3x \sum_{\alpha} P(\alpha : x) \cdot E(x)^{\perp}$$

$$- \int d^3x \sum_{\alpha} M(\alpha : x)' \cdot B(x) + \int d^3x \int d^3x' \sum_{\alpha} \underline{X}(\alpha : x, x') : B(x)B(x')$$

$$+ \left(\frac{1}{2\varepsilon_0} \right) \int d^3x \left\{ \sum_{\alpha} |P(\alpha : x)^{\perp}|^2 \right\} + \left(\frac{1}{2\varepsilon_0} \right) \int d^3x \sum_{\alpha \neq \beta} P(\alpha : x) \cdot P(\beta : x)$$

$$\tag{6-83}$$

in which the division of the hamiltonian into the form $\mathcal{H} = H_0 + V$ for use in the conventional perturbation theory is now obvious. An especially attractive feature of (6-83) is the explicit appearance of the field strengths in the interaction terms. The interaction energies are strongly reminiscent of the well-known classical expressions in macroscopic electrodynamics, but here they are fully quantum mechanical. Moreover, within the nonrelativistic approximation, (6-83) is exact. As we have indicated, it is a simple matter to recover the multipolar representations of the polarization fields if this is required. It is often remarked in the literature that since this hamiltonian is expressed in terms of the field strengths, there is no problem with gauge-invariance difficulties; the remark is ingenuous, since as we have seen the hamiltonian is intimately connected with the family of gauge conditions shown in (1-9). In these gauges there is an arbitrariness in the choice of the paths z and the reference point r; different choices are unitarily equivalent, however, which means that the hamiltonian (6-83) admits of a class of unitary transformations that are very similar to the familiar gauge transformations of the potentials in the conventional theory. In particular it should be understood that attempts to assign a special physical significance to a certain reference point (say, the center of mass, or the electronic charge centroid) are inappropriate and are not in accordance with the general principles of quantum theory.[88] All these unitary transformations in electrodynamics are consequences of charge conservation.

APPENDIX A. ON INTEGRALS INVOLVING THE POLARIZATION FIELDS

A.1. Matrix Elements

If for simplicity we choose the straight line paths $z = r + s(R_i - r)$, then (2-13) and (2-14) may be written

$$P(x) = \sum_{i=1}^{N} e_i(R_i - r) \int_0^1 ds\, \delta^3[s(R_i - r) + r - x] \qquad (A\text{-}1)$$

$$M(x) = \sum_{i=1}^{N} e_i(R_i - r) \times \dot{R}_i \int_0^1 ds\, s\, \delta^3[s(R_i - r) + r - x]$$

$$= \sum_{i=1}^{N} \left(\frac{e_i}{m_i}\right)(R_i - r) \times p_i \int_0^1 ds\, s\, \delta^3[s(R_i - r) + r - x] \qquad (A\text{-}2)$$

where we have used the usual relation $p_i = m_i \dot{R}_i$. The matrix elements of the polarization fields $P(x)$ and $M(x)$ therefore take the form of generalized electric and magnetic dipole moments and represent the displacement and circulation of the charge density of the system, either in a given state n or in a transition $n \to m$. In general, the quality of molecular wavefunctions as judged by the charge distribution is not very good, and quantitative values for the polarization fields are not readily available. Nevertheless, one could hope to apply symmetry arguments to these fields, perhaps avoiding the difficulties of origin dependence that hinder discussion of electromagnetic phenomena in terms of the molecular electric and magnetic multipoles. Nothing has been done along these lines, however, and it remains a topic for investigation in the future.

If we write the many-body wavefunction for state n as $\Psi_n(\{\mathbf{R}_i\})$, we have

$$\langle \Psi_m | \mathbf{P}(\mathbf{x}) | \Psi_n \rangle = \int \prod_{i=1}^{N} d^3\mathbf{R}_i \Psi_m^*(\{\mathbf{R}_i\}) \mathbf{P}(\mathbf{x}) \Psi_n(\{\mathbf{R}_i\}) \qquad \text{(A-3)}$$

If we now combine (A-1) and (A-3) and put

$$\mathbf{Q}_i = s(\mathbf{R}_i - \mathbf{r}), \qquad s = t^{-1} \qquad \text{(A-4)}$$

the matrix element becomes,

$$\mathbf{P}(\mathbf{x})_{mn} = \int_1^{\infty} dt\, t^2 \int \prod_{i=1}^{N} d^3\mathbf{Q}_i \Psi_m^*(\{t\mathbf{Q}_i + \mathbf{r}\}) \sum_{j=1}^{N} e_j \mathbf{Q}_j \delta^3 [\mathbf{Q}_j + \mathbf{r} - \mathbf{x}]$$

$$\times \Psi_n(\{t\mathbf{Q}_i + \mathbf{r}\}) \qquad \text{(A-5)}$$

The further reduction of this integral is only possible when the form of the wavefunction is given. If, however, the nuclei are assumed fixed and the electronic wavefunction is written in terms of one or more Slater determinants, the usual rules for simplifying matrix elements apply and there remains only the integration over t, which depends on the specific form of the basis functions employed. Similar arguments can be used to evaluate the matrix elements of the magnetization field.

A.2. The Fourier Transform of $\langle 1S | \mathbf{P}(\mathbf{x}) | 2P \rangle$

In the text, Section VI, we required the following integral:

$$I = \int d^3\mathbf{x} \sin^2\theta_x \left\{ \begin{matrix} \sin^2\phi_x \\ \cos^2\phi_x \end{matrix} \right\} e^{i\mathbf{k}\cdot\mathbf{x}} e^{-px} x^{-2} \left(6 + 6(px) + 3(px)^2 + (px)^3 \right)$$

$$\text{(A-6)}$$

which is the Fourier transform of $\langle 1S | \mathbf{P}(\mathbf{x}) \cdot \hat{\mathbf{E}}_{x,y}(\mathbf{k}) | 2P \rangle$ for atomic hydrogen. Because we choose $\mathbf{k}$ as the polar axis, the integral over ϕ simply gives π in both cases. Then using the usual expansion of a plane wave as a series of Legendre functions and cylinder functions[91]

$$e^{i\mathbf{k}\cdot\mathbf{x}} = \sum_L i^L (2L+1) j_L(kx) P_L(\cos\theta) \qquad \text{(A-7)}$$

we obtain

$$I = \pi \sum_L i^L (2L+1) \int_0^\pi d\theta \sin^3\theta P_L(\cos\theta) \int_0^\infty dx\, e^{-px} j_L(kx)$$

$$\times \{6 + 6(px) + 3(px)^2 + (px)^3\} \qquad \text{(A-8)}$$

The angular integral is immediate, since

$$\int_{-1}^{+1} dx\,(1-x^2)P_L(x) = \begin{cases} \dfrac{4}{3} & L=0 \\[4pt] 0 & L=1 \\[4pt] \dfrac{-4}{15} & L=2 \\[4pt] 0 & L>2 \end{cases} \qquad \text{(A-9)}$$

and so collecting terms we find,

$$I = \left(\frac{4\pi}{3k}\right) \int_0^\infty dx\, e^{-sx} [j_0(x) + j_2(x)]\{6 + 6\,(px) + 3(px)^2 + (px)^3\} \qquad \text{(A-10)}$$

where $s = p/k$. Now[92]

$$j_2(x) = \tfrac{1}{2}\left[j_0(x) - \frac{3 dj_1(x)}{dx} \right] \qquad \text{(A-11)}$$

so that

$$j_0(x) + j_2(x) = \frac{3}{2}\left[j_0(x) - \frac{dj_1(x)}{dx} \right] \qquad \text{(A-12)}$$

The integral involving $j_1(x)'$ can be reduced to an integral involving $j_0(x)$ by repeated integration by parts because $j_1(x) = -j_0(x)'$. Then, since

$$j_0(x) = \frac{\sin(x)}{x} \qquad \text{(A-13)}$$

we finally obtain I as a sum of elementary integrals

$$I = \left(\frac{2\pi}{k} \right) \int_0^\infty dx\, e^{-sx} \sin(x) \left\{ \frac{6}{x} + 6s + 3s^2(1 - s^2)x + s^3(1 + s^2)x^2 \right\}$$

(A-14)

$$= 4\pi \left\{ \frac{3}{k} \tan^{-1}\left(\frac{k}{p} \right) + \frac{3p}{k^2 + p^2} + \frac{2p^3}{(k^2 + p^2)^2} \right\}$$

(A-15)

as in the text.

Acknowledgments

The author wishes to thank ICI Ltd. and Trinity Hall, Cambridge, for financial support.

References

1. P. A. M. Dirac, *Proc. Roy. Soc. (London)*, **A114**, 243–265 (1927).
2. P. A. M. Dirac, *Proc. Roy. Soc. (London)*, **A114**, 710–728 (1927).
3. A. Salem and E. P. Wigner, Eds., *Aspects of Quantum Theory*, Cambridge University Press, Cambridge, 1972.
4. E. Fermi, *Rev. Mod. Phys.*, **4**, 87–132 (1932).
5. G. Breit, *Rev. Mod. Phys.* **4**, 504–576 (1932).
6. S. Stenholm, *Phys. Rep.*, **6C**, 1–121 (1973).
7. M. Göppert-Mayer, *Ann. Phys. (Leipzig)*, **9**, 273–294 (1931).
8. P. I. Richards, *Phys. Rev.*, **73**, 254 (1948).
9. E. A. Power and R. Shail, *Proc. Camb. Phil. Soc.*, **55**, 87 ff. (1959).
10. E. A. Power and S. Zienau, *Phil. Trans. Roy. Soc. (London)*, **A251**, 427–454 (1959).
11. P. W. Atkins and R. G. Woolley, *Proc. Roy. Soc. (London)*, **A321**, 549–563 (1970).
12. R. G. Woolley, *Ann. Inst. Henri Poincaré*, **23**, No. 4 (1975).
13. R. G. Woolley, *Proc. Roy. Soc. (London)*, **A321**, 557–572 (1971).
14. R. G. Woolley, *Mol. Phys.* **22**, 1013–1023 (1971).
15. P. A. M. Dirac, *Proc. Camb. Phil. Soc.*, **30**, 150–163 (1934).
16. R. E. Peierls, *Proc. Roy. Soc. (London)*, **A146**, 420–441 (1934).
17. V. Fock, *Phys. Z. Sow.*, **12**, 404 ff. (1937).
18. J. Schwinger, *Phys. Rev.*, **82**, 664–679 (1951).
19. J. G. Valatin, *Proc. Roy. Soc. (London)*, **A222**, 93–108 (1954).
20. D. Bohm and Y. Aharonov, *Phys. Rev.*, **115**, 485–491 (1959).
21. Y. Aharonov, H. Pendleton, and A. Petersen, *Int. J. Theor. Phys.*, **2**, 213–230 (1969).
22. W. Ehrenberg and R. E. Siday, *Proc. Phys. Soc. (London)*, **62B**, 8–21 (1949).
23. S. Mandelstam, *Ann. Phys.*, **19**, 1–24 (1962).
24. R. H. Capps and W. G. Hollday, *Phys. Rev.*, **99**, 931–943 (1955).
25. H. McManus, *Proc. Roy. Soc. (London)*, **A195**, 323–336 (1948).
26. M. Chrètien and R. E. Peierls, *Proc. Roy. Soc. (London)*, **A223**, 468–481 (1954).
27. H. F. Hameka, *Advanced Quantum Chemistry*, Addison-Wesley, Reading, Mass., 1965.
28. H. R. Reiss, *Phys. Rev.*, **A1**, 803–818 (1970).

29. R. G. Woolley, *J. Phys. B—Atoms Mol.* **7**, 488–499 (1974).
30. B. J. Howard and R. E. Moss, *Mol. Phys.*, **19**, 433–450 (1970).
31. B. J. Howard and R. E. Moss, *Mol. Phys.*, **20**, 147 ff. (1971).
32. R. E. Moss and A. J. Perry, *Mol. Phys.*, **23**, 954–962 (1972).
33. L. D. Barron and C. Gray, *J. Phys. A—Gen. Phys.*, **6**, 59–61 (1973).
34. R. G. Woolley, *J. Phys. B—Atoms Mol.*, **6**, L97–L99 (1973).
35. E. H. Kennard, *Phys. Rev.*, **39**, 435–454 (1932).
36. H. B. G. Casimir and D. Polder, *Phys. Rev.*, **73**, 360–372 (1948).
37. E. A. Power, *Introductory Quantum Electrodynamics, Longmans Green, London,* 1969.
38. H. A. Lorentz, *The Theory of Electrons,* 2nd ed., Dover, New York, 1952.
39. S. R. de Groot and L. G. Suttorp, *Foundations of Electrodynamics,* North-Holland, Amsterdam, 1972.
40. E. A. Power and T. Thirunamachandran, *Mathematika,* **18**, 240–245 (1971).
41. F. Rohrlich, *Classical Charged Particles,* Addison-Wesley, Reading, Mass., 1965.
42. J. Schwinger, *Particles, Sources and Fields,* Addison-Wesley, Reading, Mass, 1970.
43. R. E. Moss, *Advanced Molecular Quantum Mechanics,* Chapman Hall, London, 1973.
44. F. J. Belinfante, *Phys. Rev.*, **128**, 2832–2837 (1962).
45. L. D. Landau and R. E. Peierls, *Phys. Z. Sow.* (1932).
46. J. Fiutak, *Can. J. Phys.*, **41**, 12–21 (1963).
47. M. Babiker, E. A. Power, and T. Thirunamachandran, *Proc. Roy. Soc. (London),* **A332**, 187–197 (1973).
48. S. Goldstein, *Classical Mechanics,* Addison-Wesley, Reading, Mass., 1960.
49. P. A. M. Dirac, *The Principles of Quantum Mechanics,* 4th ed., Oxford University Press, Oxford, 1967.
50. P. A. M. Dirac, *Can. J. Math.* **2**, 147 ff. (1950).
51. P. A. M. Dirac, *Ann. Inst. Henri Poincaré,* **13**, 1–42 (1952).
52. P. A. M. Dirac, *Lectures on Quantum Mechanics,* Academic Press, London, 1964.
53. S. Shanmugadhasan, *Proc. Camb. Phil. Soc.*, **59**, 743–757 (1963).
54. H. P. Künzle, *Ann. Inst. Henri Poincaré,* **11A**, 393 (1969).
55. Y. Hagihara, *Celestial Mechanics,* Vol. 1, M.I.T. Press, Cambridge, Mass., 1970, Chapter 1.
56. A. Mercier, *Canonical Formalism in Classical Mechanics,* Dover, New York, 1963.
57. P. G. Bergman and I. Goldberg, *Phys. Rev.*, **98**, 531–538 (1955).
58. P. A. M. Dirac, *Lectures on Quantum Field Theory,* Academic Press, London, 1966.
59. P. Lorrain and D. Corson, *Electromagnetic Fields and Waves,* 2nd ed., Freeman, San Francisco, 1970, Chapter 11.
60. V. B. Berestetskii, E. M. Lifshitz, and L. P. Pitaevskii, *Relativistic Quantum Theory,* Part 1, Pergamon Press, Oxford, 1971.
61. P. W. Atkins and L. D. Barron, *Proc. Roy. Soc. (London),* **A304**, 303–317 (1968).
62. P. W. Atkins and L. D. Barron, *Mol. Phys.*, **16**, 453–466 (1969).
63. D. M. Lipkin, *J. Math. Phys.*, **5**, 696 (1964).
64. M. Schwartz, *Phys. Rev.*, **123**, 1903–1909 (1961).
65. H. J. Bhahba, *Phys. Rev.*, **70**, 759–760 (1946).
66. E. H. Kerner, *J. Math. Phys.*, **3**, 35–42 (1962).
67. J. R. Oppenheimer, *Phys. Rev.*, **35**, 461–477 (1930).
68. W. E. Lamb and R. C. Retherford, *Phys. Rev.*, **79**, 549–572 (1950).
69. W. E. Lamb, *Phys. Rev.*, **85**, 259–276 (1952).
70. W. E. Lamb, E. S. Dayhoff, and S. Triebwasser, *Phys. Rev.*, **89**, 98–106 (1953).
71. H. A. Kramers, *Nuovo Cimento,* **15**, 108 (1938).
72. H. A. Kramers, Rapports du 8^e Conseil Solvay, 241 (1948): reprinted in *Collected Scientific Papers of H. A. Kramers,* North-Holland, Amsterdam, 1956.

73. G. E. Uhlenbeck, *Oude en Nieuwe Vragen der Natuurkunde*, North-Holland, Amsterdam, 1955.
74. H. A. Bethe, *Phys. Rev.*, **72**, 339–341 (1947).
75. R. Schiller and M. Schwartz, *Phys. Rev.*, **126**, 1582–1588 (1962).
76. N. G. van Kampen, *Kgl. Danske Videnskab. Selskab.,   Mat.-Fys. Medd.*, **26**, No. 15 (1951).
77. H. Steinwedel, *Ann. Phys. (Leipzig)* **15**, 207 (1955).
78. J. Schwinger, *Phys. Rev.*, **75**, 651–679 (1949).
79. D. S. Kershaw, *Phys. Rev.*, **D4**, 3572–3579 (1971).
80. M. D. Crisp and E. T. Jaynes, *Phys. Rev.*, **179**, 253 ff. (1969).
81. C. R. Stroud and E. T. Jaynes, *Phys. Rev.*, **A1**, 106 ff. (1970).
82. V. Weisskopf and E. P. Wigner, *Z. Phys.*, **63**, 54 ff (1930).
83. J. M. Weisner, D. K. Andersen, and R. T. Robiscoe, *Phys. Rev. Lett.*, **29**, 1126 ff. (1972).
84. J. F. Clauser, *Phys. Rev.*, **A6**, 49 ff. (1972).
85. H. M. Gibbs, *Phys. Rev. Lett.*, **29**, 459 (1972).
86. P. W. Atkins and R. G. Woolley, *Proc. Roy. Soc. (London)*, **A314**, 251–267 (1969).
87. R. G. Woolley, *Mol. Phys.*, **22**, 555–559 (1971).
88. R. G. Woolley and J. E. Cordle, *Chem. Phys. Lett.*, **22**, 411–413 (1973).
89. A. D. Buckingham and P. J. Stiles, *Mol. Phys.*, **24**, 99–108 (1972).
90. R. E. Raab, *Mol. Phys.*, **29**, 1323–1331 (1975).
91. D. M. Brink and G. R. Satchler, *Angular Momentum*, 2nd ed., Clarendon Press, Oxford, (1968).
92. M. Abramowitz and J. A. Stegun, *Handbook of Mathematical Functions*, Dover, New York, 1965.

SPECTRAL LINE SHAPES IN GASES IN THE BINARY–COLLISION APPROXIMATION

A. BEN–REUVEN*

*Department of Chemistry,
Massachusetts Institute of Technology,
Cambridge, Massachusetts*

CONTENTS

Abstract

Linear-response theory is used to obtain general expressions for the shape of spectral lines of ordinary gases of neutral molecules, under stationary and homogeneous initial conditions, weakly coupled to an arbitrary external field. The susceptibility is related (by using Green's

*Present address: Department of Chemistry, Tel Aviv University, Tel Aviv, Israel.

function methods) to a self-frequency tetradic, defined on a single-molecule-excitation basis. The self frequency is expressed in terms of the Fano double-space transition matrix, simply related to the Lippmann-Schwinger T matrix in the binary-collision approximation. Properties of the binary-collision self frequency are discussed, with generalization to gas mixtures to incorporate both self broadening (including resonance exchange) and foreign-gas broadening. The chapter deals with various applications, such as impact approximation (light perturbers), velocity-dependent self frequencies (heavy perturbers) with non-Markovian long-time tails, line-wing transient collision effects, and "statistical" broadening by slow particles. The extension to collision-induced polarizations is discussed.

I. INTRODUCTION

Several methods have been suggested in recent years for the evaluation of the spectral density of resonance lines of a molecular gas system responding to electric-dipole radiation.[1-14] Most of these methods apply the direct relation[15,16] existing between the spectral density and a retarded Green's function pertaining to the propagation of a particle excited from the "thermal bath" (of the molecular system at thermal equilibrium), together with the accompanying propagation of the "hole" left in the thermal bath.

An important feature of this particle-hole duality is that their propagation may not be treated independently. They interfere with each other, since the gas molecules may collide with neighboring molecules *while* absorbing (or emitting) the radiation. As a result, it is not adequate to concentrate on the evolution of the particle and hole states a and b separately, but on the pair (a,b) as one entity. The description of the propagation of these entities requires, therefore, the introduction of tetradic quantities labeled by four sets of labels $(ab; cd)$ relating the pair a,b to a pair c,d.

This tetradic (or superoperator, or two-particle) notation is characteristic of two of the recent developments in line shape theory. One of these developments (Ross,[3] Bezzerides[4,5]) uses the diagrammatic techniques of quantum field theory for calculating temperature (imaginary-time) Green's functions. In this approach, the spectral distribution is related to a two-particle Green's function, corresponding to the propagation of the particle-hole pair. In terms of a complete one-particle basis, these functions have a tetradic labeling.

The other approach (Fano,[1] Ben-Reuven[9]) is based on the idea of expanding dynamical variables in a Hilbert space of operators (double space, or Liouville space) and projecting onto a subspace of single-molecule excitation modes. In this representation, vectors have diadic labeling and operators tetradic labeling.

Each of the two aforementioned methods has advantages and disadvantages. One of the most helpful aspects of the *diagrammatic approach* is the one-to-one correspondence between algebraic expressions and Feynman diagrams. The generality of this method is most powerfully extended, however, by unified treatment of statistical and dynamical correlations with the introduction of temperature Green's functions, using the Wick-Matsubara theorem.[17] This natural incorporation of the statistical (or initial) correlations, resulting from the effects of intermolecular forces on the statistical density operator in a canonical distribution, has an advantage over the double-space formalism. Although it does not completely remove the problem of statistical correlations (they reenter through the back door in the evaluation of the frequency sums appearing in the two-particle self-energy diagrams), it relegates it to the evaluation of the proper self-energy operators appearing in the denominator of the Green's function.

The existence of other algorithms for incorporating the statistical correlations was pointed out by De Dominicis,[18] using a real-time diagrammatic technique developed by Bloch,[19] and by Albers and Oppenheim,[13] using a double-space approach. In such algorithms, the exact retarded Green's function is expressed in terms of a related one, in which no interactions are included in the density matrix, multiplied by a correction factor.

We are not concerned here with these corrections. Fortunately they are usually unimportant for thermal dilute gases. It can be shown that the effects of statistical correlations can be neglected around the line centers if $\Delta t \gg \tau$ and $\Delta t \gg \beta$, where $\beta = \hbar / k_B T$ is the inverse mean kinetic energy (in time units), τ is a measure of the duration of a binary collision, and Δt is the relaxation time (inverse half-width). Although these inequalities hold in most gases under ordinary conditions, there may be some questions regarding the relative unimportance of the corrections when applied to the wings of a line at off-resonance frequency differences comparable to β^{-1}.

The diagrammatic imaginary-time approach is adapted to canonical distributions. In certain interesting experimental situations, however, the initial distribution is stationary but not canonical—for example, some molecular-beam, fluorescence, and laser-induced spectra. The method of temperature Green's functions then loses its particular savor.

The most attractive aspect of *double-space formalism*[1] is the direct introduction of operators (rather than contractions, as required by the Wick-Matsubara theorem). The relevant Green's functions can be treated as projections of a Green's (or resolvent) operator. The latter, in turn, can be reduced to expressions involving collision operators, such as the T (or S) matrix. In the low-density limit one may consider the binary collision as the basic "event" in the gas. It is therefore the ultimate goal of theory to reduce the Green's functions into expressions involving the two-particle T matrix. In the double-space formalism we actually use the explicit T

matrix, known from scattering theory. In the diagrammatic approach we only use so-called ladder (or T-approximation) diagrams, which are contractions of a modified (temperature-dependent) T matrix. The explicit introduction of operators has the advantage of an immediate relation to the vastly expanding work on the calculation of molecular scattering amplitudes. Moreover, introduction of operators in closed form (rather than perturbation expansions, such as ladder diagrams are made of) enables one to use approximation techniques other than perturbation methods for their evaluation (e.g., classical-trajectory, distorted-wave approximations). This is a nontrivial point in considering molecular collisions, where perturbation expansions are so frequently very poorly convergent.

A still further advantage of operator techniques is in the treatment of second-order radiation effects (such as light scattering) by an explicit introduction of many-body polarizability operators.[20] They can also be used in a systematic investigation of collision-induced polarizations (and polarizabilities), by separation of electronic and nuclear degrees of motion, both in first-order (absorption) and second-order (light-scattering) spectra. We discuss here only first-order spectra. The use of double-space techniques has been extended recently to nonlinear optics of gases.[21,22]

The foundations of the double-space approach have been laid down by Fano.[1] However, some basic questions were left without adequate study. A critical review of the appropriate definition of the projection operators, an explicit introduction of translational degrees of freedom, the admission of resonance exchange between identical molecules, a more compact derivation of the relation of the line shape parameters to the T matrix, and a deductive derivation of various approximations, are among the topics treated here. The generality of the method is stressed: its applicability to any kind of resonance spectrum in the linear-response approximation of weak fields, starting from any stationary distribution of the gas as initial condition. One can thus use this formalism in a unified treatment of such phenomena as electric and magnetic multipole spectra,[23] Raman scattering,[24] and paramagnetic resonance.[25,26] Explicit attention is paid to the binary-collision limit (where the line shape parameters are linear in the density) neglecting three-body effects.[13,27] Translationally invariant systems (in the absence of the external fields)—that is, homogeneous gas systems whose dimensions are larger than the wavelength of the radiation in the relevant spectral range—are considered.

As a final corollary, it should be stated that both methods just mentioned treat the molecules as "elementary" (i.e., stable molecules, though with a compound internal structure). In their present form they are not made to include photodissociation, photoionization, chemically reactive collisions, and like phenomena. Applications, though, have been made to ionized plasmas (Smith and Hooper,[28] Smith,[29] Klein[6]). A manner of their extension to chemically reactive systems is suggested by Berrondo's work.[30]

This chapter is a theoretical progress report rather than a review. For monographs, review articles, and other papers of general interest concerning line shapes, consult Margenau and Lewis,[31] Traving,[32] Breene,[33,34] Tsao and Curnutte,[35] Griem,[36] Gordon,[37] Birnbaum,[38] Berman and Lamb,[39] Cooper,[40] Futrelle,[41] Hindmarsh and Farr,[42] and Rabitz.[43] A comprehensive bibliography on atomic spectra (including selected references on molecular spectra) was compiled by the National Bureau of Standards.[44]

II. UNITS AND NOTATION

All quantities describing energy, momentum, mass, angular momentum, and polarization (e.g., dipole moments) appear here in ordinary (cgs) units *divided by* $\hbar$. Temperature is multiplied by Boltzmann's constant k_B and divided by $\hbar$ (i.e., T represents $k_B T/\hbar$ in cgs units). In this fashion $\hbar$ and k_B are practically eliminated.

Ordinary quantum states, as vectors in a Hilbert space (H space), are denoted by the Dirac brackets. The lower case italics $|a\rangle$, $|b\rangle$, $|c\rangle$, and $|d\rangle$ are used to denote a complete set of quantum numbers of a *single* molecule. When we want to explicitly separate the translational degrees of freedom of the molecular center of mass from its internal degrees of freedom (including rotation), we use the notation $|a\rangle = |\alpha\mathbf{p}_a\rangle$, and so on, where $\mathbf{p}_a$ is the translational momentum and the Greek α stands for all internal degrees of freedom.

States of two or more particles are denoted $|i\rangle$, $|j\rangle$, and so on; for example,

$$|i\rangle = |a\rangle|g\rangle = |\alpha\mathbf{p}_a\rangle|\zeta\mathbf{p}_g\rangle$$

is a two-particle state, where $|g\rangle$ or $|h\rangle$ denotes the second-particle states.

Double, or Liouville, space is a Hilbert space on which quantum-mechanical operators are defined[1] (L space). For example, in the expansion of an operator A in a complete N-molecule basis

$$A = \sum_{ij} A_{ij} |i\rangle\langle j| \tag{1a}$$

the products $|i\rangle\langle j|$ can form a basis for L space. The corresponding vectors are denoted by Baranger's double brackets[45] $|ij\rangle\rangle$, with which (1a) can be written as

$$|A\rangle\rangle = \sum_{ij} A_{ij}|ij\rangle\rangle \tag{1b}$$

The scalar product of two vectors $|A\rangle\rangle$ and $|B\rangle\rangle$ is defined as

$$\langle\langle B|A\rangle\rangle = \text{tr}\{wB^\dagger A\}, \qquad (B_{ij}^\dagger = B_{ji}^*) \tag{2}$$

which requires an introduction of a metric (or weight) operator w. In dealing with statistical ensemble averages, it is customary to choose the statistical (density) operator ρ as the metric.

Superoperators operating on L-space vectors are written with a script capital letter. Their matrix elements, in double-space notation are

$$\langle\langle ij|\mathcal{L}|kl\rangle\rangle = \text{tr}\{w|j\rangle\langle i|(\mathcal{L}|k\rangle\langle l|)\} \tag{3}$$

Quite generally, L-space operators can be expressed in terms of bilinear forms of H-space operators of the kind $\mathcal{L} = AB^*$, defined by

$$(\mathcal{L}X)_{ij} = \sum_{kl} \mathcal{L}_{ij;\,kl} X_{kl} = \sum_{kl} A_{ik} B_{jl}^* X_{kl}$$

$$= (AXB^\dagger)_{ij} \tag{4}$$

The *asterisk* implies both *complex conjugation* and operation on the *columns* of the matrix X. The nonasterisked operators operate on the rows of X.

The two forms of tetradic matrix elements defined in (3) and (4) are not necessarily identical. They become such only if an orthonormal metric is used—that is, if

$$\text{tr}\{w|j\rangle\langle i||k\rangle\langle l|\} = \delta_{ik}\delta_{jl}$$

Generally, the two forms are related to each other by

$$\langle\langle ij|\mathcal{L}|kl\rangle\rangle = \sum_{mn} \langle\langle ij|mn\rangle\rangle \mathcal{L}_{mn;\,kl}$$

From the definition (4), the product of two bilinear forms can be easily worked out:

$$(AB^*)(CD^*)X = ACB^*D^*X$$

$$= ACXD^\dagger B^\dagger$$

That is, asterisked and nonasterisked matrices *commute* with each other.

Our general policy is to express all L-space operators as sums (or integrals) of bilinear forms such as (4).

Some acronyms used below are defined as follows:

BCA = binary-collision approximation

SME = single-molecule excitation

SRPA = statistical random-phase approximation

III. LINEAR RESPONSE

The spectral density of energy dissipation by absorption of externally imposed radiation is known to be related to a retarded Green's function.[15-17] These Green's functions involve the molecular moments (or *polarizations*) on which the external field acts as a time-dependent *constraint*. By a polarization we mean an *extensive* field (whose density per unit volume is finite in the thermodynamic limit of infinite volume). By a constraint we imply an *intensive* (volume-independent) applied field. We follow the practice of Kubo in assuming that the coupling between the molecular system and the field is weak enough to render negligible the reaction of the system on the field itself. The constraining field can then be treated as a classical-like, externally controllable quantity.

Corresponding to a given set of constraining fields $E^j(\mathbf{x}, t)$ $(j = 1, \ldots, f)$, there usually exists a set of intensive fields $P^i(\mathbf{x}, t)$ to which the constraints are coupled, called the *polarizations*. The polarizations are intrinsic properties of the molecular system depending on the external fields (and on the thermodynamic conditions of the system). They can be generally expressed as functionals of the constraints

$$P^i(\mathbf{x}, t) = \Phi_i\big(\{E^j(\mathbf{x}, t)\}_{j=1,\ldots,f}\big) \tag{5}$$

Assume that this functional dependence behaves well enough near the limit of zero field intensity to allow expansion in functional derivatives. Then

$$P^i(\mathbf{x}, t) = \Phi_i^{(0)}(\mathbf{x}, t) + \int d^3x'\, dt'\, \Phi_{ij}^{(1)}(\mathbf{x}, \mathbf{x}'; t, t') E^j(\mathbf{x}', t')$$

$$+ \int\int d^3x'\, dt'\, d^3x''\, dt''\, \Phi_{ijk}^{(2)}(\mathbf{x}, \mathbf{x}', \mathbf{x}''; t, t', t'') E^j(\mathbf{x}', t') E^k(\mathbf{x}'', t'') + \ldots \tag{6}$$

where

$$\Phi_{ijk\cdots}^{(n)} = \left(\frac{\delta^n \Phi_i}{\delta E^j \delta E^k \cdots}\right)_{\{E^j\}=0}$$

are the functional derivatives of Φ_i at zero constraints. These derivatives are properties of the molecular system, independent of the external fields. Considering stationary properties of a large sample of gas (neglecting surface effects), the quantities $\Phi^{(n)}$ generally involve molecular properties averaged over a statistical distribution that is invariant under time and space translation. They then depend only on coordinate and time *differences*, $\mathbf{x} - \mathbf{x}'$, $t - t'$, and so on. The exact type of statistical averaging to use depends on the initial conditions under which the system is prepared and brought into interaction with the fields.

Invariance under time and space translations implies that $\Phi_i^{(0)}$ must be constant, independent of the field position $\mathbf{x}, t$. Therefore, without much loss of generality we may ignore it here. We are concerned here with the *linear response approximation*, in which only

$$\Phi_{ij}^{(1)}(\mathbf{x}, \mathbf{x}'; t, t') \equiv \chi^{ij}(\mathbf{x} - \mathbf{x}', t - t') \tag{7}$$

is retained. These functions are known as the (linear) *response functions*, and their Fourier transforms

$$\chi^{ij}(\mathbf{k}, \omega) = \int d^3\mathbf{x} \, dt \exp(-i\mathbf{k} \cdot \mathbf{x} + i\omega t) \chi^{ij}(\mathbf{x}, t) \tag{8}$$

as the (linear) *susceptibilities*.[15]

Microscopic expressions for the functional derivatives can be obtained if the macroscopic polarizations $P^i(\mathbf{x}, t)$ can be expressed as statistical averages of operators $M^i(\mathbf{x})$ explicitly constructed from the microscopic dynamical variables to which the fields are coupled by the hamiltonian

$$H(t) = H + H_1(t)$$

$$= H - \sum_i \int M^i(\mathbf{x}) E^i(\mathbf{x}, t) d^3\mathbf{x} \tag{9}$$

Here H is the time-independent translationally invariant hamiltonian of the constraint-free system. More explicitly, we deal here only with polarizations constructed from single-molecule terms. Ignoring the finite dimensions of the molecules (as we usually do at low $\mathbf{k}$ values), we can write

$$M^i(\mathbf{x}) = \sum_{A=1}^{N} \mu_A^i \delta(\mathbf{x} - \mathbf{R}_A) \tag{10}$$

where $\mathbf{R}_A$ is the position of molecule A in the gas and μ_A^i is the molecular *moment* coupled to E^i.

To this kind of constraint, whose microscopic effect can be expressed by

(9), belong the multipole interactions of neutral molecules with an external electromagnetic field. The form (10) generally applies whenever the molecules are much smaller than the wavelength $\lambda = 2\pi/k$. The μ^i's represent the various electric and magnetic multipoles; the E^i's the applied electric and magnetic fields and their spatial derivatives.

Again it should be stressed that molecular reaction fields are considered negligible. Also, this treatment implies that relativistic retardation effects are neglected and that all electromagnetic fields induced by the molecular charges are of the Coulombic (longitudinal) type, whereas the applied fields are radiative (transverse). All the Coulombic interactions are incorporated in H.

The macroscopic intensive fields $P^i(\mathbf{x}, t)$ are related to the microscopic operators $M^i(\mathbf{x})$ by a statistical averaging

$$P^i(\mathbf{x}, t) = \mathrm{tr}\left\{ \rho(t) M^i(\mathbf{x}) \right\} \tag{11}$$

Here $\rho(t)$ is the density operator (in the Schrödinger picture), which obeys the Von Neumann equation

$$i\frac{\partial}{\partial t}\rho(t) = [H(t), \rho(t)] \tag{12a}$$

with the full hamiltonian (9). To avoid the difficulties caused by the infinite phases of $\rho(t)$ in the Schrödinger picture, which may occur at the limits $t \to \pm\infty$, it is preferable to transform to the interaction picture, in which

$$i\frac{\partial}{\partial t}\rho_I(t) = [H_I(t), \rho_I(t)] \tag{12b}$$

Here

$$
\begin{aligned}
H_I(t) &= e^{iHt}H_1(t)e^{-iHt} \\
&= -\sum_i \int M^i(\mathbf{x}, t) E^i(\mathbf{x}, t) d^3\mathbf{x}
\end{aligned} \tag{13}
$$

where

$$
\begin{aligned}
M^i(\mathbf{x}, t) &= e^{iHt}M^i(\mathbf{x})e^{-iHt} \\
&= \sum_{A=1}^{N} \mu_A^i(t)\delta(\mathbf{x} - \mathbf{R}_A(t))
\end{aligned} \tag{14}
$$

are the Heisenberg operators for the polarizations in the field-free system.

Equation (11) should then be replaced by

$$P^i(\mathbf{x},t) = \mathrm{tr}\left\{\rho_I(t)M^i(\mathbf{x},t)\right\} \tag{15}$$

Equation (12b) can be solved by iterative integration if its value at some initial time t_0 is given:

$$\rho_I(t) = \rho_I(t_0) - i\int_{t_0}^{t}\left[H_I(t'),\rho_I(t_0)\right]dt'$$

$$+(-i)^2\int_{t_0}^{t}\int_{t_0}^{t'}dt'\,dt''\left[H_I(t'),\left[H_I(t''),\rho_I(t_0)\right]\right]+\cdots \tag{16}$$

We now require that $\rho_I(t_0)$, at some early time t_0, to be stationary, translationally invariant, and independent of the external fields, obeying

$$[H,\rho_I(t_0)]=0, \quad [\mathbf{P},\rho_I(t_0)]=0$$

where $\mathbf{P}$ is the total momentum operator. This statement means that the fields E^i are "switched on" only after t_0. Serious limitations would be imposed on the allowed forms of time dependence of the field unless we let $t_0\to-\infty$. For then we can choose a much wider class of time-dependent functions (e.g., monochromatic fields), add a convergence factor $e^{\varepsilon t}$, and let $\varepsilon\to0$ *after* the limit $t_0\to-\infty$ is taken. Thus we have

$$\rho_I(t) = \rho(-\infty) - i\int_{-\infty}^{\infty}\theta(t-t')[H_I(t'),\rho(-\infty)]dt'+\cdots \tag{17}$$

where $\rho_I(-\infty)$ was replaced by the Schrödinger operator, since it commutes with H. The upper limit in (17) was extended to $+\infty$ by introducing the Heaviside step function $\theta(t)$, which vanishes at negative values of its argument.

Inserting (17) into (15), and retaining only the linear terms in H_I (neglecting higher-order terms as well as the "spontaneous" zero-order term), the macroscopic polarization obtained is of the expected form

$$P^i(\mathbf{x},t) = \sum_j \int d^3x'\,dt'\chi^{ij}(\mathbf{x}-\mathbf{x}',t-t')E^j(\mathbf{x}',t')$$

with

$$\chi^{ij}(\mathbf{x}-\mathbf{x}',t-t') = i\theta(t-t')\mathrm{tr}\left\{M^i(\mathbf{x},t)[M^j(\mathbf{x}',t'),\rho(-\infty)]\right\}$$

$$= i\theta(t-t')\langle[M^i(\mathbf{x},t),M^j(\mathbf{x}',t')]\rangle \tag{18}$$

The angular brackets denote a statistical averaging over the stationary distribution $\rho(-\infty)$. In many applications we consider the effects of small perturbations on a system in thermal equilibrium, where $\rho(-\infty)$ can be replaced by the canonical-distribution density matrix

$$\rho(-\infty) = \rho_e = Z^{-1}e^{-\beta H} \qquad \left(\beta = \frac{1}{T}; Z = \operatorname{tr}\{e^{-\beta H}\}\right) \qquad (19)$$

However, this is seldom called for in a linear-response formulation of the theory. All that we require here is that ρ be stationary and independent of the applied fields. Also for simplicity's sake, let ρ be translationally invariant. There is, however, no conceptual difficulty in extending the theory to nonuniform systems. The specification of the initial state $\rho(-\infty)$ may to a large extent determine the response of the system to weak probing fields.

IV. GREEN'S FUNCTIONS

The limitation of the response function (18) to positive values of $t - t'$ establishes a causal relationship between cause (E^j) and effect (P^i), with the polarization being a *retarded* effect of the past history of the field. From a formal point of view (considering the microscopic reversibility of the laws of motion), an *advanced* relationship could be derived between P^i and later-time values of E^j by imposing the a priori knowledge of $\rho(+\infty)$ as "initial" conditions. However, only the first case is physically realizable by a proper preparation of the system. Hence the irreversible character of the temporal variation of the macroscopic polarizations. The retarded (advanced) Green's function for a pair of polarizations (i,j) is defined as

$$G_{ij}^{R(A)}(\mathbf{x} - \mathbf{x}', t - t') = \pm i\theta(\pm t \mp t')\langle[M^i(\mathbf{x},t), M^j(\mathbf{x}',t')]\rangle \qquad (20)$$

where the upper (lower) sign pertains to the retarded (advanced) function. The angular brackets imply averaging over a translation-and-time-invariant distribution. The Fourier transforms of (20) in space and time are defined by

$$G(\mathbf{k},\omega) = \int e^{i\omega t}G(\mathbf{k},t)\,dt = \int e^{i\omega t - i\mathbf{k}\cdot\mathbf{x}}G(\mathbf{x},t)\,d^3\mathbf{x}\,dt \qquad (21)$$

In the spatial Fourier analysis we use the so-called box normalization, imposing periodic boundary conditions in a large box of volume V, to introduce a denumerable basis. Thus, for example, we put

$$M^i(\mathbf{x},t) = V^{-1}\sum_{\mathbf{k}} e^{i\mathbf{k}\cdot\mathbf{x}}M^i(\mathbf{k},t) \qquad (22)$$

The thermodynamic limit $V \rightarrow \infty$ may eventually be taken by substituting

$$V^{-1} \sum_{\mathbf{k}} \longrightarrow (2\pi)^{-3} \int d^3\mathbf{k} \tag{23a}$$

$$V\delta_{\mathbf{kk'}} \longrightarrow (2\pi)^3 \delta(\mathbf{k} - \mathbf{k'}) \tag{23b}$$

provided the results can be expressed as volume-independent. Let us formally introduce now a complete denumerable basis of eigenfunctions of H, say, $|m\rangle$, $|n\rangle$,..., with respective eigenvalues $\varepsilon_m, \varepsilon_n, \dots$. Owing to the invariance properties of ρ, we can write

$$G_{ij}^{R(A)}(\mathbf{k}, t) = \mp i\theta(\pm t)V^{-1} \sum_{mn} (\rho_m - \rho_n)\langle m|M^i(\mathbf{k})|n\rangle \langle n|M^j(-\mathbf{k})|m\rangle e^{-i\omega_{nm}t}$$

$$\tag{24}$$

where $\omega_{nm} = \varepsilon_n - \varepsilon_m$, and $\rho_m = \langle m|\rho(-\infty)|m\rangle$. The temporal Fourier transform of (24) is

$$G_{ij}^{R(A)}(\mathbf{k}, \omega) = V^{-1} \sum_{mn} (\rho_m - \rho_n)\langle m|M^i(\mathbf{k})|n\rangle\langle n|M^j(-\mathbf{k})|m\rangle(\omega - \omega_{nm} \pm i\eta)^{-1}$$

$$\tag{25}$$

where the limit $\eta \rightarrow +0$ is implied; that is,

$$\lim \frac{1}{\omega - \omega_{nm} \pm i\eta} = \mathcal{P}\left(\frac{1}{\omega - \omega_{nm}}\right) \mp i\pi\delta(\omega - \omega_{nm})$$

thus resolving the Green's functions into terms involving Cauchy principal parts and Dirac delta functions. The nonanalytic character of the introduction of the Heaviside step function thus results in nonanalytic behavior of the Fourier trnasforms as functions defined on the complex-ω plane. Its expression is a cut along the line $\omega = \mp i\eta$ parallel to the real axis, leaving the retarded (advanced) functions analytic only in the upper (lower) half-plane.

The Green's functions can be related to the analytic *correlation functions* defined by

$$G_{ij}^{>}(\mathbf{x}, t) = \langle M^i(\mathbf{x}, t)M^j(0, 0)\rangle$$

$$G_{ij}^{<}(\mathbf{x}, t) = \langle M^j(0, 0)M^i(\mathbf{x}, t)\rangle \tag{26a}$$

and their Fourier transforms

$$G_{ij}^{>(<)}(\mathbf{k},\omega)=2\pi V^{-1}\sum_{mn}\rho_{m(n)}\langle m|M^i(\mathbf{k})|n\rangle$$

$$\langle n|M^j(-\mathbf{k})|m\rangle\delta(\omega-\omega_{nm}) \tag{26b}$$

through the integral relation

$$G^{R(A)}(\mathbf{k},\omega)=\frac{1}{2\pi}\int_{-\infty}^{\infty}\frac{G^>(\mathbf{k},\omega')-G^<(\mathbf{k},\omega')}{\omega-\omega'\pm i\eta}d\omega'$$

$$=-\frac{1}{2\pi}\int_{-\infty}^{\infty}[G^>(\mathbf{k},\omega')-G^<(\mathbf{k},\omega')]\mathscr{P}\left(\frac{1}{\omega'-\omega}\right)d\omega$$

$$\mp\left(\frac{i}{2}\right)[G^>(\mathbf{k},\omega)-G^<(\mathbf{k},\omega)]$$

$$=G'(\mathbf{k},\omega)\pm iG''(\mathbf{k},\omega) \tag{27}$$

The polarization operators are hermitian

$$\langle n|M^j(-\mathbf{k})|m\rangle=\langle m|M^j(\mathbf{k})|n\rangle^*$$

and therefore the matrices $G^>$, $G^<$, G', and G'' in (26b) and (27) are all hermitian in the indices i, j. Considering also that $G_{ij}^{R(A)}(\mathbf{x},t)$ is real, we get

$$G_{ij}^{R(A)}(\mathbf{k},\omega)=\left[G_{ji}^{A(R)}(\mathbf{k},\omega)\right]^*=G_{ji}^{A(R)}(-\mathbf{k},-\omega) \tag{28a}$$

or, equivalently,

$$G_{ij}^{R(A)}(\mathbf{x},t)=G_{ji}^{A(R)}(-\mathbf{x},-t) \tag{28b}$$

These symmetry properties are closely related to parity and time-reversal symmetry. If ρ is invariant under both operations, and μ^i and μ^j have definite signature (± 1) under either, we write

$$G_{ij}^{R(A)}(\mathbf{x},t)=\sigma(i,j)G_{ij}^{A(R)}(-\mathbf{x},-t)$$

where $\sigma(i,j)=\pm 1$ is the combined signature of μ^i and μ^j under parity and time reversal. Hence Onsager's reciprocity relations

$$G_{ij}^{R(A)}=\sigma(i,j)G_{ji}^{R(A)} \tag{29}$$

follow. From (29) it turns out that $G'_{ij}(\mathbf{k},\omega)$ and $G''_{ij}(\mathbf{k},\omega)$ are real or imaginary, depending on whether $\sigma(i,j)$ is even or odd.

It is generally sufficient to calculate $G^>$ or $G^<$ to obtain the Green's functions. The two correlation functions are related by

$$G^<_{ij}(\mathbf{k},\omega) = G^>_{ji}(-\mathbf{k},-\omega)$$

If ρ is a *canonical distribution* (19) then, simply,

$$G^<(\mathbf{k},\omega) = e^{-\beta\omega}G^>(\mathbf{k},\omega) \tag{30}$$

The susceptibility (in our units) is simply given by

$$\chi^{ij}(\mathbf{k},\omega) = -G^R_{ij}(\mathbf{k},\omega) \tag{31}$$

In conventional (cgs) units, an extra factor $\hbar^{-1}$ should be inserted into the right-hand side of (31).

The change in the entropy S is given by

$$T dS = -V^{-1}\sum_{\mathbf{k}}\sum_{i}E^i(-\mathbf{k},t)dP^i(\mathbf{k},t) \tag{32}$$

Assume that E^i is a monochromatic field

$$E^i(\mathbf{k},t) = E^i(\mathbf{k})\cos\omega t$$

By averaging over time and going to the thermodynamic limit, we find the net heat flow into the gas

$$\left\langle\frac{dQ}{dt}\right\rangle_\omega = T\left\langle\frac{dS}{dt}\right\rangle_\omega = \frac{\omega}{16\pi^3}\sum_{ij}\int d^3k\, E^i(-\mathbf{k})\chi''^{ij}(\mathbf{k},\omega)E^j(\mathbf{k}) \tag{33}$$

depending on χ'' only, where $i\chi''$ is the antihermitian part of χ, corresponding to G''. From

$$\chi'' = \frac{(G^> - G^<)}{2}$$

and (26b) it follows that (33) is positive (energy is dissipated) if $\rho_m > \rho_n$ for $\omega_{nm} > 0$. To an incoming flux of plane-wave, plane-polarized radiation, given by

$$I = \frac{c(E^i)^2}{8\pi}$$

as in electric-dipole radiation, there is a corresponding absorption coefficient

$$\alpha_{ii}(\mathbf{k}, \omega) = \left(\frac{4\pi\omega}{c} \right) \chi''^{ii}(\mathbf{k}, \omega) \tag{34}$$

In the following analysis we concentrate on the one-sided correlation functions $G^>$ or, rather, on the corresponding one-sided Green's functions

$$F_{ij}^{R(A)}(\mathbf{k}, t) = \mp i\theta(\pm t)V^{-1}\langle M^i(\mathbf{k}, t)M^j(-\mathbf{k}, 0)\rangle \tag{35}$$

to which $G^>$ is related by

$$G_{ij}^>(\mathbf{k}, \omega) = i\{ F_{ij}^R(\mathbf{k}, \omega) - F_{ij}^A(\mathbf{k}, \omega)\}$$

$$= i\{ F_{ij}^R(\mathbf{k}, \omega) - [F_{ji}^R(-\mathbf{k}, -\omega)]^*\} \tag{36}$$

The reason for this choice is a practical one. In canonical distributions, the relation (30) between $G^<$ and $G^>$ is rigorous, whereas in actual calculations we sometimes use approximate methods for obtaining the matrix elements of ρ. In taking the difference of populations (as in G^R) rather than the populations themselves (as in F^R) we may introduce larger errors, particularly at low-frequency (far-infrared or microwave) spectra.

V. DOUBLE–SPACE FORMALISM

Owing to the invariance of ρ under time translations, we can write

$$F_{ij}^R(\mathbf{k}, t) = -i\theta(t)V^{-1}\langle M^i(\mathbf{k})e^{-iHt}M^j(-\mathbf{k})e^{iHt}\rangle \tag{37}$$

where, from now on, M^i, and so on, denote the time-independent (Schrödinger-picture) operators. We can now consider the polarizations M^i as vectors in a Hilbert space of operators (L space), as discussed in Section II. Thus we may write, in the limit $t \to +0$,

$$F_{ij}^R(\mathbf{k}, t = +0) = -iV^{-1}\langle\langle M^i(-\mathbf{k})|M^j(-\mathbf{k})\rangle\rangle$$

provided we identify the metric w with ρ:

$$w \to \rho = \rho(-\infty) \tag{38}$$

The generator of time evolution of L-space vectors is the double-space *Liouville operator* (or *liouvillian*) defined by[1]

$$\mathcal{K} = HI^* - IH^*, \qquad (I = \text{unit operator}) \tag{39}$$

in double-space notation. Since ρ commutes with H, it turns out that $\mathcal{H}$ is a hermitian operator in the double-bracket notation; that is,

$$\langle\langle A|\mathcal{H}B\rangle\rangle = \langle\langle \mathcal{H}A|B\rangle\rangle$$

From the definition (4) of L-space operators as bilinear forms, it is easy to see that

$$e^{-iHt}M^j(-\mathbf{k})e^{iHt} = [e^{-iHt}(e^{-iHt})^*]M^j(-\mathbf{k})$$
$$= e^{-i\mathcal{H}t}M^j(-\mathbf{k}) \tag{40}$$

The convenience of this formalism is in the possibility of introducing the Fourier decomposition of (37) as a matrix element of a Green's function operator (or *resolvent operator*)

$$F_{ij}^R(\mathbf{k},\omega) = V^{-1}\langle\langle M^i(-\mathbf{k})|\mathcal{G}(\omega)|M^j(-\mathbf{k})\rangle\rangle$$
$$= V^{-1}\langle M^i(\mathbf{k})\mathcal{G}(\omega)M^j(-\mathbf{k})\rangle \tag{41}$$

where

$$\mathcal{G}(\omega) = (\omega - \mathcal{H} + i\eta)^{-1} \tag{42}$$

The general properties of this L-space (retarded) operator have been studied in detail by Zwanzig,[46] Fano,[1] Mori,[47] and others. We summarize here its properties, mostly following Fano's work. In particular, we stress its relation to the analogous operator

$$G(\varepsilon) = (\varepsilon - H + i\eta)^{-1} \tag{43}$$

in ordinary H space. The latter has been studied extensively in relation to the Lippmann-Schwinger equation in scattering theory.[48,49]

The inverse Fourier transform of (42) can be written, in double-space fashion, as

$$\mathcal{G}(t) = -i\theta(t)e^{-i\mathcal{H}t} = -iG(t)G^*(t) \tag{44}$$

where

$$G(t) = -i\theta(t)e^{-iHt} \tag{45}$$

is the analogous H-space operator. From the convolution theorem of Fourier transforms, it follows that

$$\mathcal{G}(\omega) = (2\pi i)^{-1}\int_{-\infty}^{\infty} d\varepsilon\, G(\varepsilon + \omega)G^*(\varepsilon) \tag{46}$$

This relation can be verified by contour integration, in the complex-ε plane, of the integral.

$$\int d\varepsilon [\varepsilon + \omega - H + i(\eta - \eta')]^{-1} [\varepsilon - H^* - i\eta']^{-1}$$

taking the zero limit in the order $\eta > \eta' > 0$. The integral along the real axis may be replaced by an integral around a loop containing the cut, corresponding to positive (unbound) eigenvalues of H, and the poles, corresponding to negative (bound) ones, on either side of the real axis, as in Fig. 1.

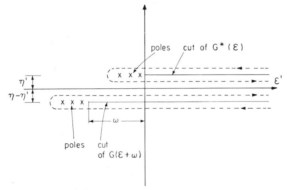

Fig. 1. Two alternative contours (dashed lines) for evaluating the energy transforming a product of Green's (resolvent) operators into the corresponding double-space operator.

We generally assume that H can be separated into

$$H = H_0 + V \tag{47}$$

where H_0 is the sum of independent free-molecule hamiltonians and V is a sum of pairwise interactions. Introduce now G_0 (or $\mathcal{G}_0$) as the operator corresponding to replacing H by H_0 (or $\mathcal{H}$ by $\mathcal{H}_0$) in the proper definition. The N-molecule T operator (in H space) can be introduced by

$$G(\varepsilon) = G_0(\varepsilon)[1 + T(\varepsilon)G_0(\varepsilon)] \tag{48}$$

Hence T obeys the Lippmann-Schwinger equation

$$T = V + VG_0T = V \sum_{n=0}^{\infty} (G_0V)^n = V + TG_0V = V + VGV \tag{49}$$

The expansion in power series of V in (49) is the renowned Born series. The T is a nonhermitian operator; its antihermitian part obeys on the energy shell the "optical theorem"

$$i(T - T^\dagger) = 2\pi T \delta(\varepsilon - H_0) T^\dagger \tag{50}$$

The analogous L-space operators obey equivalent theorems. Thus splitting $\mathcal{H}$ into

$$\mathcal{H} = \mathcal{H}_0 + \mathcal{V} \tag{51}$$

where

$$\mathcal{H}_0 = H_0 I^* - I H_0^*, \qquad \mathcal{V} = VI^* - IV^*$$

we can introduce the L-space $\mathcal{T}$ operator by

$$\mathcal{G}(\omega) = \mathcal{G}_0(\omega)[\mathcal{J} + \mathcal{T}(\omega)\mathcal{G}_0(\omega)] \tag{52}$$

where $\mathcal{J} = II^*$ is the L-space unit operator. The analog of the Lippmann-Schwinger equation is

$$\mathcal{T} = \mathcal{V} + \mathcal{V}\mathcal{G}_0\mathcal{T} = \mathcal{V} + \mathcal{T}\mathcal{G}_0\mathcal{V}$$

$$= \mathcal{V} + \mathcal{V}\mathcal{G}\mathcal{V} \tag{53}$$

The analytic properties of $\mathcal{G}(\omega)$ in the complex-ω plane, are, however, different from those of $G(\varepsilon)$ in the complex-ε plane. The latter has a cut at $\varepsilon > 0$ and, possibly, poles at $\varepsilon < 0$ corresponding to true bound states of the many-body system. The $\mathcal{G}(\omega)$ is expressed in terms of energy *differences* and therefore has a cut all along the real axis with, possibly, poles superimposed on it. To avoid complicating matters, we ignore true bound states of the gas system and retain just the cut. We thus turn $F^R(\mathbf{k}, \omega)$ into a smooth, "well-behaved" function of ω on the real axis.

As we show later on, in the low-density limit, the line shape problem reduces to the calculation of certain expectation values of the $t(\omega)$ matrix, the two-body analogue of the N-body $\mathcal{T}(\omega)$. It is therefore desirable to obtain an expression for $\mathcal{T}(\omega)$ in terms of the more familiar $T(\varepsilon)$, thus tying up the line shape problem with the theory of molecular scattering. Such an expression has been derived by Fano.[1] We derive it here in a more compact form, by using the bilinear expansion (46) of the resolvent operator. From (46) and (53) we get (using the definition of $\mathcal{V}$)

$$\mathcal{T}(\omega) = (VI^* - IV^*)\left\{ II^* - \frac{i}{2\pi}\int d\varepsilon\, G(\varepsilon + \omega)G^*(\varepsilon)(VI^* - IV^*) \right\} \tag{54}$$

It is easy to show that since V does not depend on ε,

$$\int d\varepsilon\, VG^*(\varepsilon) = 2\pi i \int d\varepsilon\, V\delta(\varepsilon - H^*) = 2\pi i V$$

with a similar relation for V^*. It can also be verified by contour integration, noting that the singularities of T and G_0^* lie on opposite sides of the real axis, that

$$\int d\varepsilon\, T(\varepsilon + \omega) G_0^*(\varepsilon) = 2\pi i \int d\varepsilon\, T(\varepsilon + \omega)\delta(\varepsilon - H_0^*)$$

$$= 2\pi i T(H_0^* + \omega)$$

This means that

$$[T(H_0^* + \omega)]_{ij;\, kl} = T_{ik}(\varepsilon_j + \omega)\delta_{jl}$$

where eigenfunctions of H_0 are used as a basis. A similar relation can be derived for T^*. Hence (54) can be rewritten

$$\mathcal{T}(\omega) = \int \frac{d\varepsilon}{2\pi i} \{ V[1 + G(\varepsilon + \omega)V]G^*(\varepsilon)$$

$$+ G(\varepsilon + \omega)V^*[I^* + G^*(\varepsilon)V^*]$$

$$- T(\varepsilon + \omega)G_0(\varepsilon + \omega)G_0^*(\varepsilon)T^*(\varepsilon) - G_0(\varepsilon + \omega)T(\varepsilon + \omega)T^*(\varepsilon)G_0^*(\varepsilon) \}$$

Using the last identity in (49), and recalling that asterisked and non-asterisked operators commute, we finally get

$$\mathcal{T}(\omega) = \int d\varepsilon \big[(T\Delta_0^* - \Delta_0 T^*)$$

$$+ (2\pi i)^{-1}\, \mathcal{D}\, TT^*\, \mathcal{D}\, \big] \tag{55}$$

where

$$\mathcal{D} = G_0 I^* - IG_0^* \tag{56}$$

and

$$\Delta_0(x) = \delta(x - H_0)$$

All arguments in (55) have been suppressed, keeping in mind that all nonasterisked operators T, G_0, and Δ_0, should have $\varepsilon + \omega$ as argument, whereas the asterisked T^*, G_0^*, and Δ_0^*, have ε as argument. Equation (55) is equivalent (in a more compact form) to (55) of Fano,[1] except that here it refers to the N-body T matrix.

A tetradic-form expression can be written for (55) in the complete basis of eigenfunctions of H_0:

$$\mathcal{T}_{ij;\,kl}(\omega) = T_{ik}(\varepsilon_j + \omega)\delta_{jl} - \delta_{ik} T_{jl}^*(\varepsilon_i - \omega)$$

$$+ \int \frac{d\varepsilon}{2\pi i} d_{ij} T_{ik}(\varepsilon + \omega) T_{jl}^*(\varepsilon) d_{kl} \qquad (57)$$

where

$$d_{ij}\delta_{ii'}\delta_{jj'} = \mathcal{D}_{ij;\,i'j'}$$

$$d_{ij} = \left[(\varepsilon + \omega - \varepsilon_i + i\eta)^{-1} - (\varepsilon - \varepsilon_j - i\eta')^{-1} \right] \qquad (58)$$

are the matrix elements of the (diagonal) supermatrix (56), consisting of Dirac delta functions and Cauchy principal parts.

As is well known from scattering theory,[48] the asymptotic time behavior of the wavefunctions is determined only by the on-the-energy-shell matrix elements of T or, equivalently, by the S matrix

$$S_{ik} = \delta_{ik} - 2\pi i \delta(\varepsilon - \varepsilon_i) T_{ik}(\varepsilon), \qquad (\varepsilon_i = \varepsilon_k) \qquad (59)$$

In a similar fashion, the asymptotic behavior of operators is determined by on-the-frequency-shell elements of (57), with

$$\omega = \omega_{ij} = \omega_{kl}$$

The factors d_{ij} defined by (58) then become

$$d_{ij}(\omega = \omega_{ij}) = -2\pi i \delta(\varepsilon - \varepsilon_j)$$

Therefore, the integral in (57) vanishes unless $\varepsilon_j = \varepsilon_l$; that is, on-the-frequency-shell elements of $\mathcal{T}(\omega)$ contain only on-the-energy-shell elements of $T(\varepsilon + \omega)$ and $T^*(\varepsilon)$, confined to the energy shells $\varepsilon_i = \varepsilon + \omega$ and $\varepsilon_j = \varepsilon$, respectively. Using (59), it immediately follows that

$$[\mathcal{T}_{ij;\,kl}(\omega)]_{\omega = \omega_{ij} = \omega_{kl}} = \int \frac{d\varepsilon}{2\pi i} [\delta_{ik}\delta_{jl} - S_{ik} S_{jl}^*]$$

$$= \int \frac{d\varepsilon}{2\pi i} [\mathcal{I} - \mathcal{S}]_{ij;\,kl} \qquad (60)$$

where the integration is carried across the energy shells. Here

$$\mathcal{S} = SS^* \qquad (61a)$$

is the L-space analogue of the S matrix, related to $\mathfrak{T}$ by

$$\mathcal{S}_{ij;\,kl} = \delta_{ik}\delta_{jl} - 2\pi i\delta(\omega-\omega_{ij})\,\mathfrak{T}_{ij;\,kl}(\omega), \qquad (\omega_{ij}=\omega_{kl}) \qquad (61b)$$

One may interpret ω_{ij} and ω_{kl} as representing the oscillation frequencies of two unperturbed "modes" of the ideal (noninteracting) gas. Equation (60) implies that the asymptotic amplitude for mixing of the modes ($\mathcal{S}_{ij;\,kl}$) by collisions will vanish not only if $\omega_{ij}\neq\omega_{kl}$ but also if $\varepsilon_i\neq\varepsilon_k$. Thus mixing of modes with different initial or final energies is transient only. The range of energy mismatch $|\varepsilon_i-\varepsilon_k|$ allowed will be determined by the variation of T and T^* off the respective energy shells. If T has a range of variation $\Delta\varepsilon$ beyond which it rapidly falls off, $T_{ij;\,kl}(\omega)$ will be generally significant only if $|\omega-\omega_{ij}|$, $|\varepsilon_i-\varepsilon_k|$, and $|\varepsilon_j-\varepsilon_l|$ all lie within $\Delta\varepsilon$ approximately.

As we see later on, cross relaxation between neighboring resonance lines in the spectrum requires such matrix elements of T. The impossibility of using the $\mathcal{S}$ matrix (except in exact degeneracy) may seem to be a disadvantage, since almost all numerical evaluation of line shape parameters is based on the S matrix. It is nevertheless possible, in approximate calculations, to introduce an analogue of the S matrix that is not strictly confined to the energy shell and is almost equal to the S matrix within the range $\Delta\varepsilon$ of energy variation. This matrix may be replaced by the S matrix in these approximate calculations (e.g., classical-trajectory calculations).

VI. PROJECTION OPERATORS

So far, we have dealt with general expressions for the N-molecule operators $\mathcal{G}(\omega)$ and $\mathfrak{T}(\omega)$. However, in calculating $F^R(\mathbf{k},\omega)$ we are only interested in very special double-space matrix elements of these operators. We only consider their projection on a restricted subspace to which the polarizations M^i are confined. This subspace is constructed only of single-molecule excitation modes—that is, $|i\rangle\langle j|$ in which i and j differ only in the state of one molecule.[9] The interaction $\mathcal{V}$ has the effect of transferring the single-molecule modes into two-molecule modes, in which two particles are simultaneously excited. Repeated application of $\mathcal{V}$ will excite two-or-more-particle modes, depending on whether we consider the same pair of molecules, or different pairs, in this act. We thus form the basis for expanding F^R in contributions of two-body collisions, three-body collisions, and so on (density expansions[13]).

There may, obviously, be ambiguity in the choice of the subspace to which the M^i's belong. For example, we could consider each M^i as a *single* vector (i.e., a single "mode"), rather than treating all $|i\rangle\langle j|$ (of the type discussed previously) as defining the subspace. It would seem that the small the basis chosen, the better off we are. However, it is shown later that we can do this only at the cost of hopelessly confusing the conceptual structure of the theory, with the simple relation of density expansions to collisions of two or more molecules completely lost. Only by retaining the full subspace of all single-molecule modes of the type mentioned, can we retain the conceptually simple structure of the theory.

Let $\{|a\rangle\langle b|\}_A$ describe an operator defined on the degrees of freedom of molecule A, where $a, b, \ldots$, are one-molecule quantum-numbers. From (10) it follows that

$$M^j(-\mathbf{k}) = \sum_{ab} \langle a| \mu^j e^{i\mathbf{k}\cdot\mathbf{R}} |b\rangle x^{ab} \tag{62a}$$

where

$$x^{ab} = \sum_{A=1}^{N} \{|a\rangle\langle b|\}_A \tag{63a}$$

Translational invariance furthermore restricts the class of operators (63a) in which (62a) can be expanded. Let us, by using $|a\rangle = |\alpha, \mathbf{p}_a\rangle$, explicitly introduce internal and translational degrees of freedom. Then

$$M^j(-\mathbf{k}) = \sum_{\alpha\beta, \mathbf{p}_b} \mu_{\alpha\beta}{}^j x^{ab}(\mathbf{k}) \tag{62b}$$

where

$$x^{ab}(\mathbf{k}) = \sum_{A=1}^{N} \{|\alpha, \mathbf{p}_b + \mathbf{k}\rangle\langle \beta, \mathbf{p}_b|\}_A \tag{63b}$$

Here we assumed that μ^i depends only on internal coordinates (including rotation). Let us denote the subspace of all *single-molecule excitation* modes (63a) (the SME subspace) as L_1, and the kth invariant subspace (63b) as $L_1(\mathbf{k})$. Corresponding to them we can define *projection operators*, P and $P_{\mathbf{k}}$, respectively.

This definition of the subspace on which we project has the following important properties:

1. It treats all identical molecules on an equal basis. In this it differs from Fano's definition[1] of the projection operator, which distinguished between the "absorbing" molecule and the rest of the gas as a "thermal bath." This distinction, which may be proper for dealing with foreign-gas broadening (of dilute solutions in a nonactive buffer gas), is inadequate for dealing with self-broadening (in a pure gas), as is shown later.

2. At low densities, the set L_1 is a "good" set, with a short "memory" time, in the sense of the Zwanzig-Mori[46,47] transport theory. In the Mori[47]

hierarchy, L_1 is a good initial set because *all* the one-molecule quantum numbers are left virtually unchanged between collisions, whereas the memory time is essentially the duration of the collision. Smaller subsets of L_1 (other than symmetry-invariant subsets), may not be "good" in the Mori sense.

3. As already hinted, the low-density contributions to the line shape correspond to the binary-collision contributions of the $\mathfrak{T}(\omega)$ matrix *only* if P is defined on this basis.

All that was just said is good *provided the basis vectors* (63) *form an orthogonal subspace*. However, with ρ as the metric, this is not generally true. This is a major weakness of this formalism. Although, formally the theory can be fully expounded with nonorthogonal metric, the only *simple* way to overcome this difficulty is to introduce an approximate density operator as a metric. A sufficient condition to ensure orthogonality of the basis vectors is to impose the *statistical random-phase approximation* (SRPA) in which off-diagonal matrix elements of ρ (in the basis of H_0) are neglected; that is,

$$\langle i|\rho_{\text{SRPA}}|j\rangle = \rho_i\delta_{ij} \tag{64}$$

The subsets (63b), with $\mathbf{k}\neq 0$, form invariant orthogonal subsets with (64) as the metric. These invariant subsets are not mixed up by the interaction $\mathcal{V}$, which depends only on *relative* positions of the molecules. Let us use for the *row* vectors the hermitian conjugates of

$$\bar{x}^{ab}(\mathbf{k}) = N^{-1}\rho_b^{-1}x^{ab}(\mathbf{k}) \tag{65}$$

where

$$\rho_b = \text{tr}\{|b\rangle\langle b|\rho\} \tag{66}$$

is the one-molecule density-matrix element. With (65) we form an orthonormal basis, since

$$\langle\langle\bar{x}^{ab}(\mathbf{k})|x^{cd}(\mathbf{k})\rangle\rangle = \text{tr}\{\rho_{\text{SRPA}}\bar{x}^{ab\dagger}(\mathbf{k})x^{cd}(\mathbf{k})\}$$

$$= \delta_{ac}\delta_{bd}, \qquad (\mathbf{k}\neq 0) \tag{67}$$

The restriction to $\mathbf{k}\neq 0$ was necessary, since the $\mathbf{k}=0$ subspace includes the subset x^{aa} (with *all* one-molecule quantum numbers equal), which obeys

$$\langle\langle\bar{x}^{aa}(\mathbf{0})|x^{bb}(\mathbf{0})\rangle\rangle = N\delta_{ab}$$

rather than (67).

An even stronger condition on ρ (used, e.g., by Fano[1]) is to neglect in ρ *all* statistical correlations between molecules and choose

$$\rho \to \rho_0 = \prod_{A=1}^{N} \rho_A, \quad [\rho_A, H_A] = 0 \tag{68}$$

for a metric. This choice has some convenient properties. It further breaks $L_1(\mathbf{k})$ into smaller invariant subspaces, forming bases for the irreducible representations of the symmetry group of the one-molecule hamiltonian H_A. For example, if H_A is rotation-and-inversion invariant, the multiple moments (electric and magnetic) then form invariant subspaces.[50] Collisions will not "mix" an electric-dipole mode, say, with a magnetic-dipole mode. The choice of (68) is approximately justified in some conditions (viz., the impact approximation, discussed in Section VIII). For the sake of a general introduction of the formalism, however, we stick to the more widely valid SRPA.

When the SRPA breaks down, the diagrammatic temperature Green's function method has a clear advantage (provided only we deal with *canonical* ensembles; otherwise both methods are handicapped). We discuss some consequences of the nonorthogonality of the SME basis in the following section.

The SME set, defined by (63), is clearly a set of eigenvectors of $\mathcal{H}_0$ (in the SRPA). For sufficiently heavy molecules (where the recoil energy $k^2/2m_1$ can be neglected, m_1 being the mass of the molecule),

$$\mathcal{H}_0 x^{ab}(\mathbf{k}) = \omega_{ab} x^{ab}(\mathbf{k}) = \left(\omega_{\alpha\beta} + \frac{\mathbf{k} \cdot \mathbf{p}_b}{m_1}\right) x^{ab}(\mathbf{k}) \tag{69}$$

where $\omega_{\alpha\beta}$ is the (discrete) molecular resonance frequency, and $\mathbf{k} \cdot \mathbf{p}_b / m_1$ is the Doppler shift owing to the motion of the molecule relative to the field. Let P project on L_1, and $Q = 1 - P$ on the complementary subspace $\bar{L}_1$ (with $L_1 \oplus \bar{L}_1 = L$, the complete Hilbert space). The $\mathcal{H}_0$ is reduced in the two subsets, with no matrix elements connecting them; that is,

$$P \mathcal{H}_0 Q = Q \mathcal{H}_0 P = 0$$

Also $\mathcal{G}_0(\omega)$ is therefore reduced into

$$\mathcal{G}_R(\omega) = (\omega R - \mathcal{H}_R + i\eta)^{-1}, \quad (R = P \text{ or } Q)$$

where

$$\mathcal{H}_R = R \mathcal{H}_0 R$$

Following Zwanzig,[46] and Fano,[1] we define the *connected* part of $\mathcal{T}(\omega)$ by allowing only $\bar{L}_1$ vectors to appear as intermediate vectors in the Born expansion

$$\mathcal{T}_c(\omega) = \mathcal{V} + \mathcal{V} \mathcal{G}_Q(\omega) \mathcal{T}_c(\omega) \tag{70}$$

On inspection, it is easy to see that $\mathcal{T}_c(\omega)$ is related to $\mathcal{T}(\omega)$ by the Dyson equation

$$\mathcal{T}(\omega) = \mathcal{T}_c(\omega) + \mathcal{T}_c(\omega) \mathcal{G}_P(\omega) \mathcal{T}(\omega) \tag{71}$$

Hence

$$P \mathcal{G}(\omega) P = \mathcal{G}_P [1 + \mathcal{T}(\omega) \mathcal{G}_P]$$

$$= [\mathcal{G}_P^{-1} - P \mathcal{T}_c(\omega) P]^{-1} = [\omega P - \mathcal{K}_P - P \mathcal{T}_c(\omega) P]^{-1} \tag{72}$$

In (72), the denominator contains only operators defined on L_1 (otherwise it is not allowed to invert a projection operator). Thus instead of dealing with $\mathcal{G}(\omega)$ as an operator in N-molecule space, we are now confined to the much simpler SME. Zwanzig[46] has proved (72) for L-space projection operators in a somewhat more general form, in which $\mathcal{K}_0$ is not necessarily reduced by P and Q. Analogous theorems for resolvent operators in H space have also been in use.[51]

Translational invariance further reduces the denominator in (72) into the kth invariant subspace. Let $P_{\mathbf{k}}$ project on $L_1(\mathbf{k})$ and $Q_{\mathbf{k}}$ on the complementary subspace $\bar{L}_1(\mathbf{k})$, where

$$L_1(\mathbf{k}) \oplus \bar{L}_1(\mathbf{k}) = L(\mathbf{k})$$

is the complete kth invariant subspace. Let us introduce

$$\Sigma(\mathbf{k}, \omega) = P_{\mathbf{k}} \mathcal{T}(\omega) P_{\mathbf{k}} \tag{73a}$$

and

$$\Sigma_c(\mathbf{k}, \omega) = P_{\mathbf{k}} \mathcal{T}_c(\omega) P_{\mathbf{k}} \tag{73b}$$

where in (73b) we can use (70) with $Q_{\mathbf{k}}$ replacing Q, as operators defined on the invariant subspace $L(\mathbf{k})$. Following a similar practice in dealing with H-space (one-particle) Green's functions, we call (73a) and (73b), respectively, the "self-frequency" and "proper self-frequency" matrices. Both matrices are defined on the SME basis; therefore their rows and columns are labeled by pairs of one-molecule quantum numbers, all other quantum numbers being "averaged out" by application of the projection operator.

Using the orthonormalized set (63), (65), we get, in the SRPA

$$F_{ij}^R(\mathbf{k}, \omega) = \left(\frac{N}{V}\right) \sum_{abcd} \rho_b (\mu_{ab}^i)^* \mu_{cd}^j$$

$$\langle\langle \bar{x}^{ab}(\mathbf{k}) | [\omega - \Omega(\mathbf{k}) - \Sigma_c(\mathbf{k}, \omega)]^{-1} | x^{cd}(\mathbf{k}) \rangle\rangle \qquad (74)$$

where

$$\Omega(\mathbf{k}) = P_{\mathbf{k}} \mathcal{H}_0 P_{\mathbf{k}} \qquad (75)$$

is the (diagonal) matrix of resonance frequencies (69). The line shape problem thus reduced to the calculation of the proper self-frequency matrix $\Sigma_c(\mathbf{k}, \omega)$. This matrix is generally not hermitian, since $\mathcal{T}_c(\omega)$ is not. It can therefore be split into a hermitian part and an antihermitian part,

$$\Sigma_c = \Sigma_c' + i\Sigma_c'' \qquad (76)$$

The two parts of (76) constitute, respectively, the "line-shifting" and "line-broadening" matrices, acting as perturbations on the resonance-frequency matrix (75).

VII. BINARY–COLLISION APPROXIMATION

Consider the first Born approximation to $\Sigma(\mathbf{k}, \omega)$ in the basis (63). The interaction

$$\mathcal{V} = \sum_{B<C} \mathcal{V}_{BC}$$

is composed of pairwise interactions. Applied to the Ath term in the expansion (63), $\mathcal{V}_{BC}$ will have a nonvanishing contribution only if $A = B$ or C. There are $N(N-1) \approx N^2$ such terms. Therefore [recalling the factor N^{-1} from (65) required by normalization], we have

$$\langle\langle \bar{x}^{ab}(\mathbf{k}) | P \mathcal{V} P | x^{cd}(\mathbf{k}) \rangle\rangle$$

$$= N\rho_b^{-1} \sum_{gh} \rho_{bg}^{(2)} [\mathcal{V}_{ag,bg;\,ch,dh} + \mathcal{V}_{ag,bg;\,hc,hd}] \qquad (77a)$$

Here ag labels a two-molecule state, and $\rho_{bg}^{(2)}$ is the (diagonal) two-molecule density matrix (in the SRPA). The first matrix element in (77a) signifies the *direct scattering* of an absorbing (or radiating) molecule by a second one that is not directly coupled to the radiation field (the "perturber"). In the second element, the roles of absorber (or radiator) and perturber are exchanged by the scattering process, thus representing *resonance exchange*. The latter terms should be considered only if both molecules are capable of resonance with the radiation within the same narrow spectral range (although they do not have to be identical).

Using the H-space operators V_{AB}, (77a) may be rewritten as

$$(77a) = N\rho_b^{-1} \sum_g \rho_{bg}^{(2)} \left[\left(V_{ag,cg} \delta_{bd} - V_{bg,dg} \delta_{ac} \right) \right.$$

$$\left. + V_{ad,bc} \left(\delta_{gd} - \delta_{gc} \right) \right] \tag{77b}$$

Let

$$NV_{ab,cd} = N \langle \alpha\mathbf{p}_a, \beta\mathbf{p}_b | V_{AB} | \gamma\mathbf{p}_c, \delta\mathbf{p}_d \rangle$$

$$= n \int d^3\mathbf{R} \, V_{\alpha\beta,\gamma\delta}(\mathbf{R}) e^{-i\mathbf{u}\cdot\mathbf{R}}$$

$$= n V_{\alpha\beta,\gamma\delta}(\mathbf{u})$$

where $\mathbf{R} = \mathbf{R}_1 - \mathbf{R}_2$ is the relative position of the molecules and

$$\mathbf{u} = \mathbf{p}_a - \mathbf{p}_c = -\mathbf{p}_b + \mathbf{p}_d$$

is the momentum gained by molecule 1 owing to the interaction (the total momentum being conserved). The $V(\mathbf{u})$ is a spatial Fourier component of $V(\mathbf{R})$. Notice that in the direct-scattering part of (77) only forward-scattering elements ($\mathbf{u}=0$) appear, and the resonance-exchange part is not limited to the forward direction. The first Born expression (77) is proportional to the gas density $n = N/V$, provided the space integrals of $V(\mathbf{R})$ converge. In dealing with neutral molecules, only R^{-3} potentials may cause any difficulty, since their forward-scattering ($\mathbf{u}=0$) integrals diverge. However, such an interaction appears only in the form of a dipole-dipole interaction, whose first Born approximation vanishes by averaging over orientations, owing to its tensorial character.

Consider now the second Born terms. Two types of products, namely, $V_{BC}V_{BC}$ and $V_{BC}V_{CD}$, now contribute. In the first (two-body) case we get a factor N^2, as before, from summing over B and C, a factor V^{-2} from the two-space integrations, a factor V from the summation over the intermediate momentum states, and N^{-1} from normalization—all together a factor $N/V = n$. In the second (three-body) case we get N^3 from summing over the three labels (B, C, D), and *no* summation over intermediate momentum states—all together n^2. Following this line of argument it becomes obvious that in any perturbation order, terms involving $m+1$ distinct molecules will contribute a factor n^m. Therefore we can write

$$\Sigma(\omega) = n\Sigma^{(1)} + n^2\Sigma^{(2)} + \cdots \tag{78}$$

where $\Sigma^{(m)}$ involves $(m+1)$ molecule integrals. In particular, $\Sigma^{(1)}$ involves just a binary collision, considering only one pair of molecules at a time to any perturbation order. The first term in (78) constitutes the *binary-collision approximation* (BCA).

Density expansions of the type (78), including higher-than-binary terms, have been studied recently by Albers and Oppenheim.[13] It may well happen that (78) does not converge owing to the possible occurrence of logarithmic-type divergences. However, such divergences do not affect the $\Sigma^{(1)}$ term, which therefore predominates at sufficiently low densities.

To get an idea of the range of validity of the BCA, consider the second Born terms. In the thermodynamic limit $V \to \infty$, their contribution to $\Sigma^{(1)}$ involves convolution integrals of Fourier components of the potential,

$$\int d^3v\, V_{AB}(\mathbf{u}-\mathbf{v})\, V_{AB}(\mathbf{v})$$

averaged with $\rho^{(2)}$ (which depends only on the momentum of the *relative* motion of the pair). In contrast, $\Sigma^{(2)}$ contains products of the type

$$V_{AB}(\mathbf{u})V_{BC}(\mathbf{v})$$

averaged with $\rho^{(3)}$. The spatial transforms $V(\mathbf{u})$ usually have a characteristic range of variation $\Delta u \sim d^{-1}$, where d is the "range" of the potential (typically of the order of 1). Beyond this range $V(\mathbf{u})$ becomes diminishingly small. The ratio of the second Born contributions to $\Sigma^{(2)}$ and to $\Sigma^{(1)}$ is, therefore, roughly of the order of nd^3. The BCA is thus quite generally valid if

$$nd^3 \ll 1 \tag{79}$$

Condition (79) usually holds at gas densities below $\sim 10^2$ amagats.

We next want to establish the identity

$$\Sigma^{(1)} = \Sigma_c^{(1)} \tag{80}$$

which enables us to use the *full* $\mathcal{T}(\omega)$ matrix, hence also its explicit operator form, without referring to perturbation expansions, in the BCA calculation of the proper self-frequency matrix.

Equation (80) becomes obvious by consulting the Dyson equation (71), since $\mathcal{G}_P$ was made independent of the density by the way we have defined our projection operator (recalling that $\mathcal{G}_P$ depends only on the unperturbed matrix $\mathcal{H}_P$ of SME resonance frequencies). This conclusion critically depends on the choice of $P_\mathbf{k}$. Consider, as an illustration, a projection operator $P_\mathbf{k}'$ projecting onto a subspace $L_1'(\mathbf{k}) \in L_1(\mathbf{k})$, of smaller dimensions than $L_1(\mathbf{k})$. Let $P_\mathbf{k}'' = P_\mathbf{k}Q_\mathbf{k}'$ project on the nonempty subspace

$$L_1''(\mathbf{k}) = L_1(\mathbf{k}) \ominus L_1'(\mathbf{k})$$

(i.e., $P = P' + P''$). Clearly,

$$Q_\mathbf{k}'\mathcal{H}_0 P_\mathbf{k}' = Q_\mathbf{k}\mathcal{H}_0 P_\mathbf{k}' + P_\mathbf{k}''\mathcal{H}_0 P_\mathbf{k}'$$

may have nonvanishing elements, since

$$P_\mathbf{k}''\mathcal{H}_0 P_\mathbf{k}' \in P_\mathbf{k}\mathcal{H}_0 P_\mathbf{k}$$

(unless $P_\mathbf{k}'$ further reduces $L_1(\mathbf{k})$ into invariant subspaces). Then $\mathcal{H}_0$ may not be reduced into $L_1'(\mathbf{k})$ and its complementary subspace. We can still retain the form of (71), redefining

$$\mathcal{H}_0' = P_\mathbf{k}'\mathcal{H}P_\mathbf{k}' + Q_\mathbf{k}'\mathcal{H}Q_\mathbf{k}' \tag{81a}$$

and

$$\mathcal{H}_1' = P_\mathbf{k}'\mathcal{H}Q_\mathbf{k}' + Q_\mathbf{k}'\mathcal{H}P_\mathbf{k}' \tag{81b}$$

as the "unperturbed" and "perturbative" parts of the liouvillian $\mathcal{H}$. But then the simple correspondence of $\mathcal{H}_0$ to one-molecule operators and of $\mathcal{H}_1$ to two-molecule operators is lost, and the whole structure of relations, between $(m+1)$-body terms and mth power of the density, collapses. This simple correspondence is retained at the price of defining P on the entire SME basis (or an invariant subspace thereof, if symmetry allows). Only then does (80) hold.

This simple structure requires that we adhere to the orthogonal basis provided by the SRPA. We are, again, in trouble if we use the full density matrix ρ as a nonorthogonal metric. Although the basis (63) is now nonorthogonal, it still spans a subspace L_1, for which we can formally define a projection operator P and accordingly split $\mathcal{H}$ as in (81). However, as in the previous example, the two parts of $\mathcal{H}$ will not correspond exactly to $\mathcal{H}_0$ and $\mathcal{V}$, and the simple correspondence of density expansions to m-body terms is lost once more.

To be on the rigorous side, we should properly add here that $\Sigma^{(1)}$ is strictly density independent only if we neglect in $\rho^{(2)}$ three-body effects. Maintaining in $\rho^{(2)}$ many-body correlations, we may still use (78), with $\Sigma^{(m)}$ referring to $(m+1)$-body terms of the $\mathcal{T}$ matrix; and (80) may still be justified, though at the price of making $\Sigma^{(m)}$ density dependent because of the statistical correlations (even in the SRPA). However, these corrections to Σ_c are only second order in the density; they should not affect our BCA expressions.[13]

VIII. SELF–FREQUENCY MATRIX

A. General Expressions

In the last section it was shown that the proper self-frequency matrix in the BCA–namely, $\Sigma^{(1)}(\mathbf{k}, \omega)$–is obtained by considering one collision pair at a time. It therefore remains to express it in terms of the two-molecule t matrix, defined by

$$T(\varepsilon) \to T^{(1,2)}(\varepsilon) = \frac{8\pi^3}{V} \Delta_{\mathbf{p}} t(\varepsilon) \tag{82}$$

We have used the customary separation of the motion of the center of mass of the pair in the definition of the two-body t matrix. In (82) we introduced

$$\Delta_{\mathbf{p}} = \frac{V}{8\pi^3} I_{\mathbf{p}}$$

where $I_{\mathbf{p}}$ is the unit operator on the degrees of freedom of the center-of-mass motion.

Let

$$|ab\rangle = |\alpha\mathbf{p}_a, \beta\mathbf{p}_b\rangle$$

$$= |\alpha\beta; \mathbf{p}_{ab}\rangle|\mathbf{p}_a + \mathbf{p}_b\rangle \tag{83}$$

represent the separation of coordinates, with

$$\frac{\mathbf{p}_{ab}}{m} = \frac{\mathbf{p}_a}{m_1} - \frac{\mathbf{p}_b}{m_2}$$

as the relative velocity and

$$m = \frac{m_1 m_2}{m_1 + m_2}$$

as the reduced mass of the pair. In the limit $V \to \infty$, the center-of-mass motion is separated out by substituting

$$\langle ab|\Delta_{\mathbf{p}} t(\varepsilon)|cd\rangle = t_{ab,cd}(\varepsilon)\delta(\mathbf{p}_a + \mathbf{p}_b - \mathbf{p}_c - \mathbf{p}_d) \tag{84a}$$

where

$$t_{ab,cd}(\varepsilon) = \langle \alpha\beta; \mathbf{p}_{ab}|t(\varepsilon)|\gamma\delta; \mathbf{p}_{cd}\rangle \tag{84b}$$

The matrix elements of $t(\varepsilon)$, defined in this fashion, are made independent of the volume when in the limit $V \to \infty$, we replace the box normalization by a plane-wave normalization.

Let us define, in analogy to (84),

$$\langle ab|\Delta_{\mathbf{p}}|cd\rangle = \Delta_{ab,cd}\delta(\mathbf{p}_a + \mathbf{p}_b - \mathbf{p}_c - \mathbf{p}_d) \tag{85a}$$

where

$$\Delta_{ab,cd} = \langle \alpha\beta; \mathbf{p}_{ab}|\gamma\delta; \mathbf{p}_{cd}\rangle \tag{85b}$$

are the delta symbols for the relative motion. We can now retain (84b) and (85b), either in the denumerable box-normalized form or in the plane-normalization form, throughout the following discussion, completely dropping out the center-of-mass degrees of freedom.

An L-space two-body $t(\omega)$ matrix can now be introduced by using (57), (84), and (85):

$$t_{ag\,bg';ch\,dh'}(\omega) = t_{ag,ch}(\varepsilon_{bg'} + \omega)\Delta^*_{bg',dh'}$$

$$- \Delta_{ag,ch}t^*_{bg',dh'}(\varepsilon_{ag} - \omega)$$

$$+ \int \frac{d\varepsilon}{2\pi i} d_{agbg'}t_{ag,ch}(\varepsilon + \omega)t^*_{bg',dh'}(\varepsilon)d_{ch,dh'} \tag{86}$$

Here

$$\varepsilon_{ab} = \varepsilon_\alpha + \varepsilon_\beta + \frac{p_{ab}^2}{2m} \tag{87}$$

is the pair energy in the center-of-mass coordinates, and

$$d_{ag,bh} = (\varepsilon + \omega - \varepsilon_{ag} + i\eta)^{-1} - (\varepsilon - \varepsilon_{bh} - i\eta')^{-1}$$

Using (82) for the T matrix, the following expression for the BCA self-frequency matrix results:

$$n\Sigma^{(1)}_{ab;cd}(\mathbf{k},\omega) = 8\pi^3 n\rho_b^{-1}\sum_{gh}\rho_{bg}^{(2)}[t_{agbg;chdh}(\omega) + t_{agbg;hchd}(\omega)] \tag{88}$$

where $\rho_{bg}^{(2)}$ depends only on the relative velocity $\mathbf{p}_{bg}/m$. The two matrix elements in (88) constitute, respectively, the *direct-scattering* and *resonance-exchange* contributions.

Recall that the **k** dependence enters (88) by the requirement that

$$\mathbf{p}_a = \mathbf{p}_b + \mathbf{k}, \qquad \mathbf{p}_c = \mathbf{p}_d + \mathbf{k}$$

The ω dependence results from the finite thickness of the energy layer on which the t matrix is defined. Its form depends on transient effects of the collisions. The range of frequencies $\Delta\omega$ over which (88) changes appreciably is determined by the range of energies $\Delta\varepsilon$ over which $t(\varepsilon)$ changes. This range defines a *duration-of-collision time*[48] $\tau \approx |\Delta\varepsilon|^{-1}$.

Consider, again, the matrix elements (86) contributing to (88). The elements of $t(\omega)$ can be split into two types of contribution: those linear in t (or t^*) in which scattering takes place only in the upper (lower) level—the "outer" terms, in the terminology of Anderson[52]—and those bilinear in t and t^* (the "middle" terms) in which scattering occurs simultaneously in both levels.

The outer terms are simpler in structure, owing to delta symbols:

$$\left[\Sigma^{(1)}_{ab;cd}(\mathbf{k},\omega)\right]_{\text{outer}} = 8\pi^3 \rho_b^{-1}$$

$$\times \left\{ \sum_g \rho_{bg}\left[\delta_{bd}t_{ag,cg}(\varepsilon_{bg}+\omega) - \delta_{ac}t^*_{bg,dg}(\varepsilon_{ag}+\omega)\right] \right.$$

$$\left. + \rho_{bd}t_{ad,bc}(\varepsilon_{bd}+\omega) - \rho_{bc}t^*_{bc,ad}(\varepsilon_{ac}-\omega) \right\} \qquad (89)$$

In the direct-scattering terms (where a summation over g appears), only forward-scattering amplitudes appear. More particularly, in the antihermitian part $i\Sigma''$ pertaining to line broadening, only the antihermitian part of the forward-scattering amplitudes appears. Using the optical theorem, it follows that these contributions to the diagonal element $\Sigma''_{ab;ab}$ are equal to the mean value of the total scattering rates in the initial and final levels.[45] The "middle" terms[52] and the "resonance-exchange" terms[9] are not confined to forward scattering.

The resonance-exchange terms are not to be confused with the exchange integrals required by the Pauli exclusion principle. (In fact, the latter can be neglected in low-energy collisions of heavy molecules.) In particular, the "exchange" can occur between two molecules of different species (as is discussed below in Section VIII.F.).

B. Diagonal Elements

The diagonal elements $\Sigma^{(1)}_{ab;ab}(\mathbf{k},\omega)$ can be split, according to (76), into a real part and an imaginary part. The first describes a pressure shift of the resonance frequency ω_{ab}, and the latter the rate of its damping by collisions. Since $\mathfrak{I}(\omega)$ obeys the optical theorem, the diagonal elements of

Σ'' are negative-definite, which means damped oscillations.

The ω dependence of $\Sigma^{(1)}$ is governed by τ^{-1}, where τ is the duration of collision. If

$$-\tau\Sigma''_{ab;ab}(\mathbf{k},\omega)\ll 1 \qquad (90)$$

(i.e., the damping time is much longer than the duration of a collision), we can replace, within a range $\Delta\omega$ comparable to Σ'' (but smaller than τ^{-1}), the ω-dependent rates by their values at $\omega=\omega_{ab}$. Then by virtue of (60) the damping rates can be expressed in terms of the on-the-energy-shell two-molecule S matrix. Condition (90) shall be called the *first impact condition*.

Even when we consider a single isolated resonance transition $\alpha\rightarrow\beta$, the structure of $\Sigma^{(1)}$ is complicated by explicit dependence on the translational relative motion. However, conditions may exist under which this dependence may be approximately removed, and we can write

$$\Sigma^{(1)}_{ab;cd}=\Sigma^{(1)}_{\alpha\beta,\gamma\delta}\delta_{\mathbf{p}_b,\mathbf{p}_d} \qquad (91)$$

The remaining active part of $\Sigma^{(1)}$ is then defined on the discrete set of resonance transitions $\alpha\rightarrow\beta$, and so on, and an isolated line is characterized by a single damping rate. We call (91) the *second impact condition*.

The matrix of resonance frequencies $\Omega(\mathbf{k})$ is also dependent on the relative motion, owing to the Doppler shift. However, if

$$-\Sigma''_{ab;ab}\gg\left\langle\left|\frac{\mathbf{k}\cdot\mathbf{p}_b}{m_1}\right|\right\rangle \qquad (92)$$

(where the angular brackets denote a mean value of the Doppler shift, i.e., the Doppler width), we can neglect the Doppler shifts and retain only the discrete resonance frequency $\omega_{\alpha\beta}$. Condition (92) is called the *third impact condition*. Under the three impact conditions, which together constitute the *impact approximation*, the line shape of an isolated line $\alpha\rightarrow\beta$ attains the simple Lorentz formula

$$F^R_{\alpha\beta;\alpha\beta}(\mathbf{k},\omega)=[\omega-\omega_{\alpha\beta}-\Delta_{\alpha\beta}+i\Gamma_{\alpha\beta}]^{-1} \qquad (93)$$

where $\Delta_{\alpha\beta}$ and $\Gamma_{\alpha\beta}$ are, respectively, the real and minus-imaginary parts of $\Sigma_{\alpha\beta;\alpha\beta}(\omega_{\alpha\beta})$.

Conditions (90) and (92) simply require that the inverse duration of collision τ^{-1}, the damping rate Γ, and the Doppler width $\Delta\omega_D$ be of respectively decreasing orders of magnitude

$$\tau^{-1} \gg \Gamma \gg \Delta\omega_D$$

This can be attained in a wide range of experimental situations by choosing the proper range of densities and temperatures. Condition (91), unlike the other two, depends on the intrinsic dynamics of the collision pair, not on controllable thermodynamic conditions. There may, however, be several *model* cases under which (91) holds. Most typical is the *Brownian-particle model*.

Suppose the absorbing (or radiating) molecules form a dilute solution in a gas of much lighter nonabsorbing molecules. If the mass ratio m_1/m_2 is exceedingly large, the momentum of the relative motion $\mathbf{p}_{12}$ is practically identical to $\mathbf{p}_2$, and the center-of-mass momentum $\mathbf{p}_1 + \mathbf{p}_2$ is practically identical to $\mathbf{p}_1$. In the separation of coordinates, t is then a function of $\mathbf{p}_2$ only. But the calculation of (88) implies an averaging over $\mathbf{p}_2$, leaving $\Sigma^{(1)}$ a function of $\mathbf{p}_1$ only. Hence $\Sigma^{(1)}$ is independent of $\mathbf{p}_1$, in accordance with (91).

The simple Lorentzian shape corresponding to exponential decay at asymptotically long times, characteristic of the Brownian-particle model, may not be valid for smaller values of m_1/m_2. In fact, one can show that in the opposite case ($m_1/m_2 \ll 1$), assuming also that the interactions are of very short range, the correlation functions decay asymptotically as an inverse power of the time ($\sim t^{-3}$) rather than exponentially (see Section IX.B).

C. Analyticity

The exponentiallike decay characteristic of the impact approximation is, we should keep in mind, only an idealized limit. In adopting (91) and (92) we confine our resolvent to the discrete set of resonance-transition pairs α, β. In this limiting case, the Green's function $F^R(\mathbf{k}, \omega)$ will have poles on the lower half of the complex-ω plane, corresponding to the roots of the secular equation

$$|\omega - \Omega(\mathbf{k}) - \Sigma_c(\mathbf{k}, \omega)| = 0 \tag{94}$$

defined on the discrete basis. The correct basis to use, however, is continuous, owing to the translational degrees of freedom. The integration over a continuum generally has the effect of "smearing" out the poles and making $F^R(\mathbf{k}, \omega)$ holomorphic in the complex-ω plane. In a simplified version, this is equivalent to replacing a meromorphic function

$$F_1(\omega) = A[\omega - B(\omega)]^{-1}$$

with poles at the roots of the equation $\omega = B(\omega)$, by

$$F_2(\omega) = \int A(v)[\omega - B(\omega, v)]^{-1} dv$$

If $A(v)$ and $B(\omega, v)$ are "well-behaved" functions of a real parameter v, defined on a continuous domain, and $B(\omega, v)$ has a range of nonzero measure (i.e., is not independent of v), then $F_2(\omega)$ is generally holomorphic, with no poles in the ω plane. If the range of $B(\omega, v)$ degenerates into a point (i.e., if it is independent of v, as in the impact approximation), then again we may have a pole.

This discussion is in some ways reminiscent of a problem encountered recently in the study of correlation functions of ordinary fluids. Substantial progress has been achieved in evaluation of the correlation functions by use of a discrete set of "hydrodynamic" variables, $A^i(\mathbf{k}, t)$, constructed from the additive constants of the motion of the fluid, as a discrete set on which to project the resolvent operators (see, e.g., Ref. 53). It turned out later on, however, that this discrete set is not a sufficiently complete one, and it should be augmented by a continuum composed of products of hydrodynamic variables of the type $A^i(\mathbf{k} - \mathbf{q}, t)\, A^j(\mathbf{q}, t)$, with $\mathbf{q}$ as the continuous variable.[54] The incentive for this modification came from the discovery by Alder and Wainwright that correlation functions extended to long times (beyond the range of exponentiallike decay) carry a slowly decaying "tail."

True poles may appear if the set we project on has discrete modes in it (such as transitions between true bound states of the many-body system). The SME basis, which seems to be a natural choice for dilute neutral gases, does not have discrete modes (at $\mathbf{k} \neq 0$). It could be argued that the choice of basis is rather arbitrary and we might as well have started from a discrete basis in L space. (For example, the complete operator $M^i(\mathbf{k})$ could be treated as a single vector, or it could be expanded in the discrete set of $|\alpha><\beta|$.) However, in all cases we would end up with the same expression for the Green's function. It is the self-frequency function $\Sigma_c(\mathbf{k}, \omega)$ that will be radically affected by the choice of initial basis. Therefore the existence of poles should not depend on the choice of basis, and if the SME basis does not support poles, the secular equation in alternative discrete bases should have no roots.

To understand this conjecture, we should recall that our "best" choice of initial basis is dictated by seeking to satisfy Mori's condition of short "memory" times, requiring that the ω-dependence of $\Sigma_c(\mathbf{k}, \omega)$ be very slow over a range comparable to the damping rate, as in (90). Consider an idealized situation in which we project on a single discrete mode. In the first instance, assume that Mori's condition holds, and Σ_c can be replaced by its value at $\omega = \Omega$ over a wide domain. The imaginary part of the secular equation will then approximately read

$$\omega'' - \Sigma_c''(\Omega) = 0 \tag{95a}$$

where ω'' is the imaginary part of ω. Equation (95a) always has a root. Its value is negative, since Σ_c'' is negative-definite on the real axis (Fig. 2a). In the second instance, suppose Mori's condition does not hold and $\tau \Sigma_c''$ is not much smaller than one. For simplicity's sake, let us assume that Σ_c' is negligible. The imaginary part of the secular equation will now read

$$\omega'' - \operatorname{Im} \Sigma_c(\Omega + i\omega'') = 0 \tag{95b}$$

which may not have a solution if τ is too long. This may be seen, for example, if we let $-\Sigma_c''$ be a Gaussian of bandwidth τ^{-1} around Ω (Fig. 2b).

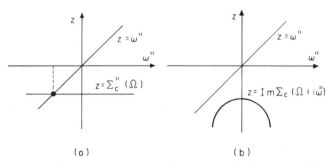

Fig. 2. The existence of true poles in the retarded Green's function projected on a discrete basis of (a) "good" modes, with short memory times, and (b) "bad" modes, with long memory times.

From the preceding illustration we learn that a discrete basis quite probably may not produce any poles if Mori's condition is not satisfied. It is, of course, only a heuristic argument, which may bear exceptions.

We should note in passing that $\Sigma_c(\mathbf{k}, \omega)$ is also a retarded Green's function. Poles in this function will appear as zeros of the original $F^R(\mathbf{k}, \omega)$. If $\Sigma_c(\mathbf{k}, \omega)$, too, is holomorphic then F^R will have no zeros (except at infinity).

D. Off-Diagonal Elements

The off-diagonal elements

$$n\Sigma_{ab;cd}^{(1)}, \qquad (a, b \neq c, d)$$

relate to the mixing of distinct resonance modes by collisions. Consider them as an off-diagonal perturbation to the matrix Ω of resonance frequencies. As is well known from perturbation theory, an off-diagonal hermitian perturbation tends to push the eigenvalues away from each other. An antihermitian perturbation will have the opposite effect of pulling together the eigenvalues of a hermitian matrix. Off-diagonal elements of Σ'' thus provide a mechanism for mixing distinct resonance modes into degenerate modes. The damping (per unit density) of the degenerate mode may be smaller than the damping rates of the individual modes. It is as if the radiation does not distinguish between the modes (once they become degenerate by a sufficiently strong mixing), and any mechanism of randomly transferring from one mode to the other will not affect the line shape.[55]

Qualitatively, this mixing mechanism may have several manifestations in gas spectra:

1. Collapse of Doppler Broadening.[56,57]

A strong mixing of the translational states of the molecule by collisions may make the Doppler profile of a line collapse into a single (discrete) mode with an impact-approximation Lorentzian shape near the center. If the ratio of scattering to line-broadening rates is particularly large, this mechanism may even actually *narrow* the Doppler-broadened line as the pressure is increased, before reaching the pressure domain where collision broadening takes over.[58]

2. Line Overlapping[45,37,59–62]

The existence of a significant mixing between two lines requires that their resonance frequencies lie close to each other (within a range $\sim \tau^{-1}$),

$$|\omega_{ab} - \omega_{cd}| \lesssim \tau^{-1} \tag{96}$$

where ω_{ab} and ω_{cd} belong to two distinct resonance lines α, β and γ, δ, respectively. It is also required that the scattering amplitudes in (89), for states g with a substantial population, be nonnegligible. The Δ and d factors in (86) impose additional conditions on the energies. Since $t(\varepsilon)$ varies smoothly with the energy, most of the contribution from the bilinear ("middle") terms will come from the even delta functions in d, and not from the odd Cauchy principal parts. As a result, contributions will be highly favored if one of the following conditions is approximately observed:

$$(a) \quad \varepsilon_{ag} = \varepsilon_{ch}, \qquad (b) \quad \varepsilon_{bg} = \varepsilon_{dh}$$

$$(c) \quad \varepsilon_{ag} = \varepsilon_{dh} + \omega, \qquad (d) \quad \varepsilon_{ch} = \varepsilon_{bg} + \omega$$

On the frequency shell, all four conditions coincide. Near the frequency shell, only terms obeying both

$$\varepsilon_{ag} \approx \varepsilon_{ch} \quad \text{and} \quad \varepsilon_{bg} \approx \varepsilon_{dh} \tag{97}$$

would contribute significantly. This seems to leave a broad range of possibilities, since (96) and (97) may both be satisfied with $|\varepsilon_a - \varepsilon_c|$ much larger than τ^{-1}. However, the scattering amplitudes contributing to the mixing usually have a much more limited range. In the classical-trajectory limit (which is usually a good approximation at high values of the translational angular momentum) $t_{ag,ch}$ is insignificant unless the change in *internal* energy $\Delta \varepsilon_{int}$ is small compared to τ^{-1}. In fact, the scattering amplitudes are inverse exponential functions of $\tau |\Delta \varepsilon_{int}|$. In foreign-gas broadening, where only the internal degrees of freedom of the absorber are affected, (97) should be replaced by the much stronger condition

$$|\varepsilon_\alpha - \varepsilon_\gamma| \lesssim \tau^{-1}, \qquad |\varepsilon_\beta - \varepsilon_\gamma| \lesssim \tau^{-1} \tag{98}$$

In self broadening, (98) should be replaced by a similar condition involving sums of internal energies of the two molecules.

A strong mixing rate will pull the two lines toward each other. At high densities, where the rates are larger than the frequency separation $|\omega_{\alpha\beta} - \omega_{\gamma\delta}|$, we can observe again the collapse into a single mode with a reduced decay rate per unity density. At lower densities, where the two lines are still resolved, the pulling will cause enhancement of the intensity in the trough between the lines at the expense of the outer wings.

3.   Transition from Resonant to Nonresonant Line Shape[61]

A phenomenon peculiar to certain low-frequency spectra (microwave or radio frequency) is the mixing of the mode $\omega_{\alpha\beta}$ with its "mirror image" $\omega_{\beta\alpha}$, which though of negative resonance frequency, extends a tail to the positive-ω domain. This mixing is strictly due to inelastic transitions between the two levels α, β, caused by collisions. As in the line overlapping case, this mixing will cause a pulling toward a mean (zero) resonance frequency, accompanied by a reduction in the damping rates per unit density. This effect was most dramatically observed in the ammonia inversion spectrum.[63]

E.   Transient Effects

We have so far implied that the t matrix varies smoothly with energy near the frequency shell and that the variation is hardly detectable over a range $\Delta\omega$ smaller than τ^{-1}, where τ is the "duration" of the collision. Transient effects, and consequently ω-dependent features in $\Sigma(\mathbf{k}, \omega)$, may appear (*a*) if $\Delta\omega$ is not smaller than τ^{-1} (i.e., at the far wings of lines), or (*b*) if a considerable contribution to the damping comes from resonance-scattering states, in which the colliding molecules may stay together for a period of time much longer than the mean collision duration τ.

From (70) and (73) it follows that $\Sigma(\mathbf{k},\omega)$ is a retarded Green's function, differing from $F^R(\mathbf{k},\omega)$ in that $[V, M^i]$ replaces M^i, and so on. In the far wings, where $\Delta\omega \gtrsim \rho^{-1}$ is much larger than the damping rates, the ω dependence of $\Sigma(\mathbf{k},\omega)$ is essentially governed by the time dependence of the potential V along the collision trajectories. We discuss the line-wing problem with more detail in Section IX.C.

Effects of resonance scattering depend on the interaction of the "absorber" molecule in the upper state $a = \alpha$, $\mathbf{p}_a$, or the lower state $b = \beta$, $\mathbf{p}_b$, with the "perturber" in state $g = \zeta$, $\mathbf{p}_g$. In the adiabatic case, where the motion of the molecules is rather slow, a typical "true" resonance scattering occurs if $V_{\alpha\zeta,\alpha\zeta}(\mathbf{R})$ can support bound states, and these states are weakly coupled to a lower level (either by off-diagonal elements of V or by nonadiabatic effects).

At higher translational energies, another source of resonance scattering is represented by the quasibound states formed by the centrifugal barrier ("orbiting" states). We should recall, however, that the calculation of $\Sigma(\mathbf{k},\omega)$ involves an averaging over the translational states $\mathbf{p}_g$, which is essentially equivalent to an averaging over the relative-motion momentum $\mathbf{p}_{ag}$ (or $\mathbf{p}_{bg}$). In most molecular spectra, this averaging implies taking into account a very large number of relative angular-momentum states ($\sim 10^2$). This summation will tend to spread over the effect of the quasibound states and destroy any possible ω dependence. Moreover, this summation can also destroy the effect of true resonances unless the potential well is deep enough to make the bound states rather insensitive to the centrifugal forces. Effects of true resonances are therefore more likely to appear in electronic spectra than in vibration-rotation spectra.

In dealing with electronic states, it is usually convenient to separate the potential into a part diagonal in the electronic states (adiabatic forces) and an off-diagonal part (dispersive forces). A special kind of transient effect, known as "collision-induced absorption" (cf. Van Kranendonk[27] and references therein), is associated with the latter kind. This phenomenon, too, is discussed in the following section.

To sum up, we note that unlike the line-center problem, which requires only the knowledge of the asymptotic amplitudes (S matrix), the problem of transient effects requires a much fuller study of the dynamics of the collision and the related t matrix. Therefore, detailed theoretical studies outside the domain of the impact approximation are very scarce and incomplete.

F. Gas Mixtures

The formalism developed earlier can be easily extended to a mixture of several gas species. Without loss of generality, it suffices to discuss here a two-component mixture.

Two distinct cases should be recognized. One in which only one component actively interacts with the radiation field in the relevant spectral range, the other gas only acting as a perturber gas, broadening (and shifting) the lines of the first gas. The second, in which the two gases may directly interact with the radiation, as well as with each other.

In the first case we distinguish between the "absorbing" gas (with density n_A) and the "perturbing" gas (with density n_B). The leading density factor in (74), which affects the integrated intensity, will then be equal to n_A. The expression for Σ in the BCA will be composed of two terms: a *self-broadening* part, proportional to n_A, in which the binary system consists of identical "absorber" molecules, and a *foreign-gas-broadening* part, proportional to n_B, owing to heterogeneous binary systems. Obviously, resonance-exchange contributions will appear only in the self-broadening part.

In the other case, where the two components are optically active, we should split our SME set (L_1) in L space into two distinct subsets, $L_1(A)$ and $L_1(B)$, corresponding to the single-molecule modes of each gas species. The Green's function can be written as a 2×2 contraction,

$$F_{ij}{}^R(\mathbf{k}, \omega) = \sum_{A,B}^{1,2} n_A \rho_A \mu_A{}^i [\omega - \Omega - \Sigma_c]_{AB}^{-1} \mu_B{}^j \qquad (99)$$

where

$$\Omega = \begin{pmatrix} \Omega_A & 0 \\ 0 & \Omega_B \end{pmatrix}$$

and

$$\Sigma_c = \begin{pmatrix} \Sigma_{AA} & \Sigma_{AB} \\ \Sigma_{BA} & \Sigma_{BB} \end{pmatrix} \qquad (100)$$

Here, Ω_A is the matrix of resonance frequencies of species A, $\mu_B{}^j$ is the column vector of all SME-mode amplitudes of species A, and so on. The "diagonal" part Σ_{AA} in the BCA is composed, as in the previous case, of two terms

$$\Sigma_{AA} = n_A \Sigma_A^{(1)}(\text{self}) + n_B \Sigma_A^{(1)}(\text{foreign}) \qquad (101)$$

The foreign-gas term is calculated as if species B is optically inactive. Similarly, Σ_{BB} is calculated by interchanging the roles of the two species. The "off-diagonal" part

$$\Sigma_{AB} = n_B \Sigma_{AB}^{(1)} \tag{102}$$

signifies the transfer of excitation from species B to species A by a near-resonance exchange, likely to occur if resonance frequencies of the two species nearly coincide. This part will be composed solely of resonance-exchange-like terms, as in the second half of (88), in which mode a, b of species A is coupled to mode c, d of species B. Here, as in the case of line overlapping, two adjacent lines belonging to the two respective species may mingle into a degenerate line if the corresponding matrix element of Σ_{AB} is sufficiently large.

IX. SPECIAL CASES

A. Impact Approximation

As shown in Section V [(60)], the matrix elements of $\mathfrak{T}(\omega)$ on the frequency shell are determined by asymptotic scattering amplitudes (S matrix). The quantum-mechanical S matrix is strictly confined to the energy shell, and therefore may not be used to evaluate off-the-shell elements of $\mathfrak{T}(\omega)$. When studying a single discrete mode α, β (as in the impact approximation), well isolated from other lines in the spectrum, we can replace $\mathfrak{T}_{\alpha\beta;\,\alpha\beta}(\omega)$ by $\mathfrak{T}_{\alpha\beta;\,\alpha\beta}(\omega_{\alpha\beta})$ within a range $|\omega - \omega_{\alpha\beta}| \ll \tau^{-1}$, and use therefore only S-matrix elements. However, in dealing with overlapping between two lines, say α, β, and γ, δ, with

$$|\omega_{\alpha\beta} - \omega_{\gamma\delta}| \ll \tau^{-1} \tag{103}$$

we may not strictly speaking use the S matrix, even though we do not expect the elements of $\mathfrak{T}(\omega)$ to vary with ω over a range including the two line centers. We can, however, introduce a *modified* S matrix, defined by

$$S(\varepsilon) = I - G_0(\varepsilon)T(\varepsilon) + T(\varepsilon)G_0^\dagger(\varepsilon) \tag{104}$$

This matrix is equal to the S matrix on the energy shell but does not vanish off the shell.

Let $\mathfrak{T}_{ij;\,kl}(\omega)$ vary only slightly with ω over a range $\Delta\omega$ around the frequencies ω_{ij} and ω_{kl}. Without much error we may substitute $G_0^\dagger(\varepsilon + \omega)$ for $G_0^*(\varepsilon)$, or $G_0^{\dagger*}(\varepsilon)$ for $G_0(\varepsilon + \omega)$, in expression (55) for $\mathfrak{T}(\omega)$ within this frequency range. It is easy to show that (55) can then be replaced by

$$\mathcal{T}(\omega) \approx \int \frac{d\varepsilon}{2\pi i} \left[II^* - S(\varepsilon + \omega) \tilde{S}^*(\varepsilon) \right] \tag{105}$$

where $\tilde{S}^*$ is the transposed S^* matrix. It is well known that (provided only H is invariant under time reversal) V can be made real, hence T (and S) symmetrical. Thus (105) becomes analogous to (60), without the strict confinement to the energy shell. The removal of this confinement is a necessary step in the introduction of *classical-trajectory approximations*, which have been so helpful in evaluating line shape parameters.[52,35,64] In a wide assortment of cases it is possible to approximately replace the quantum-mechanical $S(\varepsilon)$ matrix by a classical-trajectory counterpart. In angular momentum representation (partial-wave analysis) this amounts to writing

$$\langle \alpha\beta; plm | S(\varepsilon) | \gamma\delta; p'l'm' \rangle \approx \langle \alpha\beta | S(b,v) | \gamma\delta \rangle \delta_{pp'} \delta_{ll'} \delta_{mm'} \tag{106}$$

for a binary system. Here v is the asymptotic value of the relative velocity ($v = p/m$, where m is the reduced mass) and b is the impact parameter, related to l by

$$(mvb)^2 = l(l+1) \tag{107}$$

The substitution of (106) is usually considered to be a valid approximation unless

$$\tau \Delta\varepsilon_{\text{int}} = \tau |\varepsilon_\alpha + \varepsilon_\beta - \varepsilon_\gamma - \varepsilon_\delta| \gg 1 \tag{108}$$

That is, in the classical-trajectory form S is not strictly confined to the energy shell. The energy mismatch may not, however, much exceed τ^{-1} (which for typical values of b and v is roughly given by vb^{-1}).

A usual procedure in the calculation of $S(b,v)$ is to split V into

$$V = V_0 + V_1 \tag{109}$$

where V_0 is independent of all internal degrees of freedom and is only a function of the intermolecular distance R. This procedure is particularly suitable if V_1 is only a "weak" perturbation compared to V_0. Then V_0 can be used to determine the classical trajectories, along which S is evaluated:

$$S(b,v) = \text{T.O.} \exp\left\{ -i \int_{C(b,v)} e^{i\Delta\varepsilon_{\text{int}} t} V_1(R(t)) dt \right\} \tag{110}$$

Here the integration is performed along the classical trajectory $C(b,v)$ determined by V_0 for the given values of b and v; $\Delta\varepsilon_{int}$ is the energy mismatch, as in (108), corresponding to the matrix element $V_{\alpha\beta;\gamma\delta}(R)$ of V_1; T.O. signifies the time ordering of operators, with time increasing from right to left.

Various means of representing V_0 have been employed for obtaining the trajectories, such as hard spheres ("billiard-ball") and Lennard-Jones models. A common practice is to use the simple straight-path model ($V_0 = 0$ everywhere). However, this practice may sometimes lead to grossly erroneous results, particularly if the integrals in (110) do not much exceed unity at impact parameters comparable to (or smaller than) the gas-kinetic collision diameters. In that case, more realistic trajectories should be used. A billiard-ball model may be quite suitable, yet simple enough.

The impact approximation is only valid over a range $\Delta\omega$ smaller than τ^{-1}. This spectral range is sensitive only to stochastic fluctuations coarse-grained over a time scale that is large in comparison to τ. Over this time scale we can ignore fluctuations in the statistical correlations, which will then enter the calculation only as a mean perturbation of the one-particle distributions. A measure of this perturbation is given by

$$n|\langle \int [e^{-\beta V(\mathbf{R})} - 1]d^3\mathbf{R}\rangle| \equiv nR_0^3(\beta) \tag{111}$$

where we average over perturber states. Equation (111) defines a (temperature-dependent) correlation diameter R_0. The product nR_0^3 is typically only $\lesssim 10^{-3}$ at standard temperature and pressure.

Outside the impact domain, fluctuations in the statistical correlations may not be negligible. Their effect can be studied, for example, by the method of temperature Green's functions. It can be expressed by an introduction of a modified temperature-and-density-dependent T matrix (the so-called ladder diagram). However, it can be shown that if (90) holds, then in the range $\tau\Delta\omega \ll 1$ the contribution of this modified T matrix to the self frequency reduces to that of the ordinary T matrix.

We therefore conclude that in the impact domain ρ can be replaced by the noninteracting-gas density matrix ρ^0 without introducing appreciable error. This simplifies the expressions for Σ_c considerably since now the symmetry group of Σ_c is that of the noninteracting-molecule hamiltonian H_0. Thus, for example, if H_0 is invariant under the rotation-inversion group, Σ_c will not have matrix elements combining different multipole transitions. The multipole basis in L space is formed by[50]

$$|\alpha\pi_\alpha j_\alpha; \beta\pi_\beta j_\beta; \Pi J M\rangle\rangle = \sum_{m_\alpha m_\beta} (-1)^{j_\beta - m_\beta} C(j_\alpha j_\beta J; m_\alpha - m_\beta M)$$

$$\times |\alpha\pi_\alpha j_\alpha m_\alpha; \beta\pi_\beta j_\beta m_\beta\rangle\rangle \qquad (\Pi = \pi_\alpha \times \pi_\beta) \tag{112}$$

where π_α, j_α, and m_α, respectively represent the parity, the total (internal) angular momentum, and its projection along a fixed axis in state α, and C is a Clebsch-Gordan coefficient. In this basis, Σ_c is diagonal in Π, J, and M, and independent of M.[50] Thus, for example, for electric-dipole transitions, Σ_c should be calculated with $J = 1$, $\Pi = -1$.

With ρ^0 replacing ρ, we can use [see (88)]

$$p_{bg}^{(2)} = \rho_b{}^0 \times \rho_g{}^0 = \rho_b{}^0 \times \rho_{p_g}{}^0 \times \rho_\zeta{}^0$$

to calculate the BCA self frequency. Suppose the second impact condition (91) holds and $\mathbf{p}_g$ can be associated with the relative-motion momentum $\mathbf{p}$. In the classical-trajectory approximation (106) the kinetic energy of the relative motion is conserved and identified with ε. Therefore we can replace the integration over ε in (105) by an integration over the kinetic energy. Using (23a) to convert the integration into a summation, we can substitute

$$N \int \frac{d\varepsilon}{2\pi} \rightarrow \left(\frac{\pi N}{Vm} \right) \sum_{\mathbf{p}} p^{-1} = \left(\frac{\pi n}{m} \right) \sum_p p^{-1} \sum_l (2l+1) \tag{113a}$$

By further letting $\sum_p \rho_{\mathbf{p}}{}^0 \rightarrow \int f(v)\, dv$, where

$$f(v) = \left(\frac{m}{2\pi T} \right)^{3/2} 4\pi v^2 \exp\left(\frac{-mv^2}{2T} \right) \tag{114}$$

is the Maxwell-Boltzmann distribution of relative velocities $v = p/m$, and (in the classical limit)

$$\sum_l (2l+1) \rightarrow 2m^2 v^2 \int b\, db$$

we can finally substitute

$$\left(\frac{N}{2\pi} \right) \int d\varepsilon\, \rho_{\mathbf{p}}{}^0 \rightarrow n2\pi \int b\, db \int v f(v)\, dv \tag{113b}$$

in the calculation of the proper self frequency. Thus, in the classical-trajectory impact limit the self-frequency contribution to species A from binary collisions with species B is

$$\Sigma_{\alpha\beta;\,\gamma\delta}^{(1)} = 2\pi \int b\, db \int v f(v)\, dv\, \Phi_{\alpha\beta;\,\gamma\delta}(b,v) \tag{115}$$

where

$$\Phi_{\alpha\beta;\,\gamma\delta}(b,v) = \delta_{\alpha\gamma}\delta_{\beta\delta} - \sum_{\zeta\eta} \rho_\zeta^B [\langle \alpha\zeta | S(b,v) | \gamma\eta \rangle \langle \beta\zeta | S(b,v) | \delta\eta \rangle^*$$

$$+ \langle \alpha\zeta | S(b,v) | \eta\gamma \rangle \langle \beta\zeta | S(b,v) | \eta\delta \rangle^* \delta_{AB}] \tag{116}$$

the last term in the brackets appearing only in self broadening $(A = B)$. Equation (115) has the usual form of an integration over a (classically defined) differential cross-section, with a "broadening efficiency" factor Φ, as in Anderson's work[52] (except for the extra self-broadening term).

The preceding analysis leading to (116) was carried out under the assumption that (106) holds; that is, the S matrix was assumed to be invariant under rotations of the radius vector between the molecules (leaving all other degrees of freedom intact). The correct S matrix is invariant only under a full rotation of the binary system and is diagonal in the total angular momentum $\mathbf{J} = \mathbf{l} + \mathbf{j}_1 + \mathbf{j}_2$ (where $\mathbf{j}_1$ and $\mathbf{j}_2$ are the angular momenta of the molecules, each with respect to its center of mass), rather than in $\mathbf{l}$ alone. In this case, the sum over $\mathbf{p}$ in (113a) should be replaced by a sum over p, l, and m_l, and in (113b) an integration $\int\int\int d\psi\, d(\cos\theta)\, d\phi / 8\pi^2$ should be added over the Euler angles relating the "collision-fixed" frame [in which $S(b,v)$ was calculated] to a space-fixed frame. Thus (116) should contain only terms invariant under rotations of the collision-fixed frame.

Space degeneracy can now be taken into account by explicitly writing $S(b,v)$ as a matrix in the projection quantum numbers $(m_\alpha,$ etc.) and calculating the matrix elements of $\mathcal{S} = SS^*$ in the appropriate multipole basis, constructed with (112).

A substitution similar to (113) can be made in the classical limit even if condition (91), removing the velocity dependence of the self frequency, is relaxed. However, (115) should be modified by adding a statistical factor that depends on the two momenta $\mathbf{p}_b$ and $\mathbf{p}_g$. Carrying out the summation over $\mathbf{p}_g$, the self frequency still remains dependent on $\mathbf{p}_b$, as we show below.

B. Velocity Dependence

Velocity effects enter the line shape if the second and third impact conditions (91) and (92) are relaxed, requiring only that the duration of collision be short compared with the damping time. Removal of (92) introduces the Doppler broadening as an *inhomogeneous-broadening* effect, in which the extra broadening comes from a continuum of eigenvalues of Ω, not from the damping part of the self-frequency matrix. Removal of (91) makes the self frequency velocity dependent; thus (even if the Doppler broadening can be neglected), the shape of an isolated line is not Lorentzian near the line center (see, e.g., Mizushima,[65] Smith et al.,[66] and Berman[67]).

The impact-approximation expression (105) will be extended here to include explicit dependence on the momentum of molecule 1 in the laboratory frame of coordinates. It is again assumed that the free-molecule density matrix can be used. Also, we deal here with an isolated line (α, β), and drop out explicit reference to the second-molecule states (which can be easily incorporated in the usual manner). Also, for simplicity's sake, we deal only with foreign-gas-broadening terms. Under these assumptions (still using a discrete momentum representation), the self frequency is a matrix in the momentum states of the molecule:

$$\Sigma_{\alpha\beta}(\mathbf{p}_1,\mathbf{p}_1') = N \int \frac{d\varepsilon}{2\pi i} \left\{ \delta_{\mathbf{p}_1,\mathbf{p}_1'} - \sum_{\mathbf{p}_2} f_2(\mathbf{p}_2) \langle \mathbf{p}_1,\mathbf{p}_2|S_\alpha|\mathbf{p}_1',\mathbf{p}_2'\rangle\langle \mathbf{p}_1,\mathbf{p}_2|S_\beta|\mathbf{p}_1',\mathbf{p}_2'\rangle^* \right\}$$

$$(117)$$

where $\mathbf{p}_2' = \mathbf{p}_1 + \mathbf{p}_2 - \mathbf{p}_1'$. We use here S_α, and so on, as shorthand for the matrix elements of $S^{(2)}$ in the internal states; f_2 is the Maxwell-Boltzmann distribution of perturbers with mass m_2. The S matrix can be transformed into the center-of-mass system of coordinates of the collision pair, where it depends only on the relative velocities

$$\langle \mathbf{p}_1,\mathbf{p}_2|S|\mathbf{p}_1',\mathbf{p}_2'\rangle = \langle \mathbf{p}|S|\mathbf{p}'\rangle \delta_{\mathbf{P},\mathbf{P}'} \qquad (118)$$

where $\mathbf{P} = \mathbf{p}_1 + \mathbf{p}_2$, and (see Fig. 3)

$$\frac{\mathbf{p}}{m} = \mathbf{v} = \frac{\mathbf{p}_2}{m_2} - \frac{\mathbf{p}_1}{m_1}$$

is the relative velocity (m being the reduced mass). Because of momentum conservation, the scattering angle in the two coordinate frames is related by

$$\mathbf{p}' - \mathbf{p} = -(\mathbf{p}_1' - \mathbf{p}_1) \qquad (119)$$

The summation over $\mathbf{p}_2$ in (117) can be replaced by a summation over $\mathbf{p}$, giving

$$\Sigma_{\alpha\beta}(\mathbf{p}_1,\mathbf{p}_1') = N \int \frac{d\varepsilon}{2\pi i} \left\{ \delta_{\mathbf{p}_1,\mathbf{p}_1'} - \sum_{\mathbf{p}} K(\mathbf{p}_1,\mathbf{p}) \right.$$

$$\left. \times f(\mathbf{p})\langle \mathbf{p}|S_\alpha|\mathbf{p}-\mathbf{p}_1'+\mathbf{p}_1\rangle\langle \mathbf{p}|S_\beta|\mathbf{p}-\mathbf{p}_1'+\mathbf{p}_1\rangle^* \right\} \qquad (120)$$

Here $f(\mathbf{p})$ is the velocity distribution for the relative motion (using the reduced mass m) and

$$K(\mathbf{p}_1;\mathbf{p}) = \frac{f_2(m_2[\mathbf{v}+(\mathbf{p}_1/m_1)])}{f(\mathbf{p})} = \frac{f_2(\mathbf{p}_2)}{f(\mathbf{p})} \qquad (121)$$

In the quasielastic classical-trajectory limit, S depends only on v and b or,

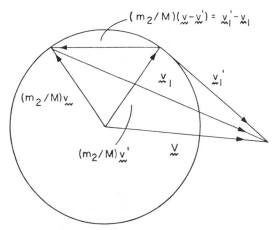

Fig. 3. The transformation of velocities before (unprimed) and after (primed) an elastic collision from laboratory to barycentric coordinates.

alternatively, on v and θ (the scattering angle);

$$\langle \mathbf{p} | S_\alpha | \mathbf{p} - \mathbf{p}_1' + \mathbf{p}_1 \rangle \rightarrow S_\alpha(v, \theta) \tag{122}$$

where

$$\sin^2\left(\frac{\theta}{2}\right) = \frac{\mathbf{p} \cdot (\mathbf{p}_1' - \mathbf{p}_1)}{2p^2}$$

The deviation of the statistical factor K from unity causes the departure from the simple impact-approximation result for, with $K = 1$, the summation over the orientations of $\mathbf{p}$ would turn into an integration over θ (or b). This limit is obviously approached whenever $m_1 \gg m_2$ (Brownian-particle model), since $f_2(\mathbf{p}_2)$ can then be identified with $f(\mathbf{p})$. Consider, however, the opposite limit

$$m_2 \gg m_1 \tag{123}$$

(*heavy perturbers*). In this limit $\mathbf{p}_1$ can be identified with $-\mathbf{p}$; therefore the summation over $\mathbf{p}_2$ becomes redundant and we get

$$\Sigma_{\alpha\beta}(\mathbf{p}_1, \mathbf{p}_1') = N \int \frac{d\varepsilon}{2\pi i} \{ \delta_{\mathbf{p}_1, \mathbf{p}_1'} - \langle \mathbf{p}_1 | S_\alpha | \mathbf{p}_1' \rangle \langle \mathbf{p}_1 | S_\beta | \mathbf{p}_1' \rangle^* \} \tag{124}$$

The sum

$$\sum_{\mathbf{p}_1'} \Sigma_{\alpha\beta}(\mathbf{p}_1, \mathbf{p}_1') = \Gamma_{\alpha\beta}(\mathbf{p}_1) \tag{125}$$

can be interpreted as the total (complex) damping coefficient at a given initial state $\mathbf{p}_1$.

Consider, further, an example in which the scattering is elastic and isotropic with equal amplitudes to all directions, making (124) independent of the direction of $\mathbf{p}'_1$. In that case, to any power of the matrix Σ, we have

$$\sum_{\mathbf{p}'_1} \Sigma_{\alpha\beta}{}^m(\mathbf{p}_1, \mathbf{p}'_1) = \sum_{\mathbf{p}'_1 \cdots \mathbf{p}_1^{(m)}} \Sigma_{\alpha\beta}(\mathbf{p}_1, \mathbf{p}_1^{(m)}) \cdots \Sigma_{\alpha\beta}(\mathbf{p}''_1, \mathbf{p}'_1)$$

$$= \left[\Gamma_{\alpha\beta}(\mathbf{p}_1)\right]^m$$

That is, $\Gamma_{\alpha\beta}(\mathbf{p}_1)$ can be treated as a diagonal representation of $\Sigma_{\alpha\beta}(\mathbf{p}_1, \mathbf{p}'_1)$. Such an example may approximately describe the scattering from a very narrow-ranged potential (a *nearly hard sphere model*). In this situation the scattering amplitudes are also approximately independent of the initial momentum. The only dependence on the momentum comes from the integration over $d\varepsilon$ which, in the classical limit, introduces a factor v, as follows from (113b). We can then write

$$i\Gamma_{\alpha\beta}(\mathbf{p}_1) \approx n_2 v_1 \sigma_{\alpha\beta}, \qquad v_1 = \frac{p_1}{m_1} \tag{126}$$

where the cross-section $\sigma_{\alpha\beta}$ is independent of velocities in this case. Another property of the nearly hard sphere model is that phase shifts are small, hence the imaginary part of Γ (pressure shift) can be neglected; that is,

$$\Gamma_{\alpha\beta}(\mathbf{p}_1) \approx -i\gamma_{\alpha\beta}(v_1) \tag{127}$$

where γ is a real damping rate.

Going to the continuum limit, the line shape for the nearly hard sphere, heavy-perturber case is given by

$$F_{\alpha\beta}{}^R(\omega) = n_1 \rho_\beta |\mu_{\alpha\beta}|^2 \int_0^\infty dv_1 f_1(v_1)[\omega - \omega_{\alpha\beta} + i\gamma_{\alpha\beta}(v_1)]^{-1} \tag{128}$$

The dissipative part $F''_{\alpha\beta}(\omega)$ is thus a "superposition" of Lorentzian terms with linewidth $\gamma(v_1)$ and weight $f(v_1)$. With the former given by (126) and the latter by (114) (with m_1 replacing m), the integral in (128) can be expressed in terms of familiar functions

$$-F''_{\alpha\beta}(\mathbf{k}, \omega) = n_1 \rho_\beta |\mu_{\alpha\beta}|^2 \left(\frac{2}{\pi}\right)^{1/2} \bar{\gamma}^{-1} A(v) \tag{129}$$

where

$$\bar{\gamma} = n_2 \sigma_{\alpha\beta} \left(\frac{T}{m_1} \right)^{1/2}$$

$$v = \frac{|\omega - \omega_{\alpha\beta}|^2}{\left(\sqrt{2}\,\bar{\gamma} \right)^2}$$

and

$$A(v) = 1 + ve^v Ei(-v) \tag{130}$$

using the exponential integral function

$$Ei(-v) = -\int_v^\infty t^{-1} e^{-t}\, dt$$

The inverse Fourier transform of (129) is the correlation function

$$G > (\mathbf{k}, t) e^{i\omega_{\alpha\beta} t} = n_1 \rho_\beta |\mu_{\alpha\beta}|^2 B(\tau) \tag{131}$$

where

$$\tau = \frac{\dot{\bar{\gamma}} t}{\sqrt{2}}$$

and

$$B(\tau) = -\left(\frac{\tau}{\sqrt{\pi}} \right) + (\tfrac{1}{2} + \tau^2) e^{\tau^2} [1 - \Phi(\tau)] \tag{132}$$

using the error function

$$\Phi(\tau) = \left(\frac{2}{\sqrt{\pi}} \right) \int_0^\tau e^{-t^2}\, dt$$

An interesting property of (129) and (132) is their asymptotic behavior. Far out in the wings ($v \gg 1$) the line shape behaves like a tail of a Lorentzian with a linewidth

$$\gamma_{\text{asym}} = \sqrt{\left(\frac{8}{\pi} \right)}\,\bar{\gamma} \tag{133}$$

At long times ($\tau \to \infty$), the correlation function decays as an inverse power of the time

$$B(\tau) \sim \left(\frac{1}{2\sqrt{\pi}}\right)\tau^{-3} \tag{134}$$

rather than exponentially. Such a behavior should be approximately expected in spectra of light molecules perturbed by heavy nonpolar atoms.

C. Line Wings

Far out in the wings of a line (α, β), where

$$|\Sigma''_{\alpha\beta;\,\alpha\beta}(\mathbf{k}, \omega)| \ll |\omega - \omega_{\alpha\beta}| \tag{135}$$

the ω dependence of $F^R(\mathbf{k}, \omega)$ is mostly determined by the ω dependence of $\Sigma_c(\mathbf{k}, \omega)$. Expanding (74) in a power series of Σ_c, and retaining only first-order terms, owing to (135), we get for the line wings (in the SRPA)

$$F_{ij}^R(\mathbf{k}, \omega) = \left(\frac{N}{V}\right) \sum_{abcd} \rho_b (\mu_{ab}{}^i)^* \mu_{cd}{}^j$$

$$\times \left[(\omega - \omega_{ab})^{-1} \delta_{ac}\delta_{bd} + (\omega - \omega_{ab})^{-1} \Sigma_{ab;\,cd}(\mathbf{k}, \omega)(\omega - \omega_{cd})^{-1} \right] \tag{136}$$

where, we should recall,

$$\Sigma_{ab;\,cd}(\mathbf{k}, \omega) = N \langle\langle \bar{x}^{ab}(\mathbf{k}) | \mathfrak{T}_c(\omega) | x^{cd}(\mathbf{k}) \rangle\rangle \tag{137}$$

This expression, since it is linear in Σ_c, can be corrected for the nonorthogonality of the SME set (forsaking the SRPA). This nonorthogonality has only a minor effect on the matrix Ω of resonance frequencies. By either neglecting it in comparison to $|\omega - \omega_{\alpha\beta}|$ or incorporating the correction in Σ', we can still retain an expression like (136), with $\rho_b \Sigma_{ab;\,cd}$ replaced by an average over the exact density matrix. Thus, for example, for the dissipative part F'' we can write (in the BCA)

$$F_{ij}''(\mathbf{k}, \omega) = \frac{N_1}{V} \sum_{abcd} (\mu_{ab}{}^i)^* \mu_{cd}{}^j (\omega - \omega_{ab})^{-1} (\omega - \omega_{cd})^{-1}$$

$$\times N_2 \sum_S \mathrm{tr} \left\{ \rho^{(2)} |b_1\rangle\langle a_1| \mathfrak{T}''^{(2)}(\omega) |c_S\rangle\langle d_S| \right\} \tag{138}$$

where $S = 1$ only in foreign-gas broadening, and $S = 1$ or 2 in self broadening ($N_1 = N_2$). Here $\rho^{(2)}$ and $\mathfrak{T}''^{(2)}$ are, respectively, the two-particle density matrix and (antihermitian part of) the double-space T matrix. In (138), $|c_S\rangle\langle d_S|$ is an excitation mode of molecule S, with $\mathbf{p}_c - \mathbf{p}_d = \mathbf{k}$, and so on.

Considering again that ω is quite far from the resonance frequencies of the binary system, we can approximately write

$$\mathcal{J}''^{(2)}(\omega) \approx -\pi \mathcal{V}_1 \delta(\omega - \mathcal{H}_0 - \mathcal{V}_0)\mathcal{V}_1 \qquad (139)$$

Here $\mathcal{H}_0$ is the liouvillian of the noninteracting pair and $\mathcal{V} = \mathcal{V}_1 + \mathcal{V}_0$ is the pair interaction where, following (109), we have separated from the potential the part $\mathcal{V}_0$ that determines the trajectory in a binary collision. It is easy to see now that (138) can be reexpressed as a Fourier transform of a correlation function:

$$F_{ij}''(\mathbf{k}, \omega) = -\frac{1}{2}\left(\frac{N_1}{V}\right) N_2 \sum_S \int_{-\infty}^{\infty} \mathrm{tr}\left\{\rho y_1^{i^\dagger}(\mathbf{k}, t) y_S^j(\mathbf{k}, 0)\right\} e^{i\omega t}\, dt \qquad (140)$$

where

$$y_S(\mathbf{k}, 0) = i\mathcal{V}_1(\mu_S e^{i\mathbf{k}\cdot\mathbf{R}_S}) \qquad (141)$$

The time evolution of $y_S(\mathbf{k}, t)$ is along the trajectory determined by $\mathcal{V}_0$.

Consider, for example, a case of foreign-gas broadening where $\mathcal{V}_1(\mathbf{R})$ is diagonal in the internal quantum numbers

$$\langle\langle \alpha\zeta; \beta\eta|\mathcal{V}_1(\mathbf{R})|\alpha'\zeta'; \beta'\eta'\rangle\rangle = \delta_{\alpha\alpha'}\delta_{\beta\beta'}\delta_{\zeta\zeta'}\delta_{\eta\eta'}$$

$$\times \left[\langle\alpha\zeta|V_1(\mathbf{R})|\alpha\zeta\rangle - \langle\beta\eta|V_1(\mathbf{R})|\beta\eta\rangle\right] \qquad (142)$$

Assume also that the second and third impact conditions (91) and (92) hold and that the classical-trajectory approximation is valid. Furthermore, suppose that the statistical correlations can be neglected, and ρ can be approximated by ρ^0. In the classical-trajectory limit, the number of collisions with relative velocity v to $v + dv$ and impact parameter b to $b + db$, and with perturber molecules in state ζ, occurring between t_0 and $t_0 + dt_0$ (where t_0 is the moment of closest approach) is

$$n_2\rho_\zeta v f(v) 2\pi b\, db\, dv\, dt_0$$

with $f(v)$ defined in (114). Because of (92) we can drop the exponential $\mathbf{k}$-dependent factor. Let the variation of $\mathbf{R}$ along the classical trajectory, as a function of time, be measured from the moment of closest approach as the origin on the time scale. Then we can write

$$F_{ij}''(\mathbf{k}, \omega) = -n_1 n_2 \sum_{\alpha\beta} \rho_\phi b\left(\mu_{\alpha\beta}^i\right)^* \mu_{\alpha\beta}^i(\omega - \omega_{\alpha\beta})^{-2} L_{\alpha\beta}(\omega) \qquad (143a)$$

where

$$L_{\alpha\beta}(\omega) = \pi \sum_{\zeta} \rho_{\zeta} \int \int vf(v)\,dv\,b\,db\,e^{i(\omega-\omega_{\alpha\beta})t}\,dt\,dt_0$$

$$\langle\langle\alpha\zeta;\beta\zeta|\mathcal{V}_1(\mathbf{R}(-t_0))|\alpha\zeta;\beta\zeta\rangle\rangle\langle\langle\alpha\zeta;\beta\zeta|\mathcal{V}_1(\mathbf{R}(t-t_0))|\alpha\zeta;\beta\zeta\rangle\rangle$$

Noting that the two time integrations are separable, we get

$$L_{\alpha\beta}(\omega) = \pi \sum_{\zeta} \rho_{\zeta} \int \int vf(v)\,b\,db\,|V_{\alpha\zeta,\alpha\zeta}(\omega-\omega_{\alpha\beta};b,v) - V_{\beta\zeta,\beta\zeta}(\omega-\omega_{\alpha\beta};b,v)|^2$$

$$(144a)$$

where

$$V_{\alpha\zeta,\alpha\zeta}(\omega;b,v) = \int_{C(v,b)} V_{\alpha\zeta,\alpha\zeta}(\mathbf{R}(t))e^{i\omega t}\,dt \qquad (145)$$

with the integration carried along the trajectory specified by b and v. Again, as in the impact case, if (144a) depends on the orientation of the collision frame in space, it should be averaged over an isotropic distribution of collisions.

Given the $\mathbf{R}$ dependence of $V_1(\mathbf{R})$ and the trajectories, we can calculate the ω dependence of $V(\omega;b,v)$, hence of $L(\omega)$. Consider, for example, a billiard-ball trajectory model

$$R(t) = (b^2 + v^2t^2)^{1/2}, \qquad b > d$$

$$= \left[d^2 + v^2t^2 + 2v(d^2-b^2)^{1/2}|t|\right]^{1/2}, \qquad b < d \qquad (146)$$

(where d is the billiard-ball diameter), and an inverse-power interaction law

$$V_{\alpha\zeta,\alpha\zeta}(R) = -A_{\alpha\zeta}R^{-n} \qquad (147)$$

Suppose we can approximately replace the averaging over v by inserting a mean value $\bar{v}$ in the integrand in (144a). The resulting ω dependence of $L(\omega)$ will be of the form

$$L(\omega) \propto P(\kappa)e^{-\kappa} \qquad (148)$$

where

$$\kappa = \frac{|\omega-\omega_{\alpha\beta}|d}{\bar{v}}$$

and P is a polynomial.

The exponential frequency dependence in the far wings may not be very sensitive to the details of the kinematics of the collision. At close frequencies, the shape of the trajectories may be more material. For example, suppose that the molecules may be trapped in quasiellipsoidal "orbiting" trajectories for quite a while. The interaction (which otherwise may have caused a monotonous frequency dependence) may appreciably enhance the line shape around a frequency displacement from the line center equal to the orbiting frequency.

The line shape expressions can be readily extended to nondiagonal interactions by replacing (143a) with

$$F_{ij}''(\mathbf{k},\omega) = -n_1 n_2 \sum_{\alpha\beta\gamma\delta} \rho_\beta (\mu_{\alpha\beta}^{\ i})^* \mu_{\gamma\delta}^{\ j} \ (\omega - \omega_{\alpha\beta})^{-1} (\omega - \omega_{\gamma\delta})^{-1} L_{\alpha\beta;\,\gamma\delta}(\omega)$$

(143b)

Here

$$L_{\alpha\beta;\,\gamma\delta}(\omega) = \pi \sum_{\substack{\zeta\eta\zeta'\eta' \\ \alpha'\beta'}} \rho_\zeta \int \int v f(v) \, dv \, b \, db$$

$$\times \langle\langle \alpha'\zeta'\beta'\eta' | \mathcal{V}_1(\omega - \omega_{\alpha'\beta'} - \omega_{\zeta'\eta'}; b, v) | \alpha\zeta\beta\zeta \rangle\rangle^*$$

$$\times \langle\langle \alpha'\zeta'\beta'\eta' | \mathcal{V}_1(\omega - \omega_{\alpha'\beta'} - \omega_{\zeta'\eta'}; b, v) | \gamma\eta\delta\eta \rangle\rangle$$

(144b)

where

$$\mathcal{V}_1(\omega; b, v) = V_1(\omega; b, v) I^* - I V_1^*(\omega; b, v)$$

This second-order perturbation expression for $L(\omega)$ is valid only if the frequency mismatch appearing as the argument of $\mathcal{V}_1$ in (144b) is much larger than the corresponding matrix elements of $\mathcal{V}_1$.

D. Statistical Broadening

The applications discussed in Sections IX.A to IX.C were restricted by the first impact condition (90), requiring the duration of a collision to be much shorter than the relaxation time. The statistical-broadening approximation (Margenau;[68] Margenau and Lewis,[31] and references therein) is used in the opposite extreme case

$$\tau \Sigma_c'' \gg 1$$

(149)

where the lines are so broad that the motion of the molecules is unnoticeable during time intervals comparable to the inverse linewidth. Under these conditions it does not make sense to talk about individual collisions as events well isolated in time. A binary-collision picture may nevertheless be retained, in a sense, by neglecting the interactions among perturbers, treating the "radiating" molecule as if it interacts with each perturber independently.

In such a situation it would be more appropriate to consider the position of the molecules as a "good" quantum number, treating the kinetic energy as a "perturbation." The most appropriate single-molecule basis in this case would be

$$z^{ab}(\mathbf{k}) = \sum_{A=1}^{N} e^{i\mathbf{k}\cdot\mathbf{R}_A}|\alpha, \mathbf{R}_A\rangle\langle\beta, \mathbf{R}_A|, \qquad a = \alpha, \mathbf{R} \qquad (150)$$

The interaction potential is diagonal in $\mathbf{R}$ and dependent only on differences $\mathbf{R}_{AB} = \mathbf{R}_A - \mathbf{R}_B$. The projection of $\mathcal{V}$ (or a function thereof) on the basis (150), which involves an integration over the positions of all but one molecule, is independent of the position of *all* molecules. Therefore, *the proper self energy* constructed from (150) while neglecting the kinetic energy *is completely independent of the position variable*:

$$\Sigma(\mathbf{k}, \omega; \mathbf{R}) \equiv \Sigma(\mathbf{k}, \omega) \qquad (151)$$

Let the equation

$$\mathcal{T}(\omega; R) = \mathcal{V}(R)\left[1 + (\omega - \mathcal{H}_{\text{int}} + i\eta)^{-1}\mathcal{T}(\omega; R)\right] \qquad (152)$$

(with R denoting all positions and $\mathcal{H}_{\text{int}}$, the liouvillian for the internal degrees of freedom) define a "static" $\mathcal{T}$ matrix in the basis of eigenstates of $\mathcal{H}_{\text{int}}$. Assuming that $\rho(R)$ is diagonal in the internal quantum numbers (the present form of the SRPA), we can write, in the binary-collision approximation,

$$\Sigma_{\alpha\beta;\gamma\delta}(\mathbf{k}, \omega) = n\Sigma^{(1)}_{\alpha\beta;\gamma\delta}(\mathbf{k}, \omega)$$

$$= n\rho_\beta^{-1}\int d^3\mathbf{R}\sum_{\zeta\eta}\rho^{(2)}_{\beta\zeta}(\mathbf{R})\left[\mathcal{T}^{(2)}_{\alpha\zeta\beta\zeta;\gamma\eta\delta\eta}(\omega; \mathbf{R})\right.$$

$$\left.+ e^{i\mathbf{k}\cdot\mathbf{R}}\mathcal{T}^{(2)}_{\alpha\zeta\beta\zeta;\eta\gamma\eta\delta}(\omega; \mathbf{R})\right] \qquad (153)$$

where the second term in the brackets appears only in self broadening. Here, as usual, $\mathcal{T}^{(2)}$ is the two-body matrix, depending on the intermolecular separation $\mathbf{R} = \mathbf{R}_{BA}$. The one-molecule density-matrix element ρ_β is, to a good approximation, independent of the potential energy. Notice that $\mathbf{k}$ enters (153) only in the resonance-exchange self-broadening term; this follows from $\mathcal{V}$ being diagonal in $\mathbf{R}$.

The major source of ω dependence now is $\Sigma(\mathbf{k}, \omega)$ which, unlike the impact approximation, may vary appreciably with ω over a range comparable to the magnitude of Σ''.

Let us consider, as an illustration, the line wings in a foreign-gas-broadened line, with a potential diagonal in the internal quantum numbers and independent of those of the perturber; that is,

$$V_{\alpha\zeta,\alpha'\zeta'}(\mathbf{R}) = V_\alpha(R)\delta_{\alpha\alpha'}\delta_{\zeta\zeta'} \tag{154}$$

The line shape in the wings may be written with an expression like (143a) (provided Σ'' is sufficiently small), where now

$$L_{\alpha\beta}(\omega) = \int d^3\mathbf{R}\, g_\beta(\mathbf{R})\, \mathcal{V}_{\alpha\beta}{}^2(R)\delta\left(\omega - \omega_{\alpha\beta} - \mathcal{V}_{\alpha\beta}(R)\right) = \Sigma''{}^{(1)}_{\alpha\beta;\alpha\beta}(\mathbf{k},\omega) \tag{155}$$

Here $\mathcal{V}_{\alpha\beta} = V_\alpha - V_\beta$, and

$$g_\beta(\mathbf{R}) = \exp\left[-\frac{V_\beta(R)}{T}\right]$$

is the binary-collision pair correlation function. Let R_ω be a root of the equation

$$\omega - \omega_{\alpha\beta} - \mathcal{V}_{\alpha\beta}(R) = 0 \tag{156}$$

Then

$$L_{\alpha\beta}(\omega) = \frac{4\pi R_\omega^2 \mathcal{V}_{\alpha\beta}^2(R_\omega)}{|d\mathcal{V}_{\alpha\beta}/dR|_{R=R_\omega}} \tag{157}$$

or sum of such terms if (156) has more than one root. The derivative of $\mathcal{V}$ in the denominator plays a critical role here. Notice that $L_{\alpha\beta}(\omega)$ will diverge if R_ω happens to be an extremum on the potential curve of $\mathcal{V}_{\alpha\beta}(R)$. We cannot use (143a), with Σ'' in the numerator, in this case since Σ'' is not small. The line shape function will tend to zero in this limit. If R_ω is an absolute maximum (minimum) point on the curve of $\mathcal{V}_{\alpha\beta}(R)$, the line shape in the statistical approximation will abruptly vanish above (below) the corresponding frequency. If it is only a local extremum, sharp features may appear in the line wings, which may be identified as "satellite bands" (Jefimenko,[69] Farr and Hindmarsh;[70] other explanations of the origin of satellite bands were suggested by Klein and Margenau[71] and Breene[72]).

The true line shape will not have such a singular behavior because of the extra broadening effect of the kinetic motion neglected in this approximation. Nevertheless, if the kinetic damping effect is sufficiently weak, the sharp features may still appear. A more comprehensive theory including kinetic effects would require solving the integral equation

$$\mathcal{T}^{(2)}(\omega; \mathbf{R}, \mathbf{R}') = \mathcal{V}^{(2)}(\mathbf{R}) \left\{ \delta(\mathbf{R} - \mathbf{R}') \right.$$

$$\left. + \left[\omega - \mathcal{H}_{\text{int}}^{(2)} + \frac{1}{2m}(\nabla_{\mathbf{R}}^2 - \nabla_{\mathbf{R}'}^2) + i\eta \right]^{-1} \mathcal{T}^{(2)}(\omega; \mathbf{R}, \mathbf{R}') \right\} \quad (158)$$

(perhaps by treating the extra kinetic term as a "small" perturbation), and substituting the solution in (153), with an extra integration over $d^3\mathbf{R}'$. The self frequency in this approximation is still independent of the position.

E. Collision-Induced Spectra

Our account of line shapes in gases would not be complete without mention of the collision-induced polarization of molecules and its effects on the spectrum.[27,73–75]

Our starting point in deriving the line shape expression was the idea that the polarization is a sum of one-molecule operators. This is true, provided the polarization operators μ^i are defined on the *complete* set of degrees of freedom of the molecule (electronic as well as nuclear). However, this scheme may be cumbersome, particularly if we deal with low-frequency effects where the dependence on electronic motion should only enter in an average sense. The customary way of removing the electronic degrees of freedom is to use the *adiabatic* (Born-Oppenheimer) *approximation*. In the ideal-gas limit, the adiabatic wave functions are derived from the *intramolecular* potentials. However, in this scheme, the *intermolecular* potentials are not diagonal in the electronic states. This off-diagonal interaction is the origin of the collision-induced polarizations.

Consider expression (41) for $F_{ij}{}^R$. In most cases, the electronic excitation energies are much higher than T. Therefore we can replace the partial averaging over electronic degrees of freedom by an expectation value over the ground electronic state. An exact calculation would require to use the "dressed" ground state (depending on the intermolecular as well as intramolecular potentials). However, in dilute gases, considering the off-diagonal interaction to first order only, we can replace it by the "bare" ground state of the independent molecules (dependent only on intramolecular nuclear positions). Let

$$|0\rangle = |0(R_{\text{intra}})\rangle$$

be the bare electronic ground state and $|e\rangle$ an excited state. The partial trace over electronic levels would give

$$F_{ij}^{R}(\mathbf{k},\omega) = V^{-1}\mathrm{tr}\sum_{e}\left\{\langle 0|\rho|0\rangle\langle 0|M^{i}(\mathbf{k})|e\rangle\langle e|(\omega-\mathcal{H}+i\eta)^{-1}M^{j}(-\mathbf{k})|0\rangle\right\}$$

$$(159)$$

where the trace is now over the nuclear degrees of freedom.

The hamiltonian can be split into

$$H = H_{\mathrm{ad}} + H_{\mathrm{non}} \qquad (160)$$

where H_{ad} is a diagonal in the adiabatic electronic states and H_{non} contains the off-diagonal parts, giving

$$(\omega-\mathcal{H}+i\eta)^{-1}\approx(\omega-\mathcal{H}_{\mathrm{ad}}+i\eta)^{-1}\left[1+\mathcal{H}_{\mathrm{non}}(\omega-\mathcal{H}_{\mathrm{ad}}+i\eta)^{-1}\right] \qquad (161)$$

to first order in H_{non}. Considering the large separation of electronic levels, compared to ground-state nuclear-motion energies, we can ignore in the sum over e in (159) all terms except the ground state, because of the large denominators, leaving

$$F_{ij}^{R}(\mathbf{k},\omega)\approx V^{-1}\mathrm{tr}\left\{\langle 0|\rho|0\rangle\langle 0|M^{i}(\mathbf{k})|0\rangle\right.$$

$$\left.(\omega-\mathcal{H}_{00}+i\eta)^{-1}\langle 0|\left[1+\mathcal{H}_{\mathrm{non}}(\omega-\mathcal{H}_{\mathrm{ad}}+i\eta)^{-1}\right]M^{j}(-\mathbf{k})|0\rangle\right\} \qquad (162)$$

where $H_{00}=\langle 0|H|0\rangle$ is the ground-state hamiltionian. The last factor in (162) has two parts. The first is the independent-molecule polarization in the ground electronic state. The second contains all induced effects (to first order) of collisions on the polarization.

The "induced" terms clearly cannot be expressed as a one-molecule operator. In this first-order expression they include one-, two-, and three-molecule terms, if H contains only pair interactions. An SME basis would not be appropriate here, and other approximate means of reducing the variables to derive the line shape would be necessary. The binary-collision concept may nevertheless be valid here, in which case the induced terms will include two-molecule contributions, at most.

Acknowledgments

The author is grateful to Professor I. Oppenheim and the Chemistry Department of the Massachusetts Institute of Technology for their kind hospitality. The work was supported in part by the National Science Foundation.

References

1. U. Fano, *Phys. Rev.*, **131**, 259 (1963).
2. P. Resibois, *Phys. Rev. B*, **138**, 281 (1965).
3. D. W. Ross, *Ann. Phys. (N.Y.)*, **36**, 458 (1966).
4. B. Bezzerides, *J. Quant. Spectrosc. Radiat. Transfer*, **7**, 353 (1967).
5. B. Bezzerides, *Phys. Rev.*, **159**, 3 (1967); **181**, 379; **186**, 239 (1969).
6. L. Klein, *J. Quant. Spectrosc. Radiat. Transfer*, **9**, 199 (1969).
7. R. Dashen and S. Ma, *J. Math. Phys.*, **12**, 1449 (1971).
8. A. Tip, and F. R. McCourt, *Physica*, **52**, 109 (1971).
9. A. Ben-Reuven, *Phys. Rev. A*, **4**, 2115 (1971).
10. R. F. Snider and B. C. Sanctuary, *J. Chem. Phys.*, **55**, 1555 (1971).
11. R. W. Lee, *J. Phys. A*, **5**, 950; *J. Phys. B*, **5**, 1271 (1972).
12. A. Royer, *Phys. Rev. A*, **6**, 1741 (1972); **7**, 1078 (1973).
13. J. Albers and I. Oppenheim, *Physica*, **59**, 161; **59**, 187 (1972).
14. C. A. Mead, *Phys. Rev. A*, **5**, 1957 (1972).
15. R. Kubo, in *Lectures in Theoretical Physics*, Vol. 1, W. E. Brittin and L. G. Dunham, Eds., Interscience, New York, 1959, p. 120.
16. D. N. Zubarev, *Usp. Fiz. Nauk*, **71**, 71 (1960); English transl.: *Sov. Phys.—Uspekhi*, **3**, 320 (1960).
17. A. A. Abrikosov, L. P. Gorkov, and I. Y. Dzyaloshinski, *Quantum Field Theoretical Methods in Statistical Physics*, 2nd ed., Pergamon Press, Oxford, 1965.
18. C. De Dominicis, *C. R. Acad. Sci. (Paris)* **265 B**, 1273 (1967).
19. C. Bloch, in *Studies in Statistical Mechanics*, Part 1, J. de Boer and G. E. Uhlenbeck, Eds., North Holland, Amsterdam, 1965.
20. R. A. Pasmanter, R. Samson, and A. Ben-Reuven, *Chem. Phys. Lett.*, **16**, 470 (1972).
21. A. Ben-Reuven and L. Klein, *Phys. Rev. A*, **4**, 753 (1971).
22. L. Klein, M. Giraud, and A. Ben-Reuven, *Phys. Rev. A*, **10**, 682 (1974).
23. A. Ben-Reuven and A. Lightman, *J. Chem. Phys.*, **46**, 2429 (1967).
24. J. Fiutak and J. Van Kranendonk, *Can. J. Phys.*, **40**, 1085 (1962).
25. P. N. Argyres and P. L. Kelley, *Phys. Rev. A*, **134**, 98 (1964).
26. R. G. Gordon, *J. Chem. Phys.*, **45**, 1635; **45**, 1649 (1966).
27. J. Van Kranendonk, *Physica*, **73**, 156 (1974).
28. E. W. Smith, and C. F. Hooper, Jr., *Phys. Rev.*, **157**, 126 (1967).
29. E. W. Smith, *Phys. Rev.*, **166**, 102 (1968).
30. M. Berrondo, *Proc. Roy. Soc. (London) A*, **328**, 353 (1972).
31. H. Margenau and M. Lewis, *Rev. Mod. Phys.*, **31**, 569 (1959).
32. G. Traving, *On the Theory of Pressure Broadening of Spectral Lines*, Verlag G. Braun, Karlsruhe, 1960.
33. R. G. Breene, Jr., *The Shift and Shape of Spectral Lines*, Pergamon Press, Oxford, 1961.
34. R. G. Breene, Jr., in *Handbuch der Physik*, Vol. 27, S. Flügge, Ed., Springer-Verlag, Berlin, 1964, p. 1.
35. C. J. Tsao and B. Curnutte, *J. Quant. Spectrosc. Radiat. Transfer*, **2**, 41 (1962).
36. H. R. Griem, *Plasma Spectroscopy*, McGraw-Hill, New York, 1964.
37. R. G. Gordon, *J. Chem. Phys.*, **44**, 3083 (1966).
38. G. Birnbaum, *Advances in Chemical Physics*, **12**, 487 (1967).
39. P. R. Berman and W. E. Lamb, Jr., *Phys. Rev.*, **187**, 221 (1969).
40. J. Cooper, in *Lectures in Theoretical Physics*, Vol. 11C, S. Geltman, K. T. Mahanthappa, and W. E. Brittin, Eds., Gordon & Breach, New York, 1969, p. 241.
41. R. P. Futrelle, *Phys. Rev. A*, **5**, 2162 (1972).

42. W. R. Hindmarsh and J. M. Farr, in *Progress in Quantum Electronics*, Vol. 2, Pergamon Press, Oxford, 1972, p. 141.
43. H. Rabitz, *Ann. Rev. Phys. Chem.*, **25**, 155 (1974).
44. J. R. Fuhr, W. L. Wiese, and L. J. Roszman, "Bibliography on Atomic Line Shapes and Shifts," National Bureau of Standards (U.S.), Special Publ. No. 366, 1972; Suppl. 1, 1974.
45. M. Baranger, *Phys. Rev.*, **111**, 481; **111**, 494; **112**, 855 (1958).
46. R. Zwanzig, in *Lectures in Theoretical Physics*, Vol. 3, W. E. Brittin, B. W. Downs, and J. Downs, Eds., Interscience, New York, 1961, p. 106.
47. H. Mori, *Progr. Theor. Phys. (Tokyo)*, **33**, 423; **34**, 399 (1965).
48. M. L. Goldberger and K. M. Watson, *Collision Theory*, Wiley, New York, 1964.
49. R. D. Levine, *Quantum Mechanics of Molecular Rate Processes*, Oxford University Press, Oxford, 1969.
50. A. Ben-Reuven, *Phys. Rev.*, **141**, 34 (1966).
51. H. Feshbach, *Ann. Rev. Nucl. Sci.*, **8**, 49 (1958).
52. P. W. Anderson, *Phys. Rev.*, **76**, 647 (1949).
53. B. U. Felderhof and I. Oppenheim, *Physica*, **31**, 1441 (1965).
54. T. Keyes and I. Oppenheim, *Phys. Rev. A*, **7**, 1384 (1973).
55. A. Ben-Reuven, *Adv. At. Mol. Phys.*, **5**, 201 (1969).
56. R. H. Dicke, *Phys. Rev.*, **89**, 472 (1953).
57. L. Galatry, *Phys. Rev.*, **122**, 1218 (1961).
58. R. S. Eng, A. R. Calawa, T. C. Harman, P. L. Kelley, and A. Javan, *Appl. Phys. Lett.*, **21**, 303 (1972).
59. A. C. Kolb and H. R. Griem, *Phys. Rev.*, **111**, 514 (1958).
60. J. Fiutak, *Acta Phys. Polon.*, **27**, 753 (1965).
61. A. Ben-Reuven, *Phys. Rev.*, **145**, 7 (1966).
62. G. Birnbaum, *Phys. Rev.*, **150**, 101 (1966).
63. B. Bleaney and J. H. N. Loubser, *Proc. Roy. Soc. (London) A*, **63**, 483 (1950).
64. D. Robert, M. Giraud, and L. Galatry, *J. Chem. Phys.*, **51**, 2192 (1969).
65. M. Mizushima, *J. Quant. Spectrosc. Radiat. Transfer*, **7**, 505 (1967).
66. E. W. Smith, J. Cooper, W. R. Chappell, and T. Dillon, *J. Quant. Spectrosc. Radiat. Transfer*, **11**, 1547 (1971); J. Ward, J. Cooper, and E. W. Smith, *ibid.*, **14**, 555 (1974).
67. P. R. Berman, *J. Quant. Spectrosc. Radiat. Transfer*, **12**, 1331 (1972).
68. H. Margenau, *Phys. Rev.*, **82**, 156 (1951).
69. O. Jefimenko, *J. Chem. Phys.*, **42**, 205 (1965).
70. J. M. Farr and W. R. Hindmarsh, *Phys. Lett. A*, **27**, 512 (1968).
71. L. Klein and H. Margenau, *J. Chem. Phys.*, **30**, 1556 (1959).
72. R. G. Breene, Jr., *Phys. Rev. A*, **2**, 1164 (1970).
73. H. B. Levine and G. Birnbaum, *Phys. Rev.*, **154**, 86 (1967).
74. E. J. Allin, A. D. May, B. P. Stoicheff, J. C. Styland, and H. L. Welsh, *Appl. Op.*, **6**, 1597 (1967).
75. H. L. Welsh, *M.T.P. International Review of Science, Physical Chemistry*, Series One, Vol. 3, Butterworths, London, 1972, p. 33.

TIME–REVERSAL INVARIANCE, REPRESENTATIONS FOR SCATTERING WAVEFUNCTIONS, SYMMETRY OF THE SCATTERING MATRIX, AND DIFFERENTIAL CROSS–SECTIONS*

DONALD G. TRUHLAR, C. ALDEN MEAD, AND
MAYNARD A. BRANDT

*Department of Chemistry,
University of Minnesota,
Minneapolis, Minnesota*

CONTENTS

*Supported in part by the National Science Foundation through grant GP-28684, by the Alfred P. Sloan Foundation through a research fellowship to one of the authors (D.G.T.), and by the Graduate School of the University of Minnesota.

295

I. INTRODUCTION

In general all the information necessary to completely describe a binary collision of two (composite or elementary) particles is contained in the scattering matrix. When it is desired to express the scattering matrix for general multichannel collision processes in the absence of external fields, a useful representation is one in which the total angular momentum and one of its components are quantum numbers.[1] Our term for this is "total-angular-momentum representation." In Section II we review some examples of descriptions of scattering processes that involve total-angular-momentum representations.

Time-reversal invariance is a fundamental symmetry property that can often be used to simplify the determination of the scattering matrix. For example, Coester[2] and others have used time-reversal invariance to prove that the scattering matrix in a total-angular-momentum representation is symmetric for certain phase conventions for the wavefunctions. In Section III we prove that all symmetry operators must be unitary or antiunitary and that time-reversal is antiunitary. We review antiunitary operators, the time-reversal operator, and some associated properties. In Section IV we prove a general theorem that provides sufficient conditions under which the scattering matrix is symmetric in a total-angular-momentum representation.

In Section V we review the formula for calculating the differential cross-section $d\sigma/d\Omega$ (where Ω specifies the scattering solid angle) from the scattering matrix, and we discuss the application of the results of Section IV to some common examples. The phase conventions used in several published expressions for calculating differential cross-sections from scattering matrices are checked for the properties of the time-reversed state.

Examples are furnished of representations that give a symmetric scattering matrix even though they do not satisfy the conditions usually stated as sufficient for this symmetry. The practical problem of consistent use of a phase convention is also discussed.

II. TOTAL–ANGULAR–MOMENTUM REPRESENTATIONS

In the usual partial-wave expansion,[1a,3-10] the scattering of an elementary particle without spin from a central potential is characterized by the good quantum numbers l, m_l, and Π. Here $\vec{l}$ is the orbital angular momentum* with quantum number l, l_z is its z-component with quantum number m_l, z is a space-fixed axis, and Π is the total parity of the system. In this case the total-angular-momentum quantum number J and the quantum number M of its z-component are identical to l and m_l. Thus the usual partial-wave expansion provides the simplest example of a total-angular-momentum representation in a scattering problem.

The hamiltonian H for an elementary particle interacting with a non-central potential[11] is not rotationally invariant. In this chapter we are concerned only with systems whose total hamiltonian is rotationally invariant.

The scattering of an elementary particle with spin from a rotationally invariant potential (which may depend on the spin) may be characterized[1b,3a,7a,8a,10a] by the quantum numbers J, M, Π, l, and the internal angular momentum quantum number s, where

$$\vec{J} = \vec{l} + \vec{s} \tag{1}$$

Since J and M are included among the quantum numbers specifying the state, such a description is a total-angular-momentum representation. An alternative description of this system which is *not* a total-angular-momentum representation is characterized by the quantum numbers l, m_l, s, and m_s, where s_z is the component of $\vec{s}$ on the space-fixed axis and m_s is the quantum number associated with that component. For certain treatments of or approximations to the scattering problem, the latter representation may prove more useful. However, since rotational invariance implies

$$[\hat{H}, \hat{J}^2] = 0 \quad \text{and} \quad [\hat{H}, \hat{J}_z] = 0$$

where a hat (caret) over a symbol denotes a quantum-mechanical operator and z is the space-fixed axis, the scattering matrix is diagonal in J and M and independent of M.[1b,7b,10a] Thus generally a representation in which J

---

*Throughout the chapter, angular momentum is measured in units of $\hbar$.

and M are included among the quantum numbers will be most convenient. This property of a block diagonal scattering matrix is the general reason for preferring a total-angular-momentum representation.

The scattering of an elementary particle without spin by a rigid diatomic rotator without spin was treated by Arthurs and Dalgarno,[12] Micha,[13] and others[14] in terms of the quantum numbers J, M, Π, l, and s, where s for this problem is the rotational angular momentum quantum number of the diatom. The scattering of an elementary particle without spin from a diatomic rotator-vibrator may be treated[15] in terms of the quantum numbers J, M, Π, l, s, and n, where s is again the internal angular momentum of the target and n is the vibrational quantum number. These problems may also be treated in other total-angular-momentum representations[16] and in representations in which J is not a quantum number.[17] In fact, the last-named type of representation sometimes provides a simplification in approximate calculations. In this chapter we consider only the total-angular-momentum representations.

The scattering of an elementary particle whose spin is neglected from a rigid diatomic rotator with spin $\vec{f}$ and angular momentum $\vec{k}$ of rotation of the molecular framework may be treated[18] in terms of the quantum numbers J, M, Π, l, s, f, and k, where

$$\vec{s} = \vec{k} + \vec{f}$$

In this case s is again the total-internal-angular-momentum quantum number.

For electron-atom scattering by low-atomic-number atoms at nonrelativistic energies, it is common to neglect spin-orbit coupling. Then the total orbital angular momentum $\vec{L}$ and the total spin angular momentum* $\vec{S}$ are separately conserved. In this case spin may be removed from the problem except for its role in enforcing a specific permutational symmetry for the spatial part of the electronic wavefunction. Then $\vec{L}$ effectively plays the role of total angular momentum and it is most convenient[9a] to include it among the good quantum numbers labeling the representation.

Further examples of total-angular-momentum representations may be found in the theory of atom-atom collisions with transitions among hyperfine levels,[19] in other atomic collisions,[20] and in Section V.

III. TIME REVERSAL

The operation commonly called "time reversal" might more logically be named "motion reversal." It consists not in a true reversal of the direction

---

*Total spin $\vec{S}$ and the quantum number S and operator $\hat{\vec{S}}$ for total spin should not be confused with the scattering operator $\hat{S}$ for which we use the same capital letter.

of time (which would be unphysical), but in a reversal of all velocities and spins, with spatial coordinates being left invariant (a more precise definition is given in Section III.A). In deference to tradition, however, we use the term "time reversal" for this operation, denoting it by $\hat{T}$. Since we shall have occasion to make use of some of the properties of $\hat{T}$, and since these are not widely understood among chemists, we present in this section a self-contained treatment carrying the theory of time reversal far enough to derive the results we shall require. For more details, the reader is referred to the excellent treatment of Wigner.[21]

A. Symmetry Operators

We define a symmetry operator $\hat{N}$ as any operator that satisfies, for all $|q\rangle$, $|r\rangle$,

$$|\langle \hat{N}q | \hat{N}r \rangle| = |\langle q | r \rangle| \tag{2}$$

and the distributive law

$$\hat{N}(|q\rangle + |r\rangle) = N|q\rangle + N|r\rangle \tag{3}$$

and has an inverse $\hat{N}^{-1}$ which is also a symmetry operator. Gottfried[22] has given a discussion of how (2) corresponds to our physical ideas of a symmetry operator. Note that we cannot use any simple physical argument involving symmetry to put restrictions on the phases of the two sides of (2). This is because $|r\rangle$ and $e^{i\alpha}|r\rangle$ represent the same physical state.[22a]

The time-reversal operator $\hat{T}$ is defined as the symmetry operator that reverses the signs of all velocities and spins, while leaving spatial coordinates unaltered. In the absence of external magnetic fields, this is equivalent to the reversal of all linear and angular momenta (including spin), while leaving coordinates unchanged. We always assume in this chapter that no external magnetic field is present, which means that $\hat{T}$ may be defined as reversing momenta and spin. This does not prevent us from including the effect of magnetic fields generated by the orbital or spin motion of the particles of the system under study. In the presence of an external magnetic field, there does not appear to be a well-defined operator that reverses velocities while leaving the external field unaltered.

The rest of this subsection is a proof of Wigner's theorem:[22b,23] all symmetry operators must be unitary or antiunitary (defined below).

To explore the consequences of properties (2) and (3), let $|u\rangle$, $|v\rangle$, etc., be a complete orthonormal set of state vectors with

$$\langle u | v \rangle = \delta_{uv} \tag{4}$$

and let $\hat{N}$ be some symmetry operator. It follows immediately from (2), plus the positive definiteness of $\langle q|q \rangle$, that $|\hat{N}u\rangle, \ldots$, are also orthonormal:

$$\langle \hat{N}u|\hat{N}v\rangle = \delta_{uv} \tag{5}$$

The $|\hat{N}u\rangle$ are also complete. For if $|w\rangle$ is orthogonal to them all, then $|\hat{N}^{-1}w\rangle$ is orthogonal to all the $|u\rangle$, contrary to the hypothesis that the $|u\rangle$ are complete.

Now select one of the $|u\rangle$ arbitrarily, and consider the action of the symmetry operator $\hat{N}$ on the vector $(a|u\rangle)$, where a is a complex number. Because $\langle v|au \rangle = a\delta_{uv}$ and because of (2), we must have

$$|\langle \hat{N}v|\hat{N}au\rangle| = |a|\delta_{uv} \tag{6}$$

From the completeness and orthonormality of the $|\hat{N}v\rangle$, it follows that

$$\hat{N}a|u\rangle = a'|\hat{N}u\rangle \quad \text{with} \quad |a'| = |a| \tag{7}$$

We next consider an arbitrary vector $|r\rangle$ expanded in terms of the $|u\rangle$:

$$|r\rangle = \Sigma r_u |u\rangle$$

Because of (3) and (7), we have

$$|\hat{N}r\rangle = \Sigma r_u'|\hat{N}u\rangle, \qquad |r_u'| = |r_u| \tag{8}$$

Now for arbitrary $|v\rangle$, $|w\rangle$ from the complete set, we define the vector

$$|f_{vw}\rangle = |v\rangle + |w\rangle$$

It follows from (3) that

$$|\hat{N}f_{vw}\rangle = |\hat{N}v\rangle + |\hat{N}w\rangle \tag{9}$$

Now, making use of (9), (5), (8), (2), and (4), we find

$$|\langle \hat{N}f_{vw}|\hat{N}r\rangle| = |r_v' + r_w'| = |\langle f_{vw}|r\rangle| = |r_v + r_w| \tag{10}$$

It is evident that (8) and (10) can be satisfied only if the relative phase of r_v' and r_w' is the same in absolute value as that of r_v and r_w. Furthermore, this must remain true when r_v and r_w are varied independently, and for all choices of $|v\rangle$ and $|w\rangle$. There are thus two possibilities: either the relative phase remains the same, in which case

$$r_u' = e^{i\phi}r_u \tag{11}$$

or the relative phase changes sign, leading to

$$r'_u = e^{i\phi} r_u^*$$ (12)

In both cases (11) and (12), the common phase factor ϕ must be the same for all $|u\rangle$. Applying (11) or (12) to the case $r_u = 1$, moreover, we find

$$\hat{N}|u\rangle \equiv |\hat{N}u\rangle = e^{i\phi}|\hat{N}u\rangle$$

from which it follows that $\phi = 0$.

We are thus left with just two possibilities. If (11) holds, we have

$$\hat{N}a|u\rangle = a|\hat{N}u\rangle$$

from which and from (3) it follows by expanding arbitrary $|q\rangle$, $|r\rangle$ in the $|u\rangle$ that

$$\hat{N}(a|q\rangle + b|r\rangle) = a|\hat{N}q\rangle + b|\hat{N}r\rangle$$ (13)

and

$$\langle \hat{N}q|\hat{N}r\rangle = \langle q|r\rangle$$ (14)

In this case, the symmetry operator $\hat{N}$ is called unitary, and, in accordance with (13), linear.

On the other hand, if (12) is obeyed, we have

$$\hat{N}a|u\rangle = a^*|\hat{N}u\rangle$$

$$\hat{N}(a|q\rangle + b|r\rangle) = a^*|\hat{N}q\rangle + b^*|\hat{N}r\rangle$$ (15)

$$\langle \hat{N}q|\hat{N}r\rangle = \langle q|r\rangle^*$$ (16)

In this case, the operator is said to be antiunitary (and antilinear). This completes the proof of Wigner's theorem.*

The familiar symmetry operations such as rotations and parity are unitary.[22b] In the Section III.B we show that $\hat{T}$ is antiunitary.

B. Antiunitary Nature of the Time-Reversal Operator

In the absence of external fields producing velocity- or spin-dependent forces (of which the only important example is the magnetic field), the

---

*If one does not initially assume the distributive law (3), the proof of Wigner's theorem is slightly more difficult and the result itself slightly weaker.[22b] Without assuming the distributive law, it can be shown that an appropriate and permissible phase convention leads to all our results being true, but other phase conventions are possible in which the distributive law is not obeyed and in which the phase angle ϕ in (11) or (12) is not necessarily zero.

operation of time reversal does not change the energy of chemical systems. Thus, for such a system, if

$$\hat{H}|j\rangle = \hbar w_j |j\rangle$$

then also

$$\hat{H}|\hat{T}j\rangle = \hbar w_j |\hat{T}j\rangle$$

We will say that such a system has a time-reversal-invariant hamiltonian. Since we are using $\hat{T}$ only in such cases, and since $\hat{T}$ is not necessarily well defined otherwise, we assume this property from now on.

Now consider a simple example of a system with a time-reversal-invariant hamiltonian: a classical free particle, moving in one dimension. At time zero, let the particle be at the origin, moving with velocity v, and consider the effect of two different sequences of operations. In the first sequence, we apply $\hat{T}$ (changing v to $-v$), and then let the particle travel for a (positive) time t. The result is that the particle is at the point $(-vt)$ and has velocity $-v$. In the second sequence, we go back in time to $(-t)$, and then apply $\hat{T}$. Again, the result is that the particle is at $(-vt)$ and has velocity $-v$. In other words, time reversal followed by propagation *forward* by t has the same effect as propagation *backward* by t followed by time reversal. A little reflection shows that this is a property of all time-reversal-invariant systems, i.e., of all systems with time-reversal-invariant hamiltonians.

Quantum mechanically, this requirement is expressed as follows: for any initial state $|q\rangle$, we have

$$\hat{G}(t)\hat{T}|q\rangle = e^{i\phi}\hat{T}\hat{G}(-t)|q\rangle \tag{17}$$

where ϕ is real and $\hat{G}(t) = \exp[-(i/\hbar)\hat{H}t]$ is the time-displacement operator. We now expand $|q\rangle$ in the eigenfunctions $|j\rangle$ of $\hat{H}$, $|q\rangle = \Sigma_j q_j |j\rangle$ and see what requirements (17) puts on $\hat{T}$. First, under the assumption that $\hat{T}$ is unitary, we find, using the rules of Section III.A,

$$\hat{G}(t)\hat{T}|q\rangle = \Sigma_j q_j e^{-iw_j t}|\hat{T}j\rangle$$

$$\hat{T}\hat{G}(-t)|q\rangle = \Sigma_j q_j e^{iw_j t}|\hat{T}j\rangle$$

which is evidently not compatible with (17).

On the other hand, assuming $\hat{T}$ to be antiunitary, we find

$$\hat{G}(t)\hat{T}|q\rangle = \Sigma_j q_j^* e^{-iw_j t}|\hat{T}j\rangle$$

$$\hat{T}\hat{G}(-t)|q\rangle = \Sigma_j q_j^* e^{-iw_j t}|\hat{T}j\rangle$$

which satisfies (17) and even satisfies

$$\hat{G}(t)\hat{T}|q\rangle = \hat{T}\hat{G}(-t)|q\rangle$$

We conclude, then, that $\hat{T}$ *must* be defined as an antiunitary operator.

C. PROPERTIES OF ANTIUNITARY OPERATORS

1. General

We use ψ to denote the set of (perhaps continuous) coefficients in the expansion of an arbitrary state vector in some complete set. We also define the operator $\overline{K}$, which replaces ψ by its complex conjugate: $\overline{K}\psi = \psi^*$.

We put a bar over K instead of a hat for the purpose of emphasizing that $\overline{K}$ as defined here has no invariant physical significance and produces in general different physical effects in different representations. For example, in the ordinary Schrödinger configuration space representation for a single particle, the complete set is just that of configuration space delta functions, and ψ is just the usual wavefunction $\psi(\vec{r})$. Application of $\overline{K}$ in this representation yields $\psi^*(\hat{r})$. On the other hand, if we change representations by expanding $\psi(\vec{r}) = \Sigma a_j \phi_j(\hat{r})$, where the ϕ_j are some complete orthonormal set, then application of $\overline{K}$ in the new representation gives $\Sigma a_j^* \phi_j(\vec{r})$, NOT $\Sigma a_j^* \phi_j^*(\vec{r})$. This is, of course, physically a different state unless the ϕ_j are all real. We therefore denote $\overline{K}$ with a bar, reserving the hat for operators with an invariant physical meaning.

In view of the foregoing discussion, we can give a meaning to an expression such as $|\overline{K}r\rangle$ only if we also specify the representation in which the ket $|r\rangle$ is being expanded. With this understanding, however, it is evident that the $\overline{K}$ associated with any particular representation is a symmetry operator, that it is antiunitary, and that $\overline{K}^2 = 1$.

Now, with a representation specified, let $\hat{W}$ be an arbitrary antiunitary operator, and consider the product $\hat{W}\overline{K}$. We find using (16) that

$$\langle \hat{W}\overline{K}q | \hat{W}\overline{K}r \rangle = \langle \hat{W}(\overline{K}q) | \hat{W}(\overline{K}r) \rangle = \langle \overline{K}q | \overline{K}r \rangle^* = \langle q | r \rangle$$

It follows that $\hat{W}\overline{K}$ for the given representation is unitary. We denote the matrix of $\hat{W}\overline{K}$ in the specified representation by β where the bold face letter denotes a matrix. Since $\overline{K}^2 = 1$, we have, in the given representation

$$\hat{W} = \tilde{\beta}\overline{K} \tag{18}$$

where $\tilde{\beta}$ is the operator in the given representation whose matrix is β.

The matrix β is unitary in the sense that $\beta\beta^\dagger = 1$. It does not, however, possess the same transformation properties as matrices of unitary operators such as those for rotations of coordinate systems. To investigate the transformation properties of β, we let the state vectors undergo a unitary transformation to a new representation, denoted by a prime, with the matrix of the transformation denoted by $\mathbf{U}$:

$$\psi' = \mathbf{U}\psi$$

The transformation of $\hat{W}$ must be such that $\hat{W}'\psi' = (\hat{W}\psi)'$, which, with the aid of (18), becomes

$$\beta'\bar{K}\mathbf{U}\psi = \mathbf{U}\beta\bar{K}\psi$$

$$\beta'\mathbf{U}^*\psi^* = \mathbf{U}\beta\psi^*$$

Since this must hold for all ψ, it follows that

$$\beta'\mathbf{U}^* = \mathbf{U}\beta$$

$$\beta' = \mathbf{U}\beta\mathbf{U}^T \tag{19}$$

where $\mathbf{U}^T$ denotes the transpose of $\mathbf{U}$. Note that (19) is different from the transformation law for the matrices $\mathbf{A}$ representing physical quantities, which obey $\mathbf{A}' = \mathbf{U}\mathbf{A}\mathbf{U}^\dagger$, where $\mathbf{U}^\dagger$ denotes the hermitian conjugate. It is because β has these unusual transformation properties that we denote the operator for it in a given representation as $\tilde{\beta}$ rather than $\hat{\beta}$ or $\bar{\beta}$. Equations involving $\tilde{\beta}$ and $\bar{K}$ are often true only in given representations, although we do not repeat this warning before each such equation.

2. Involutional Antiunitary Operators

Following Wigner,[21] we call a symmetry operator an involution if, when applied twice successively, it reproduces the original physical situation. It is evident that $\hat{T}$ is an involutional antiunitary operator. We now proceed to investigate some properties of such operators.

Let $\hat{\Theta}$ be antiunitary and involutional. It follows from (18), and from the definition of an involution, that

$$\hat{\Theta}^2 = \tilde{\beta}\bar{K}\tilde{\beta}\bar{K} = e^{i\eta}\hat{1} \tag{20}$$

where η is a real number. Applying (20) to a state function, we find

$$\hat{\Theta}^2\psi = e^{i\eta}\psi = \tilde{\beta}\bar{K}\tilde{\beta}\bar{K}\psi = \tilde{\beta}\bar{K}\tilde{\beta}\psi^* = \tilde{\beta}\tilde{\beta}^*\psi$$

It follows that

$$\tilde{\beta}\tilde{\beta}^* = e^{i\eta}$$

$$\tilde{\beta} = e^{i\eta}\tilde{\beta}^T \tag{21}$$

The transpose of (21) is

$$\tilde{\beta}^T = e^{i\eta}\tilde{\beta} \tag{22}$$

Inserting (22) into (21), we find

$$\tilde{\beta} = e^{2i\eta}\tilde{\beta}$$

from which follows $e^{i\eta} = \pm 1$. Accordingly, there are two possibilities for the square of an involutional antiunitary operator:

$$\hat{\Theta}^2 = \tilde{\beta}\tilde{\beta}^* = \pm\hat{1} \tag{23}$$

Note that (23) is not the result of normalization, as would be the case with a unitary operator; that is, a unitary operator $\hat{U}$ may be replaced by $\hat{U}' = e^{i\phi}\hat{U}$ without changing any physical properties, and in such a case $\hat{U}^2$ is replaced by $(\hat{U}')^2 = e^{2i\phi}\hat{U}^2$. If one does the same thing with an antiunitary operator $\hat{W}$, one finds

$$\left(\hat{W}'\right)^2 = e^{i\phi}\hat{W}e^{i\phi}\hat{W} = e^{i\phi}e^{-i\phi}\hat{W}^2 = \hat{W}^2$$

When an involutional antiunitary operator is applied twice, any physical quantity returns to its original value. Accordingly, the physical quantities to be considered arc divided into two classes according to whether they are left unchanged [class I ($\hat{\Theta}$)] or change sign [class II ($\hat{\Theta}$)] on application of an involutionary antiunitary operator $\hat{\Theta}$. By symmetrizing and antisymmetrizing, a variable that is neither class I ($\hat{\Theta}$) nor class II ($\hat{\Theta}$) may be expressed as a sum of class I ($\hat{\Theta}$) and a class II ($\hat{\Theta}$) variable, and we lose no generality by confining our attention to these two classes. We now investigate the commutation properties with $\hat{\Theta}$ of variables of these two classes. If $\hat{A}$ is a physical quantity that is class I ($\hat{\Theta}$), this means the complete set of eigenvectors $|j\rangle$ of $\hat{A}$ where

$$\hat{A}|j\rangle = a_j|j\rangle$$

are such that

$$\hat{A}|\hat{\Theta}j\rangle = a_j|\hat{\Theta}j\rangle$$

Expanding an arbitrary vector $|r\rangle$ as

$$|r\rangle = \Sigma_j r_j|j\rangle$$

we find

$$\hat{\Theta}\hat{A}|r\rangle = \hat{\Theta}\Sigma_j r_j a_j|j\rangle = \Sigma_j r_j^* a_j|\hat{\Theta}j\rangle$$

$$\hat{A}\hat{\Theta}|r\rangle = \hat{A}\Sigma_j r_j^*|\hat{\Theta}j\rangle = \Sigma_j r_j^* a_j|\hat{\Theta}j\rangle$$

Since $|r\rangle$ was arbitrary, we conclude that the operator for the variable A which is class I ($\hat{\Theta}$) must satisfy

$$\hat{A}\hat{\Theta}=\hat{\Theta}\hat{A} \tag{24}$$

Note that for systems with time-reversal-invariant hamiltonians (see Section III.B), $\hat{H}$ is class I ($\hat{T}$); thus it commutes with $\hat{T}$.[1c,8b] If we write $\hat{\Theta}$ as $\tilde{\beta}\bar{K}$ [see (18)], (24) becomes

$$\hat{A}\tilde{\beta}=\tilde{\beta}\hat{A}^* \tag{25}$$

For a variable B that is class II ($\hat{\Theta}$) and has a complete set of eigenvectors $|j\rangle$ where

$$\hat{B}|j\rangle=b_j|j\rangle$$

and

$$\hat{B}|\hat{\Theta}j\rangle=-b_j|\hat{\Theta}j\rangle$$

we find, proceeding as before,

$$\hat{B}\hat{\Theta}=-\hat{\Theta}\hat{B} \tag{26}$$

$$\hat{B}\tilde{\beta}=-\tilde{\beta}\hat{B}^* \tag{27}$$

All class I ($\hat{\Theta}$) variables, then, commute with $\hat{\Theta}$ and satisfy the equivalent equations (24) and (25), whereas class II ($\hat{\Theta}$) variables anticommute with $\hat{\Theta}$ and satisfy (26) and (27).

It is quite easy to show that (24) is sufficient as well as necessary for a hermitian operator $\hat{A}$ to be class I ($\hat{\Theta}$): for if (24) holds and

$$\hat{A}|j\rangle=a_j|j\rangle$$

we have

$$\hat{A}\hat{\Theta}|j\rangle=\hat{\Theta}\hat{A}|j\rangle=\hat{\Theta}a_j|j\rangle=a_j|\hat{\Theta}j\rangle$$

where the last step follows because a_j is real. Thus the physical quantity $\hat{A}$ is unchanged on application of $\hat{\Theta}$, which is what was to be proved. An entirely analogous argument shows that (26) is sufficient as well as necessary for a hermitian operator $\hat{B}$ to be class II ($\hat{\Theta}$).

3. Eigenstates of Involutional Antiunitary Operators

We next consider whether a complete set of states $|u\rangle$ may be taken to be eigenstates of $\hat{\Theta}$ so that $|u\rangle=e^{i\phi}|\hat{\Theta}u\rangle$ i.e., $|u\rangle$ and $|\hat{\Theta}u\rangle$ are physically the same state. There are two cases, corresponding to $\hat{\Theta}^2 = \pm 1$.

If $\hat{\Theta}^2 = +1$, define the states

$$|u_+\rangle = |u\rangle + |\hat{\Theta}u\rangle \tag{28}$$

$$|u_-\rangle = |u\rangle - |\hat{\Theta}u\rangle \tag{29}$$

Applying $\hat{\Theta}$, we find

$$\hat{\Theta}|u_+\rangle = |\hat{\Theta}u\rangle + \hat{\Theta}^2|u\rangle = |\hat{\Theta}u\rangle + |u\rangle = |u_+\rangle$$

A similar calculation gives

$$\hat{\Theta}|u_-\rangle = -|u_-\rangle$$

In this case, therefore, the states can be chosen to be eigenvectors of $\hat{\Theta}$. The foregoing construction gives the eigenvalues ± 1, but in actuality the phase of the eigenvalue turns out to be arbitrary. To see this, we first verify that the absolute value of the eigenvalue must be unity. If

$$\hat{\Theta}|u\rangle = c|u\rangle$$

then $\hat{\Theta}^2|u\rangle = |u\rangle = \hat{\Theta}c|u\rangle = c^*\hat{\Theta}|u\rangle = c^*c|u\rangle$, from which follows $c^*c = 1$. Now suppose that

$$\hat{\Theta}|r\rangle = e^{i\lambda}|r\rangle$$

with λ real; then $e^{i\eta}|r\rangle$, with η real, is an eigenvector of $\hat{\Theta}$ with eigenvalue $e^{i(\lambda - 2\eta)}$, since

$$\hat{\Theta}e^{i\eta}|r\rangle = e^{-i\eta}\hat{\Theta}|r\rangle = e^{i(\lambda - \eta)}|r\rangle = e^{i(\lambda - 2\eta)}e^{i\eta}|r\rangle$$

If $\hat{\Theta}^2 = -1$, however, the situation is different. Let us try to construct an eigenvector of $\hat{\Theta}$ in this case. If we postulate the existence of an eigenvector $|q\rangle$, with

$$\hat{\Theta}|q\rangle = c|q\rangle$$

we find

$$\hat{\Theta}^2|q\rangle = -|q\rangle = \hat{\Theta}c|q\rangle = c^*\hat{\Theta}|q\rangle = c^*c|q\rangle$$

Since we can never have $c^*c = -1$, this is a contradiction, and it follows that the postulated state cannot exist. In this case, therefore, $|\hat{\Theta}u\rangle$ is always physically different from $|u\rangle$.

It can also be shown for the case $\hat{\Theta}^2 = -1$ that $|\hat{\Theta}q\rangle$ is always orthogonal to $|q\rangle$. To show this, we presume the contrary—that is, for some $|q\rangle$, assumed normalized to unity, we have

$$|\hat{\Theta}q\rangle = c|q\rangle + |r\rangle \tag{30}$$

where $|r\rangle$ is a ket orthogonal to $|q\rangle$. Since $|\hat{\Theta}q\rangle$ must also be normalized, we have

$$cc^* + \langle r|r\rangle = 1 \tag{31}$$

Applying $\hat{\Theta}$ to both sides of (30), we find

$$\hat{\Theta}^2|q\rangle = -|q\rangle = c^*|\hat{\Theta}q\rangle + |\hat{\Theta}r\rangle = c^*c|q\rangle + c^*|r\rangle + |\hat{\Theta}r\rangle$$

which can be solved for $|\hat{\Theta}r\rangle$ to give

$$|\hat{\Theta}r\rangle = -(1 + c^*c)|q\rangle - c^*|r\rangle$$

Taking the norms of both sides, we obtain

$$\langle \hat{\Theta}r|\hat{\Theta}r\rangle = \langle r|r\rangle = (1 + c^*c)^2 + c^*c$$

which contradicts (31) unless $c = 0$. We conclude, then, that when $\hat{\Theta}^2 = -1$, we always have $\langle q|\hat{\Theta}q\rangle = 0$.

D. Properties of the Time-Reversal Operator

1. Explicit Form

To determine explicitly the form of the time-reversal operator $\hat{T}$ for a system of particles of arbitrary spin, we proceed as follows: we classify all variables as class I ($\hat{T}$) or class II ($\hat{T}$) according to whether they are left unchanged or change sign, respectively, on application of $\hat{T}$. According to (18), the determination of the explicit form of $\hat{T}$ reduces to that of $\tilde{\beta}$; also, if β is determined in one representation, we can use (19) to determine it in all others. We choose the Schrödinger coordinate space representation, in which the operator for a coordinate x is simply multiplication by x, and that for p_x is $(\hbar/i)(\partial/\partial x)$.

All coordinates are class I($\hat{T}$), so, according to (25), since $\hat{x}^* = \hat{x}$, $\tilde{\beta}$ must commute with all of them. The momenta are class II($\hat{T}$), and $\hat{p}^* = -\hat{p}$, so it follows from (27) that $\tilde{\beta}$ also commutes with all the momenta. We conclude that $\tilde{\beta}$ must operate solely on the spin variables and that it is simply the unit operator if the particles are spinless. It is important to note, however, that even the unit operator is not invariant under (19); thus it does *not* follow that $\tilde{\beta} = 1$ in all representations for spinless particles.

We illustrate the transformation properties of the unitary operator $\tilde{\beta}$ by calculating its form in the momentum representation for a spinless particle (in one dimension) by two methods: by direct inspection, and by means of the transformation law (19).

In this representation, we have $\hat{p}=p=\hat{p}^*$, and $\hat{x}=i\hbar(d/dp)=-\hat{x}^*$. Since $\hat{x}$ is class I $(\hat{T})$, we must have

$$\hat{x}\tilde{\beta}=\tilde{\beta}\hat{x}^*=-\tilde{\beta}\hat{x}$$

and, since $\hat{p}$ is class II $(\hat{T})$,

$$\hat{p}\tilde{\beta}=-\tilde{\beta}\hat{p}^*=-\tilde{\beta}\hat{p}$$

Thus $\tilde{\beta}$ must anticommute with both $\hat{x}$ and $\hat{p}$. It is easily verified that these conditions are satisfied by the operator $\tilde{Q}$ defined as the operator that changes p to $-p$. For, with this assumption, we have

$$\hat{x}\tilde{\beta}\phi(p)=i\hbar\frac{d}{dp}\phi(-p)=-i\hbar\frac{d\phi}{dp}(-p)$$

$$\tilde{\beta}\hat{x}\phi(p)=\tilde{\beta}i\hbar\frac{d}{dp}\phi(p)=i\hbar\frac{d\phi}{dp}(-p)=-i\hbar\frac{d}{dp}\phi(-p)$$

$$\hat{p}\tilde{\beta}\phi(p)=\hat{p}\phi(-p)$$

$$\tilde{\beta}\hat{p}\phi(p)=-\hat{p}\phi(-p)$$

Also, if $\tilde{\alpha}$ is some other unitary operator anticommuting with both $\hat{x}$ and $\hat{p}$, we have

$$\tilde{\alpha}\tilde{\beta}\hat{x}=-\tilde{\alpha}\hat{x}\tilde{\beta}=\hat{x}\tilde{\alpha}\tilde{\beta}$$

that is, $(\tilde{\alpha}\tilde{\beta})$ commutes with $\hat{x}$ and, similarly, with $\hat{p}$ as well. It follows that $(\tilde{\alpha}\tilde{\beta})$ is just a constant times the unit operator and that $\tilde{\alpha}$ is a constant times $\tilde{\beta}^{-1}$. By (23), since the upper sign applies in the present case $\tilde{\beta}^{-1}=\tilde{\beta}^*$. Since $\tilde{\beta}$ is real in the present case, this argument shows that $\tilde{\alpha}$ is a constant times $\tilde{\beta}$. Thus our assumed form for $\tilde{\beta}$ is, apart from a constant phase factor, the only possible one.

If we denote the matrix for the $\tilde{\beta}$ operator in the Schrödinger representation by $\boldsymbol{\beta}_x=1$, then according to (19), the matrix for $\tilde{\beta}$ in the momentum representation is given by

$$\boldsymbol{\beta}_p=\mathbf{U}\mathbf{U}^T$$

where in this case

$$\langle p|\hat{U}|x\rangle=\langle x|\hat{U}^T|p\rangle=\hbar^{-1/2}e^{-ipx/\hbar}$$

We thus find

$$\langle p|\tilde{\beta}|p'\rangle = \int_{-\infty}^{\infty} \langle p|\hat{U}|x\rangle dx \langle x|\hat{U}^T|p'\rangle = \delta(p+p')$$

and thus

$$\tilde{\beta}\phi(p) = \int \langle p|\tilde{\beta}|p'\rangle\phi(p')dp' = \phi(-p)$$

which verifies that the two approaches give the same result, as they should.

To determine the spin part of $\tilde{\beta}$, we first consider a single spin-$\frac{1}{2}$-particle. All three components of spin are class II ($\hat{T}$), and letting σ_x, σ_y, σ_z represent the usual Pauli matrices,[5a,6a,22c,23a] we have

$$\sigma_x^* = \sigma_x; \qquad \sigma_y^* = -\sigma_y; \qquad \sigma_z^* = \sigma_z$$

According to (27), therefore, β must anticommute with σ_x and σ_z and commute with σ_y. This is satisfied by

$$\beta = e^{i\phi}\sigma_y = e^{i\phi}\begin{pmatrix} 0 & -i \\ i & 0 \end{pmatrix}$$

where ϕ is any real number. This equation does not single out the y direction as having special properties. The spin component s_y is a physical quantity, its representation is $\frac{1}{2}\sigma_y$ only in the usual representation (see Section III.F.2), and it transforms according to $s_y' = Us_y U^\dagger$ when we go to another representation, whereas β transforms according to (19). Thus $\beta = e^{i\phi}(2s_y)$ holds only in one particular representation. Using (19) it is easy to show that

$$\beta' = Ue^{i\phi}\begin{pmatrix} 0 & -i \\ i & 0 \end{pmatrix}U^T = (\det U)e^{i\phi}\begin{pmatrix} 0 & -i \\ i & 0 \end{pmatrix}$$

Since $|\det U| = 1$ for any unitary matrix U, this yields for all representations for a spin-$\frac{1}{2}$-particle[24]

$$\beta' = e^{i\eta}\sigma_y$$

with η real.

If there are several spin-$\frac{1}{2}$-particles, the spin components of them all are separately class II ($\hat{T}$), and β becomes a direct product of the σ_y matrices for all the particles. If there are particles with spin greater than $\frac{1}{2}$, they may be regarded as built up from spin-$\frac{1}{2}$-particles, and β constructed accordingly. The treatment of Section III.F makes this more explicit.

With our explicit form of β, we can determine the sign of $\hat{T}^2$ by using (23). The space part of β clearly contributes just $+1$. For each spin-$\frac{1}{2}$-

particle, the spin part, according to (23), contributes a factor

$$\beta\beta^* = \sigma_y\sigma_y^* = -\sigma_y^2 = -1$$

It follows immediately that the sign of $\hat{T}^2$ is positive or negative according to whether the total spin is integral or half-odd integral.

It is also of interest to consider the behavior of $\tilde{\beta}$ under a rotation of coordinates. The operator for an infinitesimal rotation is*

$$\hat{U} = \hat{1} + i\vec{e}\cdot\hat{\vec{J}}$$

where the direction of $\vec{e}$ gives that of the axis of rotation and its magnitude (assumed infinitesimal), the angle of rotation; $\hat{\vec{J}}$ is the vector total-angular-momentum operator, in an arbitrary coordinate system. Since $\hat{\vec{J}}$ is hermitian and $\vec{e}$ real, we have

$$\hat{U}^T = \hat{1} + i\vec{e}\cdot\hat{\vec{J}}^*$$

Also, since $\vec{J}$ is class II ($\hat{T}$), (27) yields

$$\hat{\vec{J}}\tilde{\beta} = -\tilde{\beta}\hat{\vec{J}}^*$$

The effect of the transformation on $\tilde{\beta}$, through the first order in $\vec{e}$, is given by

$$\tilde{\beta}' = \hat{U}\tilde{\beta}\hat{U}^T = \tilde{\beta} + i\vec{e}\cdot\left(\hat{\vec{J}}\tilde{\beta} + \tilde{\beta}\hat{\vec{J}}^*\right) = \tilde{\beta}$$

In other words, $\tilde{\beta}$ is not changed by a rotation of coordinates. If, therefore, the matrix form of $\tilde{\beta}$ has been worked out with respect to one Cartesian coordinate system, one can use the same form for all such coordinate systems.

Finally we note that β is changed by a unitary transformation which just multiplies by a phase factor: If $U = e^{i\phi}1$, then

$$U\beta U^T = e^{2i\phi}\beta$$

2. Kramers Degeneracy

If the hamiltonian is time-reversal-invariant (see Section III.B), it follows that each energy eigenstate $|j\rangle$ has the same energy as $|\hat{T}j\rangle$. If these eigenstates can be chosen also to be eigenstates of $\hat{T}$, so that $|j\rangle = e^{i\phi}|\hat{T}j\rangle$, (i.e., $|j\rangle$ and $|\hat{T}j\rangle$ are the same states), this does not introduce any new degeneracy. However, if $|\hat{T}j\rangle$ is necessarily physically different from $|j\rangle$,

---

*Compare the discussion at the beginning of Section III.F.1.

the time-reversal invariance brings about twofold degeneracy.

If $\hat{T}^2 = +1$, the result of Section III.C.3 is that the eigenstates of $\hat{H}$ can be chosen to be eigenvectors of $\hat{T}$. Thus consideration of $\hat{T}$ alone does not require degeneracy.

If $\hat{T}^2 = -1$, however, the result of Section III.C.3 is that $|\hat{T}j\rangle$ is always physically distinct from $|j\rangle$. Thus the time-reversal invariance introduces a twofold degeneracy known as "Kramers degeneracy."[5b,22d]

E. Minimization of the Number of Class II ($\hat{T}$) Variables in the Complete Set of Commuting Variables

Consider a system of particles with no external fields present, so that the energy eigenstates may be chosen to be eigenstates of total angular momentum $\vec{J}$ (quantum number J) and of its z-component J_z (quantum number M). Let d denote the set of scalar observables which, together with J and M, are sufficient to determine the state.* Since J_z is a class II ($\hat{T}$) variable and $\hat{J}^2$ is class I ($\hat{T}$), we certainly have

$$\hat{T}|d,J,M\rangle = e^{i\gamma(d,J,M)}|d_t,J,-M\rangle \tag{32}$$

where $\gamma(d,J,M)$ is real and d_t differs from d in that the signs of all class II ($\hat{T}$) variables have been reversed. The question arises of whether it is possible to choose d to consist only of class I ($\hat{T}$) variables, so that $d = d_t$ and J_z is the only class II ($\hat{T}$) variable used in the specification of a state.

To answer this question, we introduce the operator $\hat{R}_x$, a rotation through $180°$ about the x-axis, and $\hat{C}_x = \hat{R}_x \hat{T}$. Applying $\hat{R}_x$ to both sides of (32), we find, since the d are scalars,

$$\hat{C}_x|d,J,M\rangle = e^{i[\gamma(d,J,M)+q(d,J,-M)]}|d_t,J,M\rangle \tag{33}$$

where $q(d,J,-M)$ is real. The operator $\hat{C}_x$ is evidently antiunitary and involutional. We see from (33) that d_t can be chosen identical with d if and only if the eigenstates can be chosen to be eigenstates of $\hat{C}_x$; and this, according to the treatment of Section III.C.3, depends on whether $\hat{C}_x^2 = +1$ or -1. We now proceed to investigate this question.

The class I ($\hat{C}_x$) variables are x, p_y, p_z, s_y, s_z, but y, z, p_x, s_x are class II ($\hat{C}_x$). The space part of the $\tilde{B}$ operator [see (18)] for $\hat{C}_x$ in the Schrödinger coordinate space representation is just the usual $180°$ rotation, and contributes $+1$ to $\hat{C}_x^2$. To calculate the spin part $\tilde{B}_s$, we again consider a spin-$\frac{1}{2}$-particle, the usual representation of spin components in terms of the Pauli matrices, and use (27). We find this time that β_s must anticommute

---

*It is shown in Appendix A that the variables d can always be chosen to be scalars.

with σ_y and σ_x and commute with σ_z. This is satisfied by $\beta_s = \sigma_z$. Using (23) again, we find $\hat{C}_x^2 = \sigma_z \sigma_z^* = \sigma_z^2 = +1$.

Since $\hat{C}_x^2 = +1$, we can choose the eigenstates of $\hat{H}$ to also be eigenstates of $\hat{C}_x$, with any eigenvalues of magnitudes unity (see Section III.C.3). By (33) this means d includes no class II ($\hat{T}$) variables and a d that is specified not to include any class II ($\hat{T}$) variables will be called D. If we choose the eigenvalues to be $+1$, we have

$$\hat{C}_x|D,J,M\rangle = \hat{R}_x \hat{T}|D,J,M\rangle = |D,J,M\rangle \qquad (34)$$

The choice of phase which ensures that (34) is satisfied is derived in Section III.F.4. Applying $\hat{R}_x^{-1}$ to both sides of (34), we find

$$\hat{T}|D,J,M\rangle = \hat{R}_x^{-1}|D,J,M\rangle = e^{i\tau(D,J,M)}|D,J,-M\rangle$$

where $\tau(D,J,M)$ is real and is determined by the phase convention used in defining the states.

The conclusion of this subsection is that in the absence of external fields in systems with rotational invariance, it is always possible to choose one's quantum numbers in such a way that J_z is the only class II ($\hat{T}$) variable used. If this is done, the Kramers degeneracy in energy introduces no *new* degeneracy, since degeneracy between states with different quantum numbers M is already required by the rotational invariance. There may, of course, be a near-degeneracy such as Kramers degeneracy of the electronic state with fixed nuclei, which is only split by the interaction between electronic and nuclear motion.

It is also easily seen that the set of scalar quantum numbers D may be chosen to include the parity of the system. From the explicit form of $\hat{T}$ derived in Section III.D.1, it is evident that $\hat{T}$ does not change the parity [i.e., that parity is class I ($\hat{T}$)]. Being a rotation, $\hat{R}_x$ also leaves the parity unchanged. Thus $\hat{C}_x$ leaves the parity unchanged. Hence if the eigenstates of $\hat{C}_x$ are constructed according to the prescription of Section III.C.3, we only have to choose the state $|u\rangle$ in (28) and (29) to be an eigenstate of parity to ensure that the eigenstates of $\hat{C}_x$ are eigenstates of parity also.

It will be useful for later purposes to *summarize* the properties of $\hat{C}_x$ that we have established: $\hat{C}_x$ is involutional and antiunitary, and $\hat{C}_x^2 = +1$ always; $\hat{C}_x$ commutes with $\hat{J}^2$, $\hat{J}_z$, and the parity operator $\hat{P}$, as well as with the hamiltonian $\hat{H}$ in problems we will be considering. Hence we can always choose our complete set of states to be simultaneous eigenstates of $\hat{J}^2$, $\hat{J}_z$, $\hat{P}$, and $\hat{C}_x$. If this is done, the time-reversal operator $\hat{T}$ reverses the eigenvalue of $\hat{J}_z$, leaving all other quantum numbers unchanged. Conversely, if the only class II ($\hat{T}$) variable in our complete set of commuting variables is $\hat{J}_z$, our states are eigenstates of $\hat{C}_x$. It is also evident that all these properties hold equally well for the analogously defined operator $\hat{C}_y = \hat{R}_y \hat{T}$, in which the role of the x-axis in $\hat{C}_x$ is assumed by the y-axis. The utility of $\hat{C}_x$ and $\hat{C}_y$ in subsequent discussions is attributable to two conditions. First, since $\hat{C}_x^2 = +1$ and $\hat{C}_y^2 = +1$, we can always choose our complete set of states to be eigenstates of $\hat{C}_x$ or $\hat{C}_y$; but since $\hat{T}^2$ is sometimes -1, we cannot always choose our states as eigenstates of $\hat{T}$. Second, $\hat{C}_x$, $\hat{C}_y$, and $\hat{P}$ commute with $\hat{J}_z$ as compared, for example, with $\hat{R}_x$, $\hat{R}_y$, and $\hat{T}\hat{P}$, which each anticommute with $\hat{J}_z$.

F. Phase Conventions in Total-Angular-Momentum Representations

1. Preliminaries

The work of Section III.E, together with that of Appendix A, shows that it is always possible to choose a representation in which the states are characterized by the total-angular-momentum quantum numbers J and M, with the other quantum numbers being class I ($\hat{T}$) scalars. In such a representation, the $(2J+1)$-dimensional manifold spanned by the states $|D, J, M\rangle$ (D and J fixed, $M = -J, -J+1, \ldots, J$) is invariant under time reversal as well as under rotation, and in particular angular momentum, operators. It follows that the properties of these operators may be studied within such a subspace without reference to the rest of the Hilbert space. We now take up the question of the properties of the operators $\hat{T}$, $\hat{R}_x$, $\hat{C}_x$, the analogous operators $\hat{R}_y$ and $\hat{C}_y$, and the angular momentum operators themselves within such a manifold, as well as their relation to one another and to the phase convention used in defining the states. The states $|D, J, M\rangle$ are assumed to be normalized and, by definition, to be eigenstates of $\hat{J}^2$ and $\hat{J}_z$. This determines each state function only up to a multiplicative phase factor $e^{i\phi}$, where ϕ is real and may be chosen arbitrarily for each state. We want to concentrate on determining the effect of this choice of phase on the properties of $\hat{T}$, $\hat{C}_x$, and so on.

We begin with a brief remark on the phasing of the rotation operators themselves. The rotation of a physical system about a given axis through an angle Θ may be expressed quantum mechanically by means of the unitary operator[8c, 22e, 23b]

$$\hat{U}(\vec{\Theta}) = e^{-i\vec{\Theta}\cdot\hat{\vec{J}}}$$

where $\vec{\Theta}$ is a vector of magnitude Θ directed along the axis of rotation in the right-handed sense, and $\hat{\vec{J}}$ is the total-angular-momentum operator. As mentioned in the discussion following (23), however, the unitary operator $\hat{U}(\vec{\Theta})$ may be multiplied by an arbitrary phase factor $e^{i\phi(\vec{\Theta})}$, causing the $\hat{U}(\vec{\Theta})$ to be replaced by

$$\hat{U}_\phi(\vec{\Theta}) = e^{i\phi(\vec{\Theta})} e^{-i\vec{\Theta}\cdot\hat{\vec{J}}}$$

The physical effect of the $\hat{U}_\phi$ operators is the same as that of the $\hat{U}$; it is desirable, however, to choose the phases in such a way that the $\hat{U}_\phi(\vec{\Theta})$ form a representation (perhaps double-valued) of the three-dimensional rotation group. Since the $\hat{U}(\vec{\Theta})$ form a representation, the $\hat{U}_\phi$ will do so only if the phase factors $e^{i\phi(\vec{\Theta})}$ form a one-dimensional representation of the rotation group. Since the only such one-dimensional representation is the unit representation, we must set all phase factors equal to unity [i.e., represent all rotations by $\hat{U}(\vec{\Theta})$]. In conformity with universal usage, this is done throughout the chapter. Another desirable property of this convention is that the rotation of the *physical system* through an angle Θ is mathematically identical (including its effect on phases) to a rotation of the *coordinate system* about the same axis through $(-\Theta)$. The infinitesimal rotation of coordinates discussed in Section III.D.1 is thus carried out in the right-handed sense about the axis $\vec{e}$.

In the work to follow, we need certain commutation relations between $\hat{T}$, $\hat{C}_x$, etc. and the raising and lowering operators $\hat{J}_\pm = \hat{J}_x \pm i\hat{J}_y$. Since all components of angular momentum are class II ($\hat{T}$), we have, according to (26),

$$\hat{T}\hat{\vec{J}} = -\hat{\vec{J}}\hat{T} \tag{35}$$

Thus $\hat{T}$ anticommutes with all components of $\hat{\vec{J}}$, and, since it is antiunitary, also with the factor i. We therefore find

$$\hat{T}\hat{J}_\pm = \hat{T}\hat{J}_x \pm \hat{T}i\hat{J}_y = \hat{T}\hat{J}_x \mp i\hat{T}\hat{J}_y = -\hat{J}_x\hat{T} \pm i\hat{J}_y\hat{T} = -\hat{J}_\mp\hat{T} \tag{36}$$

The operator $\hat{C}_x$ is antiunitary (hence anticommutes with i) and reverses only J_x, leaving J_y and J_z invariant. According to (24) and (26), therefore, it commutes with $\hat{J}_y$ and anticommutes with $\hat{J}_x$. Hence we find

$$\hat{C}_x\hat{J}_\pm = \hat{C}_x\hat{J}_x \pm \hat{C}_x i\hat{J}_y = \hat{C}_x\hat{J}_x \mp i\hat{C}_x\hat{J}_y = -\hat{J}_x\hat{C}_x \mp i\hat{J}_y\hat{C}_x = -\hat{J}_\pm\hat{C}_x \tag{37}$$

The analogous operator $\hat{C}_y$ is defined as $\hat{C}_y = \hat{R}_y\hat{T}$, where $\hat{R}_y$ is a rotation through 180° about the y-axis. The operator $\hat{C}_y$ reverses J_y, leaves J_x and J_z invariant, and is antiunitary. It therefore commutes with $\hat{J}_x$ and anticommutes with i and $\hat{J}_y$. We find for $\hat{C}_y$, analogously to (37)

$$\hat{C}_y \hat{J}_\pm = \hat{J}_\pm \hat{C}_y \qquad (38)$$

We also need the analogous relations for the unitary rotation operators $\hat{R}_x$ and $\hat{R}_y$. To obtain these, we first note that the derivation of (24) and (26) does not depend on the antiunitary nature of the operator concerned; thus the results hold for unitary involutional operators as well. It follows that $\hat{R}_x$, which reverses J_z and J_y, leaving J_x invariant, must anticommute with $\hat{J}_z$, $\hat{J}_y$ and commute with $\hat{J}_x$ and (since it is unitary, not antiunitary), with numerical factors such as i. Accordingly, we obtain

$$\hat{R}_x \hat{J}_\pm = \hat{R}_x \hat{J}_x \pm \hat{R}_x i\hat{J}_y = \hat{J}_x \hat{R}_x \mp i\hat{J}_y \hat{R}_x = \hat{J}_\mp \hat{R}_x \qquad (39)$$

The analogous calculation for $\hat{R}_y$ yields

$$\hat{R}_y \hat{J}_\pm = -\hat{J}_\mp \hat{R}_y \qquad (40)$$

2. Normal Phase Conventions

It follows from the general theory of angular momentum that[6b,25]

$$|D,J,M+1\rangle = N_{JM} e^{i\phi(D,J,M)} \hat{J}_+ |D,J,M\rangle \qquad (41)$$

and

$$|D,J,M\rangle = N_{JM} e^{-i\phi(D,J,M)} \hat{J}_- |D,J,M+1\rangle \qquad (42)$$

where $N_{JM} = [(J-M)(J+M+1)]^{-1/2}$ is a real positive normalizing factor, and $\phi(D,J,M)$ is a (real) phase angle that may be chosen arbitrarily. [The identity of the phase angles in (41) and (42) follows because $|D,J,M\rangle$ is an eigenstate of $\hat{J}_-\hat{J}_+$ with a real, nonnegative eigenvalue.] We define a *normal* (n) phase convention as one in which all the $\phi(D,J,M)$ in (41) and (42) are zero. This determines the *relative* phases of the $|D,J,M\rangle$ with common D and J, leaving only a common phase factor free. In this subsection, we study the properties of the various operators under normal phase conventions, including several special cases.

Since the matrix representation of a *linear* operator depends only on the *relative* phases of the state vectors, the form of the angular momentum operators, and of all operators constructed from them, will be completely determined when one specifies a normal phase convention. For example, it is easily verified in the case of spin-$\frac{1}{2}$ that the components of spin angular momentum have their usual Pauli form $\vec{s} = \frac{1}{2}\vec{\sigma}$ if the phase convention is normal. The form of the time-reversal and other antiunitary operators will depend on the common phase factor still left unspecified.

If the phase convention is normal, we have

$$|D,J,M+1\rangle_n = N_{JM}\hat{J}_+|D,J,M\rangle_n$$

where the subscript n simply reminds us of the normality of the phase convention. The rotation $\hat{R}_x$ reverses J_z, hence we have

$$\hat{R}_x|D,J,M\rangle_n = e^{i\rho_M}|D,J,-M\rangle_n$$

where ρ_M is a real number to be determined. From (39) and (42) with $\phi(D,J,M)=0$, from the definition of N_{JM}, and from the last two equations, it follows that

$$\hat{R}_x|D,J,M+1\rangle_n = N_{JM}\hat{R}_x\hat{J}_+|D,J,M\rangle_n$$

$$= N_{JM}\hat{J}_-\hat{R}_x|D,J,M\rangle_n = N_{JM}\hat{J}_- e^{i\rho_M}|D,J,-M\rangle_n$$

$$= N_{JM}N_{J,-M-1}^{-1}e^{i\rho_M}|D,J,-M-1\rangle_n = e^{i\rho_M}|D,J,-M-1\rangle_n$$

It follows that all the ρ_M are equal, which means that we can drop the subscript M.

To determine ρ, we first specify $\hat{R}_x$ as a rotation through π in the right-handed sense, that is,

$$\hat{R}_x = e^{-i\pi\hat{J}_x}$$

Since the form of the operators $\hat{J}$ is determined by the quantum number J plus the normal phase convention, this choice will completely determine $\hat{R}_x$. It suffices, therefore, to determine ρ for any special case with total-angular-momentum quantum number J and normal phase convention. We choose the case in which a state of total angular momentum J is built up by combining $2J$ spin-$\frac{1}{2}$-particles. For each particle, we have

$$\hat{R}_x = e^{-i\pi s_x} = e^{-i(\pi/2)\sigma_x} = -i\sigma_x \sin\frac{\pi}{2} = \begin{pmatrix} 0 & -i \\ -i & 0 \end{pmatrix}$$

The state $|D,J,J\rangle$ is simply a direct product of states $|\frac{1}{2},\frac{1}{2}\rangle$, and $|D,J,-J\rangle$ is the same direct product of $|\frac{1}{2},-\frac{1}{2}\rangle$ states.

For each single spin, we evidently have

$$\hat{R}_x|\frac{1}{2},\frac{1}{2}\rangle = -i|\frac{1}{2},-\frac{1}{2}\rangle$$

Therefore the effect of $\hat{R}_x$ applied to the direct product $|D,J,J\rangle$ is

$$\hat{R}_x|D,J,J\rangle=(-i)^{2J}|D,J,-J\rangle$$

We conclude that

$$e^{i\rho}=(-i)^{2J}$$

Our result is therefore

$$\hat{R}_x|D,J,M\rangle_n=(-i)^{2J}|D,J,-M\rangle_n \qquad (43)$$

From (43) follows the well-known result[22c]

$$\hat{R}_x^2=(-1)^{2J} \qquad (44)$$

Note that for integer J, the result would have been the same if we had defined $\hat{R}_x$ as a rotation in the left-handed sense. For half-odd integer J, the two choices yield $e^{i\rho}$ phase factors that differ by a factor of (-1), corresponding to a rotation through 2π.

Proceeding analogously for $\hat{R}_y$, we find with the aid of (40)

$$\hat{R}_y|D,J,M\rangle_n=e^{i\sigma_M}|D,J,-M\rangle_n$$

with $e^{i\sigma_{M+1}}=-e^{i\sigma_M}$. We thus have

$$\hat{R}_y|D,J,M\rangle_n=e^{i\sigma}(-1)^{J+M}|D,J,-M\rangle_n$$

(We have defined $\sigma=\sigma_{-J}$ so that $J+M$ rather than simply M appears in the exponent; this is convenient because the former is always an integer whereas the latter is not.) The choice

$$\hat{R}_y=e^{-i\pi\hat{J}_y}$$

corresponds to $e^{i\sigma}=(-1)^{2J}$. Our result is therefore

$$\hat{R}_y|D,J,M\rangle_n=(-1)^{3J+M}|D,J,-M\rangle_n=(-1)^{J-M}|D,J,-M\rangle_n \quad (45)$$

from which follows

$$\hat{R}_y^2=(-1)^{2J} \qquad (46)$$

which is analogous to (44).

The time-reversal operator $\hat{T}$ reverses J_z, and D contains no class II ($\hat{T}$)

variables; thus we have

$$\hat{T}|D,J,M\rangle_n = e^{i\tau_M}|D,J,-M\rangle_n$$

where τ_M is real. Using (36), we find

$$\hat{T}|D,J,M+1\rangle_n = N_{JM}\hat{T}\hat{J}_+|D,J,M\rangle_n = -N_{JM}\hat{J}_-\hat{T}|D,J,M\rangle_n$$

$$= -N_{JM}\hat{J}_- e^{i\tau_M}|D,J,-M\rangle_n = -e^{i\tau_M}|D,J,-M-1\rangle_n$$

Successive values of $e^{i\tau_M}$ thus differ by a factor of (-1), and we can summarize the result as

$$\hat{T}|D,J,M\rangle_n = e^{i\tau}(-1)^{J+M}|D,J,-M\rangle_n \tag{47}$$

where τ is defined as τ_{-J}. The phase angle τ depends on the common phase factor that is still left unspecified in the definition of a normal phase convention. For if we change to a new, primed normal representation by means of

$$|D,J,M\rangle_{n'} = e^{i\eta}|D,J,M\rangle_n \tag{48}$$

we find

$$\hat{T}|D,J,M\rangle_{n'} = \hat{T}e^{i\eta}|D,J,M\rangle_n = e^{-i\eta}\hat{T}|D,J,M\rangle_n$$

$$= e^{-i\eta}e^{i\tau}(-1)^{J+M}|D,J,-M\rangle_n$$

$$= e^{-2i\eta}e^{i\tau}(-1)^{J+M}|D,J,-M\rangle_{n'}$$

Thus when the common phase is changed by the transformation (48), the phase factor $e^{i\tau}$ of (47) is replaced by

$$e^{i\tau'} = e^{i\tau}e^{-2i\eta} \tag{49}$$

In particular, notice that the $e^{i\tau}$ phase factor is unchanged if $e^{i\eta} = -1$. Thus a choice of a phase convention for the $e^{i\tau}$ phase factor still leaves the absolute sign of the wavefunction undefined. (All the phase conventions named below are phase conventions for the $e^{i\tau}$ phase factor; thus they leave the absolute sign of the wavefunction undefined.)

The matrix β associated with the time-reversal operator by (18) in a normal representation of this type has the elements

$$\langle D,J,-M|\tilde{\beta}|D,J,M\rangle_n = e^{i\tau}(-1)^{J+M} \tag{50}$$

with all other elements being zero. Thus specifying the $e^{i\tau}$ phase factor is equivalent to specifying β, and vice versa.

It is now useful to work out the eigenvalues of the antiunitary operators $\hat{C}_x$ and $\hat{C}_y$. For $\hat{C}_x$ we find, with the help of (43) and (47):

$$\hat{C}_x|D,J,M\rangle_n = \hat{R}_x\hat{T}|D,J,M\rangle_n = \hat{R}_x e^{i\tau}(-1)^{J+M}|D,J,-M\rangle_n$$

$$= e^{i\tau}(-i)^{2J}(-1)^{J+M}|D,J,M\rangle_n \qquad (51)$$

Proceeding in the same way for $\hat{C}_y$, and using (45) and (47), we find

$$\hat{C}_y|D,J,M\rangle_n = e^{i\tau}|D,J,M\rangle_n \qquad (52)$$

We next examine a few special cases of normal phase conventions, which, as we see presently, correspond to phase conventions that have been used in the literature. First, the spherical harmonics as defined by Condon and Shortley[26] and Messiah[23c] span a normal representation for integer J, with the common phase being fixed by the requirement that $|D,J,0\rangle$ is the real Legendre polynomial[4a] P_J and is thus left invariant under $\hat{T}$. Referring to (47), we see that this is achieved by setting $e^{i\tau} = (-1)^J$ for integer J. A normal phase convention satisfying this relation for integer J will be called a *normal Legendre* (nL) phase convention. This still leaves the choice of τ for half-odd integer J free, but the most natural generalization is to set $e^{i\tau} = (\pm i)^{2J}$ for all J. Of these, we see from (50) that the lower sign gives $\beta = \sigma_y$ for $J = \frac{1}{2}$. Accordingly, we define the *generalized normal Legendre* (gnL) phase convention as one in which

$$e^{i\tau} = (-i)^{2J}$$

For such a convention, (47), (51), and (52) become

$$\hat{T}|D,J,M\rangle_{gnL} = (-i)^{2J}(-1)^{J+M}|D,J,-M\rangle_{gnL} \qquad (53)$$

$$\hat{C}_x|D,J,M\rangle_{gnl} = (-i)^{4J}(-1)^{J+M}|D,J,M\rangle_{gnL} = (-1)^{J-M}|D,J,M\rangle_{gnL} \qquad (54)$$

$$\hat{C}_y|D,J,M\rangle_{gnL} = (-i)^{2J}|D,J,M\rangle_{gnL} \qquad (55)$$

Another choice that is sometimes convenient is $e^{i\tau} = 1$, which we call a normal positive (+) phase convention. For this choice we find

$$\hat{T}|D,J,M\rangle_+ = (-1)^{J+M}|D,J,-M\rangle_+ \tag{56}$$

$$\hat{C}_x|D,J,M\rangle_+ = (-i)^{2J}(-1)^{J+M}|D,J,M\rangle_+ \tag{57}$$

$$\hat{C}_y|D,J,M\rangle_+ = |D,J,M\rangle_+ \tag{58}$$

Another choice frequently used is $e^{i\tau} = (-1)^{2J}$, which we call a normal negative (−) phase convention. In this case we obtain

$$\hat{T}|D,J,M\rangle_- = (-1)^{J-M}|D,J,-M\rangle_- \tag{59}$$

$$\hat{C}_x|D,J,M\rangle_- = (-i)^{2J}(-1)^{J-M}|D,J,M\rangle_- \tag{60}$$

$$\hat{C}_y|D,J,M\rangle_- = (-1)^{2J}|D,J,M\rangle_- \tag{61}$$

Note that the positive and negative phase conventions are identical to each other for integer angular momenta but differ for half-integer angular momenta. For spin-$\frac{1}{2}$, the positive and negative phase conventions yield $\beta = i\sigma_y$ and $\beta = -i\sigma_y$, respectively. See also Appendix C.

We see that the phase factors $e^{i\tau}$ for the ± phase convections differ from that for the gnL phase convention by $(\pm i)^{2J}$. Referring to (49), we conclude from this that

$$|D,J,M\rangle_\pm = (\mp i)^J|D,J,M\rangle_{\mathrm{gnL}} \tag{62}$$

For integer total angular momenta, this means that the state vectors transform like Condon-Shortley spherical harmonics multiplied by $(\mp i)^J$. Since, however, all phase conventions based on the $e^{i\tau}$ phase factor leave an overall sign of the wavefunction undefined (as discussed earlier in this subsection), we see that the positive and negative normal phase conventions are the same for integer angular momenta.

3. Angular Momentum Addition Using Normal Phase Conventions and Real Clebsch-Gordan Coefficients

The typical case considered in angular momentum addition theory is that of a system consisting of two kinematically independent parts, with total-angular-momentum quantum numbers j_1 and j_2. The state of the combined system, with total-angular-momentum quantum numbers J, M (with $J = j_1 + j_2, j_1 + j_2 - 1, \ldots, |j_1 - j_2|$) is expressed

$$|j_1, j_2, J, M\rangle = \sum_{m_1, m_2} |j_1, m_1\rangle |j_2, m_2\rangle \langle j_1, j_2, m_1, m_2 | j_1, j_2, J, M\rangle \qquad (63)$$

where D is (j_1, j_2) and the coefficients $\langle j_1, j_2, m_1, m_2 | j_1, j_2, J, M\rangle$ are Clebsch-Gordan (CG) coefficients. In general, there is no necessary connection between the phase conventions used for the partial systems and that for the full system, since the CG coefficients themselves contain arbitrary phase factors. It is customary, however, to adopt normal phase conventions for both partial systems, as well as for the full system, and to choose all the CG coefficients real.[8a,25,25a,26,27] If the addition is done in this customary way, the phase conventions for the two parts and for the combined system are not independent. We now investigate the connection between them.

Since all phase conventions used in this case are normal, the only thing to be determined in this relation between the overall phase angle τ for the combined system and the phase angles τ_1 and τ_2 of the partial systems defined by*

$$\hat{T}|j_1, m_1\rangle = e^{i\tau_1}(-1)^{j_1 + m_1}|j_1, -m_1\rangle$$

$$\hat{T}|j_2, m_2\rangle = e^{i\tau_2}(-1)^{j_2 + m_2}|j_2, -m_2\rangle$$

This is determined immediately from (52), which says that the phase factor is also the eigenvalue of $\hat{C}_y$. Since each term in the sum (63) is a direct product and since all coefficients are real, it follows from (52) that the overall eigenvalue is just the product of the partial ones. In other words,

$$e^{i\tau} = e^{i\tau_1} e^{i\tau_2} \qquad (64)$$

Note that because of the antiunitary nature of $\hat{C}_y$, (64) would not hold if the CG coefficients were allowed to be complex.

With the aid of (64), we see immediately that if both partial systems are phased according to the generalized normal Legendre phase convention, the overall phase factor $e^{i\tau}$ is $(-i)^{2(j_1 + j_2)}$. Since this is not always equal to $(-i)^{2J}$, it follows that the combined system must be rephased if it is to behave according to the (gnL) convention. On the other hand, if both systems follow the normal positive phase convention, or both the normal negative convention, the overall phase factor is $(\pm 1)^{2(j_1 + j_2)} = (\pm 1)^{2J}$. It

---

*All the names for $e^{i\tau}$ phase conventions for total-angular-momentum representations may be applied by analogy to phase conventions for these partial-angular-momenta state vectors.

follows that these two phase conventions have the useful property of "invariance under customary angular momentum addition": if both partial systems are phased according to the designated convention, the combined system will automatically also be so phased if the customary form of the CG coefficients is used. This result was given previously for the normal negative phase convention by Huby[28] and for the normal positive phase convention by Edmonds.[25b]

Equation (64) also enables us to define other types of normal phase conventions in which different types of partial angular momenta (e.g., orbital and spin) are phased differently. An example is the Alder-Winther[29] convention, in which each spin is given an $e^{i\tau}$ phase factor of unity, and each single-particle orbital angular momentum (l_i) one of $(-1)^{l_i}$. When all these are combined according to (64), the overall $e^{i\tau}$ phase factor is $(-1)^{\Sigma l_i} = \Pi$, the parity of the state. It is evident that this convention also is invariant under customary angular momentum addition.

It is sometimes convenient to write the parity as $(-1)^p$. For the Alder-Winther (AW) phase convention, we then have, using (47), (51), and (52)

$$\hat{T}|D,J,M\rangle_{AW} = (-1)^{p+J+M}|D,J,-M\rangle_{AW} \tag{65}$$

$$\hat{C}_x|D,J,M\rangle_{AW} = (-i)^{2J}(-1)^{p+J+M}|D,J,M\rangle_{AW} \tag{66}$$

$$\hat{C}_y|D,J,M\rangle_{AW} = (-1)^p|D,J,M\rangle_{AW} \tag{67}$$

Since the phase factor $e^{i\tau}$ for the Alder-Winther phase convention differs from that for the normal positive phase convention by Π, (49) and (62) show that for integer total angular momenta, the state vectors of Alder and Winther transform like Condon-Shortley spherical harmonics multiplied by $\Pi^{1/2}(-i)^J = \pm i^{p-J}$. We note that for a single spinless particle, the AW phase convention is identical with the gnL convention. For more general situations, of course, the two are not necessarily equivalent. If each of two state vectors $|j_1,m_1\rangle$ and $|j_2,m_2\rangle$ for integer spatial orbital angular momenta are phased to satisfy the gnL phase convention, they will also each satisfy the AW phase convention. The state vector $|j_1,j_2,J,M\rangle$ obtained by customary angular momentum addition will not necessarily satisfy (53) but will satisfy (65). This is the advantage of using parity to define the $e^{i\tau}$ phase convention.

Notice that (58) is equivalent to

$$\hat{C}_y\hat{P}|D,J,M\rangle_{AW} = |D,J,M\rangle_{AW} \tag{68}$$

The freedom to choose our set of states to be simultaneous eigenfunctions of $\hat{J}^2, \hat{J}_z, \hat{C}_y$, and $\hat{P}$ when D contains no class II ($\hat{T}$) variables was established in Section III.E. The freedom to choose the eigenvalues of involutional antiunitary operators $\hat{\Theta}$ for which $\hat{\Theta}^2 = +1$ to be $+1$ was established in Section III.C.3. But $(\hat{C}_y \hat{P})^2 = +1$, since $\hat{C}_y^2 = +1$, $\hat{P}^2 = +1$, and $\hat{C}_y$ and $\hat{P}$ commute (see Section III.E). This verifies that we always have the freedom to choose the set of states and phases to satisfy (68) as recommended for certain purposes by Alder and Winther.

Still another example is the class of phase conventions which may be called Kramers[30] phase conventions, in which each spin of $\frac{1}{2}$ is given an $e^{i\tau}$ phase factor of $(-i)$. This means that the matrix $\boldsymbol{\beta}$ in the spin space is just the direct product of $\boldsymbol{\sigma}_y$ for each spin. The spins thus contribute a factor $(-i)^n$ to the overall $e^{i\tau}$ phase factor, where n is the number of spin-$\frac{1}{2}$ particles in the system. The Kramers phase convention leaves open the phasing of the orbital contribution. For example, Wigner[21] defines a Kramers phase convention in which the orbital angular momentum follows the negative phase convention [cf. his eqs. (26.15) and (26.43)]. The overall $e^{i\tau}$ phase factor in Wigner's phase convention is thus $(-i)^{2n}(-1)^{2L} = (-i)^{2n}$.

4. Nonnormal Phase Conventions

It is convenient to define an arbitrary phase convention by referring it to $|D, J, M\rangle_+$ of (56) to (58). Accordingly, let

$$|D, J, M\rangle = e^{i\xi(M)} |D, J, M\rangle_+$$

We find, using (43), (45), (56) to (58), plus the unitary nature of the $\hat{R}$ operators, and the antiunitary character of $\hat{C}_x$, $\hat{C}_y$, and $\hat{T}$:

$$\hat{R}_x |D, J, M\rangle = (-i)^{2J} e^{i[\xi(M) - \xi(-M)]} |D, J, -M\rangle \tag{69}$$

$$\hat{R}_y |D, J, M\rangle = (-1)^{J-M} e^{i[\xi(M) - \xi(-M)]} |D, J, -M\rangle \tag{70}$$

$$\hat{T} |D, J, M\rangle = (-1)^{J+M} e^{-i[\xi(M) + \xi(-M)]} |D, J, -M\rangle \tag{71}$$

$$\hat{C}_x |D, J, M\rangle = (-i)^{2J} (-1)^{J+M} e^{-2i\xi(M)} |D, J, M\rangle \tag{72}$$

$$\hat{C}_y |D, J, M\rangle = e^{-2i\xi(M)} |D, J, M\rangle \tag{73}$$

In particular, we see from (72) that (34) is satisfied if we choose $e^{i\xi(M)} = e^{i(\pi/2)M}$.

IV. SCATTERING THEORY

A. Symmetry of the S Matrix

Time-reversal invariance means (see Section III.C.2)

$$\hat{T}\hat{H} = \hat{H}\hat{T} \qquad (74)$$

where $\hat{H}$ is the hamiltonian for the system. As stated in Section III.B, we only consider processes for which (74) holds. This is generally assumed to include all processes in chemical physics.[31]

Let $\hat{S}$ be the unitary scattering operator[1d,6c,7b,8d,10b] which takes the system from the state u to the state u'. We show in Appendix B that, as a consequence of (74), regardless of representation and phase convention*

$$\langle u|\hat{S}|u'\rangle = \langle \hat{T}u'|\hat{S}|\hat{T}u\rangle \qquad (B-25)$$

If our representation is characterized by the quantum numbers J, M, and D, where the D are scalars and class I ($\hat{T}$) (which, we have shown in Section III.E and Appendix A, is always possible), then (B-25) becomes

$$\langle D,J,M|\hat{S}|D',J,M\rangle = \langle \hat{T}(D',J,M)|\hat{S}|\hat{T}(D,J,M)\rangle \qquad (75)$$

Now, using (46) and (73), plus the definition $\hat{C}_y = \hat{R}_y\hat{T}$, we see that

$$\hat{T}|D,J,M\rangle = (-1)^{2J}\hat{R}_y\hat{C}_y|D,J,M\rangle = (-1)^{2J}\hat{R}_y e^{-2i\xi(M)}|D,J,M\rangle \qquad (76)$$

Application of (76) to both the primed and unprimed states and insertion into (75) gives the result

$$\langle D,J,M|\hat{S}|D',J,M\rangle = e^{2i[\xi'(M)-\xi(M)]}\langle D',J,M|\hat{R}_y^\dagger\hat{S}\hat{R}_y|D,J,M\rangle \qquad (77)$$

However, $\hat{R}_y$ commutes with the hamiltonian, hence also with $\hat{S}$, and is unitary; thus (77) becomes

$$\langle D,J,M|\hat{S}|D',J,M\rangle = e^{2i[\xi'(M)-\xi(M)]}\langle D',J,M|\hat{S}|D,J,M\rangle \qquad (78)$$

In particular, if the same phase convention is used in both the primed and unprimed manifolds, the S matrix is symmetric. If $|D,J,M\rangle$ and $|D',J,M\rangle$ are formed by customary addition of partial angular momenta, the symmetry of $\mathbf{S}$ will be assured if for each angular momentum we use a normal

---

*The proof of (B-25) does, however, depend on $\hat{T}$ being antiunitary, which, as explained in Sections III.A and III.B, requires that $\hat{T}$ obey (2) and (3).

phase convection that is invariant under customary angular momentum addition.

B. Reciprocity Theorem and Discussion

The question of symmetry of the scattering matrix is usually asked and answered in specific representations with specific phase conventions.[1c] Although the proof of the previous section is more general, in this section we rederive the symmetry of the scattering matrix in a more specific representation for various phase conventions. We also prove some related results.

The quantities used to describe the scattering of A by B are defined as follows (we use the notation of Blatt and Biedenharn[32] as much as possible):

$\vec{I}, I$ the angular momentum and the angular momentum quantum number, respectively, of the target B in the channel specified by α. All angular momenta are given in the usual way "in units of $\hbar$." Unprimed and primed quantum numbers refer to the initial state and the final state, respectively.

$\vec{i}, i$ the angular momentum and the angular momentum quantum number, respectively, of A in the channel specified by α.

$\vec{s}, s$ the angular momentum and the angular momentum quantum number, respectively, of AB, excluding the angular momentum associated with the relative motion of A and B, in the channel specified by α.

$$\vec{s} = \vec{I} + \vec{i} \tag{79}$$

We call s the internal angular momentum in a given channel.

$\vec{l}, l$ the orbital angular momentum and the orbital angular momentum quantum number, respectively, for the relative motion of A and B in the channel specified by (J, s).

$\vec{J}, J$ the total angular momentum and the total-angular-momentum quantum number, respectively; $\vec{J}$ is given by (1) or by

$$\vec{J} = \vec{L} + \vec{S} \tag{80}$$

where $\vec{L}$ and $\vec{S}$ are defined in Section II.

Π the parity [i.e., eigenvalue (± 1) of the total system parity operator $\hat{P}$].

α the channel index denoting all quantum numbers, except (s, l, J, M, Π), necessary to specify the quantum state. Appendix A shows that there is no loss of generaltiy in assuming α to contain only scalars and we shall do so.

A the group of quantum numbers (α, s, l), except the energy E.

d the group of quantum numbers (α, s, l, Π).

$Z(abcd; ef)$ the Z coefficient of Blatt and Biedenharn[32] defined as $Z(abcd; ef) = i^{f-a+c}\overline{Z}(abcd; ef)$.

$\overline{Z}(abcd; ef)$ the $\overline{Z}$ coefficient of A. M. Lane and Thomas[33] defined as

$$\overline{Z}(abcd; ef) = [(2a+1)(2b+1)(2c+1)(2d+1)]^{1/2} W(abcd; ef)\langle ac00|acf0\rangle$$

(81)

where $W(abcd; ef)$ is the Racah coefficient defined[34] by (12) of Ref. 34 and $\langle ac00|acf0 \rangle$ is the Clebsch-Gordan coefficient defined in (63) and given by (5) of Ref. 34.

M the projection quantum number of total angular momentum $\vec{J}$ on an arbitrary space-fixed axis z.

m the projection quantum number of a general angular momentum $\vec{j}$ on an arbitrary space-fixed axis z.

Y_{jm} the spherical harmonic of Condon and Shortley[26] and Messiah[23c] defined by (3.4.5), (3.4.12), and (3.4.15) of Condon and Shortley, that is,

$$Y_{jm} = (-1)^m \left[\frac{2l+1}{4\pi} \frac{(l-|m|)!}{(l+|m|)!} \right]^{1/2} P_j^{|m|}(\cos\Theta) e^{im\phi}$$

$$P_j^{|m|}(\cos\Theta) = (\sin\Theta)^{|m|} \left[\frac{d}{d(\cos\Theta)} \right]^{|m|} P_j(\cos\Theta)$$

where $P_j(\cos\Theta)$ is a Legendre polynomial[4a] and Θ and ϕ are the colatitude and the longitude, respectively.

A solution in the total-angular-momentum representation in the asymptotic (separated-subsystem) region of the time-independent Schrödinger equation in the barycentric coordinate system corresponding to given values of the conserved quantum numbers J, M, and Π is called $u(\alpha s l J M \Pi)$. This choice of quantum numbers simplifies the scattering calculations; examples have been given in Section II. The asymptotic form of $u(\alpha s l J M \Pi)$ determines the subblock $\mathbf{S}^{JM\Pi}$ of the scattering matrix $\mathbf{S}$.

It is proved in Appendix B that[10c]

$$S_{\alpha's'l'; \, \alpha sl}^{JM\Pi} = \langle \hat{T}u(\alpha slJM\Pi) | \hat{S} | \hat{T}u(\alpha's'l'JM\Pi) \rangle \tag{82}$$

Using (32) in (82) yields

$$S_{\alpha's'l'; \, \alpha sl}^{JM\Pi} = e^{-i\gamma(\alpha slJM\Pi)} e^{i\gamma(\alpha's'l'JM\Pi)}$$

$$\times \langle u(\alpha_t slJ - M\Pi) | \hat{S} | u(\alpha_t's'l'J - M\Pi) \rangle \tag{83}$$

where α_t represents the same set of quantum numbers as α but with the signs of all class II ($\hat{T}$) variables changed, and the phase angle $\gamma(\alpha slJM\Pi)$ depends on the phase conventions used in defining $u(\alpha slJM\Pi)$. For simplicity we assume that M is not negative. Using (41), (42), and the invariance of $\hat{S}$ under rotations in the form $[\hat{S}, \hat{J}_-] = 0$, we can show from (83) that

$$S_{\alpha's'l'; \, \alpha sl}^{JM\Pi} = e^{-i\gamma(\alpha slJM\Pi)} e^{i\gamma(\alpha's'l'JM\Pi)}$$

$$\times \left[\prod_{M'=-M}^{M-1}{}' e^{-i\phi(\alpha slJM'\Pi)} e^{i\phi(\alpha's'l'JM'\Pi)} \right]$$

$$\times \langle u(\alpha_t slJM\Pi) | \hat{S} | u(\alpha_t's'l'JM\Pi) \rangle \tag{84}$$

where the prime in $\prod'^{M-1}_{M'=-M}$ means the product is deleted if $M = 0$. Equation (84) yields immediately

$$| S_{\alpha's'l'; \, \alpha sl}^{JM\Pi} | = | S_{\alpha, sl; \, \alpha_t's'l'}^{JM\Pi} | \tag{85}$$

which is the reciprocity theorem (also called the reciprocity relation).[1c,6c,7b,8b]

Notice that our proof of (85) is entirely independent of phase conventions. Stronger results may be derived by making various assumptions about the phase conventions. For example, if we assume

1. $\gamma(AJM\Pi E) = \gamma(A_c JM\Pi E)$, where A_c and E are the subset of (α, s, l) which is conserved
2. $M = 0$, or $\phi(AJM'\Pi E) = \phi(A_c JM'\Pi E)$ for $-M \leqslant M' \leqslant M - 1$
3. α contains no nonneglectable* class II ($\hat{T}$) variables

---

*Variables that are conserved and of which the scattering matrix is independent are called *neglectable*.

then (84) becomes

$$S^{JM\Pi}_{\alpha's'l';\,\alpha sl} = S^{JM\Pi}_{\alpha sl;\,\alpha's'l'} \tag{86}$$

and S is symmetric. Although we have given a much more general derivation of (86) in Section IV.A, the present derivation is interesting for comparison with discussions in the literature.[1b,2,7c,8b,10b,28]

Particular phase conventions designed to ensure condition (1) have played an important role in previous discussions of symmetry of the scattering matrix. It is easily seen from Section III that condition (1) is satisfied by the generalized normal Legendre, normal positive, normal negative, and Alder-Winther phase conventions [see (32), (53), (56), (59), (65)]. Furthermore, the normal positive, normal negative, and Alder-Winther phase conventions have the property of invariance under customary angular momentum addition. Thus a particularly convenient way to ensure condition (1) is to phase each of the angular momentum states $|jm\rangle$ according to *one* of these conventions and to form $u(\alpha slJM\Pi)$ using customary angular momentum addition. Thus the symmetry of the scattering matrix has often been discussed only for the case of a normal positive[1,1b,7c,8b] or a normal negative[10b,28] phase convention. Three of the articles[13,28,33] discussed in Section V use a normal negative phase convention (note: for integer angular momenta, the normal negative and positive phases conventions are the same, see Section III.F.2) and also satisfy conditions (2) and (3). Thus both our proofs of (86) are applicable to these articles. Some previous workers,[12,32,35–38] however, used Condon-Shortley spherical harmonics Y_{jm} as their states $|jm\rangle$ for integer angular momenta j, and the usual proof in terms of customary angular momentum addition of states all phased with either the normal positive or negative phase convention does not ensure that the scattering matrix so obtained is symmetric. However, we show in Section V that condition (1) is satisfied and our two proofs are still applicable. In Section V we discuss the changes that must be made in the formulas of these workers if $i^j Y_{jm}$ is substituted for Y_{jm} in their formulas, a procedure originally suggested by Huby.[28]

Condition (2) may easily be satisfied by taking $\phi(\alpha slJM\Pi)=0$. This is called a normal phase convention in Section III.F. If we take $\phi(jm)=0$ for all $|jm\rangle$ and obtain $u(\alpha slJM\Pi)$ by customary angular momentum addition (defined in Section III.F.3), it can be shown that $\phi(\alpha slJM\Pi)$ will equal zero.[25a] However, condition (2) is less restrictive than this particularly convenient choice of phase conventions.

Condition (3) has apparently not received general discussion in the past. We proved in Section III.E that it is always possible to satisfy condition (3).

The proof of (86) is much easier for systems involving no spatial or spin angular momenta, and simple proofs have been given elsewhere[3b,39] for those cases. The complications come from the angular momenta as discussed previously.

Note that condition (2) alone suffices for the scattering matrix to be independent of M [this is easily seen by comparing (83) to (84) under the assumption of condition (2)].[8c] Condition (2) holds in all cases treated in Section V, which means that M is neglectable there, and the superscript M will be suppressed.

C. Detailed Balance

We have discussed the reciprocity theorem (85) and the symmetry of the S matrix [(78) or (86)] for total-angular-momentum representations. A similar result, detailed balancing[6c,8b,40]

$$|S^{JM\Pi}_{\alpha's'l';\ \alpha sl}| = |S^{JM\Pi}_{\alpha sl;\ \alpha's'l'}|$$

may be derived from (85) by invoking condition (3) of Section IV.B, or in particular, that α contain no nonneglectable class II ($\hat{T}$) variables. Although we have shown in Section III.E that it is always possible to satisfy condition (3) it may sometimes be convenient to characterize a system using a class II ($\hat{T}$) variable in α (but it is hard to give an example). The reciprocity theorem, symmetry of S, and detailed balance can sometimes be proved in nontotal-angular-momentum representations. In some such cases it may be convenient to characterize the states using a nonconserved component of angular momentum, usually spin, as one of the quantum numbers. Then detailed balancing does not hold in general, but a weaker result, semidetailed balancing, in which probabilities are averaged over spin directions in the initial and final states, does hold.[6c,40]

The reader should be cautioned that the labels "time-reversal invariance," "the reciprocity theorem," "the principle of detailed balancing," and "the principle of microscopic reversibility"[41] are often used interchangeably in the literature. For example, Landau and Lifshitz[5c] use "the principle of detailed balancing" to designate what is here called the "reciprocity theorem." Some chemists use "detailed balance" to refer to a relationship between rates at equilibrium and use "the principle of microscopic reversibility" to label what is here called semidetailed balancing.[41]

V. APPLICATIONS

In this section we discuss the symmetry of the S matrix and proper equations for calculating the differential cross-section for several standard scattering problems of chemical physics and also in general.

Blatt and Biedenharn[32] presented a general expression, free of all sums over angular-momentum-projection quantum numbers, for calculating from the scattering matrix S the differential cross-section for the scattering of unpolarized beams.[42] Their expression is particularly useful because it gives the differential cross-section as an expansion in Legendre polynomials[4a] with real coefficients. Their derivation was based in part on a paper by Wigner and Eisenbud.[43] Huby[28] remarked that the representations and phase conventions used by Wigner and Eisenbud and Blatt and Biedenharn do not always lead to a symmetric scattering matrix. The scattering matrix in the representation of Blatt and Biedenharn and corresponding to their phase conventions is designated ^{BB}S. The assumption that S is symmetric is not required in the derivation of Blatt and Biedenharn's differential cross-section formula, and Huby correctly pointed out that their formula is correct if ^{BB}S is used. Huby gave another representation in which the normal negative phase convention of Section III.F.2 is used for all partial and total angular momentum state vectors. The scattering matrix in this representation is denoted by ^HS. Huby explained how to calculate the differential cross-section in terms of ^HS instead of ^{BB}S. The discussion of the symmetry of the scattering matrix given in Section IV shows that the phase conventions used by Huby do lead to a symmetric scattering matrix if α contains no nonneglectable class II ($\hat{T}$) variables. However, the application of these general considerations to particular cases is still often complicated. We now discuss in detail a few such applications, with emphasis on problems in chemical physics.

A. Blatt and Biedenharn; Huby

In deriving their general expression [their (3.16), (4.5), and (4.6)] for the differential cross-section, Blatt and Biedenharn[32] obtained the following intermediate expression [see their eq. (4.1)]:

$$
\begin{aligned}
d\sigma_{\alpha's';\,\alpha s} = \left[k_\alpha^2(2s+1)\right]^{-1} &\sum_{J_1}\sum_{l_1}\sum_{l_1'}\sum_{J_2}\sum_{l_2}\sum_{l_2'} i^{-l_1+l_1'+l_2-l_2'} \\
&\times \left(\delta_{\alpha'\alpha}\delta_{s's}\delta_{l_1'l_1} - {}^{BB}S^{J_1\Pi}_{\alpha's'l_1';\,\alpha s l_1}\right)^* \\
&\times \left(\delta_{\alpha'\alpha}\delta_{s's}\delta_{l_2'l_2} - {}^{BB}S^{J_2\Pi}_{\alpha's'l_2';\,\alpha s l_2}\right) K(J_1 l_1'l_1;\,J_2 l_2'l_2;\,s's;\,\Theta)\,d\Omega \qquad (87)
\end{aligned}
$$

where $\hbar k_\alpha$ is the momentum in the channel specified by α and Θ is the center-of-mass scattering angle. The factor K contains all the sums over projection quantum numbers, and Blatt and Biedernharn showed that K could be replaced by a sum over λ containing the product of two Z coefficients, a Legendre polynomial, and a weighted phase factor. That result can be combined with the definition of the $\overline{Z}$ coefficients [see(81)] and with (87) to yield

$$d\sigma_{\alpha's';\,\alpha s} = (-1)^{s'-s}\left[4k_\alpha^2(2s+1)\right]^{-1}\sum_\lambda \sum_{J_1}\sum_{l_1}\sum_{l_1'}\sum_{J_2}\sum_{l_2}\sum_{l_2'} i^{-l_1+l_1'+l_2-l_2'}$$

$$\times \left(\delta_{\alpha'\alpha}\delta_{s's}\delta_{l_1'l_1} - {}^{BB}S_{\alpha's'l_1';\,\alpha sl_1}^{J_1\Pi}\right)^*$$

$$\times \left(\delta_{\alpha'\alpha}\delta_{s's}\delta_{l_2'l_2} - {}^{BB}S_{\alpha's'l_2';\,\alpha sl_2}^{J_2\Pi}\right)\overline{Z}(l_1 J_1 l_2 J_2;\, s\lambda)$$

$$\times \overline{Z}(l_1' J_1 l_2' J_2;\, s'\lambda)P_\lambda(\cos\Theta)\,d\Omega \qquad (88)$$

Note that the derivation requires using the fact that $(l_1'l_2'00|l_1'l_2'\lambda0)$ is zero unless $(\lambda - l_1 + l_2)$ is even.[34]

The phase factor $i^{-l_1+l_1'+l_2-l_2'}$ in (88) results from Blatt and Biedenharn's use of Y_{lm_l} instead of $i^l Y_{lm_l}$ in their wavefunction. Since Y_{lm_l} was used, and the phase convention to be used for other partial-angular-momentum state vectors was not specified, the representation of Blatt and Biedenharn does not necessarily satisfy condition (1); ${}^{BB}S$ cannot be proved to be symmetric by the proofs of Section IV or by the special cases of the proof of Section IV.B which have been published elsewhere. Blatt and Biedenharn do use the reciprocity theorem after their eq. (4.6) but their formula for the differential cross-section does not depend on the scattering matrix being symmetric. If $i^l Y_{lm_l}$ is used instead of Y_{lm_l} as Huby suggested, a scattering matrix ${}^H S$ is defined whose elements are related to the elements of ${}^{BB}S$ by Huby's eq. (6), that is, by

$${}^{BB}S_{\alpha's'l';\,\alpha sl}^{J\Pi} = i^{l'-l}\,{}^H S_{\alpha's'l';\,\alpha sl}^{J\Pi} \qquad (89)$$

If we substitute this into (88), the phase factor cancels out and the $\overline{Z}$ coefficients alone should be used with ${}^H S$. Since the remaining angular momentum eigenstates χ_{sm_s} in the representation of Huby are chosen to satisfy a normal negative phase convention, conditions (1) and (2) will be satisfied and ${}^H S$ will satisfy

$${}^H S_{\alpha's'l';\,\alpha sl}^{J\Pi} = {}^H S_{\alpha sl;\,\alpha's'l'}^{J\Pi} \qquad (90)$$

If α contains no nonneglectable class II ($\hat{T}$) variables, ${}^H S$ is symmetric.

Our statements concerning the symmetry of S apply as well to the reactance matrix R related to S by

$$S = (1 - iR)^{-1}(1 + iR) \qquad (91)$$

The formulas of Blatt and Biedenharn have been used to calculate the differential cross-sections for the scattering of structureless particles (or particles assumed to be structureless) from rotator-vibrators by Henry[44] (e–H_2). Sums of such differential cross-sections have been calculated by Wagner and McKoy[15d] (Ar–H_2, He–O_2, He–I_2). Differential cross-sections for a transition among the $3p$ ^2P states of Na induced by collisions with He have been calculated by Reid.[19] Additional examples are cited in Sections V.B and V.D. Wolken et al.[15b] and McGuire and Micha[15c] have calculated differential cross-sections for the scattering of atoms (assumed structureless) from rotator-vibrators using a total-angular-momentum representation different from that used by Blatt and Biedenharn. They used the helicity representation[1b,7d,16a,45] in which $\vec{s}$ is quantized along $\vec{k}_\alpha$ and $\vec{s}'$ is quantized along $\vec{k}_{\alpha'}$.

B. Percival and Seaton; Smith

Percival and Seaton[35] have derived an expression for the differential cross-section for scattering of electrons by atomic hydrogen under the assumption (see Section II) that $\bar{L}$ and $\bar{S}$ are separately conserved. The quantum numbers used by Percival and Seaton in labeling their representation are a, l, s, L, M_L, and Π, where the set a contains the principal quantum number of the hydrogen atom, the total energy, the quantum number S [see Section II and (80)], and the quantum number M_S associated with the component S_z of $\bar{S}$ on a space-fixed axis, and s is the orbital-angular-momentum quantum number of the bound electron. Using the ideas presented by Percival and Seaton, K. Smith[37] has derived an expression for the differential cross-section for the scattering of positrons by atomic hydrogen. These expressions have been discussed elsewhere.[46] Here we summarize that discussion and relate it to the present chapter.

For all their angular momentum eigenstates $|jm\rangle$, Percival and Seaton and K. Smith chose the spherical harmonics Y_{jm} that do not satisfy normal positive or normal negative phase conventions but do satisfy condition (2). Furthermore, using the fact[8e] that Π is $(-1)^{l+s}$ for this system and the identity [eq. (3.5.17) of Ref. 25a]

$$(acef|abcd) \equiv (-1)^{e-a-c}(ace-f|a-bc-d) \tag{92}$$

we can show that condition (1) is satisfied with

$$e^{i\gamma(aslLM_L\Pi)} = \Pi(-1)^{L+M_L} \tag{93}$$

Also, since $\hat{H}$ is rotationally invariant and L and S are good quantum numbers in this approximation, ^{PS}S is diagonal in L, M_L, S, and M_S and independent of both M_L and M_S by a theorem[7b,8c] used in Section IV.B. Thus even though a contains the class II ($\hat{T}$) variable S_z and the Percival-Seaton representation is not a total-angular-momentum representation, S_z is a neglectable variable, and the proof of Section IV.B may be modified with L and M_L replacing J and M to show that ^{PS}S is symmetric. This is an example of something pointed out in Section IV and Ref. 46—namely, that the usual rule[1b,7c,8b,28] requiring normal positive or normal negative phase conventions is not a *necessary* condition for a symmetric scattering matrix.

Since Percival and Seaton and K. Smith chose the same $|lm_l\rangle$ as did Blatt and Biedenharn, the correct expressions for the differential cross-section in terms of their scattering matrices ^{PS}S are identical to the one given by Blatt and Biedenharn. Burke and co-workers[47–49] have used the phase conventions of Percival and Seaton to calculate the reactance matrix for electron–hydrogen-atom collisions in a series of successively improved approximations. Equation (91) and the formulas of Blatt and Biedenharn have been used to calculate the differential cross-sections from these reactance matrices; the results are published elsewhere.[46,50] The equations of Blatt and Biedenharn have also been used to calculate differential cross-sections for electron scattering from larger atoms.[51]

If we apply the Huby phase convention to these problems–that is, if we use $i^j Y_{jm}$ for all $|jm\rangle$, except those for spin—we find that the new representation satisfies conditions (1) and (2) [with (J,M) replaced by (L,M_L)], and the scattering matrix in this representation $^H S$ can be shown to be symmetric. The correct formula to use in calculating the differential cross-section from $^H S$ in these cases is Huby's formula.

If $\mathbf{X}$ represents one of the matrices $\mathbf{S}$, $\mathbf{R}$, or $\mathbf{V}$, where V is the interaction potential, the elements of $^H\mathbf{X}$ are related to the elements of $^{PS}\mathbf{X}$ by [46]

$$^H X^{L\Pi}_{a's'l';\,asl} = i^{l-l'+s-s'}\,^{PS}X^{L\Pi}_{a's'l';\,asl} \tag{94}$$

Note that $i^{l-l'+s-s'}$ is real by conservation of total system parity;[46] thus applying the Huby phase convention does not affect the reality of the matrix elements.

C. Lane and Thomas

A. M. Lane and Thomas[33] have derived an expression for the differential cross-section for nuclear reactions. They correctly phased their representation to satisfy conditions (1) and (2) by using a normal negative phase convention and the scattering matrix ^{LT}S defined by them is symmetric for systems for which α does not contain any nonneglectable class II ($\hat{T}$) variables. Since their representation used $i^l Y_{lm_l}$ for $|lm_l\rangle$, their differential cross-section formula is identical to Huby's formula and correctly uses the $\bar{Z}$ coefficients with ^{LT}S.

D. Arthurs and Dalgarno; Micha

Arthurs and Dalgarno[12] have derived an expression for the differential cross-section for the scattering of particles by rigid rotators with rotational angular momentum quantum number s. In this case the set of quantum numbers α includes only the total energy. Like Percival and Seaton, they chose Y_{jm} for all $|jm\rangle$, therefore their representation does satisfy condition (1), with J, M replacing L, M_L in (93). Thus ^{AD}S is symmetric by the proof given in Section IV.B, as well as by the more general proof of Section IV.A. Their formula for the differential cross-section in terms of their ^{AD}S is identical to the one given by Blatt and Biedenharn. By arguments similar to those used in discussing the formulas of Percival and Seaton, it can be seen that their formula is correct for this problem. Micha[13] applied Huby's phase convention to this problem (i.e., he used $i^j Y_{jm}$ for $|jm\rangle$), and his scattering matrix ^MS is symmetric by the proof given by Huby, as well as by the more general proofs of Sections IV.A and IV.B. His differential cross-section formula is identical to Huby's and uses the $\bar{Z}$ coefficients with his ^MS; ^MX is related to ^{AD}X by (94), with L replaced by J just as ^HX is related to ^{PS}X in Section V.B.

The formulas of Blatt and Biedenharn and Arthurs and Dalgarno have been used to calculate the differential cross-section $d\sigma_{s';s}/d\Omega$ for a transition from rotational state s to rotational state s' $(s \to s')$ many times. For scattering of an atom or molecule (assumed structureless) from a rigid rotator they have been used by Roberts[52] $(0 \to 2)$, Allison and Dalgarno[53] $(0 \to 2)$, Munn and Monchick[54] $(1 \to 1)$, Erlewein et al.[55] $(0 \to 2)$, Johnson and Secrest[56] $(0 \to 2, 1 \to 3)$, Miller[16a] $(0 \to 2)$, Hayes et al.[57] $(0 \to 0, 0 \to 2)$, Heukels and van de Ree[58] $(0 \to 0, 1 \to 1, 2 \to 2,\ 0 \to 1, 1 \to 2, 0 \to 2)$, and two of the present authors and R. L. Smith[50] $(0 \to 2, 2 \to 2)$. For scattering of an electron by a rigid rotator, they have been used by Henry and N. F. Lane[59] ("elastic," $1 \to 3, 2 \to 4, 3 \to 5$), N. F. Lane and Geltman[60] $(1 \to 1, 1 \to 3)$, Itikawa[61] $(0 \to 0, 0 \to 1)$, Crawford and Dalgarno[62] $(0 \to 0, 0 \to 1, 0 \to 2)$, and Sams et al.[63] $(1 \to 1, 1 \to 3)$. Sums of such differential cross-sections have been calculated by Burke and Chandra.[64]

E. Davison

Davison[38] has derived an expression for the differential cross-section for the scattering of two rigid rotators, one with rotational angular momentum quantum number I and the other with rotational angular momentum quantum number i. In this case the set of quantum numbers α is (I, i, E), where E is the total energy. Davison also used Y_{jm} for the angular momentum eigenfunctions. Since Davison used the same $|lm_l\rangle$ chosen by Blatt and Biedenharn, his expression for the differential cross-section correctly involves the Z coefficients with his scattering matrix ^DS. Also the total system parity, which is $(-1)^{l+I+i}$ for this system is again conserved

Once more both conditions (1) and (2) are satisfied, as they were in the Sections V.B and V.D. Since α contains no class II ($\hat{T}$) variables, both our proofs may be used to show $^D\mathbf{S}$ is symmetric. By arguments similar to those given previously in this section, if we apply the Huby phase convention to this problem, we get another symmetric $^H\mathbf{S}$ to be used with Huby's formula for the differential cross-section: $^H\mathbf{X}$ is related to $^D\mathbf{X}$ by (94) but with $i^{s-s'}$ and L replaced by $i^{I+i-I'-i'}$ and J, respectively.

Davison's formulas were used to compute differential cross-sections for inelastic scattering by Roberts.[52]

F. Alder and Winther Phase Conventions

Alder and Winther[29] have given a phase convention for angular momentum eigenfunctions that may be very convenient because it will make all the single-particle matrix elements of the electromagnetic multipole operators[65] real. Using (65) it is clear that $u(\alpha slJM\Pi)$ obtained with these phase conventions satisfies (93), with J, M replacing L, M_L in (93). Thus condition (1) is satisfied. Condition (2) is also satisfied. Thus if α contains no nonneglectable class II ($\hat{T}$) variables, both proofs of Section IV show that $\mathbf{S}$ is symmetric for this phase convention.

Two examples of the Alder-Winther phase convention for $|lm_l\rangle$ are Y_{lm_l} and $(-1)^l Y_{lm_l}$. If either of these phase conventions is used in defining $\mathbf{S}$, the Z coefficient of Blatt and Biedenharn[32] should again be used in the differential cross-section formula.

G. Summary

In general, the answer to whether to use the differential cross-section formula of Blatt and Biedenharn[32] (involving Z coefficients[32]) or that of Huby[28] and A. M. Lane and Thomas[33] (involving $\bar{Z}$ coefficients[33]) is strictly a function of which of the choices Y_{lm_l}, $(-1)^l Y_{lm_l}$, or $i^l Y_{lm_l}$ is made for the angular basis function for relative motion. The first choice implies that the formula of Blatt and Biedenharn[32] should be used. The second choice is also consistent with the formula of Blatt and Biedenharn. The third choice is consistent with the formula of Huby[28] or the identical formula of A. M. Lane and Thomas.[33] Specific choices of representation and phase convention for χ_{sm_s} and $\chi_{s'm'_s}$ (the wavefunctions for the separated reagents and products, respectively, in the notation of Blatt and Biedenharn[32]) will determine whether (86) and (90) are valid, as discussed earlier, but will not determine whether to use the Z or $\bar{Z}$ coefficients.

APPENDIX A

Proof that the Nontotal-Angular-Momentum Quantum Numbers in a Total-Angular-Momentum Representation may be Taken to Consist Entirely of Scalar Variables.

That the quantum numbers d (defined near the beginning of Section III.E) can be chosen to be scalars is simply shown as follows:

We choose our states to have quantum numbers J, M, Π, E, and enough other quantum numbers to completely specify the states where J, M, and Π have the meaning indicated in Section III and near the beginning of Section IV, and E is the eigenvalue of some hamiltonian (either exact or approximate) that is invariant under rotations. If there are no two states with the same values of all four of these quantum numbers, our task of specifying our states by J, M, Π, and scalar quantum numbers is complete (since E is of course a scalar).

Suppose, however, that there are two or more states with the same J, M, Π, and E. Choose one of them (with $M = M_0$, say), and arbitrarily denote it by $|A_1, J, M_0, \Pi, E\rangle$. By applying angular momentum raising and lowering operators to this state, we generate $(2J + 1)$ states all with the same J, Π, and E, which we denote by $|A_1, J, M, \Pi, E\rangle$, with $M = -J, -J+1, \ldots, J$. Now let $|A_2, J, M_0, \Pi, E\rangle$ be a state with the quantum numbers J, M_0, Π, and E but orthogonal to $|A_1, J, M_0, \Pi, E\rangle$. By hypothesis, such a state exists. Again applying raising and lowering operators, we generate the states $|A_2, J, M, \Pi, E\rangle$, which are orthogonal to the $|A_1, J, M, \Pi, E\rangle$. We continue this process until all the states with quantum numbers J, M_0, Π, and E are used up. If we think of A_1, A_2, and so on, as different values of some quantity A, we can say that a state is completely specified by specifying J, M, Π, E, and A. The quantity A is a physical property of the system, since different values of A correspond in general to physically different states. Moreover, by construction it is invariant under rotations about the z-axis (since it is diagonal simultaneously with J_z) as well as those about the x- and y-axes (since it is left unaltered by the raising and lowering operators $\hat{J}_x \pm i\hat{J}_y$, hence commutes with them). It also commutes with parity, since it is simultaneously diagonal with it. It follows that A is a scalar, which is what was to be proved.

This result is a special case of a more general group theoretic result.[66] Using this more general result, we may show that any vector components included in the complete commuting set of variables in addition to J_z must be a scalar times J_z. Thus we can use the scalar as the additional dynamical variable.

APPENDIX B

The Time-Reversal-Invariance Property of the Scattering Matrix

In this appendix, we use the following known properties of the time-reversal operator $\hat{T}$ (see ref. 21 and Section III):

1. $\hat{T}$ is antilinear. That is, for any state functions ϕ_1 and ϕ_2 and complex numbers a_1 and a_2, we have

$$\hat{T}(a_1\phi_1 + a_2\phi_2) = a_1^* \hat{T}\phi_1 + a_2^* \hat{T}\phi_2 \qquad \text{(B-1)}$$

2. If the total spin of the system is integral, $S = n$, then

$$\hat{T}^2\phi = \phi \qquad \text{for all state functions } \phi \qquad \text{(B-2)}$$

In this case, if the hamiltonian $\hat{H}$ is time-reversal invariant, the arguments of Section III.C.3 show that its eigenfunctions ϕ_j may be chosen to be also eigenfunctions of $\hat{T}$ with eigenvalues ± 1:

$$\hat{T}\phi_j = k_j\phi_j, \qquad k_j = \pm 1 \qquad \text{(B-3)}$$

Actually Section III.C.3 shows that we could always choose $k_j = +1$, but we shall not need to do so.

3. If the total spin is half-odd integral $S = n + \frac{1}{2}$, we have

$$\hat{T}^2\phi = -\phi \qquad \text{for all } \phi \qquad \text{(B-4)}$$

In this case, the eigenfunctions of $\hat{H}$ cannot be eigenfunctions of $\hat{T}$, but they can be grouped into degenerate pairs ϕ_j and $\bar{\phi}_j$ with the properties

$$\hat{T}\phi_j = \bar{\phi}_j \qquad \text{(B-5)}$$

$$\hat{T}\bar{\phi}_j = -\phi_j \qquad \text{(B-6)}$$

Notice that the phase conventions of (53), (56), and (59) yield

$$\hat{T}^2|jm\rangle = (-1)^{2j}|jm\rangle \qquad \text{(B-7)}$$

which is compatible with (B-2) and (B-4).

We now consider the matrix element of the time-displacement operator $\hat{G}(t) = \exp(-i\hat{H}t/\hbar)$ between an initial state u and a final state u', which we compare with the matrix element of the same operator between $\hat{T}u'$ and $\hat{T}u$.

First, in the case $S = n$, we can expand u and u' in the simultaneous eigenfunctions of $\hat{H}$ and $\hat{T}$:

$$u = \Sigma_j a_j \phi_j \tag{B-8}$$

and

$$u' = \Sigma_j b_j \phi_j \tag{B-9}$$

Obviously,

$$\hat{G}(t)u = \Sigma_j a_j e^{(-iw_j t)} \phi_j \tag{B-10}$$

where $\hbar w_j$ is the jth eigenvalue of $\hat{H}$. Using orthonormality of the $\{\phi_j\}$ (see Section III.C.3), we find

$$\langle u' | \hat{G}(t) | u \rangle = \Sigma_j b_j^* a_j e^{(-iw_j t)} \tag{B-11}$$

To study the time-reversed situation, we use (B-1) and (B-3) to obtain

$$\hat{T}u = \Sigma_j a_j^* k_j \phi_j \tag{B-12}$$

and

$$\hat{T}u' = \Sigma_j b_j^* k_j \phi_j \tag{B-13}$$

Applying $\hat{G}(t)$ to (B-13) yields

$$\hat{G}(t)\hat{T}u' = \Sigma_j b_j^* k_j e^{(-iw_j t)} \phi_j$$

and we obtain

$$\langle \hat{T}u | \hat{G}(t) | \hat{T}u' \rangle = \Sigma_j a_j b_j^* e^{(-iw_j t)} \tag{B-14}$$

Comparing (B-11) and (B-14) yields

$$\langle \hat{T}u | \hat{G}(t) | \hat{T}u' \rangle = \langle u' | \hat{G}(t) | u \rangle \tag{B-15}$$

To prove that (B-15) also holds for $S = n + \frac{1}{2}$, we proceed analogously. We expand u and u' in eigenfunctions of $\hat{H}$ chosen to satisfy (B-5) and (B-6):

$$u = \Sigma_j \left(a_j \phi_j + \bar{a}_j \bar{\phi}_j \right) \tag{B-16}$$

$$u' = \Sigma_j \left(b_j \phi_j + \bar{b}_j \bar{\phi}_j \right) \tag{B-17}$$

Using these expansions and the degeneracy of the states ϕ_j, $\bar{\phi}_j$, we immediately obtain

$$\hat{G}(t)u = \Sigma_j \left(a_j \phi_j + \bar{a}_j \bar{\phi}_j \right) e^{(-iw_j t)} \tag{B-18}$$

and

$$\langle u' | \hat{G}(t) | u \rangle = \Sigma_j \left(b_j^* a_j + \bar{b}_j^* \bar{a}_j \right) e^{(-iw_j t)} \tag{B-19}$$

For the time-reversed situation, (B-1), (B-5), and (B-6) give

$$\hat{T}u = \Sigma_j \left(a_j^* \bar{\phi}_j - \bar{a}_j^* \phi_j \right) \tag{B-20}$$

$$\hat{T}u' = \Sigma_j \left(b_j^* \bar{\phi}_j - \bar{b}_j^* \phi_j \right) \tag{B-21}$$

Equation (B-21) yields

$$\hat{G}(t)(\hat{T}u') = \Sigma_j \left(b_j^* \bar{\phi}_j - \bar{b}_j^* \phi_j \right) e^{(-iw_j t)} \tag{B-22}$$

and

$$\langle \hat{T}u | \hat{G}(t) | \hat{T}u' \rangle = \Sigma_j \left(a_j b_j^* + \bar{a}_j \bar{b}_j^* \right) e^{(-iw_j t)} \tag{B-23}$$

Comparing (B-19) and (B-23) yields (B-15) again.

Equation (B-15) shows that the time-displacement operator $\hat{G}(t)$ possesses the symmetry claimed for the scattering operator $\hat{S}$ in Section IV.A. But[1d]

$$\hat{S} = \lim_{\substack{t_1 \to \infty \\ t_2 \to -\infty}} \exp\left(i \frac{\hat{H}_0}{\hbar} t_1 \right) \hat{G}(t_1 - t_2) \exp\left(-i \frac{\hat{H}_0}{\hbar} t_2 \right)$$

Thus, if u, u' are eigenfunctions of the zero-order hamiltonian $\hat{H}_0$ with eigenvalues $\hbar w, \hbar w'$, respectively, we have

$$\langle u' | \hat{S} | u \rangle = \lim_{\substack{t_1 \to \infty \\ t_2 \to -\infty}} \exp[i(w't_1 - wt_2)] \langle u' | \hat{G}(t_1 - t_2) | u \rangle \tag{B-24}$$

The passage to infinite time causes the matrix element (B-24) to vanish unless u, u' are "on the energy shell" (i.e., unless $w = w'$). In this case, however, the phase factor on the right-hand side of (B-24) becomes

$$\exp[iw(t_1 - t_2)]$$

Because of the degeneracy between $u, \hat{T}u$, as well as between $u', \hat{T}u'$, it follows that this phase factor will be the same for $\langle \hat{T}u|\hat{S}|\hat{T}u'\rangle$, from which we conclude that [10c]

$$\langle \hat{T}u|\hat{S}|\hat{T}u'\rangle = \langle u'|\hat{S}|u\rangle \tag{B-25}$$

This result is independent of representation. In particular total-angular-momentum representations (B-25) becomes (75) and (82). It is important to point out that our proofs of (B-25), (75), and (82) are independent of phase convention, except that the representation must be compatible with

$$\hat{T}(|q\rangle + |r\rangle) = \hat{T}|q\rangle + \hat{T}|r\rangle$$

[see (3)].

APPENDIX C

For fixed D, J an arbitrary state $|\Phi\rangle$ may be expanded as

$$|\Phi\rangle = \sum_M \phi_M |D,J,M\rangle_+$$

Applying (56) we find

$$\hat{T}|\Phi\rangle = \sum_M (-1)^{J+M}\phi_M^* |D,J,-M\rangle_+$$

and

$$\phi_M^T \equiv {}_+\langle D,J,M|\hat{T}|\Phi\rangle = (-1)^{J-M}\phi_{-M}^* \tag{C-1}$$

Although (C-1) looks similar to (59), they really correspond to different phase conventions (for half-odd-integral spin). Possible ambiguity in the sign of β for spin $\frac{1}{2}$ can be caused by failure to be explicit as to whether (59) or (C-1) is meant[67] or by changing from one to the other without warning.[68]

Notes and References

1. R. G. Newton, *Scattering Theory of Waves and Particles*, McGraw-Hill, New York, 1966, pp. 491–493, (a) Chapter 11, (b) Chapter 15, (c) pp. 192–193, (d) pp. 160–164, 480–482. The spherical harmonics used in Newton's book (see p. 31) are $i^j Y_{jm}$, where Y_{jm} is the spherical harmonic of Ref. 23b and 26. Much of the work in Chapter 15 referenced here can also be found in R. G. Newton, *J. Math. Phys.*, **1**, 319 (1960).
2. F. Coester, *Phys. Rev.*, **89**, 619 (1953).
3. N. F. Mott and H. S. W. Massey, *The Theory of Atomic Collisions*, 3rd ed., Clarendon Press, Oxford, 1965, pp. 19 ff. (a) pp. 263–264, (b) pp. 369–370.
4. P. M. Morse and H. Feshbach, *Methods of Theoretical Physics*, Vols. 1 and 2, McGraw-Hill, New York, 1953, pp. 1066 ff. (a) pp. 749, 1325.

5. L. D. Landau and E. M. Lifshitz, *Quantum Mechanics: Non-Relativistic Theory*, 2nd ed., Pergamon Press, Oxford, 1965, pp. 469 ff. (*a*) p. 193, (*b*) pp. 206–208, (*c*) p. 553.

6. A. S. Davydov, *Quantum Mechanics*, I. V. Schensted, transl., NEO Press, Ann Arbor, Mich., 1966; or D. ter Haar, transl., Pergamon Press, Oxford, 1965, pp. 385 ff. (*a*) p. 221, (*b*) pp. 140–143, (*c*) pp. 417–435.

7. M. L. Goldberger and K. M. Watson, *Collision Theory*, Wiley, New York, 1964, pp. 226 ff. (*a*) pp. 358 ff., (*b*) pp. 79, 80, 113–116, 166–171, (*c*) p. 351, (*d*) pp. 882–898.

8. L. S. Rodberg and R. M. Thaler, *Introduction to the Quantum Theory of Scattering*, Academic Press, New York, 1967, pp. 26 ff. (*a*) pp. 278–289, (*b*) pp. 266–277, (*c*) pp. 261–263, (*d*) pp. 183–185, 227–234, (*e*) pp. 263–266.

9. K. Smith, *The Calculation of Atomic Collision Processes*, Wiley-Interscience, New York, 1971, pp. 10 ff. (*a*) p. 98.

10. J. R. Taylor, *Scattering Theory*, Wiley, New York, 1972, pp. 181 ff. (*a*) pp. 103 ff., (*b*) pp. 34–35, 334–335, (*c*) pp. 93–95, 354.

11. A. D. Boardman, A. D. Hill, and S. Sampanthar, *Phys. Rev.*, **160**, 472 (1967); S. Geltman, *Topics in Atomic Collision Theory*, Academic Press, New York, 1969, pp. 73–81.

12. A. M. Arthurs and A. Dalgarno, *Proc. Roy. Soc. (London)*, **A256**, 540 (1960).

13. D. A. Micha, *Phys. Rev.*, **162**, 88 (1967).

14. Additional references and reviews are given in W. A. Lester, Jr., *Methods in Computational Physics*, **10**, 211 (1971) and in D. E. Golden, N. F. Lane, A. Temkin, and E. Gerjuoy, *Rev. Mod. Phys.*, **43**, 642 (1971).

15. See, e.g., (*a*) W. Eastes and D. Secrest, *J. Chem. Phys.*, **56**, 640 (1972), (*b*) G. Wolken, W. H. Miller, and M. Karplus, *J. Chem. Phys.*, **56**, 4930 (1972), (*c*) P. McGuire and D. A. Micha, *Int. J. Quantum Chem.*, **S6**, 111 (1972), (*d*) A. F. Wagner and V. McKoy, *J. Chem. Phys.*, **58**, 2604 (1973), (*e*) J. Schaefer and W. A. Lester, Jr., *Chem. Phys. Lett.*, **20**, 575 (1973).

16. See, e.g., (*a*) W. H. Miller, *J. Chem. Phys.*, **49**, 2373 (1968), **50**, 407 (1969), and (*b*) R. T. Pack, *J. Chem. Phys.*, **60**, 633 (1974), and references therein.

17. See. e.g., K. Takayanagi, *Progr. Theor. Phys. (Kyoto)*, **8**, 497 (1952), K. P. Lawley and J. Ross, *J. Chem. Phys.*, **43**, 2930 (1965); C. F. Curtiss, *J. Chem. Phys.*, **49**, 1952 (1968), **52**, 4832 (1970), L. W. Hunter and R. F. Snider, to be published. Time-reversal properties of a representation in which total angular momentum is not a quantum number are considered in L. W. Hunter and C. F. Curtiss, *J. Chem. Phys.*, **58**, 3884 (1973).

18. See, e.g., S. Geltman and K. Takayanagi, *Phys. Rev.*, **143**, 25 (1966).

19. See, e.g., D. C. S. Allison and P. G. Burke, *J. Phys. B*, **2**, 941 (1969); R. H. G. Reid and A. Dalgarno, *Phys. Rev. Lett.*, **22**, 1029 (1969); R. H. G. Reid and A. Dalgarno, *Chem. Phys. Lett.*, **6**, 85 (1970); R. H. G. Reid, *J. Phys. B*, **6**, 2018 (1973); F. H. Mies, *Phys. Rev. A*, **7**, 942, 957 (1973); E. L. Lewis and L. F. McNamara, *Phys. Rev. A*, **5**, 2643 (1972).

20. See, e.g., D. W. Jepson and J. O. Hirschfelder, *J. Chem. Phys.*, **32**, 1323 (1960); W. R. Thorson, *J. Chem. Phys.*, **42**, 3878 (1965); and E. L. Lewis, L. F. McNamara, and H. H. Michels, *Phys. Rev. A*, **3**, 1939 (1971).

21. E. P. Wigner, *Group Theory*, J. J. Griffin, transl., Academic Press, New York, 1959, Chapter 26.

22. K. Gottfried, *Quantum Mechanics*, Vol. 1, Benjamin, New York, 1966, p. 225, (*a*) p. 213, (*b*) pp. 226–230, (*c*) pp. 275–276, (*d*) p. 322, (*e*) pp. 265 ff.

23. A. Messiah, *Mécanique Quantique*, Vol. 2, Dunod, Paris, 1959, pp. 540–543, (*a*) pp. 465–467, (*b*) pp. 455–456, (*c*) Vol. 1, pp. 420–421 [English transl.: A. Messiah, *Quantum*

Mechanics, Vol 2, Wiley, New York, 1962, pp. 633–636, (*a*) p. 545, (*b*) pp. 533–534, (*c*) Vol. 1, pp. 494–495].

24. See G. C. Wick, *Annu. Rev. Nucl. Sci.*, **8**, 1 (1958), eq. (67).
25. A. R. Edmonds, *Angular Momentum in Quantum Mechanics*, 2nd ed., Princeton University Press, Princeton, N. J., 1960, pp. 14–17. (*a*) pp. 36–42, (*b*) pp. 51–52.
26. E. U. Condon and G. H. Shortley, *The Theory of Atomic Spectra*, Cambridge University Press, Cambridge, England, 1935.
27. M. E. Rose, *Elementary Theory of Angular Momentum*, Wiley, New York, 1957, pp. 32–47.
28. R. Huby, *Proc. Phys. Soc.* (*London*), **67A**, 1103 (1954).
29. K. Alder and A. Winther, *Phys. Lett.*, **34B**, 357 (1971). In the sentence following eq. (3) of this reference, i^l can be $(\pm i)^l$. In their notation $(-1)^\Pi$, not Π, is the parity.
30. H. A. Kramers, *Koninkl. Ned. Akad. Wetenschap., Proc.*, **33**, 959 (1930).
31. There is experimental evidence that the product of charge conjugation and parity is not conserved in high-energy particle physics. One consequence of this, according to present relativistic quantum field theories, would be violation of time-reversal invariance. This might have some observable consequences in chemical physics but so far none have been observed. For further discussions of these points, see W. R. Frazer, *Elementary Particles*, Prentice-Hall, Englewood Cliffs, N. J., 1966, pp. 165–167; V. W. Hughes, *Physics Today*, 33 (February 1969).
32. J. M. Blatt and L. C. Biedenharn, *Rev. Mod. Phys.*, **24**, 258 (1952).
33. A. M. Lane and R. G. Thomas, *Rev. Mod. Phys.*, **30**, 257 (1958).
34. L. C. Biedenharn, J. M. Blatt, and M. E. Rose, *Rev. Mod. Phys.*, **24**, 249 (1952).
35. I. C. Percival and M. J. Seaton, *Proc. Camb. Phil. Soc.*, **53**, 654 (1957).
37. K. Smith, *Proc. Phys. Soc.* (*London*), **78**, 549 (1961).
38. W. D. Davison, *Disc. Faraday Soc.*, **33**, 71 (1962). The theory of the scattering between two rigid rotators has been considered in a different total-angular-momentum representation by H. Klar, *Z. Physik*, **228**, 59 (1969).
39. J. M. Blatt and V. F. Weisskopf, *Theoretical Nuclear Physics*, Wiley, New York, 1952, pp. 517–529.
40. F. Coester, *Phys. Rev.*, **84**, 1259 (1951); E. C. G. Stückelberg, *Helv. Phys. Acta*, **25**, 577 (1952); W. Heitler, *The Quantum Theory of Radiation*, 3rd ed., Clarendon Press, Oxford, 1954, pp. 412–414.
41. See, e.g., N. Davidson, *Statistical Mechanics*, McGraw-Hill, New York, 1962, pp. 230–235; J. Ross, J. C. Light, and K. E. Shuler, in *Kinetic Processes in Gases and Plasmas*, A. R. Hochstim, ed., Academic Press, New York, 1969, pp. 281–320; B. Widom, *Adv. Chem. Phys.*, **5**, 363 (1963).
42. By polarized-beam experiments we mean any experiments in which a component of internal angular momentum (possibly spin) of either scatterer is selected (rather than averaged over or summed over) either before or after the scattering event. Time-reversal considerations also play an important role in the treatment of polarized-beam experiments; see, e.g., L. Wolfenstein and J. Ashkin, *Phys. Rev.*, **85**, 947 (1952); R. H. Dalitz, *Proc. Phys. Soc.* (*London*), **A65**, 175 (1952).
43. E. P. Wigner and L. Eisenbud, *Phys. Rev.*, **72**, 29 (1947).
44. R. J. W. Henry, *Phys. Rev. A*, **2**, 1349 (1970).
45. M. Jacob and G. C. Wick, *Ann. Phys.* (*N. Y.*), **7**, 404 (1959); R. van Wageningen, *Ann. Phys.* (*N. Y.*), **31**, 148 (1965).
46. M. A. Brandt and D. G. Truhlar, *Phys. Rev. A.*, **9**, 1188 (1974).
47. P. G. Burke and K. Smith, *Rev. Mod. Phys.*, **34**, 458 (1962); P. G. Burke, H. M. Schey,

344 D. G. TRUHLAR, C. A. MEAD, AND M. A. BRANDT

and K. Smith, *Phys. Rev.*, **129**, 1258 (1963); P. G. Burke, A. J. Taylor, and S. Ormonde, *J. Phys. B*, **1**, 325 (1968).

48. P. G. Burke, S. Ormonde, and W. Whitaker, *Proc. Phys. Soc. (London)*, **92**, 319 (1967).
49. See, e.g., S. Geltman and P. G. Burke, *J. Phys. B*, **3**, 1062 (1970), and references therein.
50. M. A. Brandt, D. G. Truhlar, and R. L. Smith, *Compt. Phys. Commun.*, **5**, 456 (1973); **7**, 172, 177 (1974).
51. D. G. Truhlar, S. Trajmar, W. Williams, S. Ormonde, and B. Torres, *Phys. Rev. A*, **8**, 2475 (1973), and references therein.
52. C. Roberts, *Phys. Rev.*, **131**, 209 (1963).
53. A. C. Allison and A. Dalgarno, *Proc. Phys. Soc.*, **90**, 609 (1967). These authors and the authors of the electron-scattering applications (except Sams et al.) do not refer to the computationally efficient contracted general formula (4.7) of Blatt and Biedenharn but instead to Blatt and Biedenharn's eq. (4.6), which is equivalent but less efficient.
54. R. J. Munn and L. Monchick, *Mol. Phys.*, **16**, 25 (1969). These authors, Erlewein et al. (next reference), and Reid (Ref. 19) did not use the expression [involving eq. (4.7) of Blatt and Biedenharn] referred to at the beginning of Section V which is free of all sums over angular momentum projection quantum numbers; rather, they used the equivalent eqs. (3.13) and (3.14) of Blatt and Biedenharn. Thus they calculated an amplitude, squared it, and summed the result over projection quantum numbers. This has certain computational advantages.
55. W. Erlewein, M. von Seggern, and J. P. Toennies, *Z. Physik*, **211**, 35 (1968).
56. B. R. Johnson and D. Secrest, *J. Chem. Phys.*, **48**, 4682 (1968).
57. E. F. Hayes, C. A. Wells, and D. J. Kouri, *Phys. Rev. A*, **4**, 1017 (1971).
58. W. F. Heukels and J. van de Ree, *J. Chem. Phys.*, **57**, 1393 (1972).
59. R. J. W. Henry and N. F. Lane, *Phys. Rev.*, **183**, 221 (1969).
60. N. F. Lane and S. Geltman, *Phys. Rev.*, **184**, 46 (1969). See also N. F. Lane and S. Geltman, *Phys. Rev.*, **160**, 53 (1967).
61. Y. Itikawa, *J. Phys. Soc. Japan*, **27**, 444 (1969).
62. O. H. Crawford and A. Dalgarno, *J. Phys. B*, **4**, 494 (1971).
63. W. N. Sams, L. Frommhold, and D. J. Kouri, *Phys. Rev. A*, **6**, 1070 (1972).
64. P. G. Burke and N. Chandra, *J. Phys. B*, **5**, 1696 (1972).
65. A. Bohr and B. R. Mottelson, *Nuclear Structure*, Vol. 1, Benjamin, New York, 1969, pp. 379–394.
66. H. Weyl, *Gruppentheorie und Quantenmechanik*, S. Hirzel, Leipzig, 1928, pp. 140–143 [English transl.: *Theory of Groups and Quantum Mechanics*, translated from the second (revised) German edition by H. P. Robertson, Methuen, London, 1931, pp. 170–175, republished by Dover Publications, New York, 1931].
67. L. C. Biedenharn and M. E. Rose, *Rev. Mod. Phys.*, **25**, 729 (1953).
68. G. Breit, *Handbook of Physics*, **41**/1, 1 (1959).

TRANSITION STATE
STABILIZATION ENERGY
AS A MEASURE OF CHEMICAL
REACTIVITY

M. V. BASILEVSKY

Karpov Institute of Physical Chemistry
Moscow, USSR

CONTENTS

I. INTRODUCTION

The organic reactivity problem may be examined at various levels. The most fundamental one is a dynamical study of the elementary reaction act, that is, the evaluation of reaction cross-sections. Recent progress in this field allows us to consider the chemical dynamics of only the simplest reactions, such as that between an atom and a diatomic molecule. To deal with a dynamical treatment, moreover, reliable potential surfaces are necessary, but calculating them presents a separate problem.

On the other hand, a potential surface permits us to obtain insight into the elementary reaction processes even without subsequent dynamical treatment. Transition state theory expresses the rate constant in the Arrhenius form $A \cdot \exp(-U^{\neq}/kT)$. The dynamical information is condensed into the preexponential factor. The value of A is approximately constant if the set of reactions involves the same reaction center. This is a consequence of transition state theory. Thus the activation energy $U^{\neq}$—the interaction energy at the saddle point—is left as the only characteristic of the reactivity within the reaction series. By this means the reactivity problem reduces to the standard technique of quantum chemistry. A semiempirical calculation of potential surfaces usually allows us to make reasonable conclusions concerning the geometry and energy of transition states of rather complex reactions. Calculations of this kind correspond to the more empirical level of treatment of the problem. Such procedures are being intensively developed at present, and one may say that they represent the current situation in the field.

If one agrees with this classification, the dynamical studies represent the level of the future; hence the level of the past is also expected to exist. It is a well-known reactivity indices (RI) method indeed.

The applications of the RI method are thoroughly discussed in a number of monographs.[1-4] The method proves to be very useful in considerations of a relative reactivity within a carefully chosen limited reaction series. The RIs have a certain advantage that seems important at least for the chemist who deals with reactions in retorts, not with computer reactions. Simple to handle, RIs appeal directly to the chemical intuition. In fact, the RI method formulates in terms of quantum mechanics those elementary conceptions that have empirically generalized the extensive experience of organic chemistry. Such chemical conceptions dissolve and disappear in the course of modern complicated calculations with all valence electrons included. Nevertheless they are needed for the qualitative treatment, which often satisfies the experimentalist and always precedes the more serious theoretical investigation.

Finally, the reactivity within reaction series is often considered in terms of simple empirical relations, such as Hammett, Polanyi, Alfrey-Price, and isokinetic, which have no direct connection with quantum-chemical calculations. We refer to such approaches as empirical correlations.

The accompanying scheme is an attempt to represent the hierarchy of various approaches to the chemical reactivity problem. Our consideration will be restricted by the part of this scheme disposed under the block called "transition state theory." We dwell on the relations between different methods, not on the methods themselves, because the latter are well known. In other words, not the blocks in the diagram but the arrows linking the blocks are under discussion in this chapter.

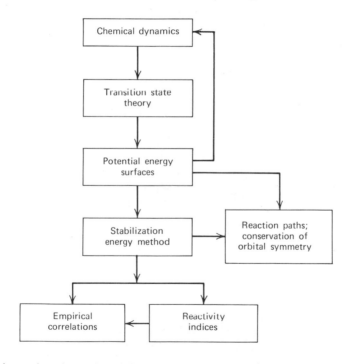

Our intention is to formulate a consistent sequence of physical arguments, becoming more approximate the further we descend along our scheme, allowing us to follow the logical deduction of various approaches and conceptions down to the most primitive ones from the more advanced levels, beginning from transition state theory.

The central point is the method suggesting the conjugation energy in a transition state as a measure of chemical reactivity, for which we assume the term "stabilization energy" (SE) method. Just as transition state theory reduces the treatment of a rate constant to a quantum-chemical calculation of the transition state by means of elimination of dynamical aspects, so the SE method eliminates a straightforward investigation of a transition state as a saddle point of a multidimensional potential surface, introducing instead several parameters that characterize it empirically. Then the research is immediately transferred at a RI level.

Our investigation of the intermolecular interaction energy is followed by the discussion of semiempirical procedures that are usually employed when the interaction energy is concerned. The subject of stabilization energy arises in this analysis among many other contributions. The results obtained present a natural background for a semiempirical method of construction of potential surfaces for addition reactions of π-electron systems. It appears that the SE is a characteristic quantity within reaction sets with a slightly varying reaction center; that is, it varies regularly when substituents at the reaction center are being changed. This important conclusion is a basis of the SE method.

The SE method is then extended over the numerous reactions involving π-electron systems. In particular, for concerted reactions with a cyclic reaction center, the interpretation in terms of SEs is a meaningful alternative to the popular Woodward-Hoffman correlations. As indicated by Dewar,[3] this interpretation contains the same qualitative deductions but can additionally furnish more quantitative information. We present a somewhat different derivation of the same conclusion.

The following questions are also considered:

1. How is it possible to extend the simple idea of electron delocalization in the transition state to cover the reactions with reaction centers formed by σ-bonds?
2. Why do a number of various, usually very primitive, estimations of the stabilization energy often give the same predictions of the relative reactivities in a reaction series?
3. What is the rationale behind the Hammett-like $\sigma\rho$ correlations between relative reaction rates in two reaction series?

Our treatment is based on the simple Hückel theory. We suppose the more complicated methods to be unreasonably refined for the discussion at the RI level. The treatment of question 1 requires a certain generalization of the Mulliken hyperconjugation theory.[5] Our study of correlations between various RIs is the development of the paper by Fukui, Yonezawa, and Nagata.[6]. The Hammett and other empirical correlations are considered as a consequence of the correlations between RIs.

II. INTERACTION ENERGY

When intermolecular forces are considerably lower in magnitude than intramolecular ones, perturbation expansion is the most appropriate tool for a study of interaction energy. Using matrix representation, the zero-order hamiltonian H_0 is constructed as follows: for a pair of interacting molecules A and A' their MOs (calculated with the interaction neglected) are taken as basis functions, in that basis the matrix H of a complete hamiltonian is calculated; then those elements of this matrix that are associated with the interaction are omitted. Formally, one should assume infinite intermolecular distance l in matrix elements of $H : H_0 = \lim_{l \to \infty} H$, the perturbation being the difference $H - H_0$.

In the framework of Hückel method this procedure was introduced by Dewar.[7] Recently Dewar's treatment has been developed to take into consideration electron interaction[8-11] and nonorthogonality effects.[12] The account in this section is based on a general procedure[13] that consideres all the physically important energy contributions and involves the results of Refs. 7 to 12 as particular cases.

A. Perturbation Theory

The MOs of noninteracting molecules φ_i (molecule A) and $\varphi_{i'}$ (molecule A') are nonorthogonal. The form of the overlap matrix for the basis set $(\varphi_i, \varphi_{i'})$ is

$$T = I + t$$

where $t_{ij} = \delta_{ij}, t_{i'j'} = \delta_{i'j'}, t_{ii'} = \langle \varphi_{i'} | \varphi_i \rangle$. The Löwdin orthogonalization gives a new set:

$$(\lambda_i, \lambda_{i'}) = (\varphi_i, \varphi_{i'}) T^{-1/2}$$

With the new basis set used to construct the hamiltonian matrix H, the nonorthogonality effects are transferred from the wavefunctions into the hamiltonian. The two versions of perturbation expansion depend on the choice of many-particle basis functions as:

1. Slater determinants Λ_M built of the orbitals $\lambda_i, \lambda_{i'}$:

$$H_{MN} = \langle \Lambda_N | H | \Lambda_M \rangle \tag{1a}$$

2. Linear combinations $\Psi_M = \Sigma \Lambda_N C_{NM}$, where coefficients C_{MN} are available from a complete configuration interaction solution for the non-interacting molecules A, A' or the ion pairs A^{+t}, A'^{-t} corresponding to an intermolecular transfer of t electrons:

$$-(\text{number of electrons in } A') \leqslant t \leqslant (\text{number of electrons in } A)$$

The hamiltonian matrix is then

$$H_{MN} = \langle \Psi_N | \mathbf{H} | \Psi_M \rangle \qquad (1b)$$

To obtain the zero hamiltonian, we assume $l \rightarrow \infty$ in (1a) or (1b) and take the diagonal part of the resulting matrix. In the first case the intramolecular electron correlation becomes an additional perturbation (HF zero approximation). In the second case it is involved in the zero hamiltonian, which means that the matrix (1b) itself becomes diagonal at $l \rightarrow \infty$.

The interaction energy based on a HF zero approximation is obtained as the following sum:

$$U = E_{\text{Coul}} + E_{\text{pol}} + E_{\text{disp}} + E_{\text{res}} + E_{\text{er}} + E_{\text{ex}} \qquad (2)$$

where the subscripts are explained in Sections II.A.1 to II.A.6.

The exact expressions for the separate contributions[13] will not be written out here. We introduce at once two semiempirical assumptions: the neglect of differential overlap (NDO) for electron distributions inside the molecules, and the Mulliken approximation for intermolecular electron distributions involved in two-electron integrals. Now AOs $\chi_r, \chi_{r'}$ become orthogonal inside the molecules, the MOs $\varphi_i, \varphi_{i'}$ being related to them by a unitary transformation $\mathbf{c}$

$$(\varphi_i, \varphi_{i'}) = (\chi_r, \chi_{r'}) \mathbf{c}$$

and $\mathbf{T}^{-1/2}$ is expressed by means of the AO overlap matrix $\mathbf{S} = \mathbf{I} + \mathbf{s}$

$$\mathbf{T}^{-1/2} = \mathbf{c}^\dagger \mathbf{S}^{-1/2} \mathbf{c}$$

$$\mathbf{S}^{-1/2} = \mathbf{I} - \tfrac{1}{2}\mathbf{s} + \tfrac{3}{8}\mathbf{s}^2 - \cdots \qquad (3)$$

The two assumptions we have accepted for electron distributions $\bar{\chi}_b \chi_a \equiv \Omega_{ab}$ (indices a, b being attributed to both the molecules A and A') may be summarized as a matrix relation

$$\Omega = \mathbf{D} + \tfrac{1}{2}(\mathbf{s}\mathbf{D} + \mathbf{D}\mathbf{s}) \qquad (4)$$

where matrix $\mathbf{D}$ is diagonal: $D_{ab} = |\chi_a|^2 \delta_{ab}$.

Simplified in this way, expressions for the energy contributions in (2) take the following form:

1. Coulomb Energy

$$E_{\text{Coul}} = \sum_{rr'} (p_{rr} - N_r)(p_{r'r'} - N_{r'})\gamma_{rr'} + \sum_{R,R'} Q_{RR'} \tag{5}$$

$$Q_{RR'} = \sum_{r \in R} p_{rr} U_{R',rr} + \sum_{r' \in R'} p_{r'r'} U_{R,r'r'} + W_{RR'} - \sum_{r \in R, r' \in R'} N_r N_{r'} \gamma_{rr'}$$

Here the small indices label AOs, the capital labels, the atoms; the notation $r \in R$ means that AO r belongs to atom R. The other notations are: $p_{rs}, p_{r's'}$, elements of intramolecular bond order matrices; $N_r, N_{r'}$, occupation numbers of AOs $\chi_r, \chi_{r'}$ in the valence states of respective atoms; $W_{RR'}$, energies of nuclear repulsion; $\gamma_{rr'} \equiv (\chi_r \chi_r | \chi_{r'} \chi_{r'})$, Coulomb replusion integrals. Operators $U_R, U_{R'}$ are the potentials of neutral atoms, and they are related to the atomic core potentials $U_R^+, U_{R'}^+$

$$U_R = U_R^+ + \sum_{r \in R} N_r (\chi_r \chi_r | \quad)$$

Their matrix elements are penetration integrals.

2. Polarization Energy

$$E_{\text{pol}} = \sum_{i_1 j_2} \frac{|V_{i_1 j_2}^{A'}|^2}{\Delta \varepsilon_{i_1 j_2}} + \text{sum with the inverse arrangement of primes} \tag{6}$$

$$V_{i_1 j_2}^{A'} = \sum_{rr'} c_{ri_1} c_{rj_2} (p_{r'r'} - N_{r'}) \gamma_{rr'}$$

The subscripts 1 and 2 label occupied and vacant MOs, respectively.

3. Dispersion Energy

$$E_{\text{disp}} = \sum_{i_1 j_2 i'_1 j'_2} \frac{|\Gamma_{i_1 j_2, i'_1 j'_2}|^2}{\Delta \varepsilon_{i_1 j_2, i'_1 j'_2}} \tag{7}$$

$$\Gamma_{i_1 j_2, i'_1 j'_2} = \sum_{rr'} c_{ri_1} c_{rj_2} c_{r'i'_1} c_{r'j'_2} \gamma_{rr'}$$

4. Resonance Energy

$$E_{res} = \sum_{i_1' j_2} \frac{\left(\sum_{rr'} c_{r'i_1'} c_{rj_2} \beta_{rr'} \right)^2}{\Delta \varepsilon_{i_1' j_2}} + \text{sum with the inverse} \qquad (8)$$
$$\text{arrangement of primes}$$

The peculiarity of this formulation is that the Hückel definition is accepted for the resonance parameters $\beta_{rr'}$: β is an off-diagonal matrix element of the screened one-electron energy operator, consisting of the core hamiltonian h and effective HF potential G:

$$F = h + G; \quad h = -\frac{\hbar^2}{2\mu} \Delta + \sum_R U_R^+ + \sum_{R'} U_{R'}^+$$

where G is the usual sum of Coulomb and exchange operators over all occupied MOs and

$$\beta_{rr'} = F_{rr'} - \tfrac{1}{2}(F_{rr} + F_{r'r'})s_{rr'} \qquad (9a)$$

5. Exchange Repulsion or Orthogonalization Energy

$$E_{er} = -\frac{1}{2} \sum_{rst'} p_{rs}(s_{rt'} \beta_{t's} + \beta_{rt'} s_{t's}) + \text{sum with the inverse} \qquad (10)$$
$$\text{arrangement of primes}$$
$$+ \text{small terms}$$

6. Exchange Energy

$$E_{ex} = -\frac{1}{8} \sum_{rsr's'} p_{rs} p_{r's'} s_{sr'} s_{rs'} (\gamma_{rs} + \gamma_{r's'} + \gamma_{rs'} + \gamma_{r's}) \qquad (11)$$

The energetic denominators in the second-order perturbation formulas (6) to (8) are not simple differences of MO energies but include in addition several (mainly Coulomb) interactions.[13,14] Summation in these formulas proceeds over the spin orbitals. If we had performed the integration over the spin variable, it would have been necessary to write out different expressions for the systems with even and odd numbers of electrons. This may be easily performed when particular systems are treated. The NDO approximation inside the molecules implies that the parameters of a semiempirical calculation are ascribed to the AOs, orthogonalized by the Löwdin method, inside each of the molecules. Such basis functions are nonorthogonal if they belong to different molecules ($s_{rr'} \neq 0$). This gives rise to important effects in the interaction energy which by no means can be neglected. Our formulas use the Mulliken approximation for their description.

B. The Energy Contributions

In the following discussion we refer to semiempirical variational MO theories, indicating whether the particular energy contribution is taken into account or is neglected. As calculated by these theories, the energy cannot be expressed in an additive form like (2). However, the contributions (5) to (8), (10), and (11) are associated with certain matrix elements; thus our conclusion concerning the presence or absence of a particular contribution is based on the presence or absence of the respective matrix element in a given semiempirical procedure.

1. Coulomb-Type Interactions

The first term—Coulomb energy, (5)—is present in all semiempirical SCF theories. This is the classical electrostatic term calculated with the convention that all the Coulomb interactions (electron–electron, electron–nuclei, and nuclei–nuclei) have the same absolute value $\gamma_{rr'}$ per unit charge. Let us consider more carefully the correction terms $Q_{RR'}$. Their magnitude and relative importance depend on the distance between atoms R and R'. In the intramolecular range (close to equilibrium bond lengths), $Q_{RR'}$ expresses strong repulsion caused by the predominance of the nuclear interaction $W_{RR'}$. Penetration integrals $U_{R, r'r'}$ and $U_{R', rr}$ represent strong attraction that partly cancels the repulsion due to $W_{RR'}$. In the intermolecular range, the values of $Q_{RR'}$ are most easily estimated for the atoms having spherical electron distributions (C, H), where they damp exponentially as a square of the overlap integral s. (The notation s without indices, accepted throughout the chapter, corresponds to some characteristic value of overlap integral which can always be specified if necessary.) For nonspherical distributions, quadrupole and higher multipole interactions are present as well. However, the invariance conditions[15] of a semiempirical procedure require that the integrals $U^+_{R, r'r'}$, $U^+_{R', rr}$, and $\gamma_{rr'}$ were assumed independent of the orientation of AOs. This corresponds to a spherical approximation for individual electron distributions $|\chi_r|^2$ or $|\chi_{r'}|^2$. In the framework of this scheme, the exponential (of s^2 order) estimation for penetration integrals as well as for the whole $Q_{RR'}$ values seems natural.

Polarization energy (6) is also present in SCF procedures because the HF potential of a combined system $A + A'$ governs the form of SCF MOs.

Dispersion energy (7) is due to the intermolecular electron correlation. It can be introduced in a semiempirical MO treatment by one of the usual perturbation techniques.[16]

2. Exchange-Type Interactions

Resonance energy [see (8)] is present in every semiempirical MO theory. In our perturbation expansion it arises because of the inclusion of charge transfer states in a wavefunction. Resonance makes the single energy contribution where the Hückel theory or its perturbation version are concerned.

Contributions (10) and (11) are lacking in all the procedures based on NDO approximation. The exchange energy [see (11)] is a regular complementary term to the Coulomb energy (5). The exchange repulsion (10) arises as a result of Löwdin orthogonalization of the interacting AOs. Since the Löwdin AOs become delocalized, it seems to be inconsistent to treat the integrals associated with them as permanent characteristics of atoms or atom pairs. These AOs also depend on the number and the kind of atoms surrounding the given atom or bond. Therefore the molecular environment creates specific corrections to one-electron and two-electron matrix elements if they are treated as permanent parameters. Both corrections are combined in (10), since the βs, as defined by (9a), contain both one- and two-electron interactions. The characteristic factors of "small terms" in (10) are of the order of $[(F_{rr} - F_{r'r'})/2]s^2$, and they are relatively small indeed in systems of weak polarity.

C. Interaction of Two Hydrogen Molecules as an Example

Figure 1a shows the relative importance of different contributions to the interaction energy of the collinear system $H_2 + H_2$. This is a nonempirical [without approximation (4)] calculation using a minimal basis set of $1s$ functions.[14] It is expedient to subdivide the whole range of intermolecular distance l into exchange and Coulomb regions. For nonpolar systems, such as the H_4 system, the exchange effects, decreasing exponentially as l increases, predominate within the exchange region. In the Coulomb region, Coulomb interactions, decreasing as inverse powers of l, become the main effects. The absolute value of interaction energy gets quite small in the Coulomb region; in the scale accepted for Fig. 1a the respective energy contributions are hardly discernible.

From Fig. 1b it is clear that an HF approximation works satisfactorily in the exchange region but fails to account for the interaction in the Coulomb and even the intermediate regions. Moreover, to get a satisfactory account of interaction in the Coulomb region, one should give up the zero approximation based on the hamiltonian (1a) and include the intramol-

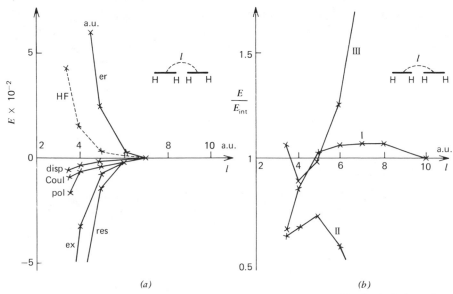

Fig. 1. Interaction energy E for the collinear $H_2 + H_2$ system (nonempirical calculation using a minimal ls basis set;[14]) the abbreviations are those used in Sections II.A.1 to II.A.6. (a) The contributions to E as defined by (2). Dotted curve represents total HF interaction energy (without dispersion). (b) Convergence of various approximations: E_{int} is the result of a variational calculation with complete account of confuguration interaction corresponding to an exact solution for the given basis set. Labels as follows: I, II, complete second-order perturbation treatments—I based on hamiltonian (1b), II based on hamiltonian (1a); III, HF component of the curve II, corresponding to the dotted curve of Fig. 1a.

ecular correlation in a zero approximation—that is, start with the hamiltonian (1b). Comparison of the calculation presented with those using extended basis sets[17,18] leads to another conclusion: the minimal set appears to be sufficient for the exchange region, but it becomes quite unsatisfactory in the Coulomb region.

Both conclusions seem to be of general value. It follows that usual semiempirical theories based on the HF approximation in a minimal basis set are not fit to deal with the Coulomb region. Subsequent discussion is therefore concerned with the exchange region only; hence we imply that the transition states for the investigated reactions are always sited in the latter region.

The behavior of potential curves in the exchange region is determined by a competition of three terms, (8), (10), and (11). For the ground state of $H_2 + H_2$ (Fig. 1a), the resonance is relatively small, whereas exchange repulsion is the main effect. If one of the H_2 molecules is excited, the absolute value of E_{res} gets much larger, but E_{er} and E_{ex} do not change significantly; thus the potential curve becomes attractive.[14]

The statement that resonance energy is very sensitive to details of the electronic structure, contrary to the insensitivity of the exchange and exchange repulsion terms, is also a general one. Resonance is a result of the interaction of the ground state of the noninteracting $A + A'$ system with charge transfer states. The strong dependence of its magnitude on the energy differences between these states is expected, and it is revealed in drastic changes of the denominators in the second-order formula (8). Since the two other exchange terms are first-order effects, they are independent of the positions of virtual states.

III. ESTIMATIONS OF INTERACTION ENERGY IN THE FRAMEWORK OF SEMIEMPIRICAL MO THEORIES

The formulas we have written out are not well suited for the quantitative treatment of molecular configurations corresponding to their equilibrium geometry. Perturbation theory is invalid for such systems because it implies that the interaction is small; in addition, expansion (3) of the overlap matrix implies s far less than unity.

Energy calculations for these systems need variational method involving all necessary matrix elements.

The perturbation treatment of interaction is also invalid for transition states in which both intra- and intermolecular interactions are of the same magnitude. This is the usual case, since rearrangement of chemical bonds occurs in a transition state, and for some bonds—namely, those forming the reaction center—it is impossible to make a clear distinction between the intra- and intermolecular ones. However, the overlap integrals of such bonds are often less than those of usual σ-bonds, which means that expansion (3) remains valid. If reactions of conjugated molecules are considered, overlap integrals are also small within the whole π-system.

Under the conditions just described, the following approach becomes meaningful.[16] Let the interaction energy be calculated by some SCF semiempirical (variational) method based on an NDO scheme. This calculation involves Coulomb, polarization, and resonance interactions but neglects contributions (10) and (11). To take the latter into account, we can again use expansion (3) and perturbation theory. This is a simple way to investigate the importance of corrections to an NDO scheme. We expect qualitative conclusions of such investigation to retain some value even for molecules with normal σ-bonds and large overlap integrals. Dispersion energy [see (7)] may be treated in the same manner, but we concentrate our attention on the exchange effects.

A. The NDO Approximation for Two-Electron Interactions

We start with some reconstruction of (10) and (11). First of all, the definition of "resonance integral" should be modified to correspond to the usual definition of semiempirical SCF theories:

$$\beta_{ab} = h_{ab} - \tfrac{1}{2}(h_{aa} + h_{bb})s_{ab} \tag{9b}$$

where h is the core hamiltonian. The part of the orthogonalization energy associated with one-electron matrix elements h_{ab} coincides with (10) except that βs are redefined according to (9b) and "small terms" are now of the form $[(h_{rr} - h_{r'r'})/2]s^2$. Then it becomes expedient to unite the exchange energy (11) and the remaining two-electron part of the orthogonalization energy [due to the HF operator G of (9a)]. Their sum is expected to be small for the following reasons. Corrections to NDO schemes are of the two types. The first type vanish after Mulliken approximation is accepted. Such is the first term of (10), since $\beta_{ab} = 0$ if h_{ab} in (9b) is calculated using this approximation. Corrections of the second type are always "small terms," in the same sense as we accepted in (10). We have applied Mulliken approximation for electron distributions involved in two-electron integrals, hence only corrections of the second type remain. The transformation of electron distributions $^{\lambda}\Omega_{ab}$ for the orthogonalized AOs with approximation (4) in mind yields

$$^{\lambda}\Omega = S^{-1/2}\Omega S^{-1/2} = D + \tfrac{1}{8}(s^2 D + D s^2 - 2sDs)$$

The second term is a desired two-electron correction to an NDO scheme. In calculation of matrix elements, it gives rise to factors

$$\frac{1}{8}\sum_{s'} s_{rs'}s_{s't}(\gamma_{ar} + \gamma_{at} - 2\gamma_{as'}); \qquad a = q \quad \text{or} \quad q'$$

They are small indeed when the intra- and intermolecular Coulomb integrals γ are of approximately the same magnitude. However, the smallness of the γ differences is necessary only for the pairs r,s' and t,s' producing the overlap integrals $s_{rs'}$ and $s_{s't}$ close to a characteristic value s, (i.e., to the pairs coming into the most close contact). For the remote atoms, the correction becomes negligibly small because of the smallness of respective overlap integrals.

These conditions seem to hold for the near exchange region, adjoining the intramolecular one ($l \sim 2$–$2,5$Å for carbon atoms), where the transition states we are interested in are disposed; then we have

$$E_{ex} + \left(\begin{array}{c} \text{orthogonalization energy due to} \\ \text{two-electron interactions} \end{array} \right) = \text{small terms}$$

The integrals β in the resonance energy equation (8) should also be redefined according to (9b). The the two-electron resonance contribution is separated out and again is proved to be a "small term."

Thus the representation of exchange contributions to interaction energy using (8), (10), and (11) with definition (9a) for resonance integrals is a reasonable approximation for the far exchange region, adjoining the Coulomb region ($l \sim 3$–$3, 5$ Å for carbon atoms), where the two-electron corrections to an NDO scheme are not negligible ($\gamma_{rs} \gg \gamma_{r's}$). For the most important near exchange region, one may take only two contributions (8) and (10), assuming the definition (9b).

This conclusion is a perturbational analogue of the results obtained in the analysis of the $S^{-1/2}$ expansion in the framework of the π-electron approximation[19, 20, 16] and their extension to the σ-electrons.[21]

In the subsequent discussion we neglect small terms due to two-electron corrections. This decision is based on the results of previous semiempirical investigations of π-electron systems.[22] The influence of the molecular environment on the electron repulsion parameters γ_{ab} is believed to be involved in some average manner in their effective empirical values. The similar corrections to the matrix elements of the core hamiltonian giving rise to the exchange repulsion term are explicitly retained in our treatment. That is, we do not use the NDO approximation when one-electron matrix elements are concerned but keep it for two-electron matrix elements.

B. Competition of Coulomb and Exchange Effects in the Intramolecular and Intermolecular Exchange Regions

Both exponentially decreasing terms $Q_{RR'}$ (5) and E_{er} (10) are usually neglected in semiempirial procedures, and we therefore investigate them in more detail. Let us separate the diagonal part of the exchange repulsion

$$\alpha_d = \sum_{RR'} \alpha_{d_{RR'}}$$

$$\alpha_{d_{RR'}} = - \sum_{r \in R, r' \in R'} (p_{rr} + p_{r'r'}) s_{r'r} \beta_{rr'}$$

It may be shown that α_d involves the main part of E_{er}. For example, in alternate π-electron systems, the nondiagonal part is of order s^3.[20, 22]

Now for any pair of atoms R, R', the values $\alpha_{d_{RR'}}$ may be combined with penetration integrals picked out of $Q_{RR'}$. If we accept that resonance integrals are proportional to s

$$\beta_{rr'} = -ks_{rr'} \tag{12}$$

and that penetration integrals are proportional to s^2 (with the coefficient k'), we arrive at the result[22]

$$\alpha = \sum_{RR'} \alpha_{d_{RR'}} + Q_{RR'} = \sum_{RR'} \alpha_{RR'} \tag{13a}$$

$$\alpha_{RR'} = \sum_{r \in R, r' \in R'} k_1(p_{rr} + p_{r'r'})s_{rr'}^2$$

$$+ \left(W_{RR'} - \sum_{r \in R, r' \in R'} N_r N_{r'} \gamma_{rr'} \right), \qquad k_1 = k - k'$$

For carbon atoms the values of k and k' are approximately equal; thus $k_1 \approx 0$.[22] That is, the exchange repulsion is almost completely canceled by the penetration integrals. Since the remaining nuclear contribution is also of order s^2, we get the net repulsion

$$\alpha_{RR'} \approx k_2 \sum_{r \in R, r' \in R'} s_{rr'}^2$$

Such estimations are valid only in the intermolecular region. In the intramolecular region, the absolute values of penetration integrals and the second (nuclear) term from $\alpha_{RR'}$ increase more rapidly than s^2 if interatomic distance decreases. The first term of (13a) becomes negative, but the positive nuclear term grows still more rapidly. The total effect is repulsion, to which the s^2 estimation is inapplicable. Its competion with the resonance attraction gives rise to an energy minimum at the equilibrium length of a chemical bond.

C. Application to π-Electron Systems

Overlap integrals are small in π-systems; thus expansion (3) is valid even at equilibrium distances. Moreover, corrections (13a) are small compared to the total π-electron energy, which makes possible their perturbation treatment. According to the procedure just formulated, we make a variational calculation, neglecting corrections (13a) (i.e., a usual PPP calculation) and add the corrections as a perturbation. The result obtained is the same expression (13a), except that the bond orders now should be taken from the PPP treatment of the combined system. The binding π-energy takes the form[16]

$$E_\pi = E_\pi^0 + \alpha \tag{14}$$

where E_π^0 is the PPP energy. We have already mentioned that for alternate systems the nondiagonal part of the exchange repulsion is of order s^3; therefore, (14) contains all the corrections to an NDO scheme having the order s^2. Furthermore, for alternate systems the π-charges in (13a) are all equal to unity, and we obtain

$$\alpha = \gamma \sum_{rs} s_{rs}^2 \qquad (13b)$$

where the sum is over the neighboring AOs of π-system and γ is a constant parameter.

Correction (13b) remains constant until geometry or charges of the π-system fail to change. Thus (13b) is the same for the ground and first excited states and makes no contribution to any respective transition energy. However, it proves to be very important when processes involving changes of molecular environment are considered. For instance, Dewar[23] and Lorquet[24] in their calculations of atomization energies had to accept the resonance parameter β_0 for aromatic π-systems ($\beta_0 \approx -1.7$ eV), which is markedly less in value than the usual spectroscopic figure ($\beta_0 \approx -2.4$ eV). After correction (13b) was brought into the calculation, the spectroscopic β_0 fitted atomization energies quite well, and this procedure furnished the empirical estimation of parameter γ ($\gamma \approx 8$ eV).[16,22]

D. NDO Procedures for All Valence Electrons

Now we are properly prepared to discuss the CNDO-2 procedure, which neglects penetration integrals and exchange repulsion but involves the second nuclear term of (13a). That could be a reliable approximation for carbon atoms in the intermolecular region. However, the parameters of CNDO-2 were calibrated at the equilibrium bond lengths. Hence because the penetration integrals have been neglected, all the attraction effects have accumulated in resonance parameters β. The net result is a very large value of k in (12): $k \approx 15$ to 20 eV. When the intermolecular range is treated, the overestimated value of k is retained but the effect it has been designed to express does not exist any more. Therefore the CNDO-2 procedure exaggerates attraction in the intermolecular range.

We point to the two characteristic illustrations.

1. The CNDO calculation of the potential surface for the Diels-Alder reaction ethylene + butadiene was absolutely unable to reveal the activation barrier,[25] the experimental activation energy being ~1 eV.

2. The calculated energy difference between cyclobutane and two ethylenes amounts to 360 kcal/mole,[26*] the experimental value being 17 kcal/mole.

---

*In Ref. 26 this defect is eliminated by means of a nonrealistic parametrization of β.

In both cases the attraction proves to be too large between the carbon atoms at "intermolecular" distances: namely, for the atoms of the reaction center of the Diels-Alder reaction or for the nonneighboring atoms of cyclobutane.

The main improvement of the MINDO-2 procedure[27,28] (transferred recently to CNDO-2[29]) is probably the artificial reduction of nuclear repulsion for the intramolecular range by means of the exponential cutting off factor:

$$(N_R N_{R'})^{-1} \times \left(\begin{array}{c} \text{repulsion of} \\ \text{nuclei } R \text{ and } R' \end{array} \right) - \gamma_{RR'} = (l_{RR'}^{-1} - \gamma_{RR'}) \exp(-\xi l_{RR'})$$

(Here integrals γ are supposed to be independent of the type and orientation of AOs and are supplied by the atomic indices.) This modification considerably reduces the absolute values of resonance integrals and so partly absorbs corrections to the NDO scheme. The price for this achievement is having to deal with an additional empirical parameter ξ, which has no physical interpretation, for every atomic pair.

Clearly the binding energy is a very subtle effect arising from the competition of numerous Coulomb and exchange forces. The straightforward way to obtain a reliable estimate of this quantity is to involve all important effects in a variational calculation. As a practical matter, the NDO procedures should be improved in two respects: a special parametrization should be introduced for core attraction integrals $U_{R,ss}^+$, independent of the parametrization for γ_{rs}, and one should explicitly involve the nonorthogonality corrections to one-electron matrix elements. A satisfactory procedure still does not exist.

IV. THE STABILIZATION ENERGY IN π–SYSTEMS

The first formulation of the idea that the transition state of a π-addition reaction is a delocalized "quasi-π-system" was apparently due to Evans, Gergely, and Seaman[30] and Bagdasaryan.[31] Fukui and his co-workers developed it to treat various organic reactions[2,32,33] by means of perturbation theory. The main point was application of the usual technique of the π-electron approximation to describe a delocalized transition state. As a measure of interaction in the transition state, the value

$$\Delta_\pi = E_\pi^{MX} - E_\pi^M - E_\pi^X \tag{15}$$

was suggested, where the π-energies E_π were attributed to the combined system (transition state) MX and reactants M and X. The calculations involve either the simple Hückel procedure or its perturbational version as presented by Dewar.[7] We refer to the Δ_π value as "stabilization energy" (SE) and call the approach that uses SE to estimate reaction rates the "SE method."

This approximation seems to be attractive because of its simplicity, but it needs justification. Indeed, the Δ_π value (15) is by no means the interaction energy of M and X. Not only do the σ-electrons make a considerable contribution, even the π-electron fraction of the interaction energy is quite unsatisfactorily accounted for, if a standard semiempirical MO theory is used to evaluate the energies in (15). It suffices to indicate that the SE in a Hückel calculation is merely a resonance energy, and all the other contributions of (2) are lacking in definition (15). The Δ_π value obtained from a PPP calculation of π-energies will additionally contain the first (electrostatic) term of Coulomb energy (5) and polarization energy (6). Since these contributions vanish for nonpolar reactants M and X, SE proves to be negative for all intermolecular distances. As a consequence, it fails to reproduce the most important region of positive energy, where attraction conquers repulsion at the point associated with a transition state.

A. Qualitative Discussion of Interactions of Various π-Systems

Interaction energy may be estimated from (13) and (14). Applying them for the calculation of π-energies in (15), we obtain

$$U = \Delta + \alpha$$

The α term involves all valence electrons of interacting atoms, not only π-electrons as in (13b). The derivation of this formula implies that the energies of the σ-bonds did not change and were included into the permanent σ-core energy, which is removed from consideration. Now we deal with the two π-systems that interact by an arrangement of the $\sigma-\sigma$ type in coplanar planes. Interactions of such σ-cores with the π-electrons and between themselves are to be considered; thus the separation of σ- and π-electrons is no longer a well-defined procedure. For the evaluation of α, (13a) should be applied in such a situation. The σ-contribution to SE is discussed below.

Let us consider the dependence of U on the intermolecular distance l for several typical cases. We begin with $l \sim 3.4$Å, which corresponds to the sum of van der Waals radii of carbon atoms. The separation being large, both quantities Δ and α are very small, and proportional to the squares of the intermolecular overlap integrals. Thus U is determined by the competition of two exponential terms, an attractive term and a repulsive one. This sum, considered as a function of l, still cannot produce a maximum. For instance, if two ethylene molecules are considered, the absolute value of Δ

is far less than α for all values of l, and repulsion therefore occurs. By contrast, the addition reaction of the methyl cation to ethylene presents a monotonic attraction until a new σ-bond is formed.

More interesting is the case when the difference between the absolute values of Δ and α is relatively small, and secondary effects become decisive. First, let us suppose that repulsion prevails at large separation. Then while the reactants get closer, a new factor appears which changes the character of the interaction. At some value of l, the inter- and intramolecular overlap integrals become approximately equal. Here the intramolecular structure of reactants is bound to change significantly, and it is not permissible to consider only one geometric parameter of the reaction—the distance l. Some new geometric variables prove to be essential, since their optimization provides a new source of the stabilization. In this region the potential curve $U(l)$ becomes an invalid means of describing the reaction path, and multidimensional potential surfaces are needed.

When the overlap integrals s are leveled off, the SE value becomes proportional to s. Since s is less than unity, the absolute Δ value begins to increase more rapidly than α. This results in a saddle point appearance on the potential surface with the interaction energy dropping after this point is passed. That means that the reaction of a new σ-bond formation proceeds. For the reaction center constructed of carbon atoms, the critical separation l is 2.3 to 2.4 Å.

Finally, it sometimes happens that the distortion of the structure of the reactants is energetically unfavorable. We mentioned at the very beginning the situation when $\Delta + \alpha > 0$, which led to a monotonous repulsion. But perhaps, although $\Delta + \alpha < 0$ at $l \sim 3.4$ Å, the attraction will convert into repulsion while the molecules come closer, producing a minimum at ~ 3 Å. The reason is that the overlap of the s–σ and s–s types becomes decisive. The $2p_z$ electrons of the reactants interact by the σ–σ type, this interaction being dominant at large separations. The mixing of $2p_z$–$2s$ and $2s$–$2s$ interactions produces quite different influences on the values of Δ or α. The former exerts only a minor change, since the resonance between both the σ- and π-electrons, as well as between the σ-electrons, is small. However, since the α is additive, it increases more rapidly than Δ decreases if new interactions are noticeable. Because we have accepted that the intramolecular distortions are energetically forbidden, the sources of the Δ decrease are exhausted and the system is incapable of passing this region. The addition reaction does not occur, but a π-complex may be formed.

B. Potential Surfaces for Addition Reactions

The investigation of potential surfaces cannot be avoided if real understanding of the mechanism of addition reactions is wanted. It is convenient to reproduce the rearrangement of intramolecular bonds by considering the hybrid AOs of the reaction center carbon atoms, which gradually pass from the trigonal to the tetrahedral state. Figure 2b presents a typical example, the reaction center of the simplest addition reaction: ethylene + methyl radical.

The σ-bond forms anew, and its hybrid orbital is

$$\chi_R = a_R(2s_R) + \sqrt{1 - a_R^2}\ (2p_{zR}) \tag{16}$$

where (2s) and (2p_z) are respective AOs of atom R. Furthermore, there are three more hybrid orbitals directed along the σ-bonds. They change from state sp^2 to sp^3, and this process is described in terms of a hybridization parameter a, changing from 0 to 0.5. Assuming that the corresponding deflection of the three σ-bonds proceeds synchronously, the relation is obtained between the deflection angle φ (Fig. 2a) and the hybridization parameter

$$a = \sqrt{2}\ tg\left(\varphi - \frac{\pi}{2}\right) \tag{17}$$

Then the interaction energy becomes

$$U = \Delta_\pi + \Delta_\sigma + \alpha \tag{18}$$

The SE Δ_π is now available from a PPP calculation with orbitals (16) for the reaction center atoms and the usual $2p_z$ AOs ($a_R = 0$) for the other atoms involved in the π-systems. The extra term Δ_σ represents the σ-bonds' energy contribution, which originates from hybridization changes. Since an explicit dependence of all three terms in (18) on the values of a_R is known,[34,22] this formula allows us to find the interaction energy in a certain range of interatomic distances and angles φ_R.

Equation (18) is a representation of a "almost" π-electron approximation[34,35] that holds for initial and intermediate stages of addition reactions. When intermolecular overlap integrals and hybridization parameters become larger, the SE cannot be reduced to a simple sum of π- and σ-electron contributions, the accuracy of (13a) for α gets worse, and the resulting formula (18) becomes a poor approximation. The important conclusion drawn from calculations of potential surfaces by means of this method is that for configurations corresponding to saddle points of radical addition and Diels-Alder reactions, the conditions of validity of the "almost" π-electron approximation are still satisfied. That is, the reacting system in its transition state still remembers that the reactants have been π-systems.

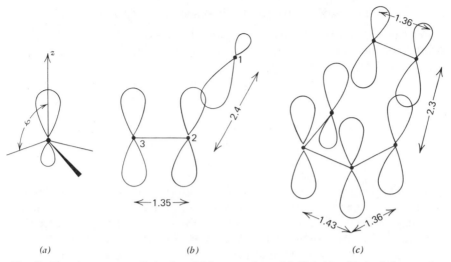

Fig. 2. Reaction centers of simple addition reactions. (*a*) Hybrid orbital of the reaction center. (*b, c*) Transition states of the reactions (*b*) methyl + ethylene (radical addition), (*c*) ethylene + butadiene (Diels-Adler). Bond lengths are in ångstrom units. Solid dots represent carbon atoms.

We would like to note that there was no necessity to represent the interaction energy in the form of (18). A variational calculation including all valence electrons of the type described in the preceding section would provide a reliable potential surface without any references to the hybridization, resonance, or exchange repulsion.* Our approach has the only following advantage: it enables us to proceed to the logical deduction of the simple SE treatment from a more advanced theory and leaves a chance for a qualitative discussion.

As calculations have shown,[34,35] the hybrid orbitals are not directed along the line drawn through the reaction center atoms (Fig. 2*b,c*)), and one may say that bent bonds are formed. In the transition state, a resonance situation arises when the values of intra- and intermolecular overlap and resonance integrals are leveled off. Thus for radical reactions the overlap integrals for the σ-bond C_1–C_2 and for the π-bond C_2–C_3 are almost identical at the saddle point and are close to 0.2—the same value found in ordinary π-electron systems. The distance between the reaction center atoms is between 2.3 and 2.4 Å.

---

*The results of MINDO calculation of radical addition reactions[36] coincided with the calculations described here.

C. Derivation of the SE Method and the Hammond Rule

We can see now that the SE Δ_π (15) should be associated with the hybrid orbitals of the "almost" π-electron approximation. Therefore (18) forms a background for further discussion.

Let us consider a set of radical addition reactions with the permanent reaction center drawn in Fig. 2b and various π-electron substituents. The structure of the transition state is

$$R_1-CH \overset{...}{=} CH_2 \cdots CH_2-R_2 \qquad (I)$$
$$\qquad (3) \qquad (2) \qquad (1)$$

The last two terms in (18)—namely, Δ_σ and α—are additive with respect to the pair exchange interactions between AOs. Hence they depend on the structure of the reaction center but not on the substituents (we assume that the overlap integrals between the AOs of R_2 and the molecule R_1CHCH_2, as well as those between R_1 and the radical R_2CH_2, are negligibly small). The only term in (18) that depends on substituents is the SE Δ_π.

Now suppose that the reaction coordinate q is the same for the whole reaction set and write

$$U(q) = \Delta_\pi(q) + C(q) \qquad (19)$$

The function C is independent of the substituents. We denote by $q_0^{\neq}$ the saddle point position for the standard reaction without substituents (ethylene + methyl) and by $q^{\neq}$ that for the other reactions. If the shift $\Delta q^{\neq} = q_0^{\neq} - q^{\neq}$ is not large, we can write for any reaction

$$U(q_0^{\neq}) = U(q^{\neq}) + \frac{\partial U}{\partial q} \Delta q^{\neq}$$

However, the derivative vanishes at the saddle point because of the extremal character of this point. Thus we arrive at the main principle of the SE method: the activation energy $U^{\neq}$ for all reactions of the given reaction set may be calculated at the same point $q_0^{\neq}$ corresponding to the transition state of the standard reaction.[37] Thus we have (the index "π" now may be omitted)

$$U^{\neq} = \Delta(\delta) + C \qquad (20)$$

The δ value is a Hückel resonance parameter for the C_1–C_2 bond in the configuration with the reaction coordinate $q_0^{\neq}$. The constants δ and C are the characteristics of reaction center I and the kinetic parameters of the whole reaction set. Their values are available from the potential surface of the standard reaction. After this surface has been constructed, Hückel calculations of Δ provide the estimates of activation energies for other reactions.

Another consequence of the main relation (19) concerns the sign of the shift $\Delta q^{\neq}$ of the saddle point within a reaction series. Using the extremal properties of the saddle point, the following expression can be easily derived:[37]

$$\Delta q^{\neq} = q_0^{\neq} - q^{\neq} = 2(p_{12} - p_{(0)12}) \frac{\partial \delta / \partial q}{\partial^2 U / \partial q^2}$$

The bond orders p_{12} and $p_{(0)12}$, as well as both derivatives, correspond to the point $q = q_0^{\neq}$. The index "0" labels the standard reaction. The last factor proves to be positive, so $\Delta q^{\neq} > 0$ for $p_{12} > p_{(0)12}$ or for $|\Delta(q_0^{\neq})| > |\Delta_0(q_0^{\neq})|$, which is the same. We sum up as follows: for the more reactive system, the saddle point locates at the earlier stage of the reaction path. This is the well-known Hammond rule.[38]

D. Radical Addition Reactions

The treatment of radical addition reactions by means of $(20)^{37}$ is illustrated by Fig. 3. The value $\delta = 0.9\beta_0$ (β_0 being the resonance parameter in benzene) was taken from the potential surface investigation. The fit of the SEs to the experimental relative rates shows satisfactory agreement for three reaction series: radical addition of methyl to olefins (Fig. 3a) and to aromatic substrates (Fig. 3b), and chain propagation in the copolymerization processes (Fig. 3c). The correlation is excellent for the

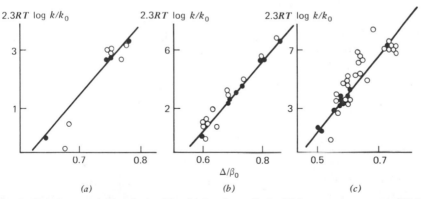

Fig. 3. Relative reactivities (in kcal/mole) for the radical addition reactions versus Hückel SEs.[37] (a) CH_3 + olefins; solid dots, hydrocarbon molecules. (b) CH_3 + aromatics; solid dots, hydrocarbon molecules. (c) Copolymerization data; solid dots, hydrocarbon polymer radicals. The distinctions in reactant bond lengths are taken into account; $\delta = 0.9\beta_0$.

hydrocarbon reactants (solid dots) and worse for the heteroatomic ones. This may be attributed to the well-known defects of the Hückel method. In the reaction series of Fig. 3*c*, the solid dots correspond to the reactions of hydrocarbon radicals. Here the experimental data are the least reliable, and steric effects may be present. The scattering is most prominent but still does not produce errors exceeding 2 kcal/mole in the calculation of activation energies.

The slope of the correlation lines (drawn through the solid dots) depends on the choice of δ and determines the value of β_0. The value $\beta_0 = -24$ kcal/mole was obtained for all three series, in close agreement with the thermochemical Hückel estimate,[4] demonstrating the internal selfconsistency of the SE method.

The plots in Fig. 3 provide the empirical estimate of the C value as well. For the addition of CH_3 to olefins and for the copolymerization reactions, C is 24 kcal/mole but the value is 27 kcal/mole for the addition reaction to aromatics. From these figures we conclude that the exchange repulsion is more pronounced for the reaction center of the series in Fig. 3*b*:

$$
\begin{array}{ccc}
& \overset{-}{\underset{\cdots}{C}} & \\
& (3) & \diagdown \\
& & C \cdots C \\
& \diagup & (2) \quad (1) \\
\overset{=}{\underset{\cdots}{C}} & & \\
(4) & &
\end{array}
$$

The natural explanation is that apart from the main interaction 1–2, two more interactions, 1–3 and 1–4 contribute to the α value, whereas there is only one such interaction in reaction center I. Thus the relatively low reactivity of aromatic compounds proves to be associated with a peculiar steric effect. It would be interesting to calculate the potential surface to check this prediction. The SE treatment of addition reactions of radicals CCl_3 and CF_3 provided similar information concerning the transition states of these processes.[39]

We do not discuss the numerous applications of the SE theory considered in Refs. 2–4, but only outline briefly the relation between SE and other RIs. This is a preliminary discussion: detailed analysis is given in a special section below.

Usually RIs, such as π-electron charge, self-polarizability, free valence, localization energy, and superdelocalizability, are nothing more than approximate estimations of SE. Depending on the type of transition state, one or another RI is more suitable. They often correlate among themselves, and then it does not seem to matter which one is used. The main distinction of the SE method is that it characterizes a pair of reactants $M + X$, whereas the other indices are associated with isolated molecules. This is an essential merit of the SE method, since it is able to describe the mutual interaction of substituents through the delocalized reaction center. A case of special importance occurs when one substituent is an electron donor and the other an electron acceptor, and charge transfer takes place in the transition state. None of the other RIs can reproduce charge-transfer effects.

A typical example is provided by the alternation of monomer links in the polymer chain produced by a copolymerization reaction.[40] If the polar properties of the monomers of a reactive pair can be distinguished, a clear-cut selectivity occurs: the polymer radical prefers addition to the "alien" monomer. Only the SE method gives an explanation for this phenomenon. In Fig. 3c, the respective radical–monomer pairs are arranged in the right order.

V. REACTION CENTERS INVOLVING σ–BONDS. HYPERCONJUGATION OF THE NONSYMMETRICAL GROUPS

Many reaction centers are linked by σ-bonds to π-bonded substituents. Extending the SE method to cover these reactions would significantly widen the scope of its application.

We consider a molecule

$$R-C\begin{array}{l} \diagup Z_1 \\ -Z_2 \\ \diagdown Z_3 \end{array} \qquad (II)$$

which involves a substituted methyl group with its bonds changing in the course of reaction, R being a π-electron substituent. We want to investigate how the interaction of the fragments R and $CZ_1Z_2Z_3$ changes along the reaction path. Since it is desirable to avoid consideration of all valence electrons, we separate the π-component out of the three σ-bonds C–Z_r, this procedure being the essence of the hyperconjugation theory.[5]

A. Generalization of the Hyperconjugation Theory

Usually a C_{3v} symmetry is accepted for the σ-electron fragment CZ_3. We have to treat the case where this condition does not hold even approximately: the C–Z_1 bond breaks and atom Z_1 is brought to infinite distance from atom C. It is obvious that conjugation of the fragments in system II takes place independently of whether the $CZ_1Z_2Z_3$ group is symmetrical.

Hence a formal description of this effect should exist. Indeed, it can be shown that the separation of the π-component is feasible in a rather wide range of such groups, the symmetrical situation being only a particular case.

Let the AO basis (C_μ, z_r) be used to describe the bonds between the valence carbon AOs (C_μ) and those of the Z_r atoms (z_r). The Hückel energy matrix is

$$F = \left(\begin{array}{c|c} \alpha_C & \beta \\ \hline \beta^\dagger & \alpha_Z \end{array} \right)$$

where α_C and α_Z are the diagonal matrixes of the Coulomb parameters and β the matrix of resonance parameters. The separation conditions are the following:[41]

1. The matrix α_Z is a scalar one, that is,

$$\alpha_Z = \alpha I = \begin{pmatrix} \alpha & & 0 \\ & \alpha & \\ 0 & & \alpha \end{pmatrix} \tag{21a}$$

2. The matrix β is a product of a diagonal and unitary matrices, that is,

$$\beta = BU; \quad B = \begin{pmatrix} B_x & & 0 \\ & B_y & \\ 0 & & B_z \end{pmatrix}; \quad UU^\dagger = U^\dagger U = I \tag{21b}$$

When relations (21) are true, the following linear combinations $Z_\mu = \Sigma U_{\mu r} z_r$ exist, and in the basis (C_μ, Z_μ) $(\mu = x, y, z)$ the energy matrix is

$$F = \left(\begin{array}{c|c} \alpha_C & B \\ \hline B & \alpha I \end{array} \right)$$

The π-component is formed by the pair of orbitals $C_z = (2p_z)$ and Z_z, which do not interact with the others. This is evident from the form of matrix F.

B. Radical Abstraction of Hydrogen Atoms

Conditions (21) hold for the class of systems in which we are interested. For instance, consider a hydrogen abstraction (Fig. 4) in which the atom $H_{(1)}$ is taken away from the $-CH_3$ group and this group turns into the radical $-CH_2^-$. The reaction path is so chosen that the symmetry plane is conserved.

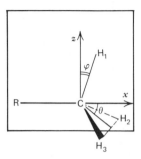

Fig. 4. Hyperconjugation of the nonsymmetrical methyl group.

We describe the R–C σ-bond by means of hybrid sp^3 orbital $C_{-x} = \frac{1}{2}(2s) - (\sqrt{3}/2)(2p_x)$. Then three carbon AOs remain for the C–H bonds:

$$C_x = \frac{\sqrt{3}}{2}(2s) + \tfrac{1}{2}(2p_x), \qquad C_y = (2p_y), \qquad C_z = (2p_z)$$

The orbitals C_x and C_z are symmetrical, and C_y is antisymmetrical respective to the x–z plane. One of the required orbitals is determined immediately by symmetry considerations. This is an antisymmetrical orbital $H_y = (1/\sqrt{2})(h_2 - h_3)$.

To satisfy conditions (21), we should put a restriction on the block of **F** matrix on the basis of four symmetrical orbitals C_x, C_z, h_1, and $H = (1/\sqrt{2})(h_2 + h_3)$:[41]

$$\frac{\langle H|\mathbf{F}|C_x\rangle}{\langle h_1|\mathbf{F}|C_x\rangle} = -\frac{\langle h_1|\mathbf{F}|C_z\rangle}{\langle h|\mathbf{F}|C_z\rangle} = k \tag{22}$$

The orbitals H_μ with the desired properties are

$$H_x = \frac{1}{\sqrt{1+k^2}}(kh_1 + H)$$

$$H_z = \frac{1}{\sqrt{1+k^2}}(h_1 - kH)$$

Condition (22) restricts the permissible configurations of the nonsymmetrical CH_3 group. It holds exactly for the tetrahedral ($k = 1/\sqrt{2}$) and trigonal ($k = 0$) states. The first case provides the well-known Mulliken orbitals H_x and H_z appearing in the regular hyperconjugation theory. The second case results in the symmetrical H orbital of the radical R–CH$_2$ and the lone AO h_1 at infinite separation.

Condition (22) requires that the angles θ and φ (Fig. 4) deflect synchronously while the distance C–H$_{(1)}$ increases. We may assume for instance $\theta = [\theta(sp^3)/\varphi(sp^3)]\varphi$, where the angles $\theta(sp^3)$ and $\varphi(sp^3)$ correspond to the tetrahedral configuration. Then we obtain a set of configurations satisfying condition (22) and blending smoothly between the sp^3 and sp^2 states. Changing in this manner, the methyl group gives away a hydrogen atom and converts to the radical. The beginning of the reaction path corresponds to the typical hyperconjugation situation, while at the end we obtain a true conjugated system. Between these terminal states there exists a continuous set of intermediate configurations, which clearly shows the relativity of the distinction between the two kinds of conjugation. In any case, there exists a conjugation of the fragments R and C–H$_z$ with the resonance parameter B_z of the latter bond changing from a considerable value ($B_z \approx 3\beta_0^1$) down to zero.

Now we are able to transfer the SE method without any change to the abstraction reaction with the reaction center

$$R_1 - C(H_2) \cdots H \cdots C(H_2) - R_2 \qquad \text{(III)}$$

The only generalization to be made is that the hydrogen AOs from both reactants should be included into H_z orbital, to obtain a continuous description of the hydrogen atom transfer from the molecule to the radical:

$$H_z = \frac{1}{\sqrt{1 + k_1^2 + k_2^2}} \left[h_1 - \frac{k_1}{\sqrt{2}}(h_2 + h_3) - \frac{k_2}{\sqrt{2}}(h_4 + h_5) \right]$$

where AOs h_1, h_2, h_3, and parameter k_1 belong to the molecule and h_4, h_5, and k_2 to the radical. While k_1 is decreasing from $1/\sqrt{2}$ to zero, k_2 increases in the inverse direction. Transition state III is characterized by the intermediate values $0 < k_1, k_2 < 1/\sqrt{2}$.

Reaction center III is quite similar formally to that of addition reaction I. The respective SEs for hydrocarbon systems have shown good correlation with the data of kinetic measurements.[4] The slope of the correlation line again corresponds to the thermochemical value $\beta_0 = -24$ kcal/mole.

The fruitfulness of this approach is supported by the experimental data on the homolytic reaction of hydrogen abstraction from amino acids.[42] The

α-amino acids were shown to be much more reactive than the β and the γ ones. This result becomes understandable if the transition state is considered:

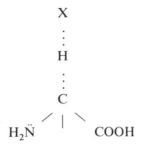

Clearly an additional contribution to SE exists in the α-isomer owing to the charge transfer from the donor group NH_2 to the acceptor CO_2H through the delocalized reaction center. The effect is absent in isolated amino acids, as well as in the transition states of the β and γ isomers.

C. General Case

The treatment of the hyperconjugation of groups with energy matrixes satisfying (21) is now straightforward. Of course, those conditions cannot be satisfied exactly if the real molecular fragments are considered. However, one can construct an effective system CZ_3, satisfying (21) and only slightly differing from the real group $CZ_1Z_2Z_3$. This procedure is described in Ref. 43. The difference between the real and effective energetic matrices should be regarded as perturbation. In the first perturbation order the effective MOs of the CZ_3 group are appropriate to describe the hyperconjugation. Using a variational technique, the effective Coulomb and resonance parameters satisfying (21) can be derived from the elements of the true energetic matrix. In particular,

$$\alpha_Z = \frac{1}{3}(\alpha_{Z_1} + \alpha_{Z_2} + \alpha_{Z_3})$$

We apply the latter result to the investigation of the Hammett correlations.

VI. CYCLIC REACTION CENTERS

Concerted reactions with cyclic transition states are naturally tractable in terms of the SE method. The potential surface calculation for the Diels-Alder reaction ethylene + butadiene[34] has shown that in its transition state, the "almost" π-electron approximation, on which the SE method is based, is quite a reasonable approach. Lately the concerted reactions have been most often discussed in terms of Woodward-Hoffman correlation diagrams.[44,45] In this section we make a brief comparison of the two methods. Note that for the reactions with open transition states already covered, the Woodward-Hoffman rules give no information except the trivial statement that they all are "allowed" ones.

A. Selection Rules Based on the SE Method

For the three main reaction types that used to be treated by the Woodward-Hoffman rules,[45] an alternative interpretation in terms of the SE method may be suggested. This has been shown by Zimmerman[46] and Dewar[3] by applying the well-known Hückel $4n+2$ rule and its anti-aromatic modification to cyclic transition states. The deduction of the same conclusions presented here is based on a standard perturbational treatment.

1.   Electrocyclic Reactions

Consider a cyclic transition state for a cyclization reaction of a conjugated molecule having $t = 2n + 2$ π-AOs (Fig. 5). We assume its initial state to be a zero approximation; thus the resonance integral δ_{1t} between the terminal atoms at first is taken to be zero. Then if the system is an alternate one, the bond order p_{1t} is positive or negative depending on whether n is even or odd. The most direct way to prove this statement is to analyze the sign of an integrand in the integral representation for p_{1t}, as introduced by Coulsom and Longuet-Higgins.[47,48] Now, taking into account the interaction δ_{1t} and dividing it into σ- and π-components, the SE is

$$\Delta = 2p_{1t}\delta_{1t} = 2p_{1t}(\delta_\sigma \cos \omega_1 \cos \omega_t + \delta_\pi \sin \omega_1 \sin \omega_t)$$

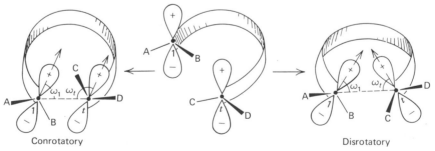

Conrotatory   Disrotatory

Fig. 5. Scheme of electrocyclic reaction (initial and transition states). The ribbon represents a conjugated system. Symmetry axes of π-electrons are disposed on its surface. Solid dots represent terminal carbon atoms.

It is evident from Fig. 5 that the two alternatives $\cos \omega_1 \cos \omega_t < 0$ or > 0 correspond to conrotatory or disrotatory cyclization processes. The δ_σ value is negative. The selection rules, which immediately follow from the conclusions concerning the sign of the σ contribution to the SE, coincide with the Woodward-Hoffman rules. The absolute values of the cosines should not be very small if a definite stereoselectivity is expected in the reaction.

2. Cycloaddition Reactions

The transition state (Fig. 6) again is assumed to involve $2n+2$ conjugated π AOs. The zero approximation corresponds to switching off the interaction: $\delta_{rr'} = \delta_{ss'} = 0$. Respective bond orders now vanish, and we appeal to a second-order perturbation theory:

$$\Delta = \pi_{rr',rr'}\delta_{rr'}^2 + \pi_{ss',ss'}\delta_{ss'}^2 + 2\pi_{rr',ss'}\delta_{rr'}\delta_{ss'} \tag{23}$$

where π are bond–bond polarizabilities.[47] This is merely another form of a usual expression [see (8)]. The two first terms are similar to SE perturbation expressions for noncyclic systems. Each of them may be combined with the respective repulsive term $\alpha_{RR'}$ or $\alpha_{SS'}$ from (13a). Therefore, if the SE had been additive, the cyclic structure of the transition state would have provided no energy gain. Preference for a concerted reaction is dictated by the sign of the cross-polarizability $\pi_{rr',ss'}$ in the interference term. Again it is easy to verify that for alternate reactants in the case at hand, this sign is minus or plus for even or odd n.

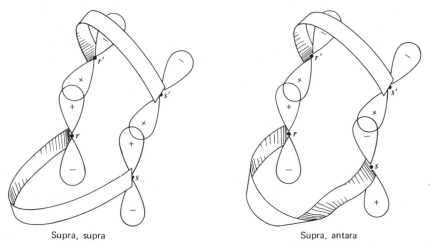

Supra, supra   Supra, antara

Fig. 6. Scheme of cycloaddition reaction (transition states). Explanations are the same as in Fig. 5.

We can propose different kinds of configuration for a transition state, and two are shown in Fig. 6: one with a normal arrangement of the π-systems of reactants (suprafacial process) and the other, in which one of the π-systems is twisted so that the axes of π-electrons form a Möbius strip (antarafacial process). To be more exact, the first case represents the *supra, supra* orientation, with normal configuration for the both reactants, the second represents the *supra, antara* process (one of the reactants in a normal configuration, the other twisted); the other two situations, which we do not consider, are obvious.[45] (It is relevant to note here that Heilbronner was the first to point out the peculiar properties of the Möbius-like π-electron systems.[49]) The different types of orientation in a transition state lead to the formation of different stereoisomers.

The first (*supra, supra*) case displayed in Fig. 6 corresponds to the integrals $\delta_{rr'}$ and $\delta_{ss'}$ having the same sign; in the second (*supra, antara*) case, the signs are opposite. The expected selection rules are derived after the sign of the cross-polarizability $\pi_{rr',ss'}$ is also taken into account. Additional consideration of the repulsion α gives a proper account of the origin of the activation energies, which prove to be rather high even for the allowed reactions.

3. Sigmatropic Reactions

The last reaction we have to investigate is the $1,k$ sigmatropic shift,[45] that is, the migration of a terminal σ (C–H) bond to a new position labeled by the number k. The following example seems sufficient to clarify all the essential points. The reacting molecule is specified as consisting of a substituted methyl group CHAB attached to a conjugated chain having $2n = k - 1$ π AO. If the counting of chain atoms begins from the C atom of the methyl group, the number of the terminal chain atom is k, and it is precisely this locus to which the hydrogen atom of the methyl group migrates.

The transition states corresponding to suprafacial and antarafacial reaction courses are exhibited in Fig. 7. According to a modified hyperconjugation theory (Section V) we can treat each of them as a cyclic $2n + 2$ system involving $2n + 1$ $2p_z$ AOs and an additional combined AO. The latter is a π-component built of the AOs of five atoms, H, A, B, C, D with the main contribution from H. We denote this orbital as Z_π. Given a suprafacial process, both the resonance integrals δ_{Z1} and δ_{Zk}, linking Z_π to the π-system, are negative, but one of them becomes positive in the case of an antarafacial process. Let us neglect the polarity of Z_π; that is, we assume that the Hückel Coulomb parameter of this orbital is the same as that of a carbon $2p$-AO. Then the energy of Z_π coincides with the energy of the nonbonding MO φ_0 of the conjugated radical, which is our $2n+1$ π-system. Applying the degenerate perturbation theory, we obtain

$$\Delta = 2(c_{01}\delta_{Z1} + c_{0k}\delta_{Zk})$$

Here c_{01} and c_{0k} are the coefficients of φ_0. They have the same or the opposite signs if n is even or odd.[3] Manipulations with the signs of c and δ again give the desired selection rules.

We would like to make two more general notes concerning the derivation of the selection rules. The first is that these rules may be automatically extended to the reverse processes proceeding through the same transition states as the direct reactions. Second, our reasoning has implied that the conjugated systems under consideration are alternate ones, and therefore the approach may appear invalid for strongly polarized transition states. However, it has not been assumed that the resonance parameters within the conjugated systems are necessarily equal.

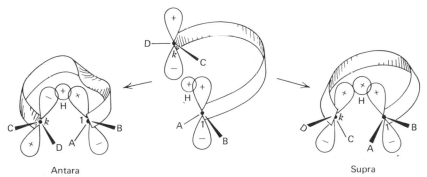

Antara   Supra

Fig. 7. Scheme of the $/1,k/$ sigmatropic reaction (initial and transition states). Explanations are the same as in Fig. 5.

B. Stability of Symmetric Transition States

If the reaction center of a concerted reaction involves a pair of equivalent bonds, say 1 and 2, then either a symmetric or an antisymmetric mode is available as a channel for the decay of a symmetric transition state. An energy profile along the decay mode exhibits a maximum at the transition state configuration, so a minimum must occur along the other mode because the configuration corresponding to maxima for both modes obviously cannot represent a transition state.

We can investigate some properties of energetic curves in the vicintiy of a symmetric point in terms of general formula (18). Neglecting the changes of σ-energy, the needed expression becomes

$$U - U_0 = \Delta - \Delta_0 + \alpha - \alpha_0$$

with index 0 labeling the symmetric reference configuration with equivalent bonds 1 and 2. Distortion of these bonds emerges as the changes Δs_1 and Δs_2 of their overlap integrals s_1 and s_2 and the changes of the resonance integrals $\Delta \delta_i = k \Delta s_i$. We consider an antisymmetric motion for which $-\Delta s_2 = \Delta s_1 = \Delta s$ and assume $\alpha = \gamma(s_1^2 + s_2^2)$. Now the linear in Δs terms cancel in the difference $\alpha - \alpha_0$. Therefore $\alpha - \alpha_0 = 2\gamma \Delta s^2$ indicating the α to have a minimum along the antisymmetric mode at the reference point. The second order perturbation formula (23) gives an estimation for $\Delta - \Delta_0$, the first order terms again vanishing due to symmetry. The result is $\Delta - \Delta_0 = 2k^2(\pi_{1,1} - \pi_{1,2})\Delta s^2$. As $\pi_{1,1} < 0$ and $|\pi_{1,1}| > |\pi_{1,2}|$, we conclude that Δ always passes through a maximum. This is an indication of a certain instability of a symmetric configuration respective to an antisymmetric motion. The magnitude of a negative force constant for Δ depends on the sign of $\pi_{1,2}$.

The simplest example of this is an exchange reaction, such as $H_2 + H$, with the collinear transition state $H \cdots H \cdots H$.[4,50] According to general theorem,[47] $\pi_{1,2}$ is positive here, so the stabilization energy decreases rapidly. The interaction energy U shows a maximum, and the antisymmetric mode proves to be a reaction coordinate, as it must.

As demonstrated by McIver[50], the Diels-Alder cycloaddition may represent another important instability case. However, this case needs more careful consideration because the sign of $\pi_{1,2}$ now alternates depending on the value $2n + 2$, the number of AOs involved in a reaction center. In most cases $n = 2$ and $\pi_{1,2}$ is negative so we expect the maximum of Δ to be very flat. The resulting behavior of the total interaction curve, as determined by the competition of Δ and α terms, may be minimum or maximum depending on the magnitudes of the parameters γ and k. The particular estimation taken from Section 3 shows that a minimum must occur.

The experiment strongly supports the idea of concerted nature of thermal Diels-Alder reactions[3,44,51]. Since the symmetric mode unquestionably goes through a maximum, we conclude that it is the repulsion α term that enables the symmetric configuration to be a transition state, although the desired minimum along the antisymmetric mode is expected to be flat.

In the recent MINDO-3 calculation[52] the Diels-Alder transition states were found to be absolutely nonsymmetric. The above analysis allows making a proposal that this is a consequence of underestimation of repulsion effects by MINDO procedure. The straightforward test using MINDO-2 parameters showed their corresponding value of the second derivative of an effective α term to be extremely small for the distance of 2.5 Å between carbon atoms. This defect was not revealed at intramolecular distances, for which MINDO method was calibrated and where $S^{-1/2}$ expansion fails. It seems that a very careful (and expensive) new calculation is needed to solve the problem quite definitely.

C. Additional Remarks

If a qualitative classification of reactions is desired, both the SE method and the Woodward-Hoffman diagrams make the same predictions. In principle, the SE method is more informative, since its predictions are not restricted by a binary code: "allowed reaction–forbidden reaction." A review of its applications in constructing correlations with quantitative kinetic measurements may be found in a recent paper by Herndon.[53] However, we encounter several difficulties that were absent when we studied the radical reactions with open transition states. Even for the simplest concerted reaction some extra contributions to the interaction energy arise, such as steric effects or changes of the ring strain energy when the reactants themselves have a cyclic structure, which is a typical case.[4,53] These factors affect the reactivity through drastic deformations of the transition state geometry and may result in changes of reaction rates that exceed by far those caused by the variation of substituents at the approximately permanent reaction center. Thus great caution is advised in choosing the reaction series, to be sure that all the effects except the Hückel SE remain constant. (A nontrivial illustration from the radical reactions field is the steric effect found for the radical addition to aromatics, as discussed in Section IV. For reactions of cyclic π-systems, it is expected to be the rule). A practical consequence is that the correct reaction series fit for the SE treatment becomes narrower and the predictions are less definite than in the case of radical reactions.

It is pertinent to mention also a special phenomenon resulting in biradical open or cyclic transition states.[54–56] Let the noninteracting reactants M and X involve odd electrons occupying the MOs φ_M and φ_X, or let two electrons be placed at one of the MOs, the other remaining unoccupied. The energy difference between the levels corresponding to φ_M and φ_X is supposed to be small.

Since the total number of electrons in a supermolecule MX is even, it might be thought that the SE method would be useful in estimating the energy of the transition state MX. However, this is not always the case. The electronic structure of MX depends on the relation between the magnitudes of two-electron Coulomb integrals associated with the MOs considered and the value of a one-electron matrix element between them. Namely, if

$$\gamma_{MM} = \left(\varphi_M \varphi_M \left| \frac{1}{r_{12}} \right| \varphi_M \varphi_M \right) \approx \left(\varphi_X \varphi_X \left| \frac{1}{r_{12}} \right| \varphi_X \varphi_X \right),$$

$$\gamma_{MX} = \left(\varphi_M \varphi_M \left| \frac{1}{r_{12}} \right| \varphi_X \varphi_X \right), \quad \text{and} \quad \beta_{MX} = (\varphi_X | h | \varphi_M)$$

the condition restricting the application of a usual MO theory is

$$\left| \frac{\gamma_{MM} - \gamma_{MX}}{4\beta_{MX}} \right| \ll 1$$

When this ratio exceeds unity, the electron correlation becomes a decisive effect. It breaks the electron pair and localizes the separated electrons at the fragments M and X. Since β_{MX} is exponentially damped when the intermolecular distance increases, the proper situation may be expected for loose transition states (recall that the difference between the energies of the MOs φ_M and φ_X should be small—namely, of the order of β_{MX}). The Hückel SE cannot give an account of reactivity for such a situation. Perturbation formula (7) is also inconsistent because we are faced by a strong correlation effect. The foregoing criterion follows from a simple configuration interaction calculation. A detailed analysis using unrestricted HF theory is given by Fukutome et al.[55,56]

VII. CORRELATIONS IN ORGANIC CHEMISTRY

A more or less close parallel between two experimental quantities in the limited reaction series is often observed in organic chemistry. The reactions to be considered are representable by a general scheme

$$\text{reactants} \rightarrow MX \rightarrow \text{products} \qquad \text{(IV)}$$

In this section we refer to fragments M and X of the transition state MX as a substituent and a reaction center, respectively. However, besides the case when M is a true substituent (i.e. does not change in the reaction), scheme IV involves bimolecular reactions with M and X being reactants. The two transition states coincide in their formal description.

Correlations of one kind are those treating the transition state MX in terms of the properties of the products, which are supposed to be known.

Typical examples are the Polanyi rule, stating the correlation between the activation energy and the heat of the reaction, or Woodward-Hoffman rules. We investigate the correlation of the theoretical quantities: the stabilization and localization energies, which fall into the same category. Such correlations have no satisfactory explanation except when the position of a transition state on the reaction coordinate is very close to products. The derivations usually presented are always based on additional assumptions needing justification themselves. For example, in the derivation of the Polanyi rule, we cannot accept uncritically the representation of the energy profile along the reaction coordinate by means of potential curves of nonreacting molecules,[3] which ignores SE altogether; and the postulate of monotonous change of SE along the reaction coordinate[40] often violated in polar transition states.

The Hammett correlations fall into a different category. According to the deduction usually adopted,[57,58] there are some interactions between the substituent M and the reaction center X. The free energy of every elementary interaction is believed to be a product $\sigma\rho$, with the first factor attributed to M and the second to X. The Hammett equation $\log(k/k_0) = \sigma\rho$ is obtained after special selection of the set of substituents has eliminated all but one interaction. A shortcoming of this formulation is that it seems incapable of explaining what the elementary interactions are like. No one has yet clearly defined such things as "polar effect" and "resonance effect." The assumption that the elementary interactions exist and are separable is again an approximation whose validity should be investigated.

More complicated empirical relations, such as that of Alfrey-Price,[59] may be regarded as a combinations of Polanyi-like and Hammett-like equations.[4] That is why it suffices to consider only correlations of the two latter types.

A. General Approach

In studying the correlations by means of the SE method, the SEs are applied to treat the relative rates k/k_0 in two reaction series, and next the correlation between SEs are considered. Thus the correlations of experimental quantities are substituted by the correlations of their mathematical models. As a result, we deal with well-defined quantities permitting accurate treatment of the subject.

The correlation of SEs is a particular case of the repeatedly discussed correlations between RIs.[1,4,6,60,61] We begin with a detailed analysis of the latter correlations for alternate hydrocarbons. The results are transferred to the Hammett reaction series, including heterolytic reactions with polar and hydrocarbon aromatic substituents.

Let the reaction center X in scheme IV be permanent. Then the set of substituents M_σ may be associated with some quantity, say the "structure variable" S. It is expedient to introduce this variable, although its mathematical nature is not yet clear. Within the limited reaction series, two RIs A and B are the functions of S: $A(S)$ and $B(S)$. It would seem sufficient to eliminate S dependence to obtain the relationship $A(B)$. However, this apparently obvious conclusion, is wrong. The actual situation is much more complicated, as can be seen by recalling that we always obtain correlations, but no true functional dependences. The question "why correlations?" needs a clear-cut answer, since we cannot refer to experimental errors in this case.

The general answer is that S is not an elementary variable. It is impossible to represent the structure of M_σ with a number s_σ; thus the set of such numbers would form a usual variable s. The structure is associated with some function. Most generally it is a Schrödinger wavefunction, but sometimes simpler quantities can be constructed. These "structure functions" vary while the structure of substituents ranges along the reaction series. Thus S is a variable in the Hilbert space, the points of this space being functions. Integration of S_σ gives expressions for RIs (or some other quantities) which depend on the form of these functions (i.e., they are functionals). Such dependence is immeasurably more complex than the simple parametric dependence $A(s)$ or $B(s)$, which yields $A(B)$ when the parameter s is eliminated. Strictly speaking, structural dependence cannot be eliminated. In the limited reaction set, however, it is possible to reduce the dependence on the form of the structure function S to that on the numerical parameter s. Then desired relationships between RIs would be obtained immediately. This, however, is not an exact procedure because the reduction to the parametric dependence involves a rather crude approximation. In other words, the correlations obtained would be less or more satisfactory, depending on the reaction set selected.

B. Correlation Between Stabilization and Localization Energies: The Structure Functions

We consider the reactions of addition of the active reactant X (cation, anion, or radical) to the substituted olefinic or aromatic molecule M:

$$M + X \rightarrow M \cdots X \rightarrow products \qquad (V)$$

Our purpose is to investigate the relationship between the SEs Δ^{MX} and sums of localization energies $L^{MX} = L^M + L^X$ for these reactions. Their correlation is a π-electron analogue of the Polanyi rule.

Both quantities Δ^{MX} and L^{MX} are supposed to be calculated by the

Hückel method. The structure of the reactants M and X is given by the matrices of their Hückel parameters α and β. We accept the same letters **M** and **X** to denote these matrices. The secular equations must be solved to reveal the dependence of RIs on the energetic matrices. Such indirect dependence is inconvenient for analytical treatment. An appropriate formulation is given in terms of the integral representations introduced by Coulson and Longuet-Higgins.[47]

Such representation for the total energy of the π-system M is

$$E^M = \frac{1}{\pi i} \int_C \left\{ z \frac{\partial}{\partial z} \ln M(z) - n^M \right\} dz$$

with a complex variable $z = x + iy$ as the argument of a characteristic polynomial $M(z)$ of the matrix **M** of order n^M. The contour of integration C should be selected so that all doubly occupied levels can be disposed within it, the singly occupied ones on it, and the vacant ones outside. The contour that is the imaginary axis from $-\infty i$ to ∞i and the infinite semicircle to the left of the y-axis obeys these conditions for even and odd alternate systems M and for any system with the doubly occupied levels $\varepsilon_\nu < 0$ as well. With this contour the integral reduces to

$$E^M = \sum \alpha_r + \frac{1}{\pi} \int_{-\infty}^{\infty} \left[y \frac{\partial}{\partial y} \ln M(iy) - n^M \right] dy$$

Based on this expression, the integral representations for various RIs were obtained. Fukui, Yonezawa, and Nagata[6] have shown that various RIs for position k of molecule M have the general form $\int f[G_k^M(y), y] dy$, where f is a function characteristic of the given RI and the following contraction is used:

$$G_k^M(y) = \frac{1}{iy} \frac{M_{kk}(iy)}{M(iy)}$$

Here $M_{kk}(iy)$ is the minor obtained from the determinant $M(iy)$ by striking out the kth row and column.

Exactly the same expressions are valid for the system X with odd number n^X of AOs. The only modification is the change in the integration contour for the polar radicals, as explained below. The RIs for position l in X are expressed in terms of the function

$$G_l^X(y) = \frac{1}{iy} \frac{X_{ll}(iy)}{X(iy)}$$

where the designations for the determinants X_{ll} and X are obvious.

The integral representations for the quantities Δ^{MX} and L^{MX} characterizing the pairs of reactants are quoted in the literature,[4,62] they make use of the functions $G_{kl}^{MX} = G_k^M \cdot G_l^X$:

$$\Delta_{kl}^{MX} = E^{MX} - E^M - E^X = -\frac{1}{\pi} \int_{-\infty}^{\infty} \ln[1 + \delta^2 y^2 G_{kl}^{MX}(y)] \, dy$$

$$L_{kl}^{MX} = L_k^M + L_l^X = \frac{1}{\pi} \int_{-\infty}^{\infty} \ln[y^4 G_{kl}^{MX}(y)] \, dy$$

(24)

In the first integrand δ is a resonance parameter of the forming bond $k–l$.

Such are the functions G_k^M, G_l^X, and G_{kl}^{MX} representing the structure of the reactants or their pairs in the reaction series. Our investigation of them enables us to understand several regularities in the behavior of the respective RIs.[6,47] For alternate M and X, they are real rational functions of y.

C. Derivation of Correlations Between RIs

Now we are ready to derive the relationship between Δ^{MX} and L^{MX}. The correlations of other RIs can be obtained in analogous fashion.

Let us give an index σ to the functions G for every particular reaction: $G_\sigma \equiv G_{kl}^{MX}$. Then we introduce a continuous parameter s and an effective function $G(y,s)$. We require that this function reproduce the functions G_σ as well as possible for the various values $s = s_\sigma$. Therefore the values of s_σ should be found from the condition that the integral

$$\int_{-\infty}^{\infty} |G_\sigma(y) - G(y,s)|^2 \, dy$$

be minimum with respect to s. The form of the function $G(y,s)$ is supposed to be chosen to give the best fit for the all functions $G_\sigma(y)$. Such selection leads to a variational procedure that is feasible in principle for the limited reaction set. Then by substituting $G_\sigma(y)$ by $G(y,s_\sigma)$ in the (24) we obtain the desired parametric dependence for the "average" RIs $\bar{\Delta}(s_\sigma)$ and $\bar{L}(s_\sigma)$, which differ from the exact values Δ_σ and L_σ because the true integrands have been changed.

This approach admits consideration of any set of reactions V with both M and X varying. But the correlations are often unsatisfactory when various X are compared. Therefore we restrict our work to the reaction series with permanent X. Then the structure functions will be $G_k^M \equiv G_\sigma$, whereas G_l^X will be the characteristic of the functional (i.e., of the respective RI). The values of L^{MX} calculated for various X differ only by a constant term L_i^X but Δ^{MX} for various X may be thought of as various RIs of molecule M.

Now consider the reaction with methyl radical ($X = CH_3$; $G_i^X = 1/y^2$). We shall not perform the complicated variational procedure to find $G(y,s)$. We assume instead that $G(y,s) = 1/(s^2 + y^2)$. This is an exact G_1 function for "ethylene" having the resonance parameter $\beta = s$. Integration in (24) yields the average functions

$$\overline{\Delta}(s) = 2(\sqrt{s^2 + \delta^2} - s)$$

$$\overline{L}(s) = 2s$$

After s is eliminated, the relationship $\overline{\Delta}(\overline{L})$ is obtained (see curve I in Fig. 8). The same relationship for the allyl radical (curve II) practically coincides with I. For both cases $\delta = \beta_0$.

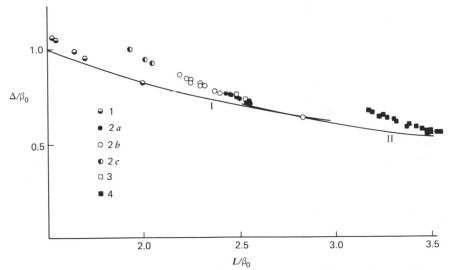

Fig. 8. Correlation of localization and stabilization energies; 1, CH_3 + olefins, 2, CH_3 + aromatics: (a) subclass (0); (b) subclass (1); (c) subclass (2); 3, allyl + olefins; 4, allyl + aromatics. For all molecules $\beta = \beta_0$, $\delta = \beta_0$.

The calculated points are distributed among four groups corresponding to the following pairs: CH_3 + olefins, CH_3 + aromatics, allyl + olefins, and allyl + aromatics. The "theoretical" relationship I–II is not satisfactory for the whole set.

D. Degeneracy of Reaction Series and Degree of Physical Resemblance

If we had performed a variational procedure to find $G(y,s)$, the "theoretical" correlation curves would have provided a much better fit. The I–II curves give much better fit for the olefinic M. Probably a more suitable curve for the aromatic M would have been obtained if we had used an exact G_k^M function for benzene or naphthalene with the effective parameter $\beta = s$ for the two bonds adjacent to the reaction center.

Thus for the special classes of functions $G_\sigma(y)$ it is possible to find an optimum function $G(y,s)$ far more appropriate than for the whole set. We can say this because the correlations within classes were considerably better. The molecules of different classes differ also in their structural features: they have a different number of carbon atoms attached to the reaction center. Koutecky, Zahradnik, and Cizek[60] found an even more detailed classification by considering the number of neighbors for the atoms attached to the reaction center in aromatic systems. [For example, the 1, 2, and 9 anthracene positions belong to the different subgroups designated as (1), (0), and (2), since they do not have the same number (1, 0, and 2) of tertiary carbon atoms attached to them.] The correlation of RIs becomes still better within such subgroups, as can be observed in Fig. 8. Nevertheless we shall not discuss this secondary effect any more because it seems to be specific only for the Hückel calculations, vanishing when the PPP method is applied;[63,64] indeed, the long-range Coulomb interactions are likely to abolish the influence of the secondary topological factors.

We arrive at the conclusion that special selection allows us to form sets of reactions that are more or less close to one-parameter set. A more explicit formulation of such proximity is presented in Appendix A, where it is also shown that the olefins and the aromatic compounds actually fall into two distinct groups, and the functions within each group comprise a better fit to one-parameter set than those from the different groups. The extreme case would be an exact one-parameter set in which an exact functional dependence for any pair of functionals were found. We refer to such reaction series as the degenerate ones.

Now we move on to the variation of reactant X, presenting in Fig. 9 the plots for the reactions of the polar anion, radical, and cation with aromatic substrates.[61] The polarity of X is determined by $\alpha_X = \alpha_C + \beta_0$. The combined system MX is not alternate any more. Thus G_{kl}^{MX} are now complex functions, since an imaginary component depending on α_X arises. The new sets of functions G^{MX} are quite different from the set with $\alpha_X = \alpha_C = 0$. Combining Figs. 8 and 9 shows that the two sets form two nonoverlapping groups of points. Hence only the discussion with X kept fixed is reasonable.

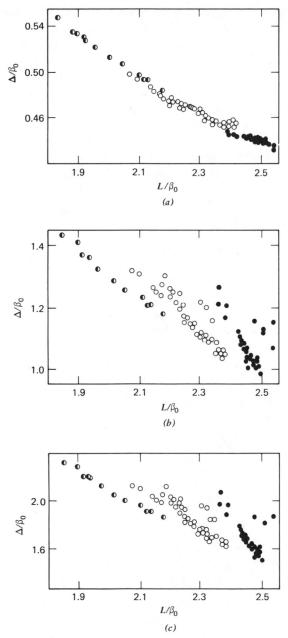

Fig. 9. Localization and stabilization energies for reactions with polar $X(\alpha_X = \delta = \beta_0)$.[61] (a) $X = $ anion, (b) $X = $ radical, (c) $X = $ cation.

For the anion, the correlation between Δ^{MX} and L^{MX} is satisfactory (Fig. 9a). It breaks down in the radical and cation cases (Fig. 9b, 9c) because the former integration contour is not suited for the last two cases, since the singly occupied or vacant level $\alpha_X < 0$ (in the isolated radical or cation) as well as the "frontier" level of the combined system $\varepsilon_f^{MX} < 0$ appear. The latter is either singly occupied in the radical MX or vacant in the cation MX^+. These points should be placed outside the integration contour or on it, in the cases of cation and radical, respectively. Additionally, the contour should be common for all three energy terms contributing to expression (24) for Δ^{MX}. Such contour for the cation is presented in Fig. 10. It surrounds α_X to fit E^X and ε_f^{MX} to fit E^{MX}. Finally we get

$$\Delta_\sigma^{MX} = -\frac{1}{\pi} \int_{-\infty}^{\infty} \ln\left[1 + \delta^2 y^2 G_\sigma^{MX}(y)\right] dy + 2\left(\alpha_X - \varepsilon_f^{MX}\right)$$

The factor 2 in the last term should be substituted by 1 for the radical case.

Expression (24) for L^{MX} does not change when the new contour is used. Thus it is the extra term $-2\varepsilon_f^{MX}$ that disturbs the correlation, because the integral—corresponding to the anionic SE—shows satisfactory correlation with L^{MX}. For the alternate systems we have $\varepsilon_f^{MX} = \alpha_X = \alpha_C = 0$. This equality holds approximately when the difference $|\alpha_X - \alpha_C|$ is not large.

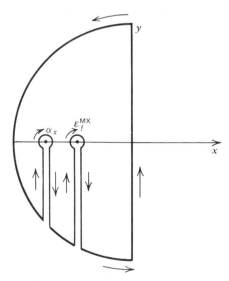

Fig. 10. Integration contour for systems with charge transfer.

But when it is large, $\varepsilon_f^{MX} \approx \varepsilon_f^M$, the last value being the highest occupied level of M. That means that the destruction of the correlations is directly related to the magnitude of $|\alpha_X - \varepsilon_f^M|$. In other words, the correlations between two RIs are the worse when the charge transfer between the reactants is more pronounced. This is true providing the effect is present in one of them (Δ^{MX}) and absent in the other (L^{MX}).

Our main conclusion is that the correlations of functionals have a dual nature. On the one hand they reflect the degree of physical resemblance between the quantities compared. For example, by comparing Figs. 8 and 9 we have learned that the degree of physical resemblance between Δ^{MX} and L^{MX} was much more visible when $X = CH_3$ than when X was a polar cation. Such comparisons should be performed with a permanent set of molecules (aromatic M in both cases). On the other hand, the correlations improve in specially selected "nearly degenerate" sets of M, reflecting the properties of the sets but having nothing to do with the physical resemblance of the functionals. The two extreme cases should be mentioned:

1. If one functional is a function of the other, an exact functional dependence is bound to be observed in any reaction series.

2. In a degenerate reaction series, an exact functional dependence is bound to be observed for any pair of functionals.

Correlations occur in other cases and we cannot separate the two effects. It is possible to estimate their relative importance only if a pair of correlations is compared within a fixed reaction series or if the quality of a single correlation is examined in different reaction series.

E. The Hammett-Type Correlations

In this section we reduce to the SE method the description of the reactions comprising a usual object for testing various $\sigma\rho$ schemes. Let us again consider reaction series IV. The reaction center may be formed by σ-bonds, but the substituents are π-systems as before. The interaction of M and X may be investigated with the hyperconjugation theory by separating of the π-component out of the σ-bond system of the reaction center.

We confine the consideration to reaction centers of the form $CZ_1Z_2Z_3$ (II). For these groups the π-component is a formal π-bond:

$$C-Y \qquad\qquad (VI)$$

with the Hückel parameters α_Y and β_{CY} depending on Z_1, Z_2, Z_3. Even if the reaction center is a π-system, we shall reduce it to the form of scheme VI.

We take the standard Hammett reaction as an illustration:

The reduction is accomplished automatically for the carboxyl anion. Defining O_1 and O_2 as the $2p_z$ AOs of the respective atoms and utilizing symmetry, we construct the orbitals $Y_1 = 1/\sqrt{2}\ (O_1 + O_2)$ and $Y_2 = 1\sqrt{2}$ $(O_1 - O_2)$ and make sure that Y_2 does not interact with M. The two electrons that occupy this orbital are excluded from our consideration.

The acid molecule is not symmetrical because the parameters α and β for O_1 and O_2 are different. Nevertheless it is possible to construct an effective symmetrical model with the two "oxygen atoms" alike. This results in a fragment similar to the carboxyl anion with the changed parameters

$$\alpha_O = \frac{1}{2}(\alpha_{O_1} + \alpha_{O_2}); \qquad \beta_{CO} = \frac{1}{\sqrt{2}}(\beta_{CO_1} + \beta_{CO_2})$$

This approximate procedure again represents the carboxyl group of the acid in a standard form (scheme VI). We can write the following scheme for the dissociation process:

$$M - C - Y \rightleftarrows M - C - Y' \qquad \text{(VII)}$$

Actually, fragment VI is a model for the reaction center specified by the effective parameters α_Y, $\alpha_{Y'}$, β_{CY}, $\beta_{CY'}$, as well as the ethylene fragment specified by the effective parameter $\beta = s$, has been accepted as a model for the hydrocarbon substituent M.

Scheme VII imitates the reaction in a simple way as a change of the Hückel parameters of atom Y. In particular one may expect $\alpha_{Y'}/\alpha_Y < 1$, since the negative charge of the anion should produce a more efficient screening of the nuclear potential.

Quite similarly, the hydrolysis of aromatic esters reduces to the following scheme:[4]

$$M - C - Y \rightarrow M - C - Y' \rightarrow \text{products} \qquad \text{(VIII)}$$

It can be shown that "atom" Y' is different for the cases of base, neutral, and acid hydrolysis, this distinction naturally explaining the peculiarities of the three reactions. Scheme VIII is also suited to describe the reactions of solvolysis of arylmethylchlorides

$$M{-}CH_2Cl \quad \xrightarrow{(n+2)H_2O} \quad M{-}CH_2 \qquad (H_2O)_n \quad \rightarrow \quad \text{products}$$

and ionic exchange

$$M{-}CH_2Cl + I^- \rightarrow \left[M{-}\ CH_2 \right]^- \rightarrow \text{products}$$

Here the nonsymmetrical version of the hyperconjugation theory should be applied. The two possible mechanisms for these reactions are bimolecular substitution of the chlorine atom (S_N2) or monomolecular heterolytic fission of the C–Cl bond as a rate-determining step (S_N1), followed by a rapid recombination of the forming cation with the solvent anion. The transition from S_N2 to S_N1 for such reactions is imitated by changing $\beta_{CY'}$ from some value to zero, as in the case of hydrogen abstraction. It should be noted that Daudel and Chalvet were the first to propose model VIII for ionic exchange reactions.[65]

We shall not enter into the details of these processes. It suffices to state that a great number of reactions obeying $\sigma\rho$ correlations may be represented by type VII or VIII schemes. Further comparison shows that "reactions" VII and VIII are similar to the addition reaction V, the reaction centers CY and CY' playing the part of the reactant X. Developing this analogy, one may calculate the SEs separately for the reactants (Δ) and for the transition state (Δ'). Their difference is directly related to the difference of the π-energies of the reactants (E) and the transition state (E'):

$$\Delta E = E' - E = \Delta' - \Delta - (E^{X'} - E^X)$$

The last term is the energy difference between the isolated reaction centers X' (i.e., CY') and X (i.e., CY). Being the characteristic of the reaction centers only, it does not depend on substituents. Hence the quantity $D = \Delta' - \Delta$ is the RI for the reactions discussed.

When treating a pair of reactions (with the upper index "0" representing the standard reaction) and studying the influence of the substituents M_σ (with $\sigma = 0$ for the standard substituent), let us compare the two values: $D_\sigma^{(0)}$ and D_σ. The correlation between them can be derived in the same manner as in the preceding section: this is in fact the correlation between the two types of SE. Except from the different factors $G_l^X, G_l^{X'}$ for the reaction investigated and $G_l^{X^0}, G_l^{X^{0'}}$ for the standard reaction, which are specific for the reaction centers, the integrands in (24) are alike. If D values are good RIs, then their correlation explains the Hammett correlation. Assuming ΔE to be the measure of the activation energy change within the reaction series, we find: $\Delta E_\sigma - \Delta E_0 = D_\sigma - D_0$. Then, making use of the linear dependence for the average (one-parameter) Ds within the relatively small interval of the structural change $\overline{D}_\sigma = \rho \overline{D}_\sigma^{(0)} + \text{const}$, we obtain

$$\Delta E_\sigma - \Delta E_0 \approx \rho \left(\overline{D}_\sigma^{(0)} - \overline{D}_{(0)}^{(0)} \right)$$

The difference $\overline{D}_\sigma^{(0)} - \overline{D}_0^{(0)}$ can differ from the Hammett σ constant by nothing more than a constant factor.

F. Discussion

To make the quantity D a really good RI it is always possible to refine its calculation without changing the general line of reasoning. First, the simple form VI is not at all necessary for reaction centers X and X'. The forms of the functions G_l^X or $G_l^{X'}$ are the only important ones; they may happen to be very close for systems that differ by the number of AOs involved and seemingly have quite different structures. Then, since the Hückel SEs do not take into account the long-range Coulomb interactions (5) and (6) ("field effect") and the σ-core polarization ("inductive effect transferred through the σ-bonds"), we could calculate the SEs by the PPP method to include the first effect or by an SCF method with all valence electrons to include both. Then the functions G are to be substituted by some new structure functions. They are not known yet, but the manner in which we deduced correlations was independent of the form of the structure functions. The manipulations with the Hückel G functions were nothing more than illustration. The calculations of localization energies and SEs by the PPP and CNDO-2 methods[63,64] have shown that even though the RIs refined in this way provided a better fit to the experiment, the improvement was not sufficiently impressive to warrant discarding the results of the Hückel calculations.

We present below two examples in which the Hückel treatment gives a reasonable interpretation of the experimental observations concerning $\sigma\rho$ relationship.

1. Usually the reaction series with the set of substituents $M = X - \langle \rangle$ are considered. We shall call them Hammett series. In the Hammett series the $\sigma\rho$ scheme reproduces well the relative rates of the side-chain reactions when the interaction of the reaction center X with the aromatic system M is not large. If this interaction is large (usually in the transition state), as in electrophilic substitution reactions or in solvolytic S_N1 reactions, the correlations break down and a special set of σ^+ constants is needed. [57,58]

In the first case we deal with the reaction centers X and X′ having only doubly occupied levels with their energies less than α_C. In the second case the cationic reaction center X′$^+$ has a vacant level and usually $\alpha_X < \alpha_C$. Thus the charge-transfer effect is expected in the second case, and as was shown previously (discussion of Fig. 9), correlations of the two types of SEs are bound to be destroyed. Good correlations can occur only for reaction centers of the same type. This conclusion explains the necessity for two sets of σ-constants.

More generally, we can speak of the physical resemblance between the reaction centers X, X′ and X$^{(0)}$, X$^{(0)′}$, the measure of which is the distance between the corresponding $G_1^X(y)$ functions (see Appendix A).

2. Streitwieser[1] applied the $\sigma\rho$ treatment for benzenoid substituents M: including phenyl, 1-and 2-naphthyl, and 1-, 2-, and 9-anthryl. We refer to this set as a Streitwieser series. These substituents have been studied only in electrophilic substitution and solvolytic reactions, and the respective σs are to be regarded as σ^+ constants. We would like to emphasize that the correlations in the Streitwieser series are worse by far than those in the Hammett series. The scattering of points is rather large (as opposed to the usual Hammett correlations) even if two solvolytic reactions in different media are compared,[1,64] whereas the solvolytic σ^+ constants are quite different from those derived from the electrophilic substitution.[64,66] Note that the range of the σ^+ constants in the Streitwieser set is approximately three times as small as that of the Hammett set* (0.5 and 1.5 σ^+ units, respectively). Thus the improvement of correlation cannot be explained by means of the decrease of "the range of structural changes" as was suggested by Leffler and Grünwald.[57]

The natural explanation that follows from the preceding discussion is that the Hammett set is degenerate to a larger extent (i.e., is closer to the one-parameter set) than the Streitwieser set. This statement seems reasonable because many more parameters are needed to reproduce exactly the

*We mean the σ^+ constants derived from acetolysis data.[64]

structural changes in the Streitwieser set. We therefore accept it as a very probable hypothesis.*

One more piece of evidence to support this point of view concerns the comparison of the experimental σ^+ constants with SEs calculated by CNDO-2 procedure.[64†] The respective plots again show a considerable dispersion for the Streitwieser set and a smaller one for the Hammett set. The correlation coefficients are .937 and .967, respectively. There are no physical reasons to think that the SE method is better justified in the latter case. Therefore we suppose that good correlations between various quantities, which are typical of the Hammett series, are caused not by any physical reason but by the "near degeneracy" of the Hammett set of substituents.

G. Conclusion

Finally we would like to raise the problem of the physical resemblance between SEs and the values of $\log k/k_0$ without any reference to the $\sigma\rho$ scheme. Such resemblance is exhibited much better for homolytic reactions in nonpolar media than for heterolytic reactions in polar solvents. We should not expect good correspondence in the latter case because the solvation effects and the variety of reaction mechanisms are not taken into account in the SE method. (Probably the narrowing of the reaction series could partly eliminate the influence of these factors, producing a "physical" improvement of correlations.) As far as the Hückel SEs are concerned, the best degree of physical resemblance is expected for the nonpolar reactants. Accordingly, correlation for the radical addition to aromatics (Fig. 3b, solid dots) is much better than that for the solvolytic reactions in the Streitwieser series.[64] Similarly, we expect that the correlations are more "physical" for the radical addition reactions than for the reactions with cyclic transition states, as discussed at the end of Section VI.

Manipulations with correlations cannot give an absolute estimate of physical resemblance. The calculation presented in Appendix A shows the Streitwieser set to be degenerate to a considerable extent, although probably less so than the Hammett set. It is impossible to eliminate nonphysical correlations at the RI level of investigation. For further progress to be

---

*Note that when the effect of the substituent Z in Z— is represented by varying the Coulomb parameter α for the carbon atom adjacent to Z, this purely inductive model corresponds to one-parameter set.

†In Ref. 64 the SEs $\Delta^{MX'}$ calculated for the cation MC^+ were adopted to estimate σ^+ values. The differences of the values Δ^{MX} for the initial molecules were neglected.

made, absolute calculations of the reaction rates based on the more explicit physical models of the elementary reaction act are necessary. This leads to the construction of potential surfaces, and after all, to chemical dynamics studies.

APPENDIX A

The Nearly Degenerate Substituent Sets

In the Hilbert space, where functions are points, the distance between the points $\sigma_1 \equiv G_{\sigma_1}(y)$ and $\sigma_2 \equiv G_{\sigma_2}(y)$ is defined as:[67]

$$\rho(\sigma_1, \sigma_2) = \left[\int |G_{\sigma_1}(y) - G_{\sigma_2}(y)|^2 dy \right]^{1/2}$$

For the given set of points σ, we select two basic points 0 and m and define the function $G(y,s)$ as the straight line in the Hilbert space drawn through these points :

$$G(y,s) = G_0(y) + s[G_m(y) - G_0(y)]$$

For every point σ, an optimum value s_σ of the parameter s should be found by minimizing the distance between the points σ and $s_\sigma \equiv G(y, s_\sigma)$

$$\frac{d}{ds}[\rho^2(\sigma, s)]_{s=s_\sigma} = 0$$

For real σ, the direct calculation yields

$$\rho^2(\sigma,s) = \rho^2(\sigma,0) + s^2\rho^2(m,0) - 2sa_\sigma; a_\sigma = \int (G_\sigma - G_0)(G_m - G_0) dy$$

After minimization respective to s is performed, we have

$$\rho^2(\sigma, s_\sigma) = \rho^2(\sigma, 0) - \frac{a_\sigma^2}{\rho^2(m,0)}$$

If we use the notation $\Delta_\sigma = \rho^2(m, \sigma) - [\rho(m,0) - \rho(\sigma,0)]^2$, the following identity is valid: $a_\sigma = \frac{1}{2}\Delta_\sigma - \rho(m,0)\rho(\sigma,0)$. Simple transformations result in

$$\rho^2(\sigma, s_\sigma) = \frac{\rho(\sigma,0)}{\rho(m,0)}\Delta_\sigma - \frac{\Delta_\sigma^2}{4\rho^2(m,0)}$$

The value of $\rho(\sigma, s_\sigma)$ vanishes and the one-parameter description becomes exact when $\Delta_\sigma = 0$, which is equivalent to $\rho(m,\sigma) + \rho(\sigma,0) = \rho(m,0)$

(i.e., to the additivity of distances). This is an analytical expression of the statement that the points σ form a one-dimensional array (or one-parameter set) in the Hilbert space.

The measure of the proximity to the one-parameter set is the dispersion

$$d^2 = \frac{1}{n} \sum_\sigma \rho^2(\sigma, s_\sigma)$$

with n being a number of points.

For several alternate systems the absolute ρ values are listed in Table A-I. All resonance parameters are accepted as equal.

TABLE A-I

The absolute distances ρ, in units $\beta_0^{-3/2}$, for alternate systems

	Bu	S	H	Bz	N-1	P-9	Anthracene-9 (A-9)
Ethylene (E)	0.830	0.635	1.488	0.283	0.164	0.176	1.036
Butadiene (Bu)		0.195	0.672	1.098	0.711	0.700	0.313
Styrene (S)			0.865	0.905	0.521	0.511	0.459
Hexatriene (H)				1.744	1.355	1.345	0.484
Benzene (Bz)					0.391	0.401	1.278
Naphthalene-1 (N-1)						0.000	0.892
Phenanthrene-9 (P-9)							0.881

For different sets of substituents, the relative deflections $r = d/\rho\,(0, m)$, following the notation of Table A-I, are:

1. E (point 0), Bu, S, H (point m): $r = 0.046$
2. Bz (point 0), N-1, P-9, A-9 (point m): $r = 0.027$
3. Combined sets $1 + 2$ with different selection of basic points:

   (a) $0 = E$, $m = H$: $r = 0.065$
   (b) $0 = Bz$, $m = A-9$: $r = 0.109$
   (c) $0 = Bz$, $m = H$: $r = 0.054$

Sets 1 and especially 2 are nearly degenerate indeed. The combined set $3c$ deflects from the one-parameter set more considerably. The comparison of $3a$ and $3b$ with 1 and 2 additionally confirms the statement that the G-functions for olefins and for aromatics fall into distinct classes.

It should be noted that the simple one-parameter description we have used is not the best one because the form of the function $G(y, s)$ was not optimized. For more consistent treatment, we could have begun with arbitrary functions 0 and m and determined them from the condition of minimum d^2. This would have been the best linear treatment. The nonlinear treatment might give even better approximation. However, such variational procedures need complicated calculations.

References

1. A. Streitwieser, *Molecular Orbital Theory for Organic Chemists*, Wiley, New York, 1961.
2. K. Fukui, in *Molecular Orbitals in Chemistry, Physics and Biology*, Academic Press, New York, 1964, p. 513, ed. by P. -O. Löwdin and B. Pullmann.
3. M. J. S. Dewar, *The Molecular Orbital Theory of Organic Chemistry*, McGraw-Hill, New York, 1969.
4. M. V. Basilevsky, *Molecular Orbital Method and Reactivity of Organic Molecules*, Khimia, Moscow, 1969.
5. R. S. Mulliken, C. A. Rieke, and W. G. Brown, *J. Am. Chem. Soc.*, **76**, 41 (1941).
6. K. Fukui, T. Yonezawa, and C. Nagata, *J. Chem. Phys.*, **26**, 831 (1957).
7. M. J. S. Dewar, *J. Am. Chem. Soc.*, **74**, 3341 (1952).
8. K. Fukui and H. Fujimoto, *Bull. Chem. Soc. Japan*, **41**, 1989 (1968).
9. A. Devaquet and L. Salem, *J. Am. Chem. Soc.*, **91**, 3793 (1969).
10. A. Devaquet, *Mol. Phys.*, **18**, 233 (1970).
11. R. Sustman and G. Binsch, *Mol. Phys.*, **20**, 1, 9 (1971).
12. L. Salem, *J. Am. Chem. Soc.*, **90**, 543 (1968).
13. M. V. Basilevsky and M. M. Berenfeld, *Int. Jn. Quant. Chem.*, **6**, 23, 555 (1972).
14. M. V. Basilevsky and M. M. Berenfeld, *Int. Jn. Quant. Chem.*, **8**, 467 (1974).
15. J. A. Pople, D. A. Santry, and G. A. Segal, *J. Chem. Phys.*, **43S**, 129 (1965).
16. M. V. Basilevsky, *Theor. Chim. Acta*, **13**, 409 (1969).
17. E. Kochanski, *J. Chem. Phys.*, **58**, 5823 (1973).
18. C. F. Bender and H. F. Schaefer, *J. Chem. Phys.*, **57**, 217 (1973).
19. M. G. Veselov, and M. M. Mestechkin, *Lith. Phys. Collect.*, **3**, 269 (1963) (in Russian).
20. I. Fisher-Hjalmars, *J. Chem. Phys.*, **42**, 1962 (1965).
21. R. D. Brown, and K. R. Roby, *Theor. Chim. Acta.*, **16**, 175 (1970).
22. M. V. Basilevsky, and V. A. Tikhomirov, *Teor. Exp. Khim.*, **8**, 723, 728 (1972) (in Russian).
23. A. L. H. Chung, and M. J. S. Dewar, *J. Chem. Phys.*, **42**, 756 (1965); M. J. S. Dewar, and C. deLlano, *J. Am. Chem. Soc.*, **91**, 789 (1969).
24. A. J. Lorquet, *Theor. Chim. Acta.*, **5**, 192 (1965).
25. O. Kikuchi, *Tetrahedron*, **27**, 2791 (1971).
26. H. Fisher, and H. Kollmar, *Theor. Chim. Acta*, **13**, 213 (1969).
27. M. J. S. Dewar, and E. Haselbach, *J. Am. Chem. Soc.*, **92**, 590 (1970).
28. M. J. S. Dewar, *Twenty-third International Congress of Pure and Applied Chemistry*, Vol. 1, Butterworths, London, 1971, p. 1.
29. R. J. Boyd and M. A. Whitehead, *J. Chem. Soc., Dalton Trans.*, **73**, (1972).
30. M. Evans, J. Gergely, and E. Seaman, *J. Polym. Sci.*, **3**, 866 (1948).
31. K. S. Bagdasaryan, *Zh. Fiz. Khim.*, **23**, 1375 (1949) (in Russian).
32. T. Yonezawa, K. Hayashi, C. Nagata, S. Okamura, and K. Fukui, *J. Polym. Sci.*, **14**, 312 (1954).

33. K. Hayashi, T. Yonezawa, C. Nagata, S. Okamura, and K. Fukui, *J. Polym. Sci.*, **20**, 537 (1956).
34. M. V. Basilevsky, V. A. Tikhomirov, and I. E. Chlenov, *Theor. Chim. Acta*, **23**, 75 (1971).
35. M. V. Basilevsky and V. A. Tikhomirov, *Twenty-third International Congress of Pure and Applied Chemistry*, Vol. 1, Butterworths, London, 1971, p. 109.
36. J. R. Hoyland, *Theor. Chim. Acta*, **22**, 229 (1971).
37. M. V. Basilevsky, in *Radiospectroscopic and Quantum Chemical Methods in Structural Research*, Moscow, 1967, pp. 33–58 (in Russian).
38. G. S. Hammond, *J. Am. Chem. Soc.*, **77**, 334 (1955).
39. M. V. Basilevsky, and I. E. Chlenov, *Zh. Strukt. Khim.*, **8**, 993 (1967) (in Russian).
40. K. S. Bagdasaryan, *Theory of Radical Polymerization*, Nauka, Moscow, 1966, Chapter 8, 13 (in Russian).
41. M. V. Basilevsky, *Zh. Strukt. Khim.*, **5**, 455, 461 (1964) (in Russian).
42. G. Moger, dissertation, Moscow State University, 1972 (in Russian).
43. M. V. Basilevsky, *Zh. Strukt. Khim.*, **7**, 781 (1966) (in Russian).
44. R. B. Woodward, and R. Hoffman, *J. Am. Chem. Soc.*, **87**, 395, 2046, 2511 (1965).
45. R. B. Woodward, and R. Hoffman *The Conservation of Orbital Symmetry*, Verlag Chemie–Academic Press, London, 1970.
46. H. E. Zimmerman, *J. Am. Chem. Soc.*, **88**, 1564 (1966).
47. C. A. Coulson, and H. C. Longuet-Higgins, *Proc. Roy. Soc.*, **A191**, 39 (1947); **A192**, 16 (1947).
48. C. A. Coulson, and H. C. Longuet-Higgins, *Proc. Roy. Soc., (London)*, **A193**, 447 (1948).
49. E. Heilbronner, *Tetrahedron Lett.*, **1964**, 1923.
50. G. W. McIver, *Accts. of Chem. Research*, **7**, 72 (1974).
51. G. R. McCabe, and C. A. Eckert, *Accts. of Chem. Research* **7**, 251 (1974).
52. M. J. S. Dewar, A. C. Griffin, and S. Kirschner, *J. Am. Chem. Soc.*, **96**, 6225 (1974).
53. W. C. Herndon, *Chem. Rev.*, **72**, 157 (1972).
54. L. Salem, and C. Rowland, *Angew. Chem. Intn. Ed.*, **11**, 92 (1972).
55. H. Fukutome, *Progr. Theort. Phys.*, **47**, 1156 (1972); **49**, 22 (1973).
56. K. Yamaguchi, T. Fueno, and H. Fukutome, *Chem. Phys. Letts.*, **22**, 461, 466 (1973).
57. J. E. Leffler, and E. Grunwald, *Rates and Equilibria of Organic Reactions*, Wiley, New York, 1963.
58. V. A. Palm, *Foundations of the Quantitative Theory of Organic Reactions*, Khimia, Leningrad, 1967 (in Russian).
59. T. A. Alfrey, J. J. Bohrer, and H. Mark, *Copolymerization*, Wiley, New York, 1952.
60. J. Koutecky, R. Zahradnik, and J. Cizek, *Trans. Faraday Soc.*, **57**, 169 (1961).
61. R. Zahradnik, and J. Koutecky, *Collect. Czech. Chem. Commun.*, **28**, 904 (1963).
62. C. A. Coulson, and H. C. Longuet-Higgins, *Proc. Roy. Soc. (London)*, **A195**, 188 (1948).
63. M. J. S. Dewar, and C. C. Thompson, *J. Am. Chem. Soc.*, **87**, 4414 (1965).
64. A. Streitwieser, H. A. Hammond, R. H. Jagow, R. M. Williams, R. G. Jesaitis, C. J. Chang, and R. Wolf, *J. Am. Chem. Soc.*, **92**, 5141 (1970).
65. R. Daudel, and O. Chalvet, *J. Chim. Phys.*, **53**, 943 (1956).
66. B. G. Van Leuwen, and R. J. Ouellette, *J. Am. Chem. Soc.*, **90**, 7056 (1968).
67. A. N. Kolmogorov, and S. V. Fomin, *Elements of the Theory of Functions and Functional Analysis*, Nauka, Moscow, 1968, Chapter 2 (in Russian).

THE THERMODYNAMICS OF
EVOLVING CHEMICAL SYSTEMS
AND THE APPROACH TO
EQUILIBRIUM

DOMINIC G. B. EDELEN

Center for the Application of Mathematics
Lehigh University
Bethlehem, Pennsylvania

CONTENTS

Abstract

Every solution of the equations governing the dynamics of a finite material body is shown to evolve toward an equilibrium solution when the body is subject to chemical reactions, diffusion, heat conduction, viscous and elastic stresses, body forces, and the following conditions: (*1*) the body adheres to some part of a fixed boundary; (*2*) the applied boundary tractions admit a potential functional; (*3*) the body force applied to each constituent of the body admits a potential; (*4*) the body is compatible with immersion in an external environment that is at constant temperature and in "equilibrium" with respect to a given system of conservative body forces; (*5*) all processes of the body satisfy the entropy-production inequality with an entropy-production function that vanishes only when all of the thermodynamic forces vanish; (*6*) the sum over all constituents of the product of the mass supply and the potential of the body force of each constituent vanishes; and (*7*) the equilibrium solution which satisfies the same boundary conditions is unique and minimizes the total free energy at constant temperature and constant total mass. The condition of compatibility with immersion in an external environment permits nonzero heat flux and diffusion fluxes across the boundary. A complete solution of the entropy-production inequality is obtained, which includes reaction rates of the Marcelin–de Donder type. However, no specific assumptions are made about the reaction rates other than (*5*) and (*6*) above. The stability results do not depend on the assumption of either a linear or a nonlinear Onsager theory. In the absence of bulk flow, it is shown that any system with nonlinear Onsager fluxes tends to a static state in which there can be imposed time-independent temperature and chemical potential distributions on the boundary.

I. INTRODUCTION

One of the fundamental properties of real material systems that is always subsumed in classical thermodynamics can be stated loosely as follows:

Any real material system of finite spatial extent, satisfying "reasonable" boundary conditions that do not depend explicitly on time, will ultimately evolve to an equilibrium state. Clearly if this property were not realized, such basic thermodynamic quantities as internal energy, free energy, entropy, temperature, etc. could not be obtained from experiments or given precise definitions within the theory. This latter remark is seen to be true both from the assumptions involved in the construct of macroscopic (phenomenological) thermodynamics and from the assumptions underlying statistical derivations of thermodynamic relations. It also seems evident that a theoretical construct that attempts to prove adherence to this fundamental property must include consideration of the irreversible process which occur in the evolution of real material systems from arbitrary initial states. Our purpose is to establish this fundamental property for a wide class of theories in the form of an asymptotic stability theorem for equilibrium processes. This will be accomplished within the arena of irreversible thermodynamics as governed by the entropy production inequality and the laws of conservation and balance satisfied by the physical components.

The analysis is confied to multiple-component material bodies that undergo chemical reactions and diffusion processes in the presence of thermal conduction and bulk motion. Significant progress has been made in recent years in all of the areas on which our analysis impings. In the area of chemical reactions, we note the works of Aris[1], Feinberg[2], Horn and Jackson[3], Horn[4], Bowen[5], and the numerous works of others cited in these papers, together with modern texts on chemical kinetics. In the area of chemically reacting mixtures of fluids that undergo bulk motions in the presence of diffusion and thermal conduction, we note, although not exclusively, the works of Eckart[6], Meixner[7], Meixner and Reik[8], Eringen and Ingram[9], Green and Naghdi[10], Bowen[11], Müller[12], Bowen and Garcia[13], Gurtin[14], Gurtin and Vargas[15], and the associated works cited in these papers. Studies in the modern theory of "linear" irreversible processes started with the pioneering works of Onsager[16], Meixner[17], Casimir[18], and Prigogine[19]. It has since seen significant and detailed development as reported in Refs. 8, 20, 21, 22, 23, and 24. Extensions of the theory, which include nonlinear phenomenological laws, have been given recently by Glansdorff and Prigogine[24], Gyarmati[25], and Edelen[26,27,28,29]. A summary through the middle 1960s of results concerning certain stability questions for chemically reacting systems is given by Gavalas[30]. Note should also be made of the more recent pioneering works of Nazarea, Rice, Nicolis, Glansdorff, and Prigogine[24,31,32,33] on the stability question, and the reveiw of results concerning chemical oscillations given by Nicolis and Portnow[34]. The approach to the stability question adopted in this paper is based on the method for single-component material bodies introduced by Ericksen[35], as extended by Coleman and Dill[36] and Naghdi and Trapp[37].

It seems to us that there are two intrinsic difficulties involved in current work on chemically reacting systems: (*1*) the requirement of the individual balance of each reaction in every dynamic process; and (*2*) the concomitant knowledge of all of the intermediate reactions, reactants, and products. Experimentally one may reasonably expect to know the initial and final compositions of the body, but knowledge of the totality of all of the individual reactions and subreactions and all of the possible species that may be present in the body prior to the final approach to an equilibrium composition would appear to ask for more than can be reasonably determined. It is also clear that any one overall reaction proceeding near equilibrium can consist of a large number of significant subreactions for large departures from equilibrium. Each of these subreactions can proceed with its own concomitant rate, and some of these rates can be very fast in comparison with others. Thus the requirement of instantaneous balance of each overall reaction in every dynamic process would appear to be in conflict with reality. We make no assumptions about the individual reaction processes in this paper other than consistency with the entropy-production inequality and the customary constraint on the species-production rates in the presence of potentials of externally applied body force fields. Accordingly if we can establish the universal approach to equilibrium in all reasonable irreversible processes without specification of the details of the individual reaction processes, as indeed we shall, then the two difficulties mentioned above can be dispensed with, at least as far as stability is concerned. On the other hand, if complete knowledge of the intermediate reaction processes and their constituents is at hand, then the analysis given below is applicable and the desired dynamic stability results are forthcoming. In this instance one would expect satisfaction of the mass balance laws, in which case the chemical reactions would proceed according to the values of the chemical affinities rather than the values of the chemical potentials themselves. A forthcoming paper[38] will show that the mass balance laws do not change the stability picture and that they provide a convenient decomposition of the dynamics into purely diffusive processes and reactive-diffusive processes.

An essential part of the analysis is the entropy-production inequality, since this inequality provides the basis whereby the necessary constitutive relations that complete the dynamical laws are obtained. A complete and general solution of this inequality is obtained, and it is shown that every system of thermodynamic fluxes admits a unique representation as the sum of fluxes—Onsager fluxes and fluxes that are nondissipative. It is further shown that if the entropy-production function vanishes only when the thermodynamic forces vanish, as is always the case in the linear Onsager theory, then the asymptotic stability results obtain whether or not the fluxes are Onsager fluxes. It has been shown elsewhere[29] that the analysis includes cases in which the reaction rates are of the de Donder type and these obtain from the full nonlinear Onsager theory with the associated nonlinear reciprocity relations.

II. NOTATION AND PRELIMINARY CONSIDERATIONS

The material bodies considered in this paper are comprised of $N+1$ constituents. We distinguish constituents by the use of lowercase Greek suffixes and assign such suffixes the range $[1, 2, \ldots, N+1]$. The mass density of the αth constituent is denoted by ρ_α and is considered to be a function of place x^i and time t. Chemical literature often uses ρ_α for the concentration of the αth constituent since this quantity is the mass per unit volume of constituent α. The concentration variables used in this paper are the absolute concentrations (mass fractions)

$$c_\alpha = \frac{\rho_\alpha}{\rho} \tag{2.1}$$

where

$$\rho = \sum_{\alpha=1}^{N+1} \rho_\alpha \tag{2.2}$$

is the total mass density. The quantities c_α and ρ are functions of place and time, and the c_α's satisfy the identity

$$\sum_{\alpha=1}^{N+1} c_\alpha = 1 \tag{2.3}$$

as follows from (2.1) and (2.2). Clearly, only $N+1$ of the quantities $\rho, c_\alpha, \alpha = 1, \ldots, N+1$ are independent. We accordingly base our analysis on the variables ρ and the first N of the c_α's; the latter variables are labeled $\hat{c}_\alpha$. Equation 2.3 then gives

$$c_{N+1} = 1 - \sum \hat{c}_\alpha \tag{2.4}$$

where Σ will always be used to denote summation from 1 through N, and hence (2.1) and (2.4) give

$$\rho_\alpha = \rho \hat{c}_\alpha \quad \alpha = 1, \ldots, N \qquad \rho_{N+1} = \rho \left(1 - \sum \hat{c}_\alpha \right) \tag{2.5}$$

Thus any function of the ρ_α's can be expressed as a function of the $N+1$ variables $\rho, \hat{c}_\alpha$. In particular, if

$$\Psi = \Psi(\rho_1, \ldots, \rho_N, \rho_{N+1}; \Lambda) \tag{2.6}$$

then we write

$$\hat{\Psi} = \hat{\Psi}(\rho, \hat{c}_\alpha; \Lambda) = \Psi \left[\rho \hat{c}_1, \ldots, \rho \hat{c}_N, \rho \left(1 - \sum \hat{c}_\alpha \right); \Lambda \right] \tag{2.7}$$

and if

$$\Psi = \Psi(c_1, \ldots, c_{N+1}; \Lambda) \tag{2.8}$$

then we write

$$\hat{\Psi} = \hat{\Psi}(\hat{c}_\alpha; \Lambda) = \Psi \left(\hat{c}_1, \ldots, \hat{c}_N, 1 - \sum \hat{c}_\alpha; \Lambda \right) \tag{2.9}$$

Of fundamental importance in thermodynamics are quantities that are defined as derivatives of scalar-valued functions with respect to the absolute concentration variables. Thus, for instance, if

$$\Psi = \Psi(c_1, \ldots, c_{N+1}; \Lambda) \qquad \hat{\Psi} = \Psi \left(\hat{c}_1, \ldots, \hat{c}_N, 1 - \sum \hat{c}_\alpha; \Lambda \right) \tag{2.10}$$

we can define quantities μ_α and $\hat{\mu}_\alpha$ by

$$\mu_\alpha = \frac{\partial \Psi(c_1, \ldots, c_{N+1}; \Lambda)}{\partial c_\alpha} \qquad \alpha = 1, \ldots, N+1 \qquad (2.11)$$

$$\hat{\mu}_\alpha = \frac{\partial \hat{\Psi}(\hat{c}_1, \ldots, \hat{c}_N; \Lambda)}{\partial \hat{c}_\alpha} \qquad \alpha = 1, \ldots, N \qquad (2.12)$$

It then follows from (2.10) through (2.12) that

$$\hat{\mu}_\alpha = \mu_\alpha - \mu_{N+1} \qquad (2.13)$$

Accordingly, if Ψ is the Helmholtz free energy function, μ_α are the chemical potentials and (2.13) shows that an analysis based upon $\hat{\Psi}$ with the independent variables $\hat{c}_\alpha$ will yield the quantities $\hat{\mu}_\alpha$, which are the independent quantities $\mu_\alpha - \mu_{N+1}$. We note that Ψ is stationary with respect to the variables $c_\alpha, \alpha = 1, \ldots, N+1$ such that $\sum_{\alpha=1}^{N+1} c_\alpha = 1$ if and only if $\hat{\Psi}$ is stationary with respect to the variables $\hat{c}_\alpha$, $\alpha = 1, \ldots, N$, in which case $\hat{\mu}_\alpha = 0$.

Let $v_\alpha^i(x^j, t)$ denote the components of the velocity field of the αth constituent with respect to a Cartesian coordinate system. The velocity vector field of the material body is then defined in the usual way by

$$v^i = \sum_{\alpha=1}^{N+1} c_\alpha v_\alpha^i = \sum \hat{c}_\alpha (v_\alpha^i - v_{N+1}^i) + v_{N+1}^i \qquad (2.14)$$

The latter equality follows by use of (2.4). The components of the diffusion flux of the αth constituent are defined by

$$J_\alpha^i = \rho c_\alpha (v_\alpha^i - v^i) \qquad \alpha = 1, \ldots, N+1 \qquad (2.15)$$

and satisfy the identity

$$\sum_{\alpha=1}^{N+1} J_\alpha^i = 0 \qquad (2.16)$$

Let $\hat{J}_\alpha^i$ designate an element of the first N of the diffusion fluxes. Then in exactly the same manner as with the c_α's we have

$$J_{N+1}^i = -\sum \hat{J}_\alpha^i \qquad (2.17)$$

as a consequence of (2.16). An elementary exercise based upon the above formulas gives

$$v_\alpha^i = v^i + (\rho \hat{c}_\alpha)^{-1} \hat{J}_\alpha^i \qquad \alpha = 1, \ldots, N$$

$$v_{N+1}^i = v^i - \rho^{-1} \left(1 - \sum \hat{c}_\alpha\right)^{-1} \sum \hat{J}_\beta^i$$

(2.18)

Hence a specification of the variables ρ, $\hat{c}_\alpha$, v^i, and $\hat{J}_\alpha^i$ serves to determine the densities and velocities of all of the constituents. This observation is very important since it allows us to dispense with equations for the determination of the velocity variables of each of the constituents if we have equations which serve to determine the quantities ρ, $\hat{c}_\alpha$, v^i, and $\hat{J}_\alpha^i$.

III. THE GOVERNING LAWS

The basic mechanical laws governing the evolution of a multiple-constituent material body are conservation of total mass, balance of constituent masses, balance of linear and moment of momentum, and balance of energy. We write these in several different, but equivalent, forms below.

A. Conservation of Total Mass

$$\frac{d}{dt} \int_{\mathscr{P}} \rho \, dV = 0 \tag{3.1}$$

$$\rho j = \rho_0 j_0 \tag{3.2}$$

$$\dot{\rho} = -\rho \partial_i v^i \tag{3.3}$$

$\mathscr{P}$ denotes the configuration of any piece of the body at time t. This configuration is obtained from a reference configuration $\mathscr{P}_0$ by the bulk motion

$$x^i = x^2(X^A, t) \tag{3.4}$$

Here X^A denotes the coordinates of a material point, in the reference configuration of the body, which has the coordinates x^i at time t. The quantity j is defined by

$$j = \det\left(\frac{\partial x^i}{\partial X^A}\right) \tag{3.5}$$

so that j is the volume ratio $dV(x)/dV(X)$, which obtains as a consequence of the motion (3.4). The quantity j is assumed to be positive for all motions. Equation 3.2 is thus the Lagrangian form of the continuity equation since $\rho_0 = \rho_0(X^A)$ is the initial mass density function of the body and j_0 is the initial value of j. Equation 3.3, where the superimposed dot denotes the material time derivative, is the Eulerian form of the continuity equation:

$$\dot{a}(x^i, t) \equiv \frac{\partial a}{\partial t} + v^i \frac{\partial a}{\partial x^i} \equiv \partial_t a + v^i \partial_i a \qquad (3.6)$$

The summation convention on lower- and uppercase Latin indices is assumed.

If $\mathcal{B}$ is the configuration of the whole body at time t, (3.1) gives $d/dt \int_{\mathcal{B}} \rho \, dV = d/dt M(\mathcal{B}) = 0$, where $M(\mathcal{B})$ is the total mass content of the material body at time t. It thus follows that the bodies under consideration are material bodies with respect to the bulk motions (i.e., the boundaries of the bodies move with the bulk motion of the body) and they are *closed* with respect to total mass. Thus, the diffusive processes of such bodies are required to preserve total mass with respect to transport across the material boundaries, which move with the bulk motion [i.e. (2.16) holds]. This point will be important later when we come to interpretations of the stability results in terms of the prohibition of sustained oscillations.

B. Balance of Constituent Mass

$$\frac{d}{dt} \int_{\mathcal{P}} \rho c_\alpha \, dV = \int_{\mathcal{P}} a_\alpha \, dV - \int_{\partial \mathcal{P}} J^i_\alpha n_i \, dS \qquad \alpha = 1, \dots, N+1 \qquad (3.7)$$

$$\rho \dot{c}_\alpha = a_\alpha - \partial_i J^i_\alpha \qquad \alpha = 1, \dots, N+1 \qquad (3.8)$$

Here a_α is the supply of mass of constituent α arising from chemical reactions, $\partial \mathcal{P}$ denotes the boundary of the material piece $\mathcal{P}$ of the body, and n_i are the components of the unit normal vector field of $\partial \mathcal{P}$ that points out of $\mathcal{P}$. Use of the divergence theorem and (3.3) shows the equivalence of (3.7) and (3.8). If we sum (3.8) from 1 to $N+1$ and use the relations $\sum_{\alpha=1}^{N+1} J^i_\alpha = 0 = \sum_{\alpha=1}^{N+1} c_\alpha - 1$, which express the mass closure conditions relative to the bulk motion, we obtain

$$a_{N+1} = -\sum \hat{a}_\alpha \qquad (3.9)$$

Accordingly only N of (3.7) and (3.8) are independent. They can thus be replaced by the equations

$$\frac{d}{dt} \int_{\mathscr{P}} \rho \hat{c}_\alpha \, dV = \int_{\mathscr{P}} \hat{a}_\alpha \, dV - \int_{\partial \mathscr{P}} \hat{J}_\alpha^i n_i \, dS \qquad \alpha = 1, \ldots, N \qquad (3.10)$$

$$\rho \dot{\hat{c}}_\alpha = \hat{a}_\alpha - \partial_i \hat{J}_\alpha^i \qquad \alpha = 1, \ldots, N \qquad (3.11)$$

C. Balance of Linear Momentum

$$\frac{d}{dt} \int_{\mathscr{P}} \rho v^i \, dV = \int_{\mathscr{P}} \left[\sum \hat{c}_\alpha (f_\alpha^i - f_{N+1}^i) + f_{N+1}^i \right] \rho \, dV$$

$$+ \int_{\partial \mathscr{P}} t^{ij} n_j \, dS \qquad i = 1, 2, 3 \qquad (3.12)$$

$$\rho \dot{v}^i = \rho \left[\sum \hat{c}_\alpha (f_\alpha^i - f_{N+1}^i) + f_{N+1}^i \right] + \partial_j t^{ij} \qquad i = 1, 2, 3 \qquad (3.13)$$

Here f_α^i are the components of the body force per unit mass acting on the αth constituent, and t^{ij} are the stress-tensor components that give the boundary tractions $t^{ij} n_j$.[39] In terms of the motion $x^i = x^i(X^A, t)$ we have

$$v^i(x^j, t) = \partial_t x^i(X^A, t) \big|_{X^A = X^A(x^j, t)} \qquad (3.14)$$

where $X^A = X^A(x^j, t)$ is the inverse of the motion that maps the current place x^j at time t onto the initial place X^A.

D. Balance of Moment of Momentum

Use of (3.3) and (3.13), expressing conservation of total mass and balance of linear momentum, reduces the balance of moment of momentum to the requirement[39]

$$t^{ij} = t^{ji} \qquad (3.15)$$

and hence the stress tensor is symmetric. If electromagnetic interactions are considered, then the analysis given by Eringen[40] shows that we no longer have a symmetric stress tensor in general. A combination of Eringen's results reported in Ref. 40, together with the complete solution of the Clausius-Duhem inequality given by Edelen[41], can be used in the manner given below to obtain the complete asymptotic stability results for such systems.

E. Balance of Energy

$$\frac{d}{dt}\int_{\mathcal{P}}\rho(\hat{e}+\tfrac{1}{2}v^iv_i)\,dV = -\int_{\partial\mathcal{P}}J_q^in_i\,dS + \int_{\partial\mathcal{P}}v^it^{ij}n_j\,dS$$

$$+\int_{\mathcal{P}}\rho v^i\Big[f_{N+1}^i+\sum\hat{c}_\alpha(f_\alpha^i-f_{N+1}^i)\Big]\,dV$$

$$+\int_{\mathcal{P}}\sum\hat{J}_\alpha^i(f_\alpha^i-f_{N+1}^i)\,dV \qquad (3.16)$$

$$\rho\dot{\hat{e}} = -\partial_iJ_q^i+t^{ij}d_{ij}+\sum\hat{J}_\alpha^i(f_\alpha^i-f_{N+1}^i) \qquad (3.17)$$

Here $\hat{e}$ is the internal energy per unit mass expressed as a function of $\hat{c}_\alpha$ by elimination of c_{N+1} through the relation $c_{N+1}=1-\sum\hat{c}_\alpha$, J_q^i are the components of the heat-flux vector, and d_{ij} are the rate of strain tensor components, which are defined by

$$2d_{ij}=\partial_iv_j+\partial_jv_i=2d_{ji} \qquad (3.18)$$

The equivalence of (3.16) and (3.17) is obtained by use of the divergence theorem, (3.4), and (3.13). The occurrence of the diffusion-flux vectors in (3.16) and (3.17) comes about by the following calculation based on (2.5):

$$\sum_{\alpha=1}^{N+1}\rho c_\alpha v_\alpha^if_\alpha^i=\sum_{\alpha=1}^{N+1}\rho c_\alpha[v^i+(v_\alpha^i-v^i)]f_\alpha^i$$

$$=\sum_{\alpha=1}^{N+1}(\rho c_\alpha v^if_\alpha^i+J_\alpha^if_\alpha^i)$$

$$=\rho v^i\sum\hat{c}_\alpha(f_\alpha^i-f_{N+1}^i)+\rho v^if_{N+1}^i+\sum\hat{J}_\alpha^i(f_\alpha^i-f_{N+1}^i)$$

The mechanical laws given above must be augmented by the thermodynamic laws, which are comprised of the balance of entropy and the entropy-production inequality.

F. Balance of Entropy

In order to obtain a statement of the balance of entropy, we have to make very specific assumptions concerning the arguments that serve to determine the specific internal energy $\hat{e}$. For the purposes of this paper we assume that

$$\hat{e} = \hat{e}\left(s, \hat{c}_\alpha, \partial_A x^i\right) = e\left(s, \hat{c}_1, \ldots, \hat{c}_N, 1 - \sum \hat{c}_\alpha, \partial_A x^i\right) \tag{3.19}$$

where s is the entropy per unit mass and

$$\partial_A x^i = \frac{\partial x^i(X^B, t)}{\partial X^A}$$

are the displacement gradients. Inclusion of these latter arguments provides for elastic properties of the body as well as for dependences on the total density [i.e., (3.2) and (3.5) give $\rho_0 j_0 = \rho j = \rho \det(\partial_A x^i)$]. It should be clearly noted that (3.19) states that the internal energy per unit mass depends on position and time only through the position and time dependence of its arguments s, $\hat{c}_\alpha$, and $\partial_A x^i$; that is, the value of the internal energy per unit mass in a dynamic process at a particular time and place is obtained by substituting the place- and time-dependent values of s, $\hat{c}_\alpha$, and $\partial_A x^i$ in the function $\hat{e}$, which is determined for equilibrium processes. Clearly this is the point that is most open to question in the dynamic case, since our procedure amount to direct extrapolation of thermostatic values to dynamic values. Arguments of microscopic reversibility[*] or analyses based on the Clausius-Duhem inequality are often introduced in attempts to circumvent this difficulty but, in the author's view, the final results seem in all cases to lead to exactly what we have assumed.[†] If we define the recoverable stress t_R^{ij} by the relation (see [39])

$$\rho_0 \frac{\partial \hat{e}}{\partial(\partial_A x^i)} = j t_R^{ij} \partial_j X^A(x^k, t) \tag{3.20}$$

and we note that $(\partial_j X^A) \partial_A v^i = \partial_j v^i$, a straightforward calculation gives

$$\rho \hat{e} = \rho T \dot{s} + r_R^{ij} d_{ij} + \rho \sum \hat{\mu}_\alpha \hat{c}_\alpha \tag{3.21}$$

where

$$T = \frac{\partial \hat{e}}{\partial s} \tag{3.22}$$

---

[*] See Ref. 24 for a careful discussion of this matter and of the meaning of entropy in nonequilibrium states.

[†] The complete solution of the Clausius-Duhem inequality, as given in Ref. 28, does permit dependence of the internal energy on the time rate of change of the temperature field, but then the internal energy does not obtain from the free energy by a Legendre transformation. The interested reader is referred to Ref. 28 for the details.

is the thermodynamic temperature and

$$\hat{\mu}_\alpha = \frac{\partial \hat{e}}{\partial \hat{c}_\alpha} = \mu_\alpha - \mu_{N+1} \qquad \left(\mu_\alpha = \frac{\partial e}{\partial c_\alpha} \right) \tag{3.23}$$

are the reduced chemical potentials [recall (2.13)]. A substitution of (3.21) into (3.17) and standard manipulations give the following statements of the balance of entropy:

$$\frac{d}{dt} \int_{\mathcal{P}} \rho s \, dV = - \int_{\partial \mathcal{P}} \frac{1}{T} \left(J_q^i - \sum \hat{\mu}_\alpha \hat{J}_\alpha^i \right) n_i \, dS + \int_{\mathcal{P}} \sigma \, dV \tag{3.24}$$

$$\rho \dot{s} = - \partial_i \left(\frac{J_q^i - \sum \hat{\mu}_\alpha \hat{J}_\alpha^i}{T} \right) + \sigma \tag{3.25}$$

Here

$$\sigma = \left(t^{ij} - t_R^{ij} \right) \frac{d_{ij}}{T} - J_q^i \frac{\partial_i T}{T^2}$$

$$- \sum \hat{J}_\alpha^i \left[\partial_i \left(\frac{\hat{\mu}_\alpha}{T} \right) - \frac{f_\alpha^i - f_{N+1}^i}{T} \right] - \sum \hat{a}_\alpha \frac{\hat{\mu}_\alpha}{T} \tag{3.26}$$

is the entropy-production function for the body.

G. Entropy-Production Inequality

The usually accepted statement of the second law of thermodynamics[16,19,20,21,22,23] is embodied in the entropy-production inequality

$$\sigma \geqslant 0 \tag{3.27}$$

As we shall see, satisfaction of this inequality is what distinguishes real material systems from arbitrary systems, which satisfy the mechanical laws of evolution. Further, an inspection of the laws of evolution (3.3), (3.11), (3.13), and (3.17) or (3.25) shows that we have equations that serve to determine the evolution of ρ, $\hat{c}_\alpha$, v^i and $\hat{e}$, s, or T provided the quantities t^{ij}, J_q^i, $\hat{J}_\alpha^i$, and $\hat{a}_\alpha$ are given. However, in view of (3.20), the terms that comprise σ involve just these undetermined quantities t^{ij}, J_q^i, $\hat{J}_\alpha^i$, and $\hat{a}_\alpha$, and the quantities T, d_{ij}, $\partial_i T$, $\hat{\mu}_\alpha$, $\partial_i \hat{\mu}_\alpha$, f_α^i are either given (as in the case of f_α^i) or else determined by ρ, $\hat{c}_\alpha$, v^i, T, and the internal energy function $\hat{e}$. It thus follows that the entropy-production inequality provides a means whereby we may determine the possible necessary constitutive relations for t^{ij}, J_q^i, $\hat{J}_\alpha^i$, and $\hat{a}_\alpha$ that are needed in order to complete the dynamical laws of the material bodies under investigation. We also note that determination of $\hat{J}^i_\alpha$ and knowledge of the evolution of ρ, $\hat{c}_\alpha$, and v^i are sufficient to determine all of the variables ρ_α and v^i_α of each of the constituents, as pointed out at the end of Section II.

IV. GENERAL SOLUTION OF THE ENTROPY-PRODUCTION INEQUALITY

We obtain the general solution of the entropy-production inequality in this section. Such a solution is clearly necessary in view of the fact that it provides the required constitutive relations that complete the dynamic description of the material bodies under investigation. Since the solution is general, it delineates all processes that satisfy the entropy production inequality. Such characterizations are an essential part of the asymptotic stability arguments given in Section VI.

A. Statement of the Problem

The well-established concept of fluxes and forces in irreversible thermodynamics[16,20,21,22,23,24,29] leads to a natural partition of the various terms that comprise the right-hand side of (3.26) so that σ becomes a bilinear form on a vector space V_{9+4N} of $9+4N$ dimensions. In particular we set

$$\mathbf{X} \equiv \left\{ \frac{d_{ij}}{T}, \frac{\partial_i T}{T^2}, \partial_i\left(\frac{\hat{\mu}_\alpha}{T}\right) - \frac{f^i_\alpha - f^i_{N+1}}{T}, \frac{\hat{\mu}_\alpha}{T} \right\} \varepsilon V_{9+4N} \qquad (4.1)$$

$$\mathbf{J} \equiv \left\{ t^{ij} - t^{ij}_R, -J^i_q, -\hat{J}^i_\alpha, -\hat{a}_\alpha \right\} \varepsilon V_{9+4N} \qquad (4.2)$$

and (3.26) assumes the form

$$\sigma = \mathbf{J} \cdot \mathbf{X} \qquad (4.3)$$

Here $\mathbf{J} \cdot \mathbf{X} = \Sigma_{\Gamma=1}^{9+4N} J_\Gamma X_\Gamma$ is the inner product in V_{9+4N}. In addition dependence of $\mathbf{J}$ on the arguments $\mathbf{X}$, we can also have a dependence of $\mathbf{J}$ on the state variables

$$\omega \equiv \left\{ T, \hat{c}_\alpha, \partial_A x^i \right\} \varepsilon V_{10+N} \qquad (4.4)$$

which we represent as a vector in V_{10+N}. It is thus necessary to consider the situation in which

$$\mathbf{J} = \mathbf{J}(\mathbf{X}; \omega) \qquad (4.5)$$

i.e., $\mathbf{J}$ is a mapping of $V_{9+4N} \times V_{10+N}$ into V_{9+4N}. In terms of these variables the entropy-production inequality becomes

$$\sigma = \sigma(\mathbf{J}, \mathbf{X}) = \sigma(\mathbf{J}(\mathbf{X}; \omega), \mathbf{X}) = \sigma^*(\mathbf{X}; \omega) = \mathbf{J}(\mathbf{X}; \omega) \cdot \mathbf{X} \geqslant 0 \qquad (4.6)$$

Clearly we have

$$\sigma^*(\mathbf{0}; \omega) = 0 \qquad (4.7)$$

and hence $\sigma^* \geqslant 0$ gives

$$\left. \frac{\partial \sigma^*}{\partial X_\Gamma} \right|_{\mathbf{X}=\mathbf{0}} = 0 \qquad \Gamma = 1, \dots, 9 + 4N \qquad (4.8)$$

A differentiation of $\sigma = \mathbf{J}(\mathbf{X}; \omega) \cdot \mathbf{X}$ with respect to X_Γ and evaluation at $\mathbf{X} = \mathbf{0}$ then gives the relations

$$\mathbf{J}(\mathbf{0}; \omega) = \mathbf{0} \qquad (4.9)$$

It thus transpires that in order to solve the entropy inequality (4.6), we have to find all functions $\mathbf{J}(\mathbf{X}; \omega)$ such that $\mathbf{J}(\mathbf{X}; \omega) \cdot \mathbf{X} \geqslant 0$ and $\mathbf{J}(\mathbf{0}; \omega) = \mathbf{0}$.

B. The Basic Decomposition Theorem

The basis upon which a complete solution of this problem is obtained is the following decomposition theorem established in [27]:

Let $\mathbf{J}(\mathbf{X}; \omega)$ be a continuous vector-valued function with a continuous derivative with respect to the vector $\mathbf{X}$. Then there exists a unique function $\phi(\mathbf{X}; \omega)$ and a unique vector $\mathbf{U}(\mathbf{X}; \omega)$ such that

$$J_\Gamma(\mathbf{X}; \omega) = \frac{\partial \phi}{\partial X_\Gamma} + U_\Gamma \qquad \Gamma = 1, \dots, 9 + 4N \qquad (4.10)$$

$$\phi(\mathbf{0}; \omega) = 0 \qquad (4.11)$$

$$\mathbf{U}(\mathbf{X}; \omega) \cdot \mathbf{X} = 0 \qquad \mathbf{U}(\mathbf{0}; \omega) = \mathbf{0} \qquad (4.12)$$

The explicit form of the phenomenological relations (4.10) and (4.12) obtained from (4.1) and (4.2) are*

$$t^{ij} = t_R^{ij} + \frac{\partial \phi}{\partial (d_{ij}/T)} + U^{ij}$$

$$J_q^i = - \frac{\partial \phi}{\partial (\partial_i T/T^2)} - U^i \qquad (4.10a)$$

$$\hat{J}_\alpha^i = - \frac{\partial \phi}{\partial \left[\partial_i \left(\dfrac{\hat{\mu}_\alpha}{T} \right) - \dfrac{f_\alpha^i - f_{N+1}^i}{T} \right]} - \hat{U}_\alpha^i$$

$$\hat{a}_\alpha = - \frac{\partial \phi}{\partial (\hat{\mu}_\alpha/T)} - \hat{U}_\alpha$$

$$U^{ij} \frac{d_{ij}}{T} + U^i \frac{\partial_i T}{T^2} + \sum \hat{U}_\alpha^i \left[\partial_i \left(\frac{\hat{\mu}_\alpha}{T} \right) - \frac{f_\alpha^i - f_{N+1}^i}{T} \right] + \sum \hat{U}_\alpha \frac{\hat{\mu}_\alpha}{T} = 0$$

$$(4.12a)$$

These relations reduce to the linear phenomenological relations of On-sager[16,20,21,22,23,24,29] when ϕ is a homogeneous positive-definite quadratic form in the variables $\mathbf{X}$ and $\mathbf{U} = \mathbf{0}$. (Modification of the relations (4.10a) that are demanded by satisfaction of the mass balance laws is given in [38].) Fluxes $\mathbf{J}$ for which $\mathbf{U} = \mathbf{0}$ are accordingly referred to as *Onsager fluxes*.[29] We show in the next paragraph that Onsager fluxes satisfy nonlinear generalizations of the Onsager reciprocity relations.

C. Reciprocity Relations and Dissipation Potentials

When (4.20) is substituted into (4.6), we obtain

$$\sigma = \sum_{\Gamma=1}^{9+4N} X_\Gamma \frac{\partial \phi}{\partial X_\Gamma} \geq 0 \qquad (4.13)$$

and

$$\sigma(\mathbf{U}, \omega) = \mathbf{U}(\mathbf{X}; \omega) \cdot \mathbf{X} = 0 \qquad (4.14)$$

*The result $\hat{a}_\alpha = - \partial \phi / \partial (\hat{\mu}_\alpha/T) - \hat{U}_\alpha$ is in many respects similar to the decompositions obtained for this case by Nazarea[31] and Nicolis[33]. A detailed discussion of the decomposition of $\hat{a}_\alpha$ into symmetric and antisymmetric parts is given in Ref. 24.

Thus every system of fluxes has a unique part that derives from a dissipation potential $\phi(X; \omega)$ and a unique part $U(X; \omega)$ that dissipates no energy. The dissipation potential $\phi(X; \omega)$ is easily seen to be a generalization of the corresponding quantity introduced for linear systems by Onsager and Machlup.[20] Nonequilibrium thermodynamics based on such functions has received intensive study in recent years (see Refs. 22, 23, 24, 25, 26, 27, 28, 29, and 41). From the mathematical point of view, there is thus no question about the existence of a dissipation potential for every process. What is in question is whether the nondissipative part $U(X; \omega)$ of a given system $J(X; \omega)$ of fluxes does or does not vanish. The answer to this question is equivalent to an answer to the question of the satisfaction or nonsatisfaction of the Onsager-Casimir reciprocity relations. This observation follows from (4.10) when we solve for $\partial \phi / \partial X_\Gamma$ and then write down the integrability conditions for the resulting system $\partial \phi / \partial X_\Gamma = J_\Gamma - U_\Gamma$ of overdetermined partial differential equations:

$$\frac{\partial}{\partial X_\Lambda}(J_\Gamma - U_\Gamma) = \frac{\partial}{\partial X_\Gamma}(J_\Lambda - U_\Lambda) \qquad (4.15)$$

We thus always have reciprocity relations. These become the nonlinear generalization of the Onsager-Casimir[8, 16, 18, 21, 22, 24] relations $\partial J_\Gamma / \partial X_\Lambda = \partial J_\Lambda / \partial X_\Gamma$, that is [see (4.1) and (4.2)],

$$\frac{\partial (t^{ij} - t_R^{ij})}{\partial (d_{kl}/T)} = \frac{\partial (t^{kl} - t_R^{kl})}{\partial (d_{ij}/T)}$$

$$\frac{\partial (t^{ij} - t_R^{ij})}{\partial (\partial_k T / T^2)} = -\frac{\partial J_q^k}{\partial (d_{ij}/T)}$$

$$\frac{\partial (t^{ij} - t_R^{ij})}{\partial \left[\partial_k \left(\dfrac{\hat{\mu}_\alpha}{T} \right) - \dfrac{f_\alpha^k - f_{N+1}^k}{T} \right]} = -\frac{\partial \hat{J}_\alpha^k}{\partial (d_{ij}/T)}$$

$$\frac{\partial (t^{ij} - t_R^{ij})}{\partial (\hat{\mu}_\alpha / T)} = -\frac{\partial \hat{a}_\alpha}{\partial (d_{ij}/T)} \qquad (4.16)$$

$$\frac{\partial J_q^i}{\partial (\partial_j T / T^2)} = \frac{\partial J_q^j}{\partial (\partial_i T / T^2)}$$

$$\vdots \qquad\qquad \vdots$$

$$\frac{\partial \hat{a}_\alpha}{\partial (\hat{\mu}_\beta / T)} = \frac{\partial \hat{a}_\beta}{\partial (\hat{\mu}_\alpha / T)}$$

only when $U(X; \omega) = 0$. Thus Onsager fluxes satisfy the Onsager-Casimir reciprocity relations identically. We have shown elsewhere[43] that $U(X; \omega) = 0$ characterizes and is characterized by appropriate behavior under time reversal.

D. Characterization of Solutions

It now remains to satisfy the entropy inequality in the form (4.13). This is a scalar differential inequality that immediately integrates with the "initial" condition (4.11) to give

$$\phi(X; \omega) = \int_0^1 p(\tau X; \omega) \frac{d\tau}{\tau} \tag{4.17}$$

where the function $p(X; \omega)$ is arbitrary to within the constraints

$$p(0; \omega) = 0 \tag{4.18}$$

$$p(X; \omega) \geqslant 0 \tag{4.19}$$

We have thus shown that *the entropy inequality is satisfied when and only when*

$$J_\Gamma(X; \omega) = \frac{\partial}{\partial X_\Gamma} \int_0^1 p(\tau X; \omega) \frac{d\tau}{\tau} + U_\Gamma(X; \omega) \tag{4.20}$$

$$U(X; \omega) \cdot X = 0 \qquad U(0; \omega) = 0 \tag{4.21}$$

and $p(X; \omega)$ satisfies the conditions (4.18) and (4.19). Incidentally it follows from (4.18) through (4.21) that $J(0; \omega) = 0$, and that $p(X; \omega) = \sigma^*(X; \omega)$. Thus if $\sigma = \sigma^*(X; \omega)$ is known as a function of the indicated arguments, everything is determined. On the other hand, if $\sigma^*(X; \omega)$ is not known, as is usually the case, then the process can be characterized by the function $p(X; \omega)$ in the same manner that equilibrium processes are characterized by the internal energy or the Helmholtz free energy or the Gibbs function. All that is required is to select for the process a function $p(X; \omega)$ that satisfies the conditions (4.18) and (4.19). We have shown elsewhere[26,27,28,29,43] that it is possible to select appropriate functions $p(X; \omega)$ and $U(X; \omega) = 0$ so as to secure agreement with the established equation governing linear and nonlinear viscous fluids, viscoelastic bodies, perfectly plastic bodies, and chemical reactions whose reaction rates are governed by the equations of reaction kinetics of the de Donder type. The remark of Anderson and Boyd[44] concerning the inappropriateness of an Onsager formalism in reaction kinetics thus retains its validity only if one is restricted to a linear Onsager formalism.

It is clear from (4.17) through (4.19) that

$$\phi(\mathbf{0}; \omega) = \min_{\mathbf{X}} \phi(\mathbf{X}; \omega) \qquad (4.22)$$

and that $\phi(\mathbf{X}; \omega)$ is a nondecreasing function on all rays in V_{9+4N} that emanate from the origin $\mathbf{0}$. Although the same properties are shared in linear theories by the function $\sigma^*(\mathbf{X}; \omega)$, in nonlinear theories $\sigma^*(\mathbf{X}; \omega)$ can oscillate rather wildly and can also have zeroes at points in V_{9+4N} other than $\mathbf{X} = \mathbf{0}$. We shall see in later sections that it is necessary to demand that $\sigma^*(\mathbf{X}; \omega) = 0$ only when $\mathbf{X} = \mathbf{0}$ in order to obtain the asymptotic stability results.[†] This in turn may provide the motivation for demanding that $\sigma^*(\mathbf{X}; \omega) = 0$ only when $\mathbf{X} = \mathbf{0}$, for otherwise it would appear that the foundations of classical thermodynamics go awry. On the other hand, in the appendix we sketch a proof of the fact that in the absence of bulk motion $[v^i(x^j, t) = 0]$, asymptotic stability can be predicted on the basis of the dissipation potential $\phi(\mathbf{X}; \omega)$ for chemically reacting and diffusing mixtures, provided that $\phi(\mathbf{X}; \omega) = \phi(\mathbf{X})$.[††] We think it important to point out this alternative in which $\sigma^*(\mathbf{X}; \omega) = \sigma^*(\mathbf{X})$ is allowed to oscillate wildly with as many zeros as desired and the asymptotic stability of the chemical system still obtains. Thus caution must be exercised in laying down a categorical denial of the possibility that $\sigma^*(\mathbf{X}; \omega) = 0$ for some $\mathbf{X} \neq \mathbf{0}$.

V. THE ENERGETIC FUNCTION

The basic idea involved in this section can be credited to Ericksen[35] and the generalizations given by Coleman and Dill,[36] although one can trace threads of the argument back to Duhem.[45] The analyses given in[35–37] do not allow for chemical reactions or diffusion processes.

A. Definition of the Energetic Function

We consider the totality of processes throughout the whole material body $\mathcal{B}$, which is assumed to be of finite spatial extent at the initial time. The body is also assumed to be immersed in an external environment in which there are processes that maintain the environment at a constant and uniform temperature T_0 in the presence of the (reduced) chemical potentials $\hat{\mu}_{\alpha 0}$ of the environment, which are "in equilibrium" with a system of conservative body forces with potential functions $V_\alpha (\alpha = 1, \ldots, N+1)$. By this latter condition we mean that

$$0 = \hat{\mu}_{\alpha 0} + V_\alpha - V_{N+1} = \mu_{\alpha 0} + V_\alpha - \mu_{N+1,0} - V_{N+1} \qquad \alpha = 1, \ldots, N$$

[†]The collection of all such $\sigma^*(\mathbf{X}; \omega)$ is characterized by (4.13) and (4.20) for all $p(\mathbf{X}; \omega)$ for which $p(\mathbf{0}; \omega) = 0$ and $p(\mathbf{X}; \omega) > 0$ for $\mathbf{X} \neq \mathbf{0}$.

[††]See Ref. 24 for further analysis of the stability question from this point of view.

holds at all points in the environment and on the boundary of the environment. In contrast with Coleman and Dill,[36] we also assume that the body $\mathscr{B}$ is not completely thermally isolated from this environment. We will have to be quite specific about the boundary conditions for both the temperature and the chemical potentials of the body later on. For the present we use these somewhat vague notions of the external environment to motivate the occurrence of the number T_0 and the functions $\hat{\mu}_{\alpha 0}$ in the following measure of the total energetics of the body:

$$R(t) = \int_{\mathscr{B}} \rho(\hat{e} - T_0 s + \tfrac{1}{2} v^i v_i) \, dV$$

$$- \int_{t_0}^{t} \left\{ \int_{\partial \mathscr{B}} v^i t^{ij} n_j \, dS + \int_{\mathscr{B}} \sum \hat{J}_\alpha^i (f_\alpha^i - f_{N+1}^i) \, dV \right.$$

$$+ \int_{\mathscr{B}} \rho v^i \left[f_{N+1}^i + \sum \hat{c}_\alpha (f_\alpha^i - f_{N+1}^i) \right] dV$$

$$\left. - \int_{\partial \mathscr{B}} \sum \hat{\mu}_{\alpha 0} \hat{J}_\alpha^i n_i \, dS \right\} dt \tag{5.1}$$

For want of a better nominative, we refer to the function $R(t)$ as the *energetic function* of the body $\mathscr{B}$. Coleman and Dill[36] use the term "canonical free energy," while one could also argue that $R(t)$ be termed the Liapunov function for $\mathscr{B}$ since it will be shown to have properties in common with the classical notion of Liapunov functions for dynamical systems. Those accustomed to thinking in terms of maximum usable work will also recognize the energetic function of a body as a generalization of $\int_{\mathscr{B}} \rho(\hat{e} - T_0 s + p_0 \rho^{-1}) \, dV$, the thermostatic availability function[46] [see (5.27) and (5.30) in this connection]. We also note that a special case of $R(t)$ has been used to advantage by Biot[47] in a recent paper on thermoelasticity and elastic instability. Finally we note that if the energy equation (3.17) is integrated formally with respect to time from t_0 to the current time and then integrated over the body, the result can be used to eliminate terms in (5.1) so as to obtain

$$R(t) = - T_0 \int_{\mathscr{B}} \rho s \, dV - \int_{t_0}^{t} \int_{\partial \mathscr{B}} \left\{ J_q^i - \sum \hat{\mu}_{\alpha 0} \hat{J}_\alpha^i \right\} n_i \, dS \, dt$$

with the obvious implications.

B. Equivalent Forms

We require several equivalent formulations of $R(t)$. Define the Helmholtz free energy per unit mass in the usual manner:

$$\hat{F} = \hat{e} - Ts \tag{5.2}$$

Here, as in previous sections, the superimposed hat indicates that c_{N+1} has been eliminated as an argument by use of the relation $c_{N+1} = 1 - \Sigma \hat{c}_\alpha$. We then have

$$\hat{F} = \hat{F}(T, \hat{c}_\alpha, \partial_A x^i) = \hat{F}(\omega) \tag{5.3}$$

$$s = -\frac{\partial \hat{F}}{\partial T} \tag{5.4}$$

$$\hat{\mu}_\alpha = \frac{\partial \hat{F}}{\partial \hat{c}_\alpha} = \mu_\alpha - \mu_{N+1} \tag{5.5}$$

and

$$\frac{\partial \hat{F}}{\partial(\partial_A x^i)} = \frac{\partial \hat{e}}{\partial(\partial_A x^i)} \tag{5.6}$$

Hence

$$\hat{e} = \hat{F} - T\frac{\partial \hat{F}}{\partial T} \tag{5.7}$$

and so

$$\frac{\partial \hat{e}}{\partial T} = -T\frac{\partial^2 \hat{F}}{\partial T^2} \tag{5.8}$$

Now $\partial \hat{e}/\partial T$ is the instantaneous heat capacity per unit mass, which is a non-negative quantity for real material bodies. We may thus conclude that

$$\frac{\partial^2 \hat{F}}{\partial T^2} \leqslant 0 \tag{5.9}$$

for all values of the arguments of $\hat{F}$. Since we wish to single out the variable T for special treatment, we write

$$\hat{F} = \hat{F}(T, \Gamma) \tag{5.10}$$

where Γ stands for the remaining arguments $\hat{c}_\alpha, \partial_A x^i$. A straightforward calculation based upon Taylor's theorem with remainder gives

$$\hat{F}(T_0, \Gamma) = \hat{F}[T + (T_0 - T), \Gamma]$$

$$= \hat{F}(T, \Gamma) + (T_0 - T)\frac{\partial \hat{F}(T, \Gamma)}{\partial T}$$

$$+ \tfrac{1}{2}(T_0 - T)^2 \frac{\partial^2 \hat{F}(T', \Gamma)}{\partial (T')^2} \tag{5.11}$$

where $\min(T_0, T) \leqslant T' \leqslant \max(T_0, T)$. If we define $\mathcal{K}(T', \Gamma)$ by

$$\mathcal{K}(T', \Gamma) = -\tfrac{1}{2}\frac{\partial^2 F(T', \Gamma)}{\partial (T')^2} \tag{5.12}$$

then by (5.9)

$$\mathcal{K}(T', \Gamma) \geqslant 0 \tag{5.13}$$

for all T' and Γ. When (5.4) and (5.12) are substituted into (5.11), we obtain

$$\hat{F}(T_0, \Gamma) = \hat{F}(T, \Gamma) - (T_0 - T)s(T, \Gamma) - (T_0 - T)^2 \mathcal{K}(T', \Gamma) \tag{5.14}$$

and hence (5.7) yields

$$\hat{e}(T, \Gamma) - T_0 s(T, \Gamma) = \hat{F}(T_0, \Gamma) + (T_0 - T)^2 \mathcal{K}(T', \Gamma) \tag{5.15}$$

A substitution of (5.15) into (5.1) thus gives

$$R(t) = \int_{\mathcal{B}} \rho \hat{F}(T_0, \Gamma) dV + \int_{\mathcal{B}} \rho \left[(T_0 - T)^2 \mathcal{K}(T', \Gamma) + \tfrac{1}{2} v^i v_i \right] dV$$

$$- \int_{t_0}^{t} \left\{ \int_{\partial \mathcal{B}} v^i t^{ij} n_j \, dS + \int_{\mathcal{B}} \sum \hat{J}_\alpha^i (f_\alpha^i - f_{N+1}^i) dV \right.$$

$$+ \int_{\mathcal{B}} \rho v^i \left[f_{N+1}^i + \sum \hat{c}_\alpha (f_\alpha^i - f_{N+1}^i) dV \right]$$

$$\left. - \int_{\partial \mathcal{B}} \sum \hat{\mu}_{\alpha 0} \hat{J}_\alpha^i n_i \, dS \right\} dt \tag{5.16}$$

We now turn to the terms that comprise the time integral in (5.16). The body force per unit mass acting on each constituent is assumed to be derived from a potential function that depends only on place:

$$f_\alpha^i = \partial_i V_\alpha(x^j) \tag{5.17}$$

The total potential $\mathcal{V}$ for the body per unit mass is defined by the relation

$$\mathcal{V} = \sum_{\alpha=1}^{N+1} c_\alpha(x^j, t) V_\alpha(x^j) = \sum \hat{c}_\alpha(V_\alpha - V_{N+1}) + V_{N+1} \tag{5.18}$$

so that

$$\rho\mathcal{V} = \sum \rho\hat{c}_\alpha(V_\alpha - V_{N+1}) + \rho V_{N+1} \tag{5.19}$$

A straightforward calculation and use of (3.11) and (3.17) give us

$$\rho\dot{\mathcal{V}} = \rho\sum\hat{c}_\alpha v^i\partial_i(V_\alpha - V_{N+1}) + \rho v^i\partial_i V_{N+1} + \sum(V_\alpha - V_{N+1})\rho\dot{\hat{c}}_\alpha$$

$$= -\rho\sum\hat{c}_\alpha v^i(f_\alpha^i - f_{N+1}^i) - \rho v^i f_{N+1}^i + \sum(V_\alpha - V_{N+1})(\hat{a}_\alpha - \partial_i\hat{J}_\alpha^i) \tag{5.20}$$

However

$$\sum(V_\alpha - V_{N+1})\partial_i\hat{J}_\alpha^i = \partial_i\left[\sum(V_\alpha - V_{N+1})\hat{J}_\alpha^i\right] - \sum\hat{J}_\alpha^i\partial_i(V_\alpha - V_{N+1})$$

$$= \partial_i\left[\sum(V_\alpha - V_{N+1})\hat{J}_\alpha^i\right] + \sum\hat{J}_\alpha^i(f_\alpha^i - f_{N+1}^i)$$

and hence

$$\rho\dot{\mathcal{V}} = -\rho\sum\hat{c}_\alpha v^i(f_\alpha^i - f_{N+1}^i) - \rho v^i f_{N+1}^i - \sum\hat{J}_\alpha^i(f_\alpha^i - f_{N+1}^i)$$

$$+ \sum\hat{a}_\alpha(V_\alpha - V_{N+1}) - \partial_i\left[\sum(V_\alpha - V_{N+1})\hat{J}_\alpha^i\right] \tag{5.21}$$

Thus since $(d/dt)\int_{\mathcal{B}}\rho\mathcal{V}\,dV = \int_{\mathcal{B}}\rho\dot{\mathcal{V}}dV$, because $(d/dt)\int_{\mathcal{B}}\rho\,dV = 0$, we have

$$-\frac{d}{dt}\int_{\mathcal{B}}\rho\mathcal{V}\,dV = \int_{\mathcal{B}}\left[\rho\sum\hat{c}_\alpha v^i(f_\alpha^i - f_{N+1}^i) + \rho v^i f_{N+1}^i + \sum\hat{J}_\alpha^i(f_\alpha^i - f_{N+1}^i)\right]dV$$

$$-\int_{\mathcal{B}}\sum\hat{a}_\alpha(V_\alpha - V_{N+1})\,dV + \int_{\partial\mathcal{B}}\sum(V_\alpha - V_{N+1})\hat{J}_\alpha^i n_i\,dS \tag{5.22}$$

Hence (5.16) is equivalent to

$$R(t) = \int_{\mathcal{B}} \rho \left[\hat{F}(T_0, \Gamma) + \mathcal{V} \right] dV + \int_{\mathcal{B}} \rho \left[(T_0 - T)^2 \mathcal{K}(T', \Gamma) + \tfrac{1}{2} v^i v_i \right] dV$$

$$- \int_{t_0}^{t} \left[\int_{\partial \mathcal{B}} v^i t^{ij} n_j \, dS + \int_{\mathcal{B}} \sum \hat{a}_\alpha (V_\alpha - V_{N+1}) \, dV \right.$$

$$\left. - \int_{\partial \mathcal{B}} \sum (\hat{\mu}_{\alpha 0} + V_\alpha - V_{N+1}) \hat{J}_\alpha^i n_i \, dS \right] dt \qquad (5.23)$$

The treatment of the term involving $v^i t^{ij} n_j$ follows Coleman and Dill (Ref. 36, Section 6). We assume that the motion of the body satisfies boundary conditions of place and traction. For this we assume that $\partial \mathcal{B}$ is the union of two disjoint regular sets $\partial \mathcal{B}_1$ and $\partial \mathcal{B}_2$, where $\partial \mathcal{B}_1$ is never empty, such that

1. $\partial \mathcal{B}_1$ is held fixed in space, i.e.,

$$x^i(X^A, t) = q^i(X^A) \qquad \text{for} \qquad X^A \in \partial \mathcal{B}_1 \qquad (5.24)$$

$$v^i(q^j, t) = 0 \qquad \text{for} \qquad X^A \in \partial \mathcal{B}_1 \qquad (5.25)$$

2. the contact forces on $\partial \mathcal{B}_2$ are such that there exists a functional $\zeta[x^i]$ such that

$$- \frac{d}{dt} \zeta[x^i] = \int_{\partial \mathcal{B}} v^i t^{ij} n_j \, dS = \int_{\partial \mathcal{B}_2} v^i t^{ij} n_j \, dS \qquad (5.26)$$

The latter equality obtains since, by (5.25), $v^i t^{ij} n_j = 0$ on $\partial \mathcal{B}_1$. Examples of loadings that satisfy this condition are given by (1) a free surface ($t^{ij} n_j = 0$), in which case $\zeta[x^i] = 0$; (2) the stress vector on $\partial \mathcal{B}_2$ is a constant uniform normal pressure p^0($t^{ij} n_j = -p^0 n_i$), in which case

$$\zeta[x^i] = p^0 \int_{\mathcal{B}} dV \qquad (5.27)$$

(3) the traction vector on $\partial \mathcal{B}_2$ admits a potential in the sense that

$$j t^{ij} \partial_j X^A n_A = - \frac{\partial l(x^k)}{\partial x^i} \qquad (5.28)$$

in which case

$$\zeta[x^i] = \int_{(\partial \mathcal{B}_2)_0} l\{x^i(X^A, t)\} \, dS_0 \tag{5.29}$$

$$\left(\text{i.e.,} \quad \frac{d\zeta}{dt} = \int_{(\partial \mathcal{B}_2)_0} v^i \partial_i l \, dS_0 = - \int_{(\partial \mathcal{B}_2)_0} v^j t^{ij} \partial_j X^A n_A \, dS_0 \right.$$

$$\left. = - \int_{\partial \mathcal{B}_2} v^i t^{ij} n_j \, dS \right)$$

This latter case includes the situation in which $\partial \mathcal{B}_2$ is dead loaded, $(jt^{ij} \partial_j X^A n_A)' = 0$, as well as a large number of other alternatives. When these conditions are met, (5.23) can be replaced by the equivalent function(al)

$$R(t) = \int_{\mathcal{B}} \rho\{\hat{F}(T_0, \Gamma) + \mathcal{V}\} \, dV + \zeta[x^i]$$

$$+ \int_{\mathcal{B}} \rho\{(T_0 - T)^2 \mathcal{K}(T', \Gamma) + \tfrac{1}{2} v^i v^i\} \, dV$$

$$- \int_{t_0}^{t} \left[\int_{\mathcal{B}} \sum \hat{a}_\alpha (V_\alpha - V_{N+1}) \, dV - \int_{\partial \mathcal{B}} \sum (\hat{\mu}_{\alpha 0} + V_\alpha - V_{N+1}) \hat{J}_\alpha^i n_i \, dS \right] dt$$

$$\tag{5.30}$$

Since $\partial \mathcal{B}$ is the boundary between the body and the external environment, $\partial \mathcal{B}$ can also be looked at as an internal boundary of the external environment. Noting that both the body and the environment are acted upon by the same system of body forces and that the potentials for such forces are continuous across a boundary, we see that $V_\alpha|_{\partial \mathcal{B}}$ is the same for both the body and for the environment. Now the external environment has (reduced) chemical potentials which are in equilibrium with the imposed body forces. We thus have

$$\hat{\mu}_{\alpha 0} + V_\alpha - V_{N+1} = 0 \qquad \alpha = 1, \ldots, N$$

at every point in the external environment. It thus follows that

$$(\hat{\mu}_{\alpha 0} + V_\alpha - V_{N+1})|_{\partial \mathcal{B}} = 0 \qquad \alpha = 1, \ldots, N \tag{5.31}$$

and hence we obtain

$$\int_{t_0}^{t} \int_{\partial \mathcal{B}} \sum (\hat{\mu}_{\alpha 0} + V_\alpha - V_{N+1}) \hat{J}_\alpha^i n_i \, dS \, dt = 0 \tag{5.32}$$

provided only that $\hat{J}_\alpha^i n_i$ is bounded.

We now assume that the supplies of mass that arise from chemical reactions are such that*

$$\sum \hat{a}_\alpha (V_\alpha - V_{N+1}) = 0 \qquad \text{throughout} \qquad \mathcal{B} \qquad (5.33)$$

This assumption is equivalent to the condition

$$\sum_{\alpha=1}^{N+1} a_\alpha V_\alpha = 0 \qquad \text{throughout} \qquad \mathcal{B} \qquad (5.34)$$

If the supplies of mass obtain from K distinct chemical reactions with stoichiometric coefficients $v_{\alpha r}$ and reaction rates J_r

$$a_\alpha = \sum_{r=1}^{K} v_{\alpha r} J_r \qquad (5.35)$$

the condition (5.34) is equivalent to the conditions

$$\sum_{\alpha=1}^{N+1} V_\alpha v_{\alpha r} = 0 \qquad r = 1, \dots, K \qquad (5.36)$$

Since these conditions are usually assumed to obtain in the course of chemical reactions, the conditions (5.32) are not unduly restrictive. Under satisfaction of (5.33) we have

$$\sum \hat{a}_\alpha \frac{\hat{\mu}_\alpha}{T} = \sum \hat{a}_\alpha \frac{\hat{\mu}_\alpha + V_\alpha - V_{N+1}}{T}$$

and hence changing $\hat{\mu}_\alpha / T$ to $(\hat{\mu}_\alpha + V_\alpha - V_{N+1})/T$ in the definition of the thermodynamic forces (4.1) preserves the relation $\sigma = \mathbf{J} \cdot \mathbf{X}$. Hence the representation of the $\hat{a}_\alpha$ given by (4.10a) can be replaced by

$$\hat{a}_\alpha = - \frac{\partial \phi}{\partial \left(\dfrac{\hat{\mu}_\alpha + V_\alpha - V_{N+1}}{T} \right)} - U_\alpha$$

*Thus, by (4.10a), ϕ and $\hat{U}_\alpha$ must satisfy

$$\sum (V_\alpha - V_{N+1}) \left[\frac{\partial \phi}{\partial (\hat{\mu}_\alpha / T)} + \hat{U}_\alpha \right] = 0$$

This constraint is easily satisfied since ϕ can depend on x^i; that is, $\phi = \phi[(\hat{\mu}_\alpha + V_\alpha - V_{N+1})/T, \dots]$ is the general solution of the above p.d.e.

Thus in the presence of external-body force fields with potentials V_α, the condition $\mathbf{X} = 0$ now gives $\hat{\mu}_\alpha + V_\alpha - V_{N+1} = 0$, which is one of the standard results for thermodynamic equilibrium. We therefore assume that $\hat{\mu}_\alpha / T$ is replaced by $(\hat{\mu}_\alpha + V_\alpha - V_{N+1})/T$ throughout the remainder of this paper. Under these conditions $R(t)$ can be replaced by the equivalent expression

$$R(t) = \int_{\mathcal{B}} \rho \{ \hat{F}(T_0, \Gamma) + \mathcal{V} \} \, dV + \zeta[x^i]$$

$$+ \int_{\mathcal{B}} \rho \{ (T_0 - T)^2 \mathcal{K}(T', \Gamma) + \tfrac{1}{2} v^i v^i \} \, dV \tag{5.37}$$

An alternative and weaker assumption would be to require the existence of a functional $\gamma[x^i, T, \hat{c}_\alpha]$ such that

$$\frac{d}{dt} \gamma[x^i, T, \hat{c}_\alpha] = \int_{\mathcal{B}} \sum \hat{a}_\alpha (V_\alpha - V_{N+1}) \, dV \geqslant 0 \tag{5.38}$$

in which case we would have

$$R(t) = \int_{\mathbf{B}} \rho \{ \hat{F}(T_0, \Gamma) + \mathcal{V} \} \, dV + \zeta[x^i] + \gamma[x^i, T, \hat{c}_\alpha]$$

$$+ \int_{\mathcal{B}} \rho \{ (T_0 - T)^2 \mathcal{K}(T', \Gamma) + \tfrac{1}{2} v^i v^i \} \, dV$$

Having noted this possibility, the physical interpretation of which is not altogether clear, we shall continue to require satisfaction of the condition (5.33).

C. Equilibrium States and Lower Bounds

The purpose of all of this is that we can now establish lower bounds on the function $R(t)$. Clearly, since $\mathcal{K}(T', \Gamma) \geqslant 0$, (5.37) implies

$$R(t) \geqslant \int_{\mathcal{B}} \rho \{ \hat{F}(T_0, \Gamma) + \mathcal{V} \} \, dV + \zeta[x^i] \tag{5.39}$$

with equality holding if and only if

$$T = T_0 \qquad v^i = 0 \qquad \text{throughout} \qquad \mathcal{B} \tag{5.40}$$

To proceed further we note that the right-hand side of the inequality (5.39), namely

$$\Psi[T_0, \Gamma] = \int_{\mathcal{B}} \rho \{ \hat{F}(T_0, \Gamma) + \mathcal{V}(x^i) \} \, dV + \zeta[x^i] \tag{5.41}$$

depends only on the state variables $\hat{c}_\alpha$, $\partial_A x^i$, and x^i, in addition to the parameter T_0; i.e., $\Psi[T_0, \Gamma]$ does not depend explicitly on the time variable. In fact, if $\zeta[x^i]$ were absent in (5.41), Ψ would be just the total free energy of the body in the presence of the external potential field $\mathcal{V}$ when the temperature throughout the body has the constant value T_0. This suggests that we make use of the fact in classical thermodynamics that distinguishes equilibrium states of the system at constant temperature T_0 from among the collection of all states in mechanical equilibrium ($\dot{v}^i = 0$) at constant temperature T_0; the equilibrium states at temperature T_0 are those states in mechanical equilibrium that minimize the total free energy of the system at constant temperature T_0 and constant total mass. Thus since $\zeta[x^i]$ is the potential functional for the applied surface tractions, we state the following definition.

The equilibrium states of a body $\mathcal{B}$ in mechanical equilibrium at constant temperature T_0 *and subject to*

1. given boundary conditions of place on $\partial\mathcal{B}_1$ *of the form (5.24),*

2. given boundary conditions of traction on $\partial\mathcal{B}_2$ *that have a potential functional* $\zeta[x^i]$ *such that (5.26) is satisfied,*

3. the reduced potentials $\hat{\mu}_{\alpha 0}$ *of the external environment satisfying the conditions (5.31),*

4. given body forces that admit potentials V_α *such that the condition (5.33) holds, are those states*

$$\Gamma_e \equiv \{\hat{c}_{\alpha e}, \partial_A x_e^i\} \tag{5.42}$$

that minimize the total free energy functional $\Psi[T_0, \Gamma]$ *at constant temperature* T_0 *and constant total mass* $M = \int_{\mathcal{B}} \rho \, dV$ *subject to the given boundary conditions and the initial conditions* $v^i(x^j, t_0) = 0$.[†] *Under these conditions, there exists at least one equilibrium state* (T_0, Γ_e^*) *that has the property that*

$$\Psi[T_0, \Gamma] \geqslant \Psi[T_0, \Gamma_e^*] \tag{5.43}$$

with equality holding if and only if $\Gamma = \Gamma_e^*$ when we assume Γ_e^* to be unique. Hence (5.39), (5.41), and (5.43) give the following lower bound for the functional $R(t)$:

$$R(t) \geqslant \Psi[T_0, \Gamma_e^*] \tag{5.44}$$

Further, since equality holds in (5.39) if any only if $T = T_0$ and $v^i = 0$, equality holds in (5.44) if and only if

$$T = T_0 \qquad v^i = 0 \qquad \Gamma = \Gamma_e^* \tag{5.45}$$

[†]See Ref. 38 for the modifications of this definition that account for satisfaction of the laws of mass balance.

D. Properties of the Equilibrium State

The universal lower bound (5.44) is what will allow us to obtain the fundamental results on the asymptotic approach to equilibrium states of chemically reacting and diffusing material bodies in the next section. Before turning to the dynamic stability question, it is useful to pause at this point to establish certain additional properties of equilibrium states of a material body at constant temperature T_0 subject to the imposed boundary conditions of place and traction, and the imposed body forces. Since an equilibrium state (T_0, Γ_e) minimizes the functional (5.41) at constant T_0 and $M = \int_{\mathcal{B}} \rho \, dV$ subject to the boundary conditions (5.24), (5.25), (5.26), and the condition (5.33), the variables $(\hat{c}_{\alpha e}, x_e^i)$ must satisfy the Euler equations[48] associated with the functional $\Psi[T_0, \Gamma]$ such that

$$\delta(j\rho) = 0 \quad \text{that is} \quad \delta\rho = -\frac{\rho}{j}\,\delta j = -\rho(\partial_i X^A)\partial_A \delta x^i$$

We thus obtain the following results [recall (3.20) and (5.6)]:

$$\partial_i t_{Re}^{ij} = \rho\partial_j \mathcal{V} \quad \text{throughout} \quad \mathcal{B} \tag{5.46}$$

$$\int_{\partial\mathcal{B}_2} (\delta x^i) t_{Re}^{ij} n_j \, dS = -\delta \zeta[x^i] \quad \text{on} \quad \partial\mathcal{B}_2 \tag{5.47}$$

$$x_e^i = q^i(X^A) \quad \text{on} \quad \partial\mathcal{B}_1 \tag{5.48}$$

from variation of the current configuration variable x^i;

$$0 = \frac{\partial\hat{F}}{\partial\hat{c}_\alpha} + \frac{\partial\mathcal{V}}{\partial\hat{c}_\alpha} = \hat{\mu}_{\alpha e} + V_\alpha - V_{N+1} = \mu_{\alpha e} - \mu_{N+1e} + V_\alpha - V_{N+1} \tag{5.49}$$

throughout $\mathcal{B}$ from variation of the $\hat{c}_\alpha$'s;

$$T_e = T_0 \tag{5.50}$$

throughout $\mathcal{B}$; and

$$\sum \hat{a}_{\alpha e}(V_\alpha - V_{N+1}) = 0 \tag{5.51}$$

The condition (5.50) implies $\partial_i T_0 = 0$, and the conditions of mechanical equilibrium, $\dot{v}^i = 0$, together with the initial conditions $v^i(x^j, t_0) = 0$, imply $v^i(x^j, t) = 0$, so that $d_{ij} = 0$. Hence (3.26), $f_\alpha^i = -\partial_i V_\alpha$, and $\sum \hat{a}_\alpha(V_\alpha - V_{N+1}) = 0$ give

$$\sigma_e = -\sum \hat{J}_{\alpha e}^i \partial_i \left(\frac{\hat{\mu}_\alpha + V_\alpha - V_{N+1}}{T_0} \right) - \sum \hat{a}_{\alpha e} \left(\frac{\hat{\mu}_\alpha + V_\alpha - V_{N+1}}{T_0} \right) \tag{5.52}$$

Thus (5.49) and (5.52) show that

$$\sigma_e = 0 \qquad (5.53)$$

Now we have assumed that $\sigma = 0$ when and only when $\mathbf{J} = \mathbf{X} = 0$,[†] and hence we obtain

$$t_e^{ij} - t_{Re}^{ij} = 0 \qquad (5.54)$$

$$J_{qe}^i = \hat{J}_{\alpha e}^i = 0 \qquad (5.55)$$

$$\hat{a}_{\alpha e} = 0 \qquad (5.56)$$

Combining (5.54) through (5.56) with (5.46) and the equations of balance obtained in Section III and noting that $\dot{v}_e^i = v_e^i = 0$, we see that every equilibrium process is a static process; i.e.,

$$v_e^i = \dot{v}_e^i = \hat{c}_{\alpha e} = \dot{e}_e = \dot{s}_e = \dot{\rho}_e = 0 \qquad (5.57)$$

It also follows from (5.49) that

$$(\hat{\mu}_{\alpha e} + V_\alpha - V_{N+1})|_{\partial \mathcal{B}} = 0 \qquad (5.58)$$

and hence (5.58) and (5.31) show that an equilibrium process is such that

$$\hat{\mu}_{\alpha e}|_{\partial \mathcal{B}} = \hat{\mu}_{\alpha 0}|_{\partial \mathcal{B}} = (V_{N+1} - V_\alpha)|_{\partial \mathcal{B}} \qquad (5.59)$$

Thus the definition we have given for the equilibrium process leads to equilibrium processes with all of the usual properties. It is interesting to note that we have been able to prove rather than assume that the equilibrium process is a static process. The proof, however, relies upon the properties of σ and hence depends in an essential way on consideration of nonequilibrium processes, for it is only in such processes that $\sigma \not\equiv 0$ is defined. Nonequilibrium is thus necessary in order to give meaning to the implications of $\sigma = 0$. We also note that the conclusion about the static nature of the equilibrium states obtained from satisfaction of the Euler equations (stationarity conditions), and hence the solutions of these equations, need not necessarily refer to stable equilibrium states.

Careful note must be taken of the fact that we have assumed that the equilibrium process (T_0, Γ_e^*) satisfying the given boundary conditions is unique. Thus, in particular, the functional $\Psi[T_0, \Gamma]$ must have the property that there is one and only one solution (T_0, Γ_e^*) of the Euler equations

---

[†]See Ref. 38 for analysis of the case where σ depends on μ_α / T only through the independent affinities of the reactions.

environment at points on $\partial \mathcal{B}$. Such a condition is clearly necessary if the boundary conditions are to preserve the second law of thermodynamics at the boundary. Similarly satisfaction of the inequality (6.10) is also necessary for preservation of the second law on the boundary. If the set $\partial \mathcal{B}_3$ is vacuous, so that there are no boundary conditions on the value of the temperature field, there is the possibility that the system is thermally isolated, that is, $J_q^i n_i = 0$ at every point of $\partial \mathcal{B}$. Such a condition is incompatible with the notion of immersion in an external environment on physical grounds since there are no perfect insulators. It is for this reason that we have demanded that the function h cannot vanish on $\partial \mathcal{B}$ if $\partial \mathcal{B}_3$ is vacuous unless $T = T_0$ at at least one point of $\partial \mathcal{B}$. Clearly if

$$J_q^i n_i = h = (T - T_0)f \qquad f > 0$$

is the form of the function h on some part $\partial \mathcal{P}$ of $\partial \mathcal{B}$, then we satisfy the inequality (6.8) and the requirement that $h = 0$ only if $T = T_0$ on $\partial \mathcal{P}$ Likewise (6.11) can be satisfied by

$$\hat{J}_\alpha^i n_i = \hat{h}_\alpha = \left(\frac{\hat{\mu}_\alpha}{T} - \frac{\hat{\mu}_{\alpha 0}}{T_0} \right) g_\alpha \qquad g_\alpha > 0$$

From the mathematical point of view, elimination of the possibility of a completely thermally isolated body appears to be necessary in order to secure the uniqueness of the equilibrium state, for otherwise the reference temperature T_0 is lost. For example, a rigid, linear, isotropic heat-conducting body comprised of a single constituent has the governing equation $c_v \partial_t T = k \nabla^2 T$. If the boundary conditions are $n_i \partial_i T = 0$ everywhere on $\partial \mathcal{B}$, different initial conditions will lead to different equilibrium temperatures, none of which need be equal to T_0. Put another way, immersion in an environment at temperature T_0 must have the capability of fixing the equilibrium temperature of the body at the temperature T_0 of the environment. It would appear that Coleman and Dill[36] overlooked this point in their analysis. We also note that the boundary conditions (6.8) and (6.9) are of sufficient generality as to encompass boundaries that are permeable to some species but not to others. For example, if $\partial \mathcal{B}_{5,1}$ and $\partial \mathcal{B}_{5,2}$ are empty while $h_1 = 0$ and $h_2 = (\hat{\mu}_\alpha / T - \hat{\mu}_{\alpha 0} / T_0) g_2$ when $g_2 > 0$, then the boundary is permeable to species two and impermeable to species one.

With satisfaction of the boundary conditions (6.5) through (6.10), we have

$$-\int_{\partial \mathcal{B}} \left(\frac{T - T_0}{T} \right) J_q^i n_i \, dS - T_0 \int_{\partial \mathcal{B}} \sum \left(\frac{\hat{\mu}_\alpha}{T} - \frac{\hat{\mu}_{\alpha 0}}{T_0} \right) \hat{J}_\alpha^i n_i \, dS \leqslant 0$$

(5.46) through (5.51), with $\dot{v}^i = v^i = 0$, which renders $\Psi[T_0, \Gamma]$ an *absolute* minimum for each choice to T_0 and the functions $q^i(x^j)$ and $\zeta[x^i]$ that set the boundary conditions.

VI. LARGE-TIME BEHAVIOR

The last section established the bound

$$R(t) \geqslant \Psi[T_0, \Gamma_e^*] = \int_{\mathcal{B}} \rho \{ \hat{F}(T_0, \Gamma_e^*) + \mathcal{V}(x_e^i) \} \, dV + \zeta[x_e^i] \qquad (6.1)$$

for all processes that (1) satisfy given boundary conditions of the form (5.24), (5.25), (5.26); (2) satisfy the condition (5.34); (3) are acted upon by given body forces which admit potentials; and (4) the surface tractions on $\partial \mathcal{B}_2$ admit the potential function $\zeta[x^i]$. For bodies of finite extent

$$\int_{\mathcal{B}} dV(x) = \int_{\mathcal{B}_0} j \, dV(X) < \infty \qquad (6.2)$$

the lower bound $\Psi[T_0, \Gamma_e^*]$ is finite. Thus if we can show that $R(t)$ is a monotone decreasing function of t on $[0, \infty]$, then to all intent and purposes we will have most of the results that are necessary in order to study the large-time behavior of the chemical systems considered previously. However the argument leading to the bound (6.1) allowed reduction to problems in which $T = T_0$, and hence there was no necessity of specifying boundary conditions for the temperature field. Clearly such conditions must be added if we are to obtain deterministic results and, in addition, these conditions should reflect the notion that the body is immersed in an external environment that is held at constant temperature T_0. If this were not the case, then T_0 would have no intrinsic association with the problem and physically unrealistic situations would be under consideration.

A. The External Environment and Bounds on dR/dt

A direct calculation of the time derivative of (5.1) gives

$$\frac{dR}{dt} = \frac{d}{dt} \int_{\mathcal{B}} \rho(\hat{e} + \tfrac{1}{2} v^i v_i) \, dV - T_0 \frac{d}{dt} \int_{\mathcal{B}} \rho s \, dV$$

$$- \int_{\partial \mathcal{B}} v^i t^{ij} n_j \, dS - \int_{\mathcal{B}} \rho v^i \left\{ f_{N+1}^i + \sum \hat{c}_\alpha (f_\alpha^i - f_{N+1}^i) \right\} dV$$

$$- \int_{\mathcal{B}} \sum \hat{J}_\alpha^i (f_\alpha^i - f_{N+1}^i) \, dV + \int_{\partial \mathcal{B}} \sum \hat{\mu}_{\alpha 0} \hat{J}_\alpha^i n_i \, dS \qquad (6.3)$$

Since equations of balance are satisfied for each dynamical process of the body, we can use (3.16) and (3.24) to eliminate a number of the terms which occur in (6.3). When this is done, we have

$$\frac{dR}{dt} = -T_0 \int_{\mathcal{B}} \sigma \, dV - \int_{\partial \mathcal{B}} \frac{T - T_0}{T} J_q^i n_i \, dS$$

$$- T_0 \int_{\partial \mathcal{B}} \sum \left(\frac{\hat{\mu}_\alpha}{T} - \frac{\hat{\mu}_{\alpha 0}}{T_0} \right) \hat{J}_\alpha^i n_i \, dS \tag{6.4}$$

The sign of the first term on the right-hand side of (6.4) is known for all processes since $\sigma \geqslant 0$. The signs of the remaining terms are fixed by the following definition which also serves to make the idea of immersion in a static external environment a precise statement.

A body $\mathcal{B}$ is said to be compatible with immersion in an external environment that has chemical $\hat{\mu}_{\alpha 0}$ in equilibrium with the imposed body forces with potentials V_α and constant temperature T_0 if and only if
1. the boundary conditions

$$T = T_0 \quad on \quad \partial \mathcal{B}_3 \tag{6.5}$$

$$\partial \mathcal{B} = \partial \mathcal{B}_3 \cup \partial \mathcal{B}_4$$

$$J_q^i n_i = h \quad on \quad \partial \mathcal{B}_4 \tag{6.6}$$

of temperature and heat flux satisfy

$$(T - T_0) J_q^i n_i \geqslant 0 \tag{6.7}$$

2. the boundary conditions

$$T_0 \hat{\mu}_\alpha = T \hat{\mu}_{\alpha 0} \quad on \quad \partial \mathcal{B}_{5,\alpha} \tag{6.8}$$

$$\partial \mathcal{B} = \partial \mathcal{B}_{5,\alpha} \cup \partial \mathcal{B}_{6,\alpha}$$

$$\hat{J}_\alpha^i n_i = \hat{h}_\alpha \quad on \quad \partial \mathcal{B}_{6,\alpha} \tag{6.9}$$

of chemical potential and diffusion flux satisfy

$$\sum \left(\frac{\hat{\mu}_\alpha}{T} - \frac{\hat{\mu}_{\alpha 0}}{T_0} \right) \hat{J}_\alpha^i n_i \geqslant 0 \tag{6.10}$$

3. if $\partial \mathcal{B}_3$ is vacuous then the given function h can vanish on $\partial \mathcal{B}$ only when $T = T_0$ at at least one point on $\partial \mathcal{B}$.
The inequality (6.7) simply states that either $T = T_0$ on $\partial \mathcal{B}$ or that heat flows into (out of) the body if the body is colder (hotter) than the

When this inequality is used in conjunction with (6.4), we obtain the upper bound

$$\frac{dR}{dt} \leqslant -T_0 \int_{\mathcal{B}} \sigma \, dV \tag{6.11}$$

with equality holding only when

$$\int_{\partial \mathcal{B}} \left(\frac{T-T_0}{T} \right) J_q^i n_i \, dS = \int_{\partial \mathcal{B}} \sum \left(\frac{\hat{\mu}_\alpha}{T} - \frac{\hat{\mu}_{\alpha 0}}{T_0} \right) \hat{J}_\alpha^i n_i \, dS = 0 \tag{6.12}$$

B. Asymptotic Approach to Equilibrium States

Let $\mathcal{I}(\mathcal{B}, \partial\mathcal{B})$ denote the collection of all finite initial data for processes of the finite body $\mathcal{B}$ for which solutions of the governing laws of evolution exist for all t in $[0, \infty)$ and such that

1. given boundary conditions of the kinds considered above are satisfied on $\partial\mathcal{B}$,

2. the given body forces are derived from potentials, and

3. $\Sigma \hat{a}_\alpha(V_\alpha - V_{N+1}) = 0$ throughout $\mathcal{B}$.

Clearly the initial data $T = T_0$, $v^i(x^j, t_0) = 0$, $\Gamma(x^i, t_0) = \Gamma_e(x^i)$ belong to $\mathcal{I}(\mathcal{B}, \partial\mathcal{B})$. For each process with initial data belonging to $\mathcal{I}(\mathcal{B}, \partial\mathcal{B})$ we have

$$R(t) \geqslant \Psi[T_0, \Gamma_e^*] \tag{6.13}$$

where $\Psi[T_0, \Gamma_e]$ is finite because $\int_{\mathcal{B}} dV$ is finite,[†] and

$$\frac{dR(t)}{dt} \leqslant -T_0 \int_{\mathcal{B}} \sigma \, dV \leqslant 0 \tag{6.14}$$

An integration of (6.14) leads to a contradiction of the lower bound (6.13) unless the integral $\int_{t_0}^{\infty} \int_{\mathcal{B}} \sigma \, dV \, dt$ converges, and hence we must have, for t greater than some positive number t_1,

$$\int_{\mathcal{B}} \sigma \, dV \leqslant A t^{-1-\gamma} \tag{6.15}$$

[†]Implied here, of course, is the standard assumption of thermostatics—that $F(T_0, \Gamma)$ is a continuous and differentiable f unction of its arguments, and hence is finite for finite values of these arguments.

for some $\gamma > 0$ and some positive number A. Using L to denote the limit process as t tends to infinity, (6.13) through (6.15) thus imply the existence and values of the limits

$$LR(t) \geqslant \Psi[T_0, \Gamma_e] \tag{6.16}$$

$$L\frac{dR(t)}{dt} = -T_0 L \int_{\mathcal{B}} \sigma \, dV = -T_0 L \int_{\mathcal{B}_0} \sigma j \, dV(X^A) = 0 \tag{6.17}$$

Further, since equality holds in (6.11) if and only if (6.12) holds, the integrands of (6.12) are non-negative, and σ is non-negative, it follows from (6.17) and the ability to convert all volume and surface integrations to integrations over the reference configuration that

$$L\left[(T - T_0)J_q^i j(\partial_i X^A) n_A\right] = \sum L\left[\left(\frac{\hat{\mu}_\alpha}{T} - \frac{\hat{\mu}_{\alpha 0}}{T_0}\right)\hat{J}_\alpha^i j(\partial_i X^A) n_A\right] = 0 \tag{6.18}$$

on ∂B_0 and

$$L(j\sigma) = 0 \tag{6.19}$$

In view of the conditions which the thermal boundary conditions must satisfy for compatability with immersion in an external environment at temperature T_0, (6.18) implies that

$$LT = T_0 \tag{6.20}$$

for at least one point on $\partial \mathcal{B}$ since $j > 0$ and continuous, and $\det(\partial_i X^A) \neq 0$ for all t.[†] Likewise, since $j > 0$ and continuous for all t, (6.19) implies

$$L\sigma = L\sigma^*[\mathbf{X}(x^i, t); \omega(x^i, t)] = 0 \tag{6.21}$$

and hence continuity and differentiability of σ^* implies

$$\sigma^*(L\mathbf{X}; L\omega) = 0 \tag{6.22}$$

[†]It is assumed here that the boundary conditions preclude the case of expansion into an unbounded region. If free expansion occurs, it must take place from some given finite region into a larger, but finite, enclosure.

Now we have assumed that $\sigma^*(\mathbf{X}; \omega) = 0$ for any ω if and only if both $\mathbf{X} = \mathbf{0}$ and $\mathbf{J}(\mathbf{0}; \omega) = \mathbf{0}$, while (3.26), (5.17), and (5.33) give

$$\sigma^* = \left(t^{ij} - t_R^{ij}\right)\frac{d_{ij}}{T} - J_q^i \frac{\partial_i T}{T^2}$$

$$- \sum_\alpha \hat{J}_\alpha^i \left[\partial_i \left(\frac{\hat{\mu}_\alpha}{T}\right) + \frac{1}{T}\partial_i(V_\alpha - V_{N+1}) \right]$$

$$- \sum_\alpha \hat{a}_\alpha \frac{\hat{\mu}_\alpha + V_\alpha - V_{N+1}}{T} \tag{6.23}$$

We thus conclude that

$$L\left(\frac{d_{ij}}{T}\right) = L\left(\frac{\partial_i T}{T^2}\right) = L\left[\partial_i \left(\frac{\hat{\mu}_\alpha}{T}\right) + \frac{1}{T}\partial_i(V_\alpha - V_{N+1}) \right]$$

$$= L\frac{\hat{\mu}_\alpha + V_\alpha - V_{N+1}}{T} = 0 \tag{6.24}$$

$$L(t^{ij}) = L(t_R^{ij}) \tag{6.25}$$

$$L(J_q^i) = L(\hat{J}_\alpha^i) = L(\hat{a}_\alpha) = 0 \tag{6.26}$$

Since $L(d_{ij}) = 0$, the only limiting motion of the body is a rigid body motion[39]. Accordingly, because boundary conditions of place are satisfied on $\partial \mathcal{B}_1$, and $\partial \mathcal{B}_1$ is not vacuous, rigid body motions are eliminated and we have

$$Lv^i = 0 \tag{6.27}$$

throughout $\mathcal{B}$. The equations of balance of linear momentum, (3.13), together with (6.21), imply that the limiting configuration of the body satisfies the static equations (5.46) and the static boundary conditions (5.47) and (5.48), and hence $L\dot{v}^i = 0$. Now $L\partial_i T = 0$ implies that LT is independent of position, and hence satisfaction of the condition (6.17) for at least one point on $\partial \mathcal{B}$ gives

$$LT = T_0 \tag{6.28}$$

throughout $\mathcal{B}$. A combination of (6.16), (6.23), and (6.28) with (6.22), (6.26), (3.3), (3.11), (3.17), and (3.25) shows that the limit solution satisfies the conditions

$$Lv^i = L\dot{v}^i = L\hat{c}_\alpha^\cdot = L\hat{e}^\cdot = L\dot{s}$$

$$= L\dot{\rho} = L(\hat{\mu}_\alpha - V_\alpha + V_{N+1}) = 0$$

$$L(T) = T_0$$

$$L\sum \hat{a}_\alpha (V_\alpha - V_{N+1}) = 0 \qquad (6.29)$$

$$L(\partial_j t_R^{ij}) = L(\rho \partial_i \mathcal{V})$$

and the given boundary conditions. The result $L\dot{v}^i = 0$ and the last of (6.29) follow from writing (3.13) in the equivalent form $\rho\dot{v}^i - \partial_j(t^{ij} - t_R^{ij}) = \partial_j t_R^{ij} - \rho\partial_i \mathcal{V}$. Since these are exactly the equations and boundary conditions that equilibrium processes satisfy, as shown by the analysis at the end of Section V, we arrive at the following conclusion.

Every process of a finite material body $\mathfrak{B}$ with initial data belonging to $\mathcal{G}(\mathfrak{B}, \partial\mathfrak{B})$ and given boundary data of the kind considered above tends to a (static) equilibrium state that satisfies the same boundary data.

Although we now know that any process starting in $\mathcal{G}(\mathfrak{B}, \partial\mathfrak{B})$ will wind up at an equilibrium state for large t, we do not know which of the equilibrium states this will be.

C. Asymptotic Stability

The analysis given at the end of Section V showed that the equilibrium states of a body (subject to the given boundary conditions) were static states and that they rendered the total free energy *stationary* in value. On the other hand we have also assumed that there is an equilibrium state (T_0, Γ_e^*) for which the total free energy attains an absolute minimum rather than just a stationary value (for satisfaction of the given boundary conditions, of course). It now follows that a material body for which (T_0, Γ_e^*) is the *only* stationarizing element of the total free energy for the given boundary conditions will have the property that it will tend to the unique equilibrium solution (T_0, Γ_e^*) for sufficiently large time. Under satisfaction of this uniqueness condition on the equilibrium state, we have

$$L\Gamma = \Gamma_e^* \qquad L\Psi[T, \Gamma] = \Psi[T_0, \Gamma_e^*]$$

$$LR(t) = \Psi[T_0, \Gamma_e^*] = \min_{t \in [0, \infty)} \Psi[T_0, \Gamma(t)] = \min_\Gamma \Psi[T_0, \Gamma] \qquad (6.30)$$

and hence we obtain the following basic asymptotic stability result.

If a material body has one and only one equilibrium state for given T_0 and given boundary conditions of the kind considered above, every process of such a body with initial data belonging to $\mathcal{I}(\mathcal{I}, \partial\mathcal{B})$ tends to the unique equilibrium soltuion that satisfies the same given boundary conditions as t tends to infinity; the state (T_0, Γ_e^*) is asymptotically stable with $\mathcal{I}(\mathcal{B}, \partial\mathcal{B})$ as its domain of attratcion.

D. Concluding Remarks

There are several remarks that need to be made about this asymptotic stability theorem. First, it is essential to assume existence of solutions of the governing equations for all t in $[0, \infty)$ and uniqueness of the equilibrium solution. Without these assumptions the theorem is clearly false. In particular, if the body is completely thermally isolated, it can cease to have a unique equilibrium solution for the given boundary data. In this event the asymptotic stability ceases to hold, since we do not know which equilibrium solution will obtain as a consequence of any given initial data belonging to $\mathcal{I}(\mathcal{B}, \partial\mathcal{B})$.

Second, the body must be of finite extent and the data must be finite. Otherwise the bounds on R and dR/dt can become infinite and the argument ceases to hold. This is not too heavy a price to pay, for the above theorem states that the flow of a one-component, heat-conducting, viscous fluid in a finite region with adherence to at least one fixed boundary will ultimately tend to a state of rest. Thus even if the flow becomes turbulent, it will eventually die out in time. Note carefully that the fluid must be heat-conducting and viscous, for otherwise σ can vanish without $\mathbf{X}$ and $\mathbf{J}$ vanishing simultaneously. In fact the essential difference between the results given in this paper and those derived in Ref. 36 can be traced directly to the fact that σ is required to vanish only when the fluxes and the thermodynamic forces both vanish.

Third, we have made no assumptions such as detailed balance or satisfaction of the laws of mass balance. Thus asymptotic stability of the equilibrium solutions obtains for the body under the given conditions whenever $\hat{a}_\alpha$ are such that $\sigma > 0$ for $(\mu_\alpha + V_\alpha - V_{N+1})/T \neq 0$. It can also be shown that the same results obtain when the laws of mass balance are satisfied and $\sigma > 0$ whenever the chemical affinities associated with the permissible reactions are nonzero.[38] Although it is known[34] that sustained oscillations of chemically reacting systems are possible if the system is open, the results established above show that satisfaction of the mild conditions imposed on the form of the boundary conditions precludes indefinitely sustained oscillations for systems that are *closed* with respect to the bulk motion (i.e., (2.17), (3.1), and (3.9) hold, and in particular $\sum_{\alpha=1}^{N+1} a_\alpha = 0$, which eliminates most of the examples considered in [34]).

This follows even when there are several equilibrium states, for every solution has been shown to proceed so as to decrease $R(t)$ and will wind up at static equilibrium states. In particular, biological clocks are precluded whenever the system is mass-closed and whenever boundary conditions and initial conditions belong to the kinds considered and to $\mathcal{I}(\mathcal{B},\partial\mathcal{B})$, respectively. Even though mass-closed chemical systems can still exhibit oscillations that persist for a very long time, ultimately they die out if the process is given sufficient time and if the external environment is maintained at constant temperature and in equilibrium with the externally applied potentials. We note that the mass closure considered in this paper does not preclude exchange of constituent masses with the environment by diffusion through the boundary, provided that the diffusion is driven by the chemical potentials of the external environment, as determined by the boundary conditions (6.9), (6.10), and $J^i_{N+1} = -\Sigma \hat{J}^i_\alpha$; in fact, it is this determination of J^i_{N+1} in terms of $\hat{J}^i_\alpha$ that reflects the diffusive mass closure with respect to the bulk motion. These considerations suggest that biological clocks (sustained chemical oscillations) are possible for mass-closed systems if the external environment undergoes sustained temperature and/ or chemical-potential oscillations, as is indeed the case in the biosphere.

Fourth, our asymptotic stability results are obtained as a consequence of the fact that

$$\Psi[T_0, \Gamma] \geqslant [T_0, \Gamma^*_e]$$

which can also be read as a statement of static stability of the equilibrium state (T_0, Γ^*_e) at constant temperature. The asymptotic stability theorem may thus be rephrased in the following way.

Static stability and uniqueness of the equilibrium state of a body $\mathcal{B}$ imply the asymptotic stability of the equilibrium state for all processes with initial data belonging to $\mathcal{I}(\mathcal{B},\partial\mathcal{B})$.

[Implied here, of course, are statements about the boundary conditions, body forces, immersion in a static external environment, and $\Sigma \hat{a}_\alpha (V_\alpha - V_{N+1}) = 0$.] This result appears to provide a full legitimization of classical thermodynamics in the real world of intrinsically dynamic, nonequilibrium processes.

Finally the established asymptotic stability of the unique equilibrium state for $\sigma = 0$ if and only if $\mathbf{X} = 0$ provides a means of computing the equilibrium state of a material body by means of an initial-value problem. If interest lies only in the equilibrium state, we can arrive at this equilibrium state for any choice of $\mathbf{J}(\mathbf{X}: \omega)$ provided only that $\sigma = 0$ if and only if $\mathbf{X} = 0$. Thus it is only necessary to select $\mathbf{U}(\mathbf{X}; \omega) = 0$ and $\phi(\mathbf{X}; \omega) = X_\Gamma L_{\Gamma\Lambda} X_\Lambda$, where $((L_{\Gamma\Lambda}))$ is a positive definite matrix with large eigenvalues, set $\mathbf{J} = \nabla_X \phi$ for a determination of the constitutive relations, and then integrate the equations of evolution forward in time with a reasonable choice of the initial conditions in order to arrive at the equilibrium configuration. Clearly the larger the eigenvalues of $((L_{\Gamma \Lambda}))$ the quicker the process will converge to the equilibrium solution. This method is particularly useful in determining the equilibrium concentrations of nonideal mixtures of liquids with nontrivial fugacities,[49] that is, mixtures that imply significant departures from the laws of mass action. In addition the quantity $dR(t)/dt$ gives a direct measure of the convergence rate since $dR/dt \leqslant -T_0 \int \sigma \, dV$ and σ has the positivity properties and dominates a convex function of the thermodynamic forces $\mathbf{X}$ in an open neighborhood of $\mathbf{X} = \mathbf{0}$.

APPENDIX. AN ALTERNATIVE ASYMPTOTIC STABILITY THEOREM

This appendix considers material bodies for which

$$v^i = 0 \tag{A.1}$$

$$J_\Lambda = \frac{\partial \phi}{\partial X_\Lambda} \tag{A.2}$$

$$\phi = \phi(\mathbf{X}) \tag{A.3}$$

Such bodies are those for which there is no mass flow, the thermodynamic fluxes are Onsager fluxes (i.e., $\mathbf{U} \equiv \mathbf{0}$), and the dissipation potential does not depend on the variables ω. Since ϕ is a dissipation potential, $\phi \geqslant 0$, and hence the function

$$P(t) = \int_{\mathcal{B}} \phi[\mathbf{X}(x^j, t)] \, dV \geqslant 0 \tag{A.4}$$

Since there is no mass flow, it follows that $\dot{a} = \partial_t a$ and we have

$$\frac{dP(t)}{dt} = \int_{\mathcal{B}} \dot{\phi} \, dV = \int_{\mathcal{B}} \sum_\Lambda \frac{\partial \phi}{\partial X_\Lambda} \dot{X}_\Lambda \, dV = \int_{\mathcal{B}} \sum_\Lambda J_\Lambda \dot{X}_\Lambda \, dV$$

$$= \int_{\mathcal{B}} \left\{ J_q^i \partial_i \left(\frac{1}{T} \right) - \sum \hat{J}_\alpha^i \left[\partial_i \left(\frac{\hat{\mu}_\alpha}{T} \right) - \frac{f_\alpha^i - f_{N+1}^i}{T} \right] \right.$$

$$\left. - \sum \hat{a}_\alpha \left(\frac{\hat{\mu}_\alpha}{T} \right) \dot{} \right\} dV \tag{A.5}$$

Since there is no bulk flow [i.e., (A.1) holds], we have $\dot{\rho}=0$ and

$$\partial_i t_R^{ij} + \rho \sum \hat{c}_\alpha (f_\alpha^i - f_{N+1}^i) + \rho f_{N+1}^i = 0 \tag{A.6}$$

from the conservation of mass and balance of linear momentum equations. Use of the remaining balance equations and the divergence theorem gives

$$\frac{dP}{dt} = \int_{\partial \mathcal{B}} \left[J_q^i \partial_t \left(\frac{1}{T} \right) - \sum \hat{J}_\alpha^i \partial_t \left(\frac{\hat{\mu}_\alpha}{T} \right) \right] n_i \, dS$$

$$+ \int_{\mathcal{B}} \left\{ -\partial_t \left(\frac{1}{T} \right) \left[-\rho T \dot{s} + \sum \hat{J}_\alpha^i (f_\alpha^i - f_{N+1}^i) - \rho \sum \hat{\mu}_\alpha \hat{c}_\alpha \right] \right.$$

$$+ \sum \hat{J}_\alpha^i (f_\alpha^i - f_{N+1}^i) \partial_t \left(\frac{1}{T} \right) + \frac{1}{T} \sum \hat{J}_\alpha^i \partial_t (f_\alpha^i - f_{N+1}^i)$$

$$\left. - \rho \sum \hat{c}_k \partial_t \left(\frac{\hat{\mu}_\alpha}{T} \right) \right\} dV$$

$$= \int_{\partial \mathcal{B}} \left[J_q^i \partial_t \left(\frac{1}{T} \right) - \sum \hat{J}_\alpha^i \partial_t \left(\frac{\hat{\mu}_\alpha}{T} \right) \right] n_i \, dS$$

$$+ \int_{\mathcal{B}} \frac{1}{T} \left[-\rho (\partial_t T)(\partial_t s) + \sum \hat{J}_\alpha^i \partial_t (f_\alpha^i - f_{N+1}^i) \right.$$

$$\left. - \rho \sum (\partial_t \hat{\mu}_\alpha)(\partial_t \hat{c}_\alpha) \right] dV$$

$$= \int_{\partial \mathcal{B}} \left[J_q^i \partial_t \left(\frac{1}{T} \right) - \sum \hat{J}_\alpha^i \partial_t \left(\frac{\hat{\mu}_\alpha}{T} \right) \right] n_i \, dS$$

$$+ \int_{\mathcal{B}} \frac{1}{T} \sum \hat{J}_\alpha^i \partial_t (f_\alpha^i - f_{N+1}^i) \, dV$$

$$- \int_{\mathcal{B}} \frac{\rho}{T} \left[-\frac{\partial^2 \hat{F}}{\partial T^2} (\partial_t T)^2 + \sum \sum (\partial_t \hat{c}_\alpha) \frac{\partial \hat{\mu}_\alpha}{\partial \hat{c}_\beta} (\partial_t \hat{c}_\beta) \right] dV \tag{A.7}$$

where we have used the relation $s = -\partial \hat{F}/\partial T$. Thus if
1. the boundary conditions are such that

$$J_q^i n_i = 0 \quad \text{on} \quad \partial \mathcal{B}_5 \quad \partial \mathcal{B} = \partial \mathcal{B}_5 \cup \partial \mathcal{B}_6 \quad \text{(A.8)}$$

$$\partial_t(T) = 0 \quad \text{on} \quad \partial \mathcal{B}_6 \quad \text{(A.9)}$$

$$\hat{J}_\alpha^i n_i = 0 \quad \text{on} \quad \partial \mathcal{B}_7 \quad \partial \mathcal{B} = \partial \mathcal{B}_7 \cup \partial \mathcal{B}_8 \quad \text{(A.10)}$$

$$\partial_t \left(\frac{\hat{\mu}_\alpha}{T} \right) = 0 \quad \text{on} \quad \partial \mathcal{B}_8 \quad \text{(A.11)}$$

2. the applied body forces are functions of position only (i.e., $\partial_t f_\alpha^i = 0$), we obtain the result

$$\frac{dP}{dt} = - \int_{\mathcal{B}} \frac{\rho}{T} \left[-\frac{\partial^2 \hat{F}}{\partial T^2} (\partial_t T)^2 + \sum \sum (\partial_t \hat{c}_\alpha) \frac{\partial \hat{\mu}_\alpha}{\partial \hat{c}_\beta} (\partial_t \hat{c}_\beta) \right] dV \quad \text{(A.12)}$$

Accordingly, since $\partial^2 \hat{F}/\partial T^2 < 0$,

$$\frac{dP}{dt} \leqslant 0 \quad \text{(A.13)}$$

for all time histories of processes for which

$$\frac{\partial \hat{\mu}_\alpha}{\partial \hat{c}_\beta} = \frac{\partial^2 \hat{F}}{\partial \hat{\mu}_\alpha \partial \hat{\mu}_\beta} \quad \text{(A.14)}$$

is a positive definite matrix for all values of the arguments of the Helmholtz free energy $\hat{F}$. We note that this condition is satisfied if $\hat{F}$ is convex in the variables $\hat{c}_\alpha$ with a minimum at $\hat{\mu}_\alpha = 0$ for constant T and $\partial_A x^i$.

Under satisfaction of these conditions, we have $P(t) \geqslant 0$, $dP/dt \leqslant 0$. Thus $L(dP/dt) = 0$, where L denotes the limit as t tends to infinity. It then follows from (A.12) that

$$L(\partial_t T) = L(\partial_t c_\alpha) = 0 \quad \text{(A.15)}$$

Every process thus tends to a static process as t tends to infinity. It also follows from (3.11), (3.25), and $s = \partial \hat{F} / \partial T$, that

$$0 = L(\rho \hat{c}_\alpha) = L(\hat{a}_\alpha) - L(\partial_i \hat{J}^i_\alpha) \tag{A.16}$$

$$0 = L(\hat{F}) = L\left(\frac{\partial \hat{F}}{\partial T} \partial_t T + \sum \frac{\partial \hat{F}}{\partial \hat{c}_\alpha} \hat{c}_\alpha \right) \tag{A.17}$$

$$0 = L(\rho \dot{s}) = -L\partial_i \left(\frac{J^i_q - \sum \hat{\mu}_\alpha \hat{J}^i_\alpha}{T} \right) + L(\sigma) \tag{A.18}$$

$$0 = L(\rho \hat{e}) \tag{A.19}$$

From these we conclude that these limiting static processes are such that

$$L\int_{\mathscr{B}} \hat{a}_\alpha \, dV = L\int_{\partial \mathscr{B}} \hat{J}^i_\alpha n_i \, dS$$

$$L\int_{\mathscr{B}} \sigma \, dV = L\int_{\partial \mathscr{B}} \frac{1}{T}\left(J^i_q - \sum \hat{\mu}_\alpha \hat{J}^i_\alpha \right) n_i \, dS$$

The processes are thus in equilibrium with the imposed boundary values of the heat flux and the diffusion fluxes.

Although bulk mass flow is excluded, the processes allowed under the conditions stated above are more general in respect to conduction, diffusion, and reaction than the corresponding results given in the body of the paper. Here nonzero diffusion flux is allowed at boundaries provided $\hat{\mu}_\alpha / T$ is held constant in time at corresponding points on the boundary [satisfaction of (A.11)]. Similarly with regard to the heat flux we have been able to eliminate the restriction imposed by immersion in an external environment at constant temperature T_0. Thus we can have whatever heat flux on $\partial \mathscr{B}_6$ that is required in order to maintain the part $\partial \mathscr{B}_6$ of the boundary at a temperature distribution that is time independent. Note should also be taken of the fact that the temperature on $\partial \mathscr{B}_6$ and $\hat{\mu}_\alpha / T$ on $\partial \mathscr{B}_8$ can depend on the position variables that describe the boundary. Also there is no requirement that $\sigma = 0$ when and only when $\mathbf{X} = \mathbf{J} = \mathbf{0}$, for

$$\phi(\mathbf{X}) = \int_0^1 \sigma^*(\tau \mathbf{X}) \frac{d\tau}{\tau} \geqslant 0$$

is satisfied for any function $\sigma^*(\mathbf{X}) \geqslant 0$ and $\phi(\mathbf{X})$ has only $\mathbf{X} = \mathbf{0}$ as a critical point for any $\sigma^*(\mathbf{X})$ that can be approximated by a homogeneous positive definite quadratic form in a sufficiently small neighborhood of $\mathbf{X} = \mathbf{0}$. Thus although $\sigma^*(\mathbf{X})$ is now permitted to oscillate with positive range and to vanish for certain values $\mathbf{X}^* \neq 0$, in contrast to the analysis in the body of the paper, we have $\phi(\mathbf{X}) = 0$ only when $\mathbf{X} = 0$.

Lastly we note that (A.16), (A.2), and $f_\alpha^i = f_{N+1}^i$ imply that the steady-state concentrations are given by

$$\nabla \cdot \frac{\partial \phi}{\partial [\nabla(\mu_\alpha/T)]} - \frac{\partial \phi}{\partial (\mu_\alpha/T)} = 0$$

which are just the Euler equations associated with $P = \int \phi \, dV$; that is, the steady state concentrations render $P[\phi]$ stationary in value with respect to the choice of μ_α/T.

The results established in this appendix generalize those given by Gyarmati (Ref. 22, pp. 125–131) for linear phenomenological relations. In many respects the analysis given above is similar to a very useful approach to the stability question originated by Glansdorff and Prigogine.[24]

Acknowledgments

The author is indebted to Professor I. Gyarmati for his generous correspondence and for the extremely useful account of nonequilibrium processes,[22] which often pointed the way during periods of difficulty with the analysis presented in this paper. Thanks are also due to Professor K. Klier, who kindled the author's interest in reaction processes and whose discussions of a number of the topics impinging on the results reported here were invaluable.

References

1. R. Aris, *Arch. Ration. Mech. Anal.*, **19**, 81 (1965); *ibid.*, **22**, 356 (1968).
2. M. Feinberg, *Arch. Ration. Mech. Anal.*, **46**, 1 (1972).
3. F. Horn and R. Jackson, *Arch. Ration. Mech. Anal.*, **47**, 82 (1972).
4. F. Horn, *Arch. Ration. Mech. Anal.*, **49**, 172 (1972).
5. R. M. Bowen, *Arch. Ration. Mech. Anal.*, **29**, 114 (1968).
6. C. Eckart, *Phys. Rev.*, **58**, 269 (1940).
7. J. Meixner, *Ann. Phys.*, **39**, 333 (1941).
8. J. Meixner and H. G. Reik, *Handbüch der Physik*, III/2 (1959).
9. A. C. Eringen and J. D. Ingram, *Int. J. Eng. Sci.*, **3**, 197 (1965).
10. A. E. Green and P. M. Naghdi, *Int. J. Eng. Sci.*, **3**, 231 (1965); *Arch. Ration. Mech. Anal.*, **24**, 243 (1967); *Q. J. Mech. Anal. Math.*, **22**, 427 (1969).
11. R. M. Bowen, *Arch. Ration. Mech. Anal.*, **24**, 370 (1967); *J. Chem. Phys.*, **49**, 1625 (1968); *Arch. Ration. Mech. Anal.*, **34**, 97 (1969).
12. I. Müller, *Arch. Ration. Mech. Anal.*, **28**, 689 (1968).
13. R. M. Bowen and D. J. Garcia, *Int. J. Eng. Sci.*, **8**, 63 (1970).
14. M. E. Gurtin, *Arch. Ration. Mech. Anal.*, **43**, 198 (1971).
15. M. E. Gurtin and A. S. Vargas, *Arch. Ration. Mech. Anal.*, **43**, 179 (1971).
16. L. Onsager, *Phys. Rev.*, **37**, 405 (1931); *ibid.*, **38**, 2265 (1931).
17. J. Meixner, *Z. Phys. Chem. B*, **53**, 235 (1943).
18. H. B. G. Casimir, *Rev. Mod. Phys.*, **17**, 343 (1945).
19. I. Prigogine, *Bull. Acad. Roy. Belg. L. Sci.*, **31**, 600 (1945).
20. L. Onsager and S. Machlup, *Phys. Rev.*, **91**, 1505, 1512 (1953).

21. S. R. De Groot and P. Mazur, *Non-Equilibrium Thermodynamics*, North-Holland, Amsterdam, 1962.

22. I. Gyarmati, *Non-Equilibrium Thermodynamics*, Springer, New York, Heidelberg, Berlin, 1970.

23. I. Prigogine, *Introduction to Thermodynamics of Irreversible Processes*, Interscience, New York, 1961.

24. P. Glansdorff and I. Prigogine, *Thermodynamic Theory of Structure, Stability and Fluctuations*, Wiley-Interscience, New York, 1971.

25. I. Gyarmati, *Z. Phys. Chem.*, **239**, 133 (1968); *Ann. Phys.*, **7**, 23 (1969).

26. D. G. B. Edelen, *Int. J. Eng. Sci.*, **10**, 481 (1972).

27. D. G. B. Edelen, *Arch. Ration. Mech. Anal.*, **51**, 218 (1973).

28. D. G. B. Edelen, *Int. J. Eng. Sci.*, **12**, 121 (1974).

29. D. G. B. Edelen, *Int. J. Eng. Sci.*, **12**, 397 (1974).

30. G. R. Gavalas, *Nonlinear Differential Equations of Chemically Reacting Systems*, Springer, Berlin, 1968.

31. A. D. Nazarea and S. A. Rice, *Proc. Nat. Acad. Sci. U.S.A.*, **68**, 2502 (1971).

32. A. D. Nazarea, *Biophysik*, **7**, 85 (1971); *ibid.*, **8**, 96 (1972); *ibid.*, **9**, 94 (1973).

33. G. Nicolis, *Adv. Chem. Phys.*, **19**, 209 (1971).

34. G. Nicolis and J. Portnow, *Chem. Rev.*, **73**, 365 (1973).

35. J. L. Ericksen, *Int. J. Solids Struct.*, **2**, 573 (1966).

36. B. D. Coleman and E. H. Dill, *Arch. Ration. Mech. Anal.*, **51**, 1 (1973).

37. P. M. Naghdi and J. A. Trapp, *Arch. Ration. Mech. Anal.*, **51**, 165 (1973).

38. D. G. B. Edelen, "Mass balance laws and the decomposition, evolution and stability of chemical systems," *Int. J. Eng. Sci.* (in press).

39. C. Truesdell and R. A. Toupin, *Handbüch der Physik*, III/1, Springer, Berlin, 1960.

40. A. C. Eringen, *J. Math. Phys.*, **14**, 733 (1973).

41. D. G. B. Edelen, *Lett. Appl. Eng. Sci.*, **1**, 497 (1973).

42. D. G. B. Edelen, *Int. J. Eng. Sci.*, **12**, 607 (1974).

43. D. G. B. Edelen, *Arch. Mech. (Warsaw)*, **26**, 251 (1974).

44. M. L. Anderson and R. K. Boyd, *Can. J. Chem.*, **49**, 1001 (1971).

45. P. Duhem, *Traité d'Énergetique ou de Thermodynamique Générale*, Gauthier-Villars, Paris, 1911.

46. G. J. Van Wylen, *Thermodynamics*, Wiley, New York, 1959.

47. M. A. Biot, *Indiana Univ. Math. J.*, **23**, 309 (1973).

48. I. M. Gelfand and S. V. Fomin, *Calculus of Variations*, Prentice-Hall, Englewood Cliffs, New Jersey, 1963.

49. J. G. Kirkwood and I. Oppenheim, *Chemical Thermodynamics*, McGraw-Hill, New York, 1961.

AUTHOR INDEX

Numbers in parentheses are reference numbers and show that an author's work is referred to although his name is not mentioned in the text. Numbers in *italics* indicate the pages on which the full references appear.

SUBJECT INDEX

3 5282 00368 6592